Rodent Pest Management

Editor

Ishwar Prakash, Ph.D., D.Sc.
Professor of Eminence
Coordination and Monitoring Center for Rodent Research and Training
Central Arid Zone Research Institute
Jodhpur, Rajasthan, India

CRC Press, Inc.
Boca Raton, Florida

Library of Congress Cataloging-in-Publication Data

Rodent pest management.

Bibliography: p.
Includes index.
1. Rodents—Control. 2. Rodents. 3. Rodents as carriers of disease. I. Prakash, Ishwar.
SB994.R6R62 1988 632'.69323 87-29901
ISBN 0-8493-6726-3

This book represents information obtained from authentic and highly regarded sources. Reprinted material is quoted with permission, and sources are indicated. A wide variety of references are listed. Every reasonable effort has been made to give reliable data and information, but the author and the publisher cannot assume responsibility for the validity of all materials or for the consequences of their use.

Direct all inquiries to CRC Press, Inc., 2000 Corporate Blvd., N.W., Boca Raton, Florida, 33431.

International Standard Book Number 0-8493-6726-3

Library of Congress Card Number 87-29901
Printed in the United States

PREFACE

Rodents have been dreaded animals since the Vedic times. An indication of the ravages done to granaries by these animals is found in one of the oldest (about 3000 B.C.) Indian scriptures: **(Atharva Veda 60.50.1)**

हतं तर्द समङ्कमाखुमश्विना
छिन्तं शिरो अपि पृष्टीः शृणीतम् ।
यवान्नेददानपि नह्यतं
मुखमथाभयं कृणुतं धान्याय ॥

"O, Ashwini. Kill the burrowing rodents which devastate our foodgrains, cut their heads, break their necks. Plug their mouths so that they can never destroy our food. Rid mankind of them".

Man's early efforts to control rodents are also evident from the pottery remains of mouse traps dating from the Turanian civilization. However, serious thought was given to rodents as carriers of diseases only during the early part of this century when rodent research concentrated around epidemiological aspects, as was particularly warranted by the spread of bubonic plague. Rodents have been considered as serious agricultural pests only since the Second World War. Drs. Dennis Chitty and H. N. Southern compiled research, carried out in the U.K., on biology, behavior, and toxicology in their three-volume treatise: *Control of Rats and Mice,* published in 1954. Although this compendium was largely restricted to commensal rodents such as *Rattus norvegicus, Rattus rattus,* and *Mus musculus,* it did succeed in creating interest among scientists to carry out more intensive work to achieve better success in the control of a large variety of rodent species. However, it is only recently that rodent pests have received the proper attention of agriculturists as serious threats to standing crops—probably as a result of the inordinate global increase in the human population thus necessitating increased production of food grains. In the past few decades larger areas have come under the plough, more irrigation facilities have been created, and high yielding varieties of crops are being raised. In a way, however, the agricultural revolution has created a more conducive environment for the survival of field rodents and their ravages have become even more revealing.

As a direct consequence of the challenge posed by field rodents, research for ensuring the protection of standing crops took shape. In several countries organizations were established, and a few nations launched national programs for rodent pest management. During the last few decades, scientists have shifted their emphasis from "eradication" to "control", and, more recently, to "management" of rodent pests based on ecological principles.

The rodenticides industry has not lagged behind. The acute toxicants are being replaced by anticoagulants, and now by the fast-acting second generation anticoagulants. The baiting system is also changing from mixing the poison in food baits to the use of wax blocks.

International organizations have also been taking greater interest in rodent pest problems and a number of seminars and training courses have been organized. Certain nations possessing the know-how of the latest proven technologies have come forward to assist the developing countries in standardizing technologies for minimizing losses due to rodents.

All these research efforts have created a great deal of knowledge about rodent pests, but the information is scattered in various periodicals and proceedings of workshops and seminars. It is time to compile the information in a single place. The objective of this multi-authored compendium is, therefore, to bring together the State of Arts Reports in one place, written by specialists in various fields of rodentology, and to suggest future lines of research.

It is also felt that this work on rodent pest management will trigger more research effort for the benefit of mankind and help certain countries and organizations in revitalizing serious work in this field which, it appears, has dampened during the last few years.

It is my privilege to express gratitude to all the authors who have very kindly contributed to this publication. Thanks are also due to Dr. K. A. Shankarnarayan, Director, Central Arid Zone Research Institute, Jodhpur, and Dr. P. K. Ghosh, Senior Animal Physiologist, who have always given unstinted support to our rodent research. The heavy load of secretarial assistance was upon Mrs. Bhavani Bhaskaran and Shri P. K. Rajan to whom I owe my indebtedness. Thanks are also due to my colleagues for all their assistance. My daughter, Dr. Mukta Mani assisted in finalizing the chapters on rodent-borne diseases. My wife Lakshmi, daughter Shubhra, and son Gaurav assisted me in many ways, and this work could have not been completed without their support. I am obliged to them.

A word to our Publisher, thanks for tolerating the delays.

Ishwar Prakash
Editor

THE EDITOR

Ishwar Prakash, Ph.D. is the Professor of Eminence at the Central Arid Zone Research Institute, Jodhpur, India. Earlier, he served as the Principal Animal Ecologist in the same Institute and as a Lecturer in Zoology at the University of Rajasthan. He was educated at Jaipur and Pilani, received his M.Sc. degree in 1952, Ph.D. in 1957, and D.Sc. in 1981 from the University of Rajasthan.

With his experience of carrying out pioneering research work on desert vertebrates for over three and a half decades, Dr. Prakash has been able to establish schools of Desert Ecology, Rodent Pest Mangement, and Wildlife Conservation in the xeric environment. He is regarded as an authority on desert mammals and rodent control. He has also led research and rodent control operations throughout India as the National Coordinator of the National Program on Rodent Pest Management.

He has published about 250 papers, and 6 monographs, besides several books on the Thar desert, vertebrate pest control, and rodents and their control.

Dr. Prakash has been the recipient of two National Awards, the Rafi Ahmed Kidwai Memorial Prize and the Professorial Chair of Eminence of the Indian Council of Agricultural Research, Government of India.

He is a member of various national and international scientific societies and is serving on a number of important policy-making national committees on environment, plant protection, wildlife conservation, science, and technology.

CONTRIBUTORS

S. A. Barnett
Emeritus Professor of Zoology
Australian National University
Canberra, Australia

E. A. Benigno, Ph.D.
National Crop Protection Center
College of Agriculture
University of the Philippines at Los Baños
College, Laguna, Philippines

Alan Peter Buckle, Ph.D.
Doctor
Vertebrate and Non-Crop Pesticides Section
ICI Plant Protection Division
Surrey, England

V. A. Bykovskii
All-Union Plant Protection Research Institute
Leningrad, Union of Soviet Socialist Republics

David E. Davis, Ph.D.
Professor Emeritus
Department of Zoology
North Carolina State University
Raleigh, North Carolina

Donald J. Elias, M.S.
Agricultural Officer (Vertebrate Pests)
Plant Production and Protection Division
Food and Agriculture Organization of the United Nations
Rome, Italy

Sam Erlinge
Department of Animal Ecology
University of Lund
Lund, Sweden

Michael W. Fall, Ph.D.
Chief, Branch of Predator Studies
Denver Wildlife Research Center
Animal and Plant Health Inspection Service
U.S. Department of Agriculture
Denver, Colorado

Lynwood A. Fiedler, M.A.
Wildlife Biologist (Research)
Section of International Programs
Denver Wildlife Research Center
U.S. Department of Agriculture
Denver, Colorado

William D. Fitzwater, M.S.
National Animal Damage Control Association
Albuquerque, New Mexico

Norman G. Gratz, D.Sc.
Director, Retired
Division of Vector Biology and Control
World Health Organization
Geneva, Switzerland

Lennart Hansson, Ph.D.
Professor
Department of Wildlife Ecology
Swedish University of Agricultural Sciences
Uppsala, Sweden

Melanda M. Hoque, Ph.D.
National Crop Protection Center
College of Agriculture
University of the Philippines at Los Baños
College Laguna, Philippines

Walter E. Howard, Ph.D.
Professor of Wildlife Biology and Vertebrate Ecologist
Department of Wildlife and Fisheries Biology
University of California at Davis
Davis, California

N. V. Kandybin
Professor
Laboratory of Microbiology and Pathology of Rodents and Insects
All Union Reseach Institute of Agriculture Microbiology
Leningrad,
Union of Soviet Socialist Republics

Mogens Lund
Rodent Control Specialist in Charge
Department of Rodent Control
Danish Pest Infestation Laboratory
Lyngby, Denmark

Rex E. Marsh, A.B.
Specialist in Vertebrate Ecology
Wildlife and Fisheries Biology
University of California at Davis
Davis, California

R. P. Mathur
Pest Control (India) Private Ltd.
Madras, India

Adrian P. Meehan, B.Sc.
Chief Biologist
Research and Development
Rentokil Ltd., Felcourt
Sussex, England

Heimo Posamentier, M.Sc.
Department of Plant Protection
Gesellschaft fur Technische
Zusammenarbeit (GTZ) GmbH
Eschborn, West Germany

Trevor David Redhead, Ph.D.
Doctor
Division of Wildlife and Rangelands
Research
CSIRO
Canberra, Australia

Brian D. Rennison, B.A., M.Sc., F.I.S.
Retired
Formerly of Tolworth Laboratory
Ministry of Agriculture, Fisheries and Food
Surrey, England

Christopher G. J. Richards, M.A., D. Phil.
Rodenticides Product Manager
Insecticides and Fungicides Department
ICI Agrochemicals
Haslemere, England

Terrell P. Salmon, Ph.D.
Extension Wildlife Specialist for Animal
Damage Control
Wildlife Extension
University of California at Davis
Davis, California

F. F. Sanchez, Ph.D.
Director
National Crop Protection Center
College of Agriculture
University of the Philippines at Los Baños
College, Laguna, Philippines

R. C. Saxena, Ph.D.
Department of Entomology
International Rice Research Institute
Manila, Philippines

D. Michael Stoddart, Ph.D.
Professor
Department of Zoology
University of Tasmania
Tasmania, Australia

Robert M. Timm, Ph.D.
Superintendent
Hopland Field Station
University of California
Hopland, California

G. I. Twigg, Ph.D.
Department of Zoology
Royal Holloway and Bedford New
College
Surrey, England

TABLE OF CONTENTS

Chapter 1

RODENT PROBLEMS ON THE NORTH AMERICAN CONTINENT

Rex E. Marsh

TABLE OF CONTENTS

I. INTRODUCTION

Rodent problems on the North American Continent are highly varied as they are on most other continents and involve both introduced as well as native species. Many of the species found in North America have counterparts of the same family or closely related genera on other continents, and the problems they cause or damage they do is often similar. Rodent problems can be categorized in many ways, but for the purpose of this chapter they will be discussed on the basis of taxonomic classification and how they affect man's interest or impact on humans.

Although certain rodent species or groups of rodents are discussed as pests in this chapter, relatively few of the numerous existing rodent species are considered pests. Few species are detrimental all the time as might be unintentionally implied by categorizing them as "pests." Some good or benefit can be attributed to most every species, if nothing more than serving as food for animals of a high trophic level.

Many factors may contribute to whether a rodent species becomes a pest, including its geographic distribution, preferred habitats, feeding habits, reproductive propensity, population dynamics, and other biological or behavioral characteristics; but most important is its adaptability to man-made or modified environments.

Rodents are considered pests in many situations and for a variety of reasons. They damage food crops, such as sugar cane, rice, sugar beets, and vegetables, and trees or vines that produce fruit, nuts, or berries; they compete with livestock for forage, or damage forage or hay crops; they damage oil and fiber crops such as cotton and flax; and they damage forest trees and hamper reforestation efforts. Rodents of one species or another may cause damage at most any stage of plant growth from the seed to the harvestable commodity. Often crop damage does not stop there, and some of the most serious damage and losses through rodent contamination occur postharvest in transportation and storage.

Our most serious postharvest or stored-food rodent pests are three introduced species: the black or roof rat (*Rattus rattus*), the Norway rat (*R. norvegicus*), and the house mouse (*Mus musculus*).

Rodents are implicated in the transmission of a variety of diseases to man and his domestic animals. In other situations, especially around habitations, rodents are nuisance problems. They cause considerable structural damage to buildings and other man-made or engineered structures.

The major pest genera or species are discussed individually in a general order of importance to provide an overview of the problems in the North American Continent, including the crops and other damage they cause. Space permits only a brief discussion of pest rodent groups and the nature of their damage. As a convenient reference, important rodent pests of North America, their taxonomic classifications, types of damage, and magnitude and severity of damage are presented in Table 1. Important pest rodent species are managed or controlled as needed to reduce animal damage, health-related problems, and other conflicts with man's activities. Control success varies with the pest species. Management and control methods are discussed in other chapters. More complete taxonomic information and distribution maps for the species discussed are found in *The Mammals of North America*.[14]

II. RODENT PESTS

A. Pocket Gophers

Pocket gophers are fossorial rodents that spend much of their lives in their extensive burrow systems. They are represented by five genera of the family Geomyidae and are found only in North America, from Central Canada to Panama. The thick-set bodies, about 6 to 12 in. (15 to 30 cm) long, show little external evidence of a neck. Highly adapted for living

Table 1
IMPORTANT RODENT PESTS OF NORTH AMERICA, TAXONOMIC CLASSIFICATION, TYPES OF DAMAGE, AND MAGNITUDE AND SEVERITY OF DAMAGE

Rodent pest			Type of damage[a]					
Common name	**Species**	**Family**	**Agric. crops**	**Rangeland grazing land**	**Forestry**	**Postharvest storage**	**Physical damage — man-made structures**	**Ref.**
Mountain beaver	*Aplodontia rufa*	Aplodontidae	—	—	MW, 1—4	—	L, 1—2	1—3
Chipmunk	*Tamias* spp.	Sciuridae	L, 1—2	—	L, 2—3	—	L, 1—2	4
Marmot or Woodchuck	*Marmota* spp.	Sciuridae	W, 2—3	W, 2—3	—	—	—	4, 5
Ground squirrel	*Spermophilus* spp.	Sciuridae	W, 2—4	W, 2—4	L, 1—3	—	MW, 1—3	6—9
Prairie dog	*Cynomys* spp.	Sciuridae	W, 2—4	W, 2—4	—	—	MW, 1—3	8, 10
Tree squirrel	*Sciurus* spp.	Sciuridae	L, 1—3	—	MW, 1—4	L, 1—2	MW, 1—2	3, 4
	Tamiasciurus spp.	Sciuridae	—	—	L, 1—3	—	L, 1—2	3, 4, 11
Pocket gopher	*Thomomys* spp.	Geomyidae	W, 2—4	W, 2—4	W, 2—4	—	W, 1—4	12, 13
	Geomys spp.	Geomyidae	W, 2—4	W, 2—4	W, 2—4	—	W, 1—4	12—14
Kangaroo rat	*Dipodomys* spp.	Heteromyidae	L, 1—3	MW, 1—4	—	—	—	4, 15, 16
Beaver	*Castor canadensis*	Castoridae	L, 1—2	—	W, 1—4	—	W, 1—4	3, 4, 17
Rice rat	*Oryzomys* spp.	Cricetidae	L, 1—4	—	—	—	—	4
Deer mouse	*Peromyscus* spp.	Cricetidae	L, 1—3	—	W, 2—4	L, 1—2	L, 1—2	18
Cotton rat	*Sigmodon* spp.	Cricetidae	MW, 1—3	MW, 2—3	—	—	—	19, 20
Wood rat	*Neotoma* spp.	Cricetidae	L, 1—2	—	MW, 1—3	L, 1—2	L, 1—2	21, 22
Vole	*Microtus* spp.	Cricetidae	W, 1—4	W, 1—4	MW, 1—3	L, 1—2	—	23—26
Muskrat	*Ondatra zibethicus*	Cricetidae	W, 1—4	—	—	—	—	27
Black rat[b]	*Rattus rattus*	Muridae	W, 1—4	—	—	W, 2—4	W, 2—4	28—30
Norway rat[b]	*R. norvegicus*	Muridae	W, 1—4	—	—	W, 2—4	W, 2—4	28—30
House mouse[b]	*Mus musculus*	Muridae	W, 1—4	—	—	W, 2—4	W, 2—4	28—30
Porcupine	*Erethizon dorsatum*	Erethizontidae	—	—	W, 1—4	—	W, 1—3	31, 32
Nutria[b]	*Myocastor coypus*	Myocastoridae	MW, 1—4	—	—	—	MW, 1—3	33

[a] The magnitude of the problem is indicated by the following letters: (L) localized; (MW) moderately widespread; (W) widespread. The severity of the problem is indicated by numbers or a range of numbers: (1) occasional damage; (2) light damage; (3) moderate damage, (4) heavy damage.
[b] Introduced species, not native.

below ground, they have small eyes and ears, a good sense of smell, and stout forelegs with strong claws. The gopher's incisors protrude through an opening in its lips to enable their use for digging. A pair of furlined external cheek pouches are used for carrying food items.[5]

Pocket gophers feed mostly below ground on seeds, roots, bulbs, tubers, and the bark of trees. After consuming the roots of a plant, they often pull the smaller plants into their burrows and devour them entirely. The various species have somewhat different preferences for plants, but they include both grasses and forbs. Many agricultural crops, ornamental plants, and newly planted forests suffer severe damage from root and seedling clipping. Damage to older trees involves basal debarking. The major economic losses occur to landscape plantings, young transplanted forest trees, fruit trees, and alfalfa production. On rangeland and pasture, pocket gophers compete for forage and greatly reduce the carrying capacity for livestock.[13]

In their burrow construction, pocket gophers push up numerous earthen mounds that may exceed 3000 per ha.[30] These mounds bury other vegetation as well as interfere with mowing of alfalfa and other hay crops. In irrigated crops or pastures, considerable water is lost because of extensive burrow systems, and in hilly terrain burrowing may accelerate soil erosion. Damage to earthen dams and levees can be extensive and costly.

B. Ground Squirrels

As a group, the ground-dwelling genera of the squirrel (Sciurid) family are of considerable economic importance in western U.S., especially as competitors of livestock for forage and damage they do to grain crops. Though the three groups, i.e., ground squirrels, prairie dogs, and woodchucks or marmots, have similarities, they differ considerably in biology, behavior, population densities, and the degree with which they affect man's livelihood.

Ground squirrels belong to the genus *Spermophilus* and range from Canada to Mexico, with 20 of the 23 species found in the U.S., mostly west of the Mississippi River. Over half of those species are considered significant pests to agriculture and forestry.[27] These species generally inhabit open grassy plains and valleys, slightly wooded areas, or openings in forests. Several species are relatively adaptable and thrive in and around a variety of agricultural crops, principally grain, alfalfa, and other forage crops, but they also attack sugar beets, vegetable, and fruit crops. In some regions of western U.S. they compete extensively with livestock for range forage[12,17,22,23] and, to a much lesser degree, cause reforestation problems by consuming tree seed and clipping small seedlings.[34] As diurnal rodents of substantial size that feed above ground, their presence and damage seldom go unnoticed. Some species colonize and reach high population densities. They are implicated in the transmission of several diseases to man, including the plague.

C. Prairie Dogs

Prairie dogs, the local name given to rodents belonging to the genus *Cynomys*, are found mostly in the Plains area of the U.S. One of the four species extends into Canada a short distance, and a fifth species is found well into Mexico. Larger than most ground squirrels and smaller than woodchucks or marmots, it is a stocky rodent with a short tail about one-fourth its total length. They are very social animals and thrive in large colonies called "towns", which may contain a few to thousands of individuals. Once widely distributed and abundant, particularly over the short-grass prairies of the Great Plains,[35] agricultural and grazing practices, plus control measures, have greatly reduced inhabited areas, yet they are still locally and regionally very abundant and compete with livestock for range forage. Grain and forage crops are also consumed.

D. Woodchucks

Marmots and woodchucks of the genus *Marmota* are the largest members of the squirrel family in North America. Five species range from Canada through much of the U.S. except for the southernmost areas, both east and west. The four species found in the western states are called marmots, while the single eastern species, *Marmota monax*, is referred to as the woodchuck. They are stocky animals that weight from 4 to 10 lb (1.8 to 4.5 kg) and live in open and closed forests or brushy areas. They rarely become very numerous, but their size more than makes up for this in the agricultural damage they can do. They can be significant pests in grain, alfalfa, clover, hay crops, and on rangeland. They will also feed on a variety of vegetable crops if near their occupied habitat.

E. Chipmunks

Chipmunks are also a group of the Sciurid family and are represented by two genera. The eastern chipmunk, *Tamias striatus*, is the only species of this genera, while the several chipmunks of western North America are of the genus *Eutamias*. The chipmunk ranges from Canada into Mexico. It generally is the smallest of the Sciurids, diurnal, and prefers as habitat semiwooded, brushy, forest edges, or open forests. It digs burrows or lives in hollow logs or other forest debris. It is sometimes considered a pest when it clips young tree seedlings and consumes seeds. Occasionally it digs up seed of crops before or during early germination and at times may take ripening fruit to obtain the seeds.[10] Regionally, chipmunks are a relatively minor economic pest; however, they are implicated in the transmission to humans of several diseases. They are frequently numerous in forested parks and recreational campgrounds where they become quite tame and feed from picnic tables or on scraps provided by people.

F. Tree Squirrels

Tree squirrels (*Sciurus* spp. and *Tamiasciurus* spp.) are arboreal members of the Sciurid family that naturally inhabit forests or wooded areas. Several members of this group cause damage to fruit and nut crops. Both genera cause damage to forest trees by stripping bark, debarking the upper bole and twig, and cone-cutting.[10,20] Although an economic pest, tree squirrel damage is generally relatively localized. Several tree squirrel species are found in urban parks and landscaped areas where they may become a pest. Insulation damage to overhead communication and electrical lines can be extensive in some regions, as is structural damage to buildings — principally wood-gnawing on homes and other buildings.

G. Voles or Meadow Mice

Voles and meadow mice are both terms used in reference to the genus *Microtus* and belong to the subfamily Microtinae of the family Cricetidae. Nineteen species are found in North America, with some ranges extending into the arctic regions on the north and others occupying regions as far south as Guatemala. Except for some desert areas, one or more species are found throughout the Continental U.S. and Canada.

Only about six of these species become pests of any significant economic importance. Voles all have the same general appearance with some variations in size and tail length. The rodent has a chunky body, blunt nose with small furry ears, and a scantily haired tail. The largest of the species may measure 4 to 5 in. (10 to 13 cm) in head and body length with a $1^3/_4$ to $2^3/_4$ in. (4 to 7 cm) tail. Voles generally prefer dense grassy habitat providing relatively good cover where they develop well-defined surface runways for movement and feeding. Other species are more subterranean in feeding activities. They are very prolific, and population fluctuations are common with peaks reached every 3 to 6 years. Periodically mouse irruptions occur and populations of plague proportions are reached which may be as high as 3000 mice per acre.[33]

The damage they do depends on the species and location. They cause extensive damage in apple and other orchards where they girdle the tree at or below the soil line. In the eastern part of the U.S. the vole is probably the single most damaging agricultural rodent, particularly to orchard trees.[3] Voles also damage grain crops and alfalfa by clipping the stems.[19] Damage to sprinkler-irrigated alfalfa in the west is particularly extensive when peak populations occur. Sugar beet fields are also especially hard hit in California, with substantial economic losses.[29] A variety of vegetable crops suffer some damage including peas, cabbage, celery, potatoes, carrots, and tomatoes. In California voles cause the death of artichoke plants and thus significantly reduce production. In forestry they may girdle trees and clip seedlings and low branches. Orchard and forest damage often occurs in the winter months under snow cover.

H. Muskrats

The muskrat, *Ondatra zibethicus*, is found throughout most of Canada and the U.S., but only in a few spots in northern Mexico. It is a semiaquatic species like the beaver, and in some habitats it builds conical lodges of tules or cattails in shallow ponds.

It is a large rodent weighting from $1^1/_2$ to 3 lb (0.7 to 1.4 kg) and has a naked tail that is flattened laterally. The muskrat is a most important fur-bearing species with more pelts being marketed than those of beaver.

The muskrat occasionally causes damage to crops such as corn and sugar beets grown adjacent to streams, drainage ditches, or ponds that serve as muskrat habitat. Because of its burrowing activities, the muskrat can be a serious pest in rice-growing areas where fields are kept flooded and water management is critical.[25] It can be a significant pest along irrigation ditches, levees, or in earthen dams where its burrowing causes washouts and diverts water.

I. Nutria

Nutria or coypu (*Myocastor coypus*) is a rodent introduced into North America from South America, originally brought to the U.S. as a fur-bearing animal to raise in captivity. It escaped or was deliberately released in a number of areas. It presently thrives in the southeastern and northwestern U.S. It is semiaquatic and occupies the same type of habitat as the native muskrat. The damage it does parallels that of muskrats except nutria is a much larger animal.[11] It is harvested from the wild for its pelt.

J. Beaver

Castor canadensis is distributed throughout most of Canada and the U.S., but its range extends only a short distance into Mexico. It is the largest rodent found in the U.S. and Canada. It weighs from 30 to 60 lb (14 to 27 kg) but occasionally approaches 70 lb (32 kg). Its tail is horizontally flattened and naked. It is known for its semiaquatic habits and for the dams and lodges it builds in streams.

Its tree-cutting habits are of concern to foresters, as is the flooding of standing trees. Burrowing damage to dikes, levees, and other water-management structures is common.[15] The damage it does may be compensated for by the benefits derived from its damming activities in some regions. It will at times feed on corn and other crops adjacent to streams and will sometimes cut down fruit trees such as pear and apple that may be growing near rivers or streams where beaver thrive. As an agricultural pest, most damage relates to damming of irrigation ditches, clogging water-control gates, and burrowing into embankments, diverting irrigation water.

On the positive side, beavers are an important fur-bearer and their pelts bring considerable revenue to trappers.

K. Mountain Beaver

Aplodontia rufa, commonly referred to as the mountain beaver, is also locally called "boomer". Only one living species exists with very limited distribution along the Pacific Coast from the very southern portion of Canada to northern California and along the Sierra Nevada mountain range.

The mountain beaver is not closely related to the common beaver, *Castor canadensis*, but is the sole surviving member of a primitive rodent family, Aplodontidae. It is a short, stocky animal about 12 in (30 cm) long with a stubby tail that is hardly visible. The adult weighs from 2 to 3 lb (1 to $1^1/_2$ kg). It is a burrowing animal found in forests, wooded, or brushy habitat, with relatively high rainfall.[16] Its above-ground feeding on a variety of vegetation, such as ferns, low-growing shrubs, as well as tree species occurs mostly at night. It is not considered a good climber, and feeding is generally within a few feet of the ground.

It is primarily a pest of conifer forests when it clips stems of small seedlings or terminal shoots of young trees. Feeding also extends to lateral branches and basal girdling.[1,20] It may undermine roads and accelerate erosion. It is rarely considered a pest of agricultural crops.

L. Porcupines

The porcupine, *Erethizon dorsatum*, one of the larger native rodents, is distinguished by its coat of bristling quills. Its range extends from Canada to Northern Mexico, but it is absent from the southeastern states.

The porcupine is a chunky, heavy-bodied animal with a small head, short legs, and a short thick tail. Sharp barbed quills, $1^1/_2$ to 4 in. (4 to 10 cm) long, cover most of the upper body surface and tail. When disturbed, the animal will take a defensive stand and raise its quills. Adults weight from 10 to 20 lb ($4^1/_2$ to 9 kg) or more and measure 2 to 3 feet (0.6 to 1 m) in length.

Porcupines live in forest habitats and are more common in coniferous or mixed forests, but are found also where no evergreens exist. They are excellent climbers and spend considerable time in trees, especially in the winter. They are generally considered a pest in forests where in winter they feed primarily on the inner bark (cambium and phloem tissue) of a wide variety of forest trees.[9,32] Top-girdling results in deformed tree crowns. This includes the sugar maple trees of the northeastern U.S. where the porcupines cause serious damage. Occasionally home orchards and vegetable gardens in or near wooded areas may be damaged by porcupines. Berries and corn are among the items consumed. Porcupines can become troublesome around homes, summer cabins, logging camps, campgrounds, and other outdoor recreational facilities in forested settings where they gnaw on wooden items and structures, often causing considerable damage.

M. Deer Mice

Deer or white-footed mice are of the genus *Peromyscus* of the family Cricetidae. More than 20 species are found in North America and, with the exception of the colder arctic regions, they are found nearly everywhere. The size and coloration of different species vary, but the general characteristics are all very similar. They range in size from that of a house mouse to about twice that size. They have large eyes and ears, and the tail is covered with short fine hair and generally distinctly biocolored. Most deer mice burrow, but burrow entrances are not easy to located. They all can climb, but some are more arboreal in their habits than others.

They are mostly nocturnal in habits and feed mainly on seeds, fruits, and insects. They occupy a large variety of natural habitats as long as food is available. Population densities of deer mice are constantly relatively low with only isolated instance of major fluctuations.

Because of their seed-eating propensity, they are of considerable economic importance in direct seeding of coniferous forests.[31] Without controlling the population or the use of effective seed protectants, direct seeding of forests in the northwestern U.S. frequently fails. They are at times destructive to grain crops and sometimes dig up planted seed of cereals, melons, and other crops. In rare situations they girdle orchard trees beneath trunk wraps.[7] Occasionally they enter buildings and may cause damage similar to that of house mice.

N. Woodrats

Woodrats, genus *Neotoma*, are represented by a number of species which are widely distributed over much of North America and are found in varied types of habitat from low, hot dry deserts to cold rocky slopes above timberline. This rat-sized rodent has large ears, large eyes, and a fairly long tail, which is sparsely covered with hair or, depending on the species, well-furred with long hair. Its fur is soft.

Some species, such as the dusky-footed woodrat (*Neotoma fuscipes*), are semiarboreal and often construct large bulky dens of sticks or nests high in trees. Woodrats are chiefly nocturnal. Their food is largely determined by the habitat and consists mainly of green vegetation including grass, twigs, leaves, fruits, nuts, small bulbs, bark, dry seeds, and fungi.

While they cause some occasional damage to citrus, nut, and other tree crops, they are most troublesome to foresters of the western U.S. where they cause damage to trees by crown-debarking, clipping of branch tips and the leaders of young conifers, and general bark-stripping.[8,26] Occasionally they are attracted to buildings in forested areas where they sometimes nest in the attics and feed on human food supplies if readily available. Their gnawing may cause structural damage.

O. Cotton Rats

The cotton rat, *Sigmodon* spp., is a rat-sized rodent with coarse, grizzled-gray fur and a sparsely haired tail about half the body length. It is generally found in the southern part of the U.S., extending south into Mexico and Central America, and is represented by seven species. The genus also exists in South America.

The Hispid cotton rat (*S. hispidus*), the most important species in the U.S., prefers dense, grassy natural habitats. They tunnel and create surface runways. They eat a variety of plants, although on uncultivated land grasses are highly preferred. Cotton rat populations fluctuate dramatically and can become most numerous. Serious damage can occur to sugar cane, cotton, alfalfa, and grain crops, as well as to melons, sweet potatoes, and tomatoes.[4,6] Depending on the plant and season, they may feed on leaves, stems, roots, and seeds.

P. Kangaroo Rats

Kangaroo rats, genus *Dipodomys*, belong to the family Heteromyidae. Some 23 species are distributed from southwestern Canada to Baja California and throughout much of Mexico. These nocturnal rodents thrive mostly in arid or semiarid regions that are grassy or brushy. They generally prefer more open-type habitat with sandy soil types.

Species of *Dipodomys* vary considerably in size and weight, but are generally smaller than black rats (*Rattus rattus*). They have a relatively delicate conformation and move about most of the time in a hopping motion, using their long hind legs. From this they derive the name of kangaroo rats. They have external cheek pouches, large eyes, and a long tail tufted with longer hair. Kangaroo rats live in burrow systems, and most are considered to be a colonizing species.

They feed on a variety of plants, eating leaves, stems, buds, some fruits, and many kinds of seed including cultivated grains. They store great quantities of seed in surface caches and burrows. Several species of kangaroo rats, when in moderate-to-high numbers, compete significantly with grazing livestock for range forage. Especially during periods of drought, kangaroo rats can be of critical importance on arid grazing land, substantially reducing the livestock-carrying capacity of the range.[28,36] Several species or subspecies not implicated as significant pests are on the endangered or threatened species list primarily because of loss of habitat.

Q. Rice Rats

Rice rats of the genus *Oryzomys* range along the eastern and southern coasts of the U.S.

from New Jersey to Florida, and as far west as Texas. They extend along both coasts of Mexico and south to Panama. *O. palustris* is the species found in the U.S.; some nine other species are found in Mexico and further south.

This small rat-like species prefers as habitat heavy or dense ground cover of coastal or inland marshes, although it may extend inland some distance if suitable cover exists. It feeds on a variety of vegetation, including rice and sugar cane. When its numbers are high, it may also damage corn, sweet potatoes, and other vegetable crops planted near its natural habitat. Fruit trees have been reported girdled.[10] Significant damage is usually very localized.

R. Norway Rats

The Norway rat, *Rattus norvegicus*, is not native, but one of three major introduced pest rodents in North America. It is usually found in areas inhabited by man; however, there are a few areas of the western U.S. and many relatively large regions of Canada, notably the Province of Alberta, where this species has not become established. It is well established in most temperate, semitropical, and tropical areas where it can cause damage to a variety of crops including sugar cane, rice, and vegetable crops.[2,18,24] The Norway rat is a serious problem in postharvest food storage and a most serious pest in poultry and hog-raising facilities as well as in dairies and in cattle feedlots. Wherever animals are maintained, conditions are generally favorable to the Norway rat. A number of diseases are transmitted to humans and domestic animals by the Norway rat. They include leptospirosis, salmonella, murine typhus, plague, and ratbite fever.

S. Black Rats

The black rat, *Rattus rattus*, sometimes referred to as the roof or ship rat, is an introduced species found along the east and west coasts of the U.S., extending along the Gulf Coast and presumably into the coastal regions of Mexico and other Central American countries. It is found throughout the Caribbean islands. Most of the U.S. and Canada are free of black rat. They are commensal rodents living in close association with man and cause losses to stored feed and food.[2,18,24] In the U.S. they cause damage to sugar cane, citrus, and avocado crops. Other fruit, nut, and berry crops are occasionally subject to some black rat damage. On poultry and hog farms the rats may compete for feed. This species is well known around the world and needs no special descriptive comments.

T. House Mice

The house mouse, *Mus musculus*, like the two species of *Rattus*, is an introduced rodent which undoubtedly arrived on the early sailing ships from Europe. It is a major pest of stored feed and food and is found most everywhere in North America that is inhabited by man. House mice probably cause more economic damage to stored products than any other rodent species in North America. While house mice are found in some growing agricultural crops, they are essentially seed eaters, causing occasional crop damage in the field in a few situations. Mice are often a serious problem where livestock are confined or fed, as they consume livestock rations and contaminate feed. They are also sometimes implicated in transmitting diseases to domestic animals as well as to humans.

U. Minor Rodent Pests

A few very minor rodent pests have been omitted from this discussion because they are not significant problems except in some very unusual situations and on rare occasions. These include harvest mice, *Reithrodontomys* spp.; pocket mice, *Perognathus* spp.; and red-backed mice, *Clethrionomys* spp.

III. NONRODENT SPECIES CAUSING SIMILAR DAMAGE

While this chapter is concerned with pest rodent species, several other pest species are worthy of mention because they, too, are implicated in damage not unlike that of rodents. Several species of hares and rabbits[21] of the order Lagomorpha and family Leporidae are important pests of agriculture and forestry in North America.

In the order Insectivora several species of shrews (family Soricidae) and several species of moles (family Talpidae) also are implicated in damage similar to that of certain rodents.

REFERENCES

1. **Borreco, J. E. and Anderson, R. J.,** Mountain beaver problems in the forests of California, Oregon and Washington, in *Proc. 9th Vertebr. Pest Conf.*, Clark, J. P., Ed., University of California, Davis, 1980, 135.
2. **Brooks, J. E. and Rowe, F. P.,** Commensal rodent control, World Health Organization, Geneva, WHO/VBC/79,726, 1979, 1.
3. **Byers, R. E.,** Control and management of vertebrate pests in deciduous orchards of the eastern United States, in *Horticultural Reviews*, Vol. 6, Jannick, J., Ed., AVI Publishing, Westport, Conn., 1984, 253.
4. **Carr, C.,** Cotton-rat control in Florida, Circular 41, Agricultural Extension Service, University of Florida, Gainesville, Fla., 1936, 1.
5. **Chase, J. D., Howard, W. E., and Roseberry, J. T.,** Pocket Gophers (Geomyidae), in *Wild Mammals of North America*, Chapman, J. A. and Feldhamer, G. A., Eds., Johns Hopkins University Press, Baltimore, Md., 1982, 239.
6. **Clark, D. O.,** The extending of cotton rat range in California — their life history and control, in *Proc. 5th Vertebr. Pest Conf.*, Marsh, R. E., Ed., University of California, Davis, 1972, 7.
7. **Cummings, M. W. and Marsh, R. E.,** Vertebrate pests of citrus, in *The Citrus Industry*, Vol. IV, Reuther, W., Calavan, E. C., and Carman, G. E., Eds., University of California, Davis, 1978, 237.
8. **Dice, L. R.,** Wood rat damage to fruit trees in eastern Oregon, *J. Mammal.*, 6, 282, 1925.
9. **Dodge, W. E.,** Porcupine (*Erethizon dorsatum*), in *Wild Mammals of North America*, Chapman, J. A. and Feldhamer, G. A., Eds., Johns Hopkins University Press, Baltimore, Md., 1982, 355.
10. **Eadie, R. W.,** *Animal Control in Field, Farm, and Forest*, Macmillan, New York, 1954, 257.
11. **Evans, J.,** Nutria, in *Prevention and Control of Wildlife Damage*, Timm, R. M., Ed., University of Nebraska, Lincoln, Neb., 1983, B61.
12. **Fitch, H. S.,** Ecology of the California ground squirrel on grazing lands, *Am. Midl. Nat.*, 39, 513, 1948.
13. **Foster, M. A. and Stubbendieck, J.,** Effects of the Plains pocket gopher (*Geomys bursarius*) on rangeland, *J. Range Manage.*, 33, 74, 1980.
14. **Hall, E. R.,** *The Mammals of North America*, Vols. 1 and 2, 2nd Ed., John Wiley & Sons, New York, 1981, 1.
15. **Hill, E. P.,** Beaver (*Castor canadensis*), in *Wild Mammals of North America*, Chapman, J. A. and Feldhamer, G. A., Eds., Johns Hopkins University Press, Baltimore, Md., 1982, 256.
16. **Hoover, E. F.,** The mountain beaver in Oregon: its life history and control, Research Paper No. 30, Oregon State University, School of Forestry, Corvallis, Ore., 1979, 1.
17. **Howard, W. E., Wagnon, K. A., and Bentley, J. R.,** Competition between ground squirrels and cattle for range forage, *J. Range Manage.*, 12, 110, 1959.
18. **Jackson, W. B.,** Evaluation of rodent depredations to crops and stored products, *EPPO Bull.*, 7, 439, 1977.
19. **Jameson, E. W.,** Consumption of alfalfa and wild oats by *Microtus californicus*, *J. Wildl. Manage.*, 22, 433, 1958.
20. **Lawrence, W. H., Kverno, W. B., and Hartwell, H. D.,** *Guide to Wildlife Feeding Injuries on Conifers in the Pacific Northwest*, Western Forestry and Conservation Association, Portland, Oregon, 1961, 1.
21. **Marsh, R. E.,** An overview of rodent and rabbit pests of agricultural crops in the United States, in *Proc. Conf. of the Organisation and Practice of Vertebr. Pest Control*, ICI Plant Protection Division, Haslemere, England, 1984, 57.
22. **Marsh, R. E.,** Ground squirrels, prairie dogs, marmots as pests on rangeland, in *Proc. Conf. of the Organisation and Practice of Vertebrate Pest Control*, ICI Plant Protection Division, Haslemere, England, 1984, 195.

23. **Marsh, R. E.,** Competition of rodents and other small mammals with livestock in the United States, in *Parasites, Pests and Predators,* Gaafar, S. M., Howard, W. E., and Marsh, R. E., Eds., Elsevier, Amsterdam, 1985, 485.
24. **Meehan, A. P.,** *Rats and Mice,* Brown, Knight, and Truscott, Tombridge, England, 1984, 1.
25. **Miller, J. E.,** Muskrat control and damage prevention, in *Proc. 6th Vertebr. Pest Conf.,* Johnson, W. V., Ed., University of California, Davis, Calif., 1974, 85.
26. **Nettleton, H. I.,** Wood rats damage young douglas-fir, *J. For.,* 55, 845, 1957.
27. **Record, C. R.,** Franklin, Richardson, and Columbian ground squirrels, in *Prevention and Control of Wildlife Damage,* Timm, R. M., Ed., University of Nebraska, Lincoln, Neb., 1983, B133.
28. **Reynolds, H. G.,** Relation of Merriam kangaroo rats to range vegetation in southern Arizona, *Ecology,* 31, 456, 1950.
29. **Salmon, T. P., Gorenzel, W. P., and Lickliter, R. E.,** Severity and distribution of rodent damage to sugar beets, *Protection Ecology,* 7, 65, 1984.
30. **Scheffer, T. H.,** Habits and economic status of the pocket gophers, *U.S. Dep. Agric. Tech. Bull.,* 224, 1931, 1.
31. **Smith, C. F. and Aldous, S. E.,** The influence of mammals and birds in retarding artificial and natural reseeding of coniferous forests in the United States, *J. For.,* 45, 361, 1947.
32. **Spencer, D. A.,** The porcupine, its economic status and control, *USDA Wildl. Leaflet,* 328, 1, 1950.
33. **Spencer, D. A.,** Biological aspects of the 1957-58 meadow mouse irruption in the Pacific Northwest, U.S. Department of the Interior, Fish and Wildlife Service, Denver Wildlife Res. Lab. Spec. Rep., 1958, 1.
34. **Stillinger, C. R.,** Damage to conifers in northern Idaho by the Richardson red squirrel, *J. For.,* 42, 143, 1944.
35. **Taylor, W. P. and Loftfield, J. V. G.,** Damage to range grasses by the Zuni prairie-dog, *U.S. Dep. Agric. Bull.,* 1227, 1, 1924.
36. **Vorhies, C. T. and Taylor, W. P.,** Life history of the kangaroo rat, *Dipodomys spectabilis spectabilis* Merriam, *U.S. Dep. Agric. Bull.,* 1091, 1, 1922.

Chapter 2

THE RODENT PROBLEM IN LATIN AMERICA

Donald J. Elias and Michael W. Fall

TABLE OF CONTENTS

I. INTRODUCTION

The collective term "Latin America" generally refers to those countries of North and South America (including the Caribbean Islands) that speak a language derived from Latin. The most frequent usage is restricted to Spanish- or Portuguese-speaking countries, but French-speaking French Guiana, the French West Indies, and Haiti may also be included. Though it would seem improbable, a natural zoogeographic region encompasses essentially the same area.[101] The Neotropical Region is South and Central America and the lowlands of Mexico. Some authors include the West Indies, Galapagos, and Falklands, even though the fauna of these island groups is not representative of a major continental pattern.[26,29,55,65] For purposes of this discussion, we exclude the island groups; references to that part of Mexico which pertains to the Nearctic zoogeographic region have been included.

This report summarizes a review of ecological, zoological, agricultural, and other literature relevant to rodents as pests in Latin America. Searches were conducted in the following computerized data bases: (1) AGRICOLA (U.S. Department of Agriculture), 1970-Nov. 1984; (2) BIOSIS (Biosciences Information Services), 1977-Feb. 1985; and (3) Wildlife Review (U.S. Fish and Wildlife Service), 1974-Oct. 1985. The Food and Agriculture Organization/World Health Organization (FAO/WHO) bibliographies on rodent pest biology and control for the periods 1950-59, 1960-69, and 1970-74,[40,41,115] and Kaukeinen's bibliography[64] on vertebrate pest biology and control for 1984 were searched manually. Other sources searched include Recent Literature in Mammalogy (American Society of Mammalogists), and conference proceedings that included the Vertebrate Pest Control Conferences, 1 through 11, the Noveno Congreso Latinoamericano de Zoologia: Zoologia Economica y Vertebrados como Plagas de la Agricultura,[34] and the Conference on the Organization and Practice of Vertebrate Pest Control.[30] Numerous books, some unpublished materials, information provided by colleagues, and personal experience have contributed to our discussion. Scientific nomenclature and taxonomic classification follow *Mammal Species of the World*.[57]

Much material addressing rodent pest problems in Latin America was, undoubtedly, missed; other information was unavailable to us and, in many cases, the available work has been insufficient for us to do any more than note the occurrence of rodent problems. Thus, we do not contend that this review is a comprehensive treatment of the subject; on the contrary, we submit that it is an initial look at a complex problem that merits more intense scrutiny.

II. THE NEOTROPICAL REGION

This zoogeographical area is, without a doubt, the most varied and diverse.[29,65] Many salient features contribute to the ecological diversity of the region. The South American continent is 7200 km long and 4800 km at its greatest width, comprising a land area of some 18 million $km.^2$ It is the only continental mass that extends so far toward the Antarctic. It is connected to North America by the narrow bridge of Central America, extending 2900 km from the Darien lowlands northwest to the Isthmus of Tehuantepec (Figure 1).[97] Of the six major faunal regions, the Neotropical Region is one of two classified as barrier-limited.[26] After Australia, South America is the most isolated continent.[29,102] Two-thirds of it is vast plains rising no more than 300 m above sea level; this feature is contrasted with a colossal mountain range unrivaled in length and exceeded in height only by the Himalayas of Asia. Numerous rivers drain the land, including the Amazon — world's largest in terms of volume and area of its watershed.

The region's environmental diversity results from climatic extremes. The area of highest rainfall in the Western Hemisphere, the world's driest desert, hot humid regions, mild areas

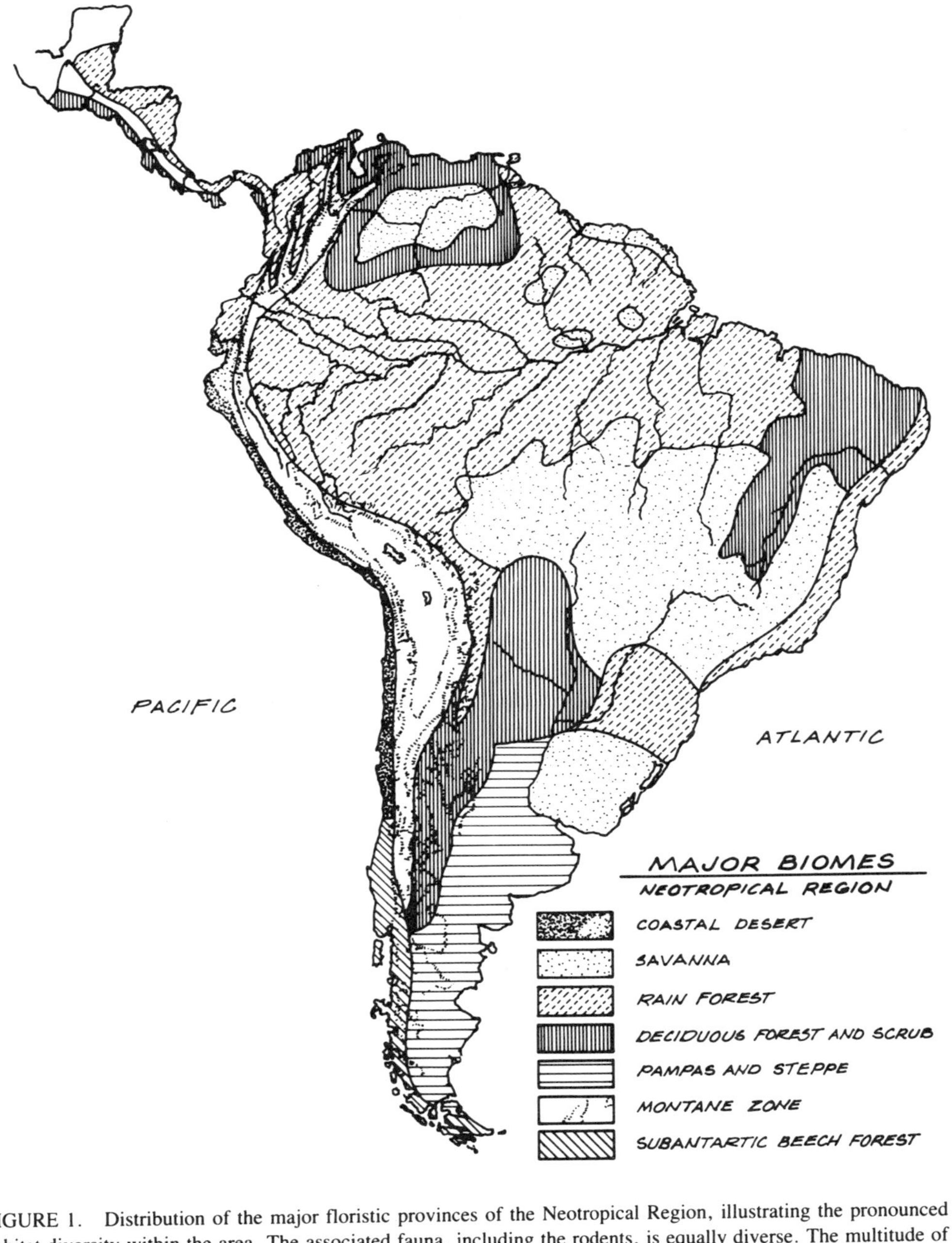

FIGURE 1. Distribution of the major floristic provinces of the Neotropical Region, illustrating the pronounced habitat diversity within the area. The associated fauna, including the rodents, is equally diverse. The multitude of ecological conditions, the variety of rodent species, and a lack of fundamental knowledge about their interactions with man has retarded progress in defining and resolving rodent pest problems. (Adapted from Dorst, J., *South America and Central America, A Natural History*, Random House, New York, 1967.)

with mediterranean climates, and vast cold regions are among the climatic variations found in the neotropics. The variety of the physical environment is reflected by the vegetation. Humid areas are covered by rain forests, that of the Amazon Basin being the most extensive forest on earth. Other vegetative types include treeless grasslands, deciduous tropical forests and scrublands, luxuriant temperate forests, the high, dry, bleak *puna* of the central Andes, and high, humid *paramos* of the northern Andes. Thus, the map of the region's major vegetative zones (Figure 1) is a mosaic with extremes of climate and vegetation found side by side.

The region's fauna is equally diverse and is as different from other areas of the world as the physical surroundings.[29,55,65] The region is characterized by many indigenous groups. South America, sometimes called the "bird continent," has the richest of all avifaunas comprising at least one-fourth of all known species; about half of the 67 families represented are indigenous.[26,29] Likewise, the mammal fauna differs notably from other zoogeographical areas; 89.5% of neotropical mammalian genera are native.[55] There are few ungulates or hooved mammals in comparison with Africa or Asia; however, other groups like the bats and primates are numerous and diversified.[21,22,29,56,65,66]

Rodents have diversified extensively in the Neotropical Region. Of the 35 families comprising the world's largest mammalian order, 19 are represented in Latin America.[24,57] Endemic neotropical rodents range in size from the tiny rice rat, *Neacomys spinosus*, to the large, semi-aquatic capybara, *Hydrochaeris hydrochaeris*.[113] The latter, the largest living rodent, measures up to 1340 mm in length and 65.5 kg in weight.[80]

While the biology of a few species is well known, knowledge of most neotropical rodents (and other wildlife) is very limited.[29,71,73] Some are known only from single museum specimens. Others are more common, but basic information on their geographic and altitudinal distributions, preferred habitats, food habits, life cycles, and taxonomic relationships is lacking.[71,73,113] This paucity of fundamental information bears directly upon our ability to define and resolve rodent pest problems. Some years ago, Hopf et al.[58] studied the existing knowledge of rodent damage to crops and stored products in tropical and subtropical areas of the world. Their report[58] concluded that, "The one single fact which emerges most clearly from the survey is the widespread ignorance of the magnitude of the rodent problem, and means to control it." With reference to Latin America, this statement is as valid today as when originally published.

III. RODENTS AS PESTS IN LATIN AMERICA

The conflicts between rodents and the economic interests or well-being of mankind in Latin America are the same as elsewhere in the world: disease transmission, food destruction, reduced timber and forage production, and physical damage. Certainly, different species are involved (Figure 2), and, taking into account the great diversity of ecological conditions found in the Neotropical Region, the resolution of rodent pest problems will require much study and understanding. It is our intent in this report to emphasize the problems of rodents in agriculture, specifically preharvest problems. However, we would be remiss not to touch on two important areas involving rodents: disease and postharvest food loss.

A. Disease

Rodents serve as reservoirs for numerous diseases or parasites that are transmissible to man and domestic animals.[2,6,8,27,44,45,71,114] They may also affect human or animal health by limiting food availability thus contributing to malnourishment.[8,51,71,84,86] It is beyond our purpose and scope to review these disease relationships in detail; it will suffice to discuss a few of the significant problems in Latin America.

The most common disease association is that of plague, the "black death" of 14th century Europe. Plague is not just a disease of history. Between 1958 and 1979 human plague occurred in 30 countries.[6] It is endemic in four Latin America countries (Bolivia, Brazil, Ecuador, and Peru) and has occurred periodically in Colombia and Venezuela.[6,7] Nearly 5000 human cases of plague were reported from South America between 1960 and 1969. Seventeen indigenous neotropical rodent genera (50 species) have been proved or strongly implicated as carriers of plague.[2,6,7,48,71]

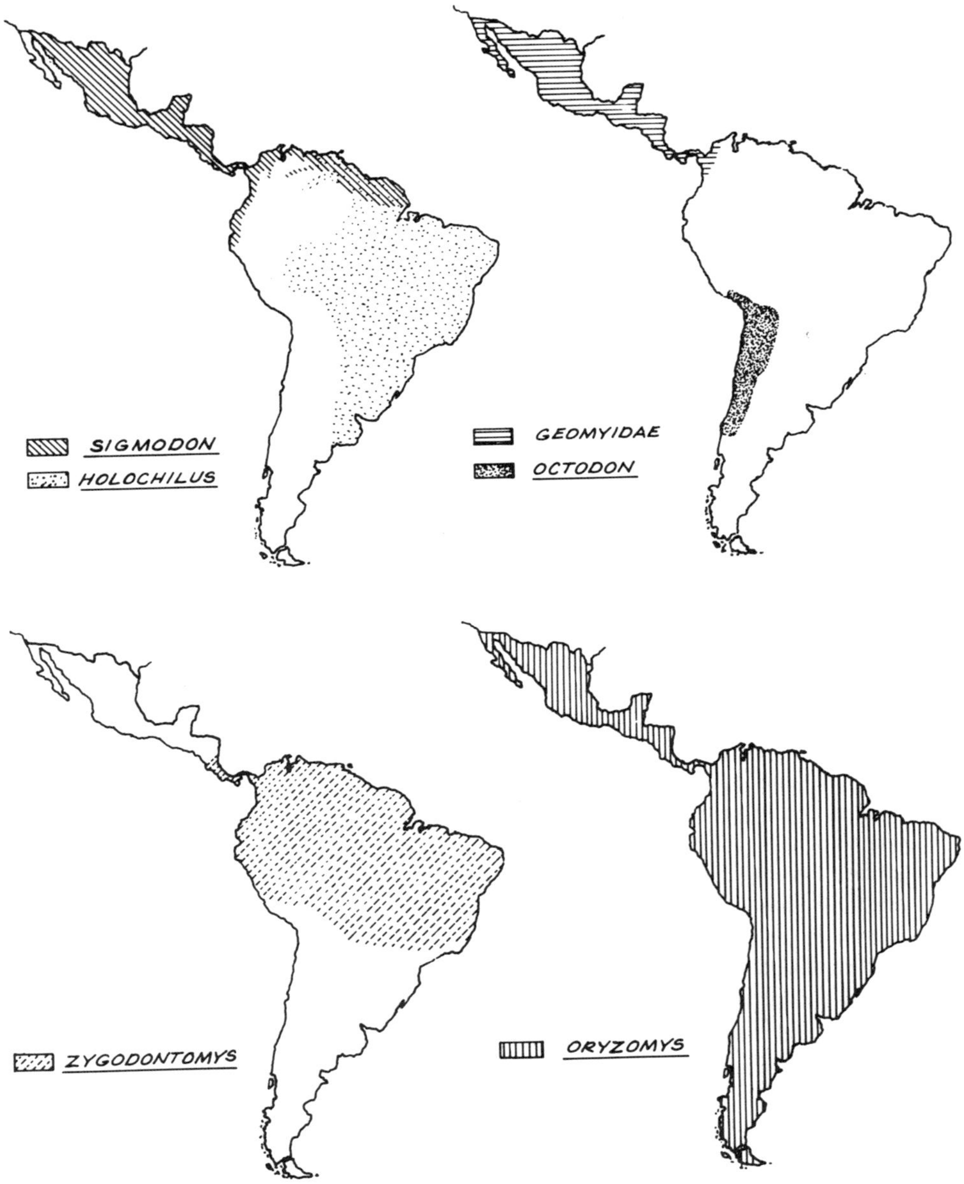

FIGURE 2. Approximate distributions of major rodent groups damaging agriculture in Latin America that are based on reports in technical literature. The cosmopolitan *Rattus* and *Mus* are major problem species throughout the region; their distribution is worldwide. *Octodon* is illustrative of several rodent groups that are not widely distributed but which have had a significant impact on a specific segment of the agricultural economy within their range. Existence of several such groups requires that rodent damage problems be examined within the context of ecological and environmental parameters.

Venezuelan encephalitis, also known as Venezuelan equine encephalitis, is a serious disease of equines and man. It is found in northern South America, Central America, and Mexico.[51] Epidemics appeared in Venezuela in 1962 and Colombia in 1967. The disease is maintained in nature by a rodent-mosquito cycle.[2,51,71]

The South American haemorrhagic fevers are associated with rodents which serve as reservoirs and probable transmitters of these diseases.[51,71,100] Argentine haemorrhagic fever

(Junin virus) epidemics have occurred primarily among wheat- and corn-field workers. The main reservoirs are *Calomys laucha* and *C. musculinus*;[51] other rodent species may be involved as well. Bolivian haemorraghic fever (Machupo virus) caused similar infections in rural Bolivia. The most notable outbreak occurred in the San Joaquin township in the department of El Beni in 1963 and had an 18% mortality rate.[71] Epidemiological studies showed that the rodent reservoir was *C. callosus* which had become commensal, and that control of epidemics could be achieved by controlling these rodents.[44,71] A complex of similar viruses have been isolated from other new-world rodents including Tamiami virus from *Sigmodon hispidus*, Pichinde virus from *Oryzomys albigularis* in Colombia, Parana virus from *O. buccinatus* in Paraguay, and Ampari virus from *O. capito* and *Neacomys guianae* in Brazil.[71]

American cutaneous or mucocutaneous leishmaniasis is found in much of Central America and large areas of Colombia, Brazil, Peru, Paraguay, Uruguay, Bolivia, and into Argentina and Chile. The reservoirs are wild rodents, including the genera *Oryzomys, Zygodontomys, Neacomys, Proechimys, Hoplomys, Coendou, Ototylomys, Heteromys, Dasyprocta, Agouti*, and *Kannabateomys*.[2,71,114]

Chagas' disease (American trypanosomiasis) is the most important cause of myocarditis in the world and is a public health problem of great importance in Brazil, Argentina, Chile, and Venezuela.[14,51,71,114] Human cases have been recorded from every South American country. Several mammal groups serve as natural reservoirs; evidence from Brazil, Panama, and Costa Rica indicated that the black rat, *Rattus rattus*, was the main intra-domiciliary reservoir.[71] *R. norvegicus, Mus musculus*, and *Sigmodon hispidus* have also been implicated as reservoirs.[16,114] It has also been established that guinea pigs (*Cavia*) kept for food in Bolivia, Peru, and Ecuador serve as reservoirs for the disease transmitted to humans by household-dwelling reduviids.[2,71]

These are a few of the diseases in which rodents are involved in affecting the health of humans and domestic animals in Latin America. Our intent is to emphasize that the public health importance of rodents is much greater than generally realized. Since knowledge of the indigenous South American rodent fauna is limited, the full extent of their public health importance is unknown.[71]

B. Postharvest Food Loss

The most lamentable food losses are those that occur after harvest; the entire investment (time, money, labor, and supplies) employed to plant, cultivate, protect, harvest, and store the crop is lost. That rodents play a significant role in postharvest food losses is a widely accepted premise.[20,28,53,58,79,89,92] That the true magnitude of the problem is unknown, is fact.[58,62] The durable commodities like cereal crops and cereal products are the most important in terms of quantity stored and are, therefore, frequently damaged by rodents.[58,79] However, other produce such as potatoes, fruits, and beans are also stored and subject to rodent damage.[58] The extent of losses in postharvest situations is unknown because there are no practical methods for obtaining loss estimates.[19,47,53,62,79,92] Several rodent groups have been implicated as pests of stored products in Latin America; the most important ones are, undoubtedly, the cosmopolitan murids, *Rattus* and *Mus*. Others named included *Oryzomys, Calomys, Akodon, Heteromys, Peromyscus, Ototylomys, Eligmodontia, Neotoma*, and *Proechimys*.[36,58,113]

Much concern about postharvest food losses in Latin America has been expressed. The United Nations Food and Agricultural Organization's Regional Office for Latin America and the Carribean (FAO/RLAC) and the Centro Nacional de Treinamento en Armazenagem (CENTREINAR) in Vicosa, Brazil, have sponsored two Latin American seminars on postharvest losses; rodent problems were included as a major topic in each.[19,32,37] FAO/RLAC has published a collection of booklets on various postharvest grain management technologies and procedures; one of these was on rodent damage and control.[36]

C. Agriculture

The development of Latin American agroecosystems covers a broad spectrum — from slash-and-burn plots, to small, multi-crop subsistence farms, to large, monoculture plantations. Although more than one-third of Latin America's population is engaged in agriculture, many parts of South America and almost all of Central America are considered to be diet-deficient areas.[21,42] There are many contributing reasons; among them are food losses to rodents. That rodents are pests of some notoriety in this part of the world is not a recent phenomenon. Depictions on Incan and pre-Incan pottery in museums in Lima, Peru, show rodents feeding on ears of corn; in the agricultural calendar of the Inca, protection of crops from rodents and other vertebrate pests was specified along with cultivation, harvest, and storage.[48,86,87]

Rodents, and other animals, become pests only because they somehow come into conflict with man; it is generally a direct result of man's activities that these conflicts arise. For example, efforts to increase agricultural production in Latin America have involved environmental changes which appear to influence the types and degree of man/rodent conflicts. A principal means for countries to increase agricultural production has been to bring new lands under cultivation by clearing forest, scrub, or marsh areas for agriculture. The newly available food sources and the disappearance of natural predators often benefit rodent populations. Other efforts to increase food production have included introduction of exotic species, improvement of native crop varieties, more intensive management practices, expansion of grazing areas for livestock, and irrigation schemes. Such activities inevitably result in radical modification of environments, beneficial to some forms of wildlife and detrimental to others.[33,39,42,72,74,81,110] The resultant changes in rodent populations may be dramatic or subtle, and the importance of any group of rodents to man's interests may change in a significant way.[27,39,50,71,77,99]

We have focused this review on the most important crops — arbitrarily defined as those produced in more than half of the 21 Latin American countries. Virtually all of these crops are subject to rodent damage; we found at least one reference to rodent damage for all but two (cabbage and peppers) of the crops listed (Table 1). The specific crop for which we found the most references to rodent damage was sugar cane. This was not surprising since Latin America is the world's leading producer of sugar cane.[42] We found numerous references to rodent damage in cereal crops; most dealt with rodent problems in rice, but maize and wheat were specified in many papers. Rodent problems in root and tuber crops, especially cassava; oil crops, especially coconut; cacao; and bananas were also discussed in a number of papers. Finally, references to rodent damage in forest and forage crops were relatively more numerous than those dealing with other major crops.

1. Sugar Cane

Sugar cane is produced in every Latin American country except Chile. Of the total area devoted to this crop in the world, 44% is in Latin America which produces 47% of the world total.[42] Twenty-two rodent genera were named as pests of sugar cane in Latin America; the most frequently mentioned groups were *Holochilus, Sigmodon, Rattus*, and several genera of the Geomyidae. The geomyids and *Sigmodon* were the major pests in Mexico, Central America, and part of Colombia.[1,16,23,38,45,49,50,58,69,95,103-105] *Holochilus* was of primary importance in South America, particularly in Venezuela, the Guianas, and Brazil.[3,4,9-11,54,58,68,75,76,84,88,107,108] Other genera that were named in a number of reports include *Peromyscus, Zygodontomys, Oryzomys*, and *Reithrodontomys*.[1,11,52,58,63,76,113]

2. Cereal Crops

Rice is produced in each of the 21 Latin American countries. Among the developing market economies of the world, Latin American rice production is second only to the Far

Table 1
REPORTS OF RODENT DAMAGE TO MAJOR CROPS IN 21 LATIN AMERICAN COUNTRIES

	No. producing countries[a]	No. of rodent damage references
Cereals		
Rice	21	28
Maize	20	19
Sorghum	15	4
Wheat	12	10
Roots and tubers		
Potato	17	7
Cassava	17	11
Sweet potato	11	5
Pulses		
Dry beans	17	3
Dry peas	11	2
Soybeans	11	1
Vegetables and melons		
Tomato	19	2
Cabbage	15	—
Dry onion	15	3
Watermelon	13	4
Cantalope	12	2
Pepper	11	—
Grapes		
Grapes	11	4
Sugar crops		
Sugarcane	20	33
Fruits		
Oranges	19	3
Banana	17	9
Pineapple	16	6
Avocado	15	3
Mango	14	2
Plantain	14	4
Papaya	11	2
Lemon/lime	11	1
Tobacco crops		
Tobacco	17	2
Fiber crops		
Cotton	14	7
Oil crops		
Cottonseed	14	7
Coconuts	14	11
Groundnuts	13	1
Coffee, cacao, tea		
Coffee	15	6
Cacao	11	10
Forest crops*	21	9
Forage*	21	9
Other or unspecified*	—	35

[a] Data from Food and Agriculture Organization, *Production Yearbook 1983,* FAO Statistics Series No. 55, Food and Agriculture Organization of the United Nations, Rome, 1984. Entries added by these authors are indicated by an asterisk.*

East[42] We found reports implicating 22 different rodent genera as pests in rice fields, but certain ones were evidently more important than others based on frequency of reference. These were *Holochilus, Sigmodon, Oryzomys, Zygodontomys*, and *Rattus*.[3,15,17,33,38,58,61,67,68-70,106,108,110,113,116,117,119]

Maize is probably the most important cereal grain in Latin America, and, according to FAO,[42] is produced in 20 of the 21 countries considered here; no production statistics were reported for Surinam. Latin American maize production is second only to that of North America and first among the developing market economies of the world.[42] More than 20 rodent genera were mentioned as problems in the literature we examined. Again, a few groups were reported with sufficient frequency to indicate them as the more important pests of this crop. The pocket gophers (family Geomyidae) were apparently the principal rodent pest of maize throughout their range.[35,58,87,103-105,113] *Sigmodon, Oryzomys, Akodon*, and *Rattus* were other rodent groups that appeared as important pests of maize in Latin America.[1,5,36,38,48,58,68,83,93,113]

Wheat is produced in 13 Latin American countries; major producers reported were Argentina, Brazil, Chile, and Uruguay.[42] Fewer rodent genera were implicated as pests in this crop; 12 were named in the literature reviewed, with only a few citations each. This was likely due to several factors. Wheat is produced in a much more restricted area than sugar cane, rice, or maize, and production areas are more temperate with relatively fewer kinds of rodents. *Oryzomys, Octodon, Rattus, Akodon, Phyllotis*, and *Mus* appear to be the principal rodent pests of wheat.[5,35,46,54,58,74,93,106]

3. Oil Crops

Major oil crops in Latin America are coconut, cottonseed, and groundnuts.[42] Oil palm has been introduced into several areas including Venezuela, Colombia,[96] Brazil, and Peru.[58] Reports of damage to young palms in newly established plantations in Brazil and Peru did not identify the rodents.[58] In Colombia, *Proechimys* caused similar damage.[111] Coconuts are produced in 15 Latin American countries; Mexico is the leading producer with Brazil, Venezuela, Ecuador, and Colombia following.[42] Rodent damage can be a limiting factor to the coconut industry. For example, rodent damage as high as 77% occurred in one coconut producing area of Colombia.[31] This high damage resulted in a problem of serious economic consequence to the local community and closure of a coconut processing plant. Of eight genera implicated in the reports we examined, those most frequently named as pests of oil crops were *Rattus* and *Sigmodon*.[1,31,35,38,58,63,106,109,112,113] Most reports on cotton did not distinguish between fiber and oilseed production and are summarized under "Forest, Forage, Fiber, and Other Crops."

4. Horticultural Crops

Root crops are important dietary staples in Latin America as they are elsewhere in the world.[59] Potatoes and cassava are each produced in 17 Latin American countries and sweet potatoes in 11.[42] The literature reviewed implicated 19 rodent genera as pests; of these, the more important were *Oryzomys, Phyllotis*, and the geomyids.[5,35,48,54,58,63,94,103-105]

Pulse crops are produced to some extent in all Latin American countries although production data for French Guiana were unavailable.[42] Brazil and Mexico were the largest producers.[42] Although pulses are an important source of dietary protein in much of Latin America, few reports of rodent damage to these crops were encountered; *Rattus* reportedly caused damage to beans in Brazil,[58] and *Pappogeomys* was said to do considerable damage to bean fields in Mexico.[113]

Vegetables and melons are produced in all countries of Latin America; specific crops and production varied considerably among the different countries but Brazil and Mexico were the leading producers.[42] Relatively little information on rodent damage to these crops was encountered. *Rattus, Sigmodon, Oryzomys, Holochilus, Phyllotis, Punomys, Andinomys, Orthogeomys, Hydrochaeris*, and *Agouti* were among the rodent genera named as causing damage to vegetable and melon crops.[1,4,5,35,38,76] The general nature of these reports was anecdotal and serious rodent problems are likely very localized in nature except in periods of high rodent populations or "ratadas".[38,48,60,71,82,108]

Latin America is a major producer of fruits, being the leading world producer of some including bananas, oranges, avocados, mangos, and papayas.[42] Damage to fruits was attributed to 18 genera of rodents in the literature examined. Geomyid rodents were noted as causing significant damage to bananas.[35,58,103-105,113] Bananas have become the most economically influential plant in Central America[63] and are of major importance as an export crop in numerous countries of tropical America. Bananas, avocados, and mangos were noted as preferred foods of pacas (*Agouti*) and agoutis (*Dasyprocta*) in South America.[13,113] Our review also indicated that pineapple was highly susceptible to rodent damage. Total losses of pineapple crops to *Punomys* and *Oryzomys* were reported in Peru.[5] *Holochilus* and *Sigmodon* damaged pineapple in Venezuela,[4,68] and *Sigmodon* was a pest of this crop in Nicaragua[12] and Mexico.[1,58] Cultivated *tuna*, the edible fruit of the prickly pear cactus, was extensively damaged by *Octodon* in Peru and Chile.[46,83,118] Fruit trees, particularly citrus and apple, were often girdled by a variety of rodents including *Oryzomys*[48] and *Sigmodon*.[98]

Grapes are produced in 11 Latin American nations; they are exported by Chile. Wine production is an important industry in several countries including Chile, Argentina, Uruguay, and Peru. Rodents implicated as vineyard pests included *Oryzomys* and *Octodon*.[46,48,83,118]

Although coffee is of African origin and is grown in practically every tropical country, coffee production is greatest in the Western Hemisphere.[59] Coffee is grown in 15 Latin American countries and is an important export crop for most of them; Brazil and Colombia were the world's leading producers.[42] Rodent groups reported to damage coffee included *Punomys, Oryzomys, Orthogeomys*, and *Rattus*.[5,36,58,87,106,113] In the Central American countries, pocket gophers (family Geomyidae) damaged root systems, often killing the entire plant. Coffee seedlings may also be destroyed by these rodents.

Cacao, in contrast to coffee, is indigenous to the American tropics, but Africa is the leading producer.[59] Brazil and Ecuador were major producers among the 11 Latin American countries where it is grown.[42] Several rodent groups were mentioned as significant pests in Latin American cacao plantations. For example, squirrels (*Sciurus*) damaged cacao fruits in Costa Rica,[63] and reports from Colombia[69,110] and Peru[5] named *Sciurus* and *Rattus*. Several rodent groups reportedly caused damage in Brazilian plantations including *Nectomys, Rattus, Thomasomys, Rhipidomys*, and *Akodon*.[25] Other reports cited *Sigmodon, Holochilus, Peromyscus, Punomys*, and *Oryzomys*.[5,93,94]

Although tobacco is a major crop and is produced in 17 Latin American countries, we found only two references to rodent damage;[1,58] both were from Mexico, but the rodent species was not named.

5. *Forest, Forage, Fiber, and Other Crops*

Forests and woodlands cover about 45% of the land area in Latin America, ranging from about 4% in Uruguay to 96% in Surinam.[42] Most of these areas are natural forest; others are intensely managed plantations of exotic species. Forests and woodlands provide a variety of products from fuelwood to lumber and paper, and are of major economic importance. For example, forestry has become the second major industry in Chile.[90] Forestry, as an industry, is important in several Latin American countries;[18,78,91] reforestation, as a necessity, is important in many areas.[14,27,43,47]

Rodents were reported as impediments to successful reforestation in several countries.[18,58,78,91,95,106] In Chile, for example, *Octodon bridgesi* was described as the primary pest in reforested areas planted to *Pinus radiata*.[85] Recent studies estimated a 43% incidence of damage and a tree mortality rate of 12.5% in young plantations.[91] Other rodents implicated in damage to Latin American forest crops include *Clyomys, Agouti, Coendou, Sciurus,* and the geomyids.[18,35,83,95,113]

Pasture or forage crops are also subject to damage by rodents. Damage may take many different forms. For example, it may occur as direct competition with livestock for available forage, as physical damage by burrowing rodents which is a hazard to livestock or contributes to soil erosion, or as destruction of forage either before or after harvest. Several different rodent groups were reported to cause such damage in Latin America. In Mexico and Central America, *Dipodomys*, the geomyids, especially *Orthogeomys*, and *Sigmodon* were reported to compete with cattle for forage and damage alfalfa crops and dairy pastures.[38,58,103,104,113] In Peru, four genera, *Oryzomys, Andinomys, Phyllotis*, and *Cavia*, were named as causing damage to natural grass pastures, and losses of 10 to 30% in alfalfa crops.[5,94] Destruction of rangeland by rodents was described as an important problem in Chile,[93] and rodents of the genera *Ctenomys* and *Lagostomus* were named as pests in pasture and forage in Argentina.[35,58]

A variety of fiber crops are grown in Latin America including flax, hemp, jute, sisal, and cotton. Cotton is, by far, the most important fiber crop, being produced in 14 countries.[42] This was the only fiber crop for which we found reports of rodent damage. Rodents implicated as pests in cotton were *Sigmodon, Oryzomys*, and *Holochilus*.[1,38,48,58,68,94]

Numerous reports which discussed rodent damage in Latin American countries did not specify the crop or crops damaged, and, all too often, did not identify the rodent species involved. The generic phrase "... rodents are an important problem in cultivated crops ..." was common in the literature. Considerable work remains to be done in defining rodent problems associated with minor crops and localized areas.

IV. DISCUSSION

The variety of rodent problems now identified in Latin America reflects substantial progress in defining the rodent species associated with major crop, disease, and postharvest problems. However, few field data have been gathered for the population ecology of most species, their behavior in man-modified environments is largely unknown, and virtually none of the problems caused in the region by rodents have been defined in economic terms.

Although most of the common rodenticides and other control materials are commercially available throughout the region, their use is largely restricted to problems involving the commensal rodent species that occur in cities, towns, and ports. Except for the few native rodent species ranging into North America, no toxicological data are available upon which to base commercial formulations for cropland use.

Numerous threatened and endangered species occur in Latin America, along with a great variety of rare fauna and flora; however, we know of no situation where questions of ecological impact have been examined in relation to rodent damage problems or control activities. The successful efforts made in many countries in recent years to develop ecologically-based crop damage control programs are, with very few exceptions, not evident in Latin America.

Clearly, new materials, techniques, and programs unique to the region will be required in some of the diverse agricultural situations where problem rodent species are encountered. We do not expect that existing rodent-damage-control technology could be adapted to deal adequately with all of the problems touched in this review — although adaptation is a starting point that should be pursued. The limits of adaptation apply to developments within the region as well. Materials and delivery systems that work well for the labor-intensive efforts on small farms will be of little practical help or economic benefit on the large mechanized farms that are common in parts of a Latin America — even when the same crops and rodent species are involved.

Long-term solutions to rodent problems in agriculture and public health will require increased emphasis on training in the public and private sectors and support for coordinated research efforts throughout the region. Initial research should examine existing and potential control materials and delivery systems for principal rodent groups causing crop damage in some of the major agroecosystems. The opportunity exists in Latin America, as in few other areas of the world, to approach this process with emphasis on the development of environmentally compatible techniques, supporting databases, and mitigation of impacts on indigenous fauna and flora.

The past decade of effort by Latin American scientists has produced a substantial increase in the knowledge of rodent problems in the region; nearly 70% of our citations were published in this period. However, most of this work deals with problem definition, serving principally to confirm what farmers already knew — that serious rodent problems occur in many parts of Latin America. We anticipate that the next decade will produce a start toward solving some of these problems in the interest of increased food production, improved human and animal health, and environmental protection.

REFERENCES

1. **Abarca, R. M.,** Rodent control in sugarcane fields (Mexico), in *Proc. 2nd Inter-American Sugar Cane Seminar,* Florida International University, Miami, Fla., October 6 to 8, 1981, 189.
2. **Acha, P. N. and Szyfres, B.,** *Zoonoses and Communicable Diseases Common to Man and Animals,* Sci. Publ. No. 354, Pan American Health Organization, Washington, D.C., 1980.
3. **Aguero, D.,** Analisis reproductive de una poblacion de *Holochilus brasiliensis* (Rodentia — Cricetidae) en cultivos de arroz del Estado Portuguesa, *Agron. Trop.,* 28, 101, 1978.
4. **Aguero, D., Espinoza, Y., and Tovar, I.,** Evaluacion de tres raticidas en condiciones de campo en cana de azucar, *Agron. Trop.,* in press.
5. **Aguilar, P. G., Beingolea, O. D., Brack, A. J., and Ceballos, I.,** Vertebrados importantes en la agricultura Peruana, *Rev. Peru. Entomol.,* 20, 25, 1977.
6. **Akiev, A. K.,** Epidemiology and incidence of plague in the world, 1958-79, *Bull. W. H. O.,* 60, 165, 1982.
7. **Almeida, C. R., Almeida, A. R., Baptiste Vieira, J., Guida, U., and Butler, T.,** Plague in Brazil during two years of bacteriological and serological surveillance, *Bull. W. H. O.,* 59, 591, 1981.
8. **Alvarez, V., Rivera, G., Neghme, A., and Schenone, H.,** Triquinosis en animales de Chile, *Bol. Chil. Parasitol.,* 25, 83, 1970.
9. **Bates, J. F.,** Rodent control in sugar cane in British Guiana, in *Proc. Br. West Indies Sugar Technol.,* 61, 1960.
10. **Bates, J. F.,** The cane field rat in British Guiana and its control, *Proc. Int. Soc. Sugar Cane Technol.,* 11, 695, 1963.
11. **Bates, J. F.,** Rodents in sugar cane — their biology, economic importance and control, in *Pests of Sugar Cane,* Williams, J. R., Metcalfe, J. R., Mungomery, R. W., and Mates, R., Eds., Elsevier, Amsterdam, 1969, 541.
12. **Bojorge, R.,** personal communication, 1971.
13. **Borrero, H. J. I.,** *Mamiferos Neotropicales,* Universidad del Valle, Departamento de Biologia, Cali, Colombia, 1967.
14. **Bucher, E. H. and Schofield, C. J.,** Economic assault on Chagas' disease, *New Sci.,* 92, 321, 1981.
15. **Cabrera, M. I.,** Habitos alimentarios de *Holochilus venezuelae* (Allen, 1904) (Rodentia, Cricetidae), en cultivos de arroz del Estado Portuguesa, Trabajo especial de grado, Universidad Central de Venezuela, 1984.
16. **Cameron, G. N. and Spencer, S. R.,** *Sigmodon hispidus, Mamm. Species,* 158, 1, 1981.
17. **Cartaya, E. and Aguilera, M.,** Estudio de la comunidad de roedores plaga en un cultivo de arroz, *Acta Cient. Venez.,* 36, 250, 1985.
18. **Carvalho, C. T. de and Bueno, R. A.,** Animais causando danos em plantios, *Silvic, Sao Paulo,* 9, 39, 1975.

19. Centro Nacional de Treinamento en Armazenagem, *Perdidas de Post-cosecha de Alimentos en Paises en Desarrollo,* Serie CENTREINAR No. 4, Centro Nacional de Treinamento en Armazenagem, Vicosa, Brazil, 1982.
20. **Charbonnier, R. and Vico, G.,** Aspectos generales de la produccion y almacenamiento de granos en Uruguay, in *Anais do 1 Seminario Latino-Americano de Perdas Pos-Colheita de Graos,* Vicosa, Brazil, 1983, 335.
21. **Cochrane, W. W.,** *The World Food Problem,* Thomas Y. Crowell, New York, 1969, 36.
22. **Cockrum, E. L.,** *Introduction to Mammalogy,* Ronald Press, New York, 1962, chap. 16.
23. **Collado, J. C. and Ruano, M. A.,** The rat problem in the sugar cane plantations of Mexico, *Proc. Int. Soc. Sugar Cane Technol.,* 11, 705, 1962.
24. **Corbet, G. B. and Hill, J. E.,** *A World List of Mammalian Species,* British Museum of Natural History, Comstock Publishing, London, 1980, 128.
25. **Cruz, P. F. N. da,** Ocorrencia e avaliacao de danos causados por roedores pragas do cacaueiro na Bahia, Brasil, *Rev. Theobroma,* 13, 59, 1983.
26. **Darlington, P. J., Jr.,** *Zoogeography: The Geographical Distribution of Animals,* John Wiley & Sons, New York, 1966, chap. 7.
27. **Daugherty, H. E.,** The impact of man on the zoogeography of El Salvador, *Biol.Conserv.,* 4, 273, 1972.
28. **Dell'Orto, T. H.,** Produccion y post-cosecha de granos en Chile, in *Anais do 1 Seminario Latino-Americano de Perdas Pos-Colheita de Graos,* Vicosa, Brazil, 1983, 211.
29. **Dorst, J.,** *South America and Central America, A Natural History,* Random House, New York, 1967.
30. **Dubock, A. C., Ed.,** *Proc. Conf. Organization and Practice of Vertebrate Pest Control,* ICI Plant Protection Division, Surrey, England, 1984.
31. **Elias, D. J. and Valencia, G. D.,** Control de roedores en el cocotero, *ICA Inf.,* 8, 5, 1973.
32. **Elias, D. J.,** Las plagas vertebradas: su papel en perdidas post-cosechas, in *Anais do 1 Seminaria Latino-Americana de Perdas Pos-Colheita de Graos,* CENTREINAR, Vicosa, Brasil, 1983, 143.
33. **Elias, D. J.,** El problema de roedores en el proyecto de riego del Bajo Rio San Francisco: status y recomendaciones, informe final al Project Manager, United Nations Food and Agriculture Organization, Proyecto BRA/78/007, Aracaju, Brazil, 1983.
34. **Elias, D. J., Ed.,** Symposium: Zoologia Economica y Vertebrados Como Plagas de la Agricultura, IX Congreso Latinoamericano de Zoologia, Arequipa, Peru, 1984.
35. **Elias, D. J. and Valencia, G. D.,** La agricultura Latinoamericana y los vertebrados plaga, *Interciencia,* 9, 223, 1984.
36. **Elias, D. J.,** *Roedores Como Plagas de Productos Almacenados: Control y Manejo,* Serie: Tecnologia Postcosecha No. 3, Oficina Regional de la Food and Agriculture Organization para America Latina y El Caribe, Santiago, Chile, 1984.
37. **Elias, D. J.,** Importancia de las plagas vertebradas en las perdidas post-cosecha, in *Anais do Mesa Redonda Latino-Americana Sobre Perdas Pos-Colheita de Graos,* CENTREINAR, Vicosa, Brazil, 1985, 13.
38. **Espinoza, H. R. and Rowe, F. P.,** Biology and control of the cotton rat *Sigmodon hispidus, PANS,* 25, 251, 1979.
39. **Fall, M. W.,** Agricultural development and the ecology of rodent control, in *Alternative Strategies for Desert Development and Management, Vol. 2,* Pergamon Press, New York, 1982, 443.
40. **Food and Agriculture Organization - World Health Organization,** *Bibliography on Rodent Pest Biology and Control, 1950-1959,* Part I WS/E2270/E/12.73/1/2600 and Part II MI/E2270/E/12.93/1/2600, FAO/WHO, Rome, 1973.
41. **Food and Agriculture Organization - World Health Organization,** *Rodent Pest Biology and Control — Bibliography 1970-1974,* FAO Plant Production and Protection Paper No. 7, FAO, Rome, 1977.
42. **Food and Agriculture Organization,** *Production Yearbook 1983,* Vol. 32, FAO Statistics Series No. 55, Food and Agriculture Organization of the United Nations, Rome, 1984.
43. **Flavin, C., Pollock, C., and Postel, S.,** Wood crisis, wood renaissance, *Interciencia,* 10, 187, 1985.
44. **Flint, M. L. and van den Bosch, R.,** *A Source Book on Integrated Pest Management,* International Center for Integrated and Biological Control, University of California at Riverside, Riverside, Calif., 1977.
45. **Fors, A. L.,** Certain philosophies related with the behaviour and control of rodents (rats and gophers) in Mexico, Central America and the Caribbean, in *Proc. 2nd Inter-American Sugar Cane Seminar,* Florida International University, Miami, Fla., October 6 to 8, 1981, 201.
46. **Fulk, G. W.,** Notes on the activity, reproduction and social behavior of *Octodon degus, J. Mammal.,* 57, 495, 1976.
47. **Giles, P. H. and Rodriguez, R. E.,** El programa del desarrollo de pequenos silos metalicos para la conservacion de granos por agricultores en Bolivia oriental, in *Anais do 1 Seminario Latino-Americano de Perdas Pos-Colheita de Graos,* CENTREINAR, Vicosa, Brazil, 1983, 173.
48. **Gilmore, P. M.,** Cyclic behavior and economic importance of the rata-muca (*Oryzomys*) in Peru, *J. Mammal.,* 28, 231, 1947.

49. **Gonzales Romero, A., Tello Sandoval, G., Serna Secundino, J., and Aguilar, V. M.,** Damages caused by cotton rat, *Sigmodon hispidus zanjonensis,* on sugar cane in San Pedro Sula, Honduras, *Proc. Vertebr. Pest Conf.,* 8, 231, 1978.
50. **Gonzalez-Romero, A., Galina-Tessaro, P., and Alvarez-Cardenas, S.,** Estudio comparativo de las comunidades de roedores en una zona agricola del sur del valle de Mexico, in *Symposium: Zoologia Economica y Vertebrados Como Plagas de la Agricultura,* IX Congreso Latinoamericano de Zoologia Arequipa, Peru, October 9 to 15, 1983, 136.
51. **Gratz, N. G.,** The global public health importance of rodents, in *Proc. Conf. Organization and Practice of Vertebrate Pest Control,* ICI Plant Protection Division, Surrey, England, 1984, 413.
52. **Hampson, S. J.,** A review of rodent damage to sugar cane with criteria for the use of rodenticides, in *Proc. of a Conference on the Organization and Practice of Vertebrate Pest Control,* ICI Plant Protection Division, Surrey, England, 1984, 227.
53. **Harris, K. L. and Linblad, C. J., Eds.,** *Postharvest Grain Loss Assessment Methods,* American Association of Cereal Chemists, St. Paul, Minn., undated, chap. 6.
54. **Hershkovitz, P.,** South American marsh rats, genus *Holochilus,* with a summary of sigmodont rodents, *Fieldiana Zool.,* 37, 639, 1955.
55. **Hershkovitz, P.,** A geographical classification of neotropical mammals, *Fieldiana Zool.,* 36, 581, 1958.
56. **Hershkovitz, P.,** The recent mammals of the neotropical region: a zoogeographic and ecological review, *Q. Rev. Biol.,* 44, 1, 1969.
57. **Honacki, J. H., Kinman, K. E., and Koeppl, J. W.,** *Mammal Species of the World: A Taxonomic and Geographic Reference,* Allen Press, Lawrence, Kansas, 1982, 345.
58. **Hopf, H. S., Morley, G. E. J., and Humphries, J. P. O.,** *Rodent Damage to Growing Crops and to Farm and Village Storage in Tropical and Subtropical Regions,* Centre for Overseas Pest Research and Tropical Products Institute, London, 1976.
59. **Hubbell, D. S.,** *Tropical Agriculture: An Abridged Field Guide,* Howard W. Sams Int., Kansas City, Mo., 1965.
60. **Hudson, W. H.,** *The Naturalist in La Plata,* D. Appleton, New York, 1892.
61. **Inchaurraga, L.,** Prevencion del ataque de ratas en arroz de riego con laminas de zinc, *Bol. Ciarco,* 3, 166, 1973.
62. **Jackson, W. B.,** Evaluation of rodent depredations to crops and stored products, *EPPO Bull.,* 7, 439, 1977.
63. **Janzen, D. H., Ed.,** *Costa Rican Natural History,* University of Chicago Press, Chicago, Ill., 1983.
64. **Kaukeinen, D. E.,** *Bibliography: Vertebrate Pest Biology and Control: The World Scientific Literature for 1984,* ICI Americas, Goldsboro, N.C., 1985.
65. **Keast, A.,** Comparisons of contemporary mammal faunas of southern continents, in *Evolution, Mammals, and Southern Continents,* Keast, A., Erk, F. C., and Glass, B., Eds., State University of New York Press, New York, 1972, 433.
66. **Kowalski, K.,** *Mammals: An Outline of Theriology,* Panstwowe Wydawnictwo, Naukowe, Warsaw, 1976.
67. **Kverno, N. B., Elias, D. J., Vogel, E., and Bojorge, R.,** Exploratory studies to reduce cotton rat damage to rice in Nicaragua, unpublished report, DWRC, 1971.
68. **Leon, M. E.,** Rodent pest species in rice and sugar cane crops in Venezuela, and applicable control methods, M.S. thesis, Colorado State University, Fort Collins, Col., 1985.
69. **Londono, V. F.,** Ratas y ratones en Latinoamerica y su control con piriminil, *Somos Rohm Hass,* 18, 20, 1977.
70. **Londono, J. F.,** Vertebrados y su control en arroz, *Arroz,* 285, 4, 1977.
71. **MacKenzie, R. B.,** Public health importance of rodents in South America, *Bull. W. H. O.,* 47, 161, 1972.
72. **Manjon, M. C., Camarasa, L., Torres, M. P., Polop, J. J., and Kravetz, F. O.,** Estudio de densidad y desplazamiento de roedores en un campo de girasol, in *Symposium: Zoologia Economica y Vertebrados Como Plagas de la Agricultra,* IX Congreso Latinoamericano de Zoologia, Arequipa, Peru, October 9 to 15, 1983, 156.
73. **Mares, M. A.,** The scope of South American mammalian biology: perspectives on a decade of research, in *Mammalian Biology in South America,* Vol. 6, Mares, M. A. and Genoways, H. H., Eds., Spec. Publ. Series, Pymatuning Lab of Ecology, University of Pittsburgh, Pittsburgh, Pa., 1982, 1.
74. **Martinez-Palacios, C., Chavez-Sanchez, R., Ituarte-Soto, R., and Marquez-Moreno, A.,** Warfarin baits bagging to control the population of cotton rats in field crops in Sinaloa, Mexico, *Proc. Vertebr. Pest. Conf.,* 8, 226, 1978.
75. **Massoia, E.,** Caracteres y rasgos bioecologicos de *Holochilus brasiliensis chacarius* Thomas de la provincia de Formosa y comparaciones con *Holochilus brasiliensis vulpinus* (Brants) (Mammalia-Rodentia-Cricetidae), *Rev. Inv. Agropec. INTA,* 8, 13, 1971.
76. **Massoia, E.,** Ataques graves de *Holochilus* y otros roedores a cultivos de cana de azucar, *IDIA,* 321, 2, 1974.

77. **Meserve, P. L., Martin, R. E., and Rodriguez, M. J.,** Feeding ecology of two Chilean caviomorphs in a central mediterranean savanna, *J. Mammal.*, 64, 322, 1983.
78. **Muller, J. A. and Macedo, J. H. P.,** Notas preliminares sobre danos causados por animais silvestres em pinhoes, *Rev. For.*, 11, 35, 1980.
79. National Academy of Sciences, *Postharvest Food Losses in Developing Countries*, National Academy of Sciences, Washington, 1978.
80. **Ojasti, J.,** *Estudio Biologica Del Chiguire o Capibara*, Fondo Nacional de Investigaciones Agropecuari, Caracas, 1973.
81. **Ojasti, J.,** Papel ecologica de mamiferos en sabanas inundables, *Bol. Soc. Venez. Cienc. Nat.*, Vol. 36, 139, 59, 1981.
82. **Pearson, O. P.,** An outbreak of mice in the coastal desert of Peru, *Mammalia*, 39, 375, 1975.
83. **Pefaur, J. E., Yanez, J. L., and Jaksic, F. M.,** Biological and environmental aspects of a mouse outbreak in the semi-arid region of Chile, *Mammalia*, 43, 313, 1979.
84. **Porquez, P. H. and Ledesma, F. I.,** Sugar losses from cane damaged by rats, *Sugar News*, 46, 364, 1970.
85. **Ramirez, O.,** personal communication, 1983.
86. **Rhoades, R. E.,** The other side of the coin: animals as pests, *Proc. Third Annual Farming Systems Research Symposium*, Kansas State University, Manhattan, Kan., 1983.
87. **Rhoades, R. E. and Bidegaray, P.,** Aspectos socio-economicos de las plagas vertebrados: tradicion y cambio, *Symposium: Zoologia Economica y Vertebrados Como Plagas de la Agricultura*, IX Congreso Latinoamericano de Zoologia, Arequipa, Peru, October 9 to 15, 1983, 49.
88. **Rincones, C.,** Observaciones sobre danos producidos por ratas en nueve variedades de cana en Piritu, Estado Portuguesa, *Bol. CIARCO*, 10, 1, 1980.
89. **Rincon, C. C. J.,** Etapas de post-cosecha en maiz en el minifundio colombiano, in *Anais do I Seminario Latino-Americano de Perdas Pos-Colheita de Graos*, CENTREINAR, Vicosa, Brazil, 1983, 201.
90. **Rodriguez, M. J.,** personal communication, 1983.
91. **Rodriguez, M. J. and Herrera, G. L. A.,** Evaluacion del dano ocasionado por *Octodon bridgesi* en plantaciones de *Pinus radiata*, unpublished manuscript, 1984.
92. **Rowe, F. P. and Taylor, K. D.,** Rodent biology, in *Food Storage Manual, Part 1*, Tropical Stored Products Center, Ministry of Overseas Development, Slough, England, 1970, chap. 3.
93. **Sanchez, F. F.,** Rodents affecting food supplies in developing countries: problems and needs, *FAO Plant Prot. Bull.*, 23, 96, 1975.
94. **Sanchez Deza, P.,** Algunos aspectos del combate de roedores daninos aplicables a nuestras zonas agricolas de selva, *Publicaciones del Comite Cafetalero del Peru*, Lima, Peru, undated.
95. **Sanchez Navarette, F.,** Rodents as agricultural pests in Mexico: National Rodent Campaign, *Proc. Vertebr. Pest Conf.*, 8, 118, 1978.
96. **Sanchez Potes, A.,** personal communication, 1973.
97. **Sauer, C. O.,** Geography of South America, in *Handbook of South American Indians*, Vol. 6, Bull. No. 143, Bureau of American Ethnology, Smithsonian Institution, Washington, D.C., 1950, 319.
98. **Sequeira, J.,** personal communication, 1976.
99. **Simonetti, J. A.,** Occurrence of the black rat (*Rattus rattus*) in central Chile, *Mammalia*, 47, 131, 1983.
100. **Simpson, D. I. H.,** Viral haemorrhagic fevers of man, *Bull. W. H. O.*, 56, 819, 1978.
101. **Simpson, G. G.,** History of the fauna of Latin America, *Am. Sci.*, 38, 361, 1950.
102. **Simpson, G. G.,** *Splendid Isolation, The Curious History of South American Mammals*, Yale University Press, New Haven, Conn., 1980.
103. **Sisk, T.,** A report on Central American pocket gopher, *Orthogeomys* sp., biology and control in Costa Rica, ACM Field Biology Program, San Jose, Costa Rica, 1982.
104. **Sisk, T. and Vaughan, C.,** Notes on some aspects of the natural history of the giant pocket gopher (*Orthogeomys*) Merriam in Costa Rica, *Brenesia*, 22, 233, 1984.
105. **Sutherland, D., Vaughn, C., and Chavez, A.,** *Control de la Taltuza en Costa Rica*, Universidad Nacional, Escuela de Ciencias Ambienteles, Hereida, Costa Rica, undated.
106. **Taylor, K. O.,** Rodent problems in tropical agriculture, *PANS*, 18, 81, 1972.
107. **Twigg, G. I.,** Rodent damage to sugar cane in British Guiana, *World Crops*, 14, 150, 1962.
108. **Twigg, G. I.,** Studies on *Holochilus sciureus berbicensis*, a cricetine rodent from the coastal region of British Guiana, *Proc. Zool. Soc. London*, 135, 263, 1965.
109. **Valencia, G. D.,** Rat control in coconut palms in Colombia, *Proc. Vertebr. Pest Conf.*, 9, 110, 1980.
110. **Valencia, G. D.,** Los roedores como plagas agricolas en Colombia, in *Symposium: Zoologia y Vertebrados Como Plagas de la Agricultura*, IX Congreso Latinoamericano de Zoologia, Arequipa, Peru, October 9 to 15, 1983, 123.
111. **Valencia, G. D.,** personal communication, 1985.
112. **Valencia, G. D. and de Finke, E. Ortiz,** *Guia Para el Control de Ratas y Ratones*, Boletin Didactico No. 10, Instituto Colombiano Agropecuario, Palmira, Colombia, 1981.

113. **Walker, E. P.,** *Mammals of the World,* Vol. 2, 3rd Ed., Johns Hopkins Press, Baltimore, Md., 1975, 569.
114. **Weber, W. J.,** *Diseases Transmitted by Rats and Mice,* Thompson Publications, Fresno, Calif., 1982.
115. **World Health Organization - Food and Agriculture Organization,** *Bibliography on Rodent Pest Biology and Control, 1960-1969,* Parts I-IV, VBC/91.9, 9a, 9b, and 9c, 1971.
116. **Williams, J. O.,** Rodents and their problems in South America and the Caribbean, in *Proc. Conf. Organization and Practice of Vertebrate Pest Control,* ICI Plant Protection Division, Surrey, England, 1984, 53.
117. **Williams, J. O. and Vega, C.,** Brodifacoum as a rodenticide to control rodent damage in rice fields in Venezuela. I. State of Portuguesa, in *Proc. Conf. Organization and Practice of Vertebrate Pest Control.* ICI Plant Protection Division, Surrey, England, 1984, 341.
118. **Woods, C. A. and Boraker, D. K.,** *Octodon degus, Mamm. Species,* 67, 1, 1975.
119. **Woronecki, P. P.,** Rodent problems in the Darien of Colombia, *Biol. Conserv.,* 5, 227, 1973.

Chapter 3

RODENT PROBLEMS IN EUROPE

M. Lund

TABLE OF CONTENTS

I. INTRODUCTION

The magnitude of the rodent problem in Europe has never been evaluated in detail due to many factors. First of all the assessment of rodent damage is very time consuming and difficult to carry out, whether the damage is inflicted to crops or to structures. The significance of rodents as carriers and transmitters of diseases has, apart from a couple of well-known bacterial diseases (leptospirosis and salmonellosis), not been thoroughly investigated. In the following, the various problems created by the most common rodent species are briefly described, with the main emphasis on field rodents.

II. FIELD RODENTS

A. Microtidae

From an agricultural, forestial, and horticultural point of view rodents belonging to this family are by far the most important in Europe; and, due to their often strictly herbivorous feeding biology they are also difficult to control by traditional poison baits.

Microtus agrestis, the short-tailed field vole, is the most important rodent pest in aforestation areas and fruit orchards in Scandinavia. In late autumn or winter, young trees and seedlings are debarked, and as this species does not climb, the trees are most often completely girdled on the lower 10 to 15 cm of the stem, or they are gnawed in two at ground level. Large areas in northern Scandinavia with larch, spruce, or pine have, in certain years, been severely damaged,[2,44,3,16] whereas oak and beech are the most exposed tree species in the southern and central part of Europe.[9] As a dense grass cover is essential for the existence of *M. agrestis*, older forests or plantations (10 to 15 years) are usually not endangered. The trees can be attacked at various stages: as seedlings in nurseries or in forest plantations, or as grafts in seed orchards.[11] In naturally renewed forests damage is less pronounced than in plantations. Plantations with rarer and more expensive species of conifer like Chamaecyparis, Cryptomeria, and Thuja, are particularly attractive to *M. agrestis*.

According to Myllymäki, a rough estimate of the losses due to *M. agrestis* since World War II is $100 million.[15,17] Due to the cyclic fluctuations of vole populations the degree of damage varies considerably from year to year.[4]

Locally, field voles may attack crops early in the growing season, but the most serious type of damage is the destruction of stored hay in barns and stocks in northern Finland and Sweden. In a single year up to 160,00 tons of hay have been destroyed by voles in Sweden.[10]

Microtus arvalis, the common vole, is mainly confined to open, grass-covered plains and meadows, but fruit orchards are sometimes invaded, and in Finland 5 to 10% of the damage to horticulture is caused by this species.[17] In France *M. arvalis* is considered a serious pest, and the damage is roughly estimated to around $20 per ha.[26] The most important damage is, however, seen in winter wheat and alfalfa crops in Central Europe.[28] In Holland, West Germany, and Denmark, in certain years, this species has created problems by perforating the grass-cover on the ditches towards the North Sea, lowering their resistance to the waves. In the marshlands the burrowing activity just beneath the grass-cover loosens the grass so that it is easily torn up by the grazing cattle that may consume fatal amounts of earth from the roots.[1,25] Damage to leys and pastures as well as to leguminous and cereal crops are reported to be important in central Europe.[22]

Clethrionomys glareolus, the bank vole, is less graminivorous than the two Microtus species, but may also damage young trees by debarking them in the winter season. Primarily in Norway and Sweden, it is considered an important pest in forests in peak years,[3] even though the trees are less often completely girdled due to the climbing ability of the species. In northern Scandinavia stored hay in barns may also be destroyed by the bank vole. In central Europe (especially Czechoslovakia) serous damage is reported to tree seeds, young shoots, and buds, as well as bark of young trees above the snow level.[9,29,30]

Voles belonging to the genus *Pitymys (savii, subterraneus, and duodecimcostatus)* are responsible for considerable damage to agricultural and horticultural crops in central and southern Europe. *P. savii* is reported to be the most serious rodent pest to horti-floricultural crops in Italy.[23,24]

Arvicola terrestris, the water vole, is found in two subspecies, an aquatic and a fossorial or subterranean form. The former is primarily abundant in northern Europe, the latter in the central and southern parts. This rat-sized vole is the most important rodent pest in southern Scandinavia and northwestern Europe (especially Denmark, West Germany, and Holland). The damage is very varied, but the gnawing of tree roots in orchards, forests, and gardens in winter time is considered the greatest problem. Broadleaf trees, especially fruit trees, are more often attacked than conifers, but almost any tree species is at risk. In arable land fields with beets, potatoes, carrots, lucerne, and grass-seeds are often invaded and seriously damaged.

A particular, but important, type of damage is the gnawing of buried telephone cables and other wires at the road side.[1] The fossorial form may cause severe damage to grassland not only because of its burrowing activity, but also due to the consumption of plants.[14] This form does not invade forests but is a problem in gardens, orchards, and fields where it may reach a density of 1000 voles per ha, causing more than 50% destruction of the vegetation. In orchards, more than 25% of the fruit trees may be destroyed during an outbreak winter.[27]

Ondatra zibethica, the musk rat, is the largest of the Microtidae and was introduced into Czechoslovakia in 1905 as a fur animal. In a few decades it had spread all over western and central Europe. Particularly in the lowlands and marshlands of the west the musk rat is considered a pest of great economic importance due to its destruction of dikes and ditches, causing stagnation of the water regime.[12] This has serious consequences with respect to flooding, yield reduction, and loss of cultivated land in Holland, Belgium, and West Germany. In spite of a large-scale trapping program, combined with the application of rodenticides locally, it has not been possible to prevent the species from increasing in numbers or from spreading to new areas.[4,5]

B. Cricetidae

Cricetus cricetus, the common hamster, is a significant pest in central and eastern Europe (East Germany, Czechoslovakia, Hungary, and Rumania). The hamster's main impact is on seeds and green crops at all stages, as well as on fields with carrots, beets, and potatoes; but locally it is also important because it penetrates ditches and dikes. Health hazards are also of concern when the hamster occurs at peak densities (e.g., 800/ha).[7,8,18]

C. Muridae

Apart from the three major commensal species the family *Muridae* only contains a few relatively unimportant pest species.

In recent years *Apodemus sylvaticus*, the long-tailed field mouse, has changed its status from being of negligable importance, invading structures locally in autumn, into becoming a serious pest to sugar beet fields shortly after sowing. This was first recognized in the U.K., but later this type of damage was reported in West Germany and Denmark.[3,21]

Apodemus flavicollis, the yellow-necked field mouse, is an important commensal pest in a few countries, especially in Denmark where locally it is as abundant indoors as the house mouse.[13]

Apodemus agrarius, the striped field mouse, is in peak years a problem in eastern and central Europe, including southern Denmark, partly by transmitting leptospirosis (*Leptospira pomona*) to grazing cattle.[13] In most European countries a comparatively high standard of living, fairly rodent-proof structures, and an effective, modern garbage-removal system, are factors that have reduced the problems caused by *Rattus norvegicus* (the brown rat) in urban areas. In most rural areas, however, this commensal species still requires substantial control efforts in order to keep it at an acceptable level. In most countries it is considered a public health and a hygienic problem more than an economic one. The diseases particularily in focus are Weil's disease caused by *Leptospira icterohaemorrhagiae* and Salmonellosis (especially *S. typhimurium*). In recent years, however, viral diseases like haemorrhagic fever with renal syndrome (HFRS) have become a matter of concern in Europe.[19,20]

Rattus rattus, the house rat, is a rare species in northern Europe, but increases in number in central and southern Europe, where it is considered an important pest in warehouses food-stores, and grain-stores.

Mus musculus, the house mouse, is now a more important pest in cities than the rat. This species, too, is regarded mainly as a hygienic and health problem, especially as a potential transmitter of Salmonellosis (*S. typhimurium*). In recent years the increasing use of computers and other electronic devices has created a new rodent problem, as the small mice invade the electronic systems and gnaw the wires and other connections, sometimes with extreme consequences.

REFERENCES

1. **Bang, P.,** Damage by small mammals in Denmark, in *Biocontrol of Rodents,* Hansson, L. and Nilsson, B., Eds., Nordic Forestry Research Committee, Lund, Sweden, Ecol. Bull./NFR 19, 1975, 13.
2. **Bärring, U.,** Damage caused by voles on seedlings of Scots pine and Norway spruce in 1961-62, Research Notes No. 1, Department of Reforestation, Royal College of Forestry, Oslo, Norway, 40, 1963.
3. **Christiansen, E.,** Field rodent damage in Norway, in *Biocontrol of Rodents,* Hansson, L. and Nilsson, B., Eds., Nordic Forestry Research Committee, Lund, Sweden, Ecol. Bull./NFR 19, 1975, 37.
4. **Christiansen, E.,** Damage by small rodents in the Norwegian forests 1971-78, Research Paper 11/81, Norwegian Forest Research Institute, Oslo, Norway, 1981, 24.
5. **Commissie Muskusrattenbestrijding,** *Mininisterie van verkeer en waterstaat, jaarverslag 1982,* Wagingen, Holland, 1983, 28.
6. **European and Mediterranean Plant Protection Organization,** Report on the second meeting of the working party on the muskrat, *EPPO Publ. Ser.*, 35, 31, 1974.
7. **Gorecki, A.,** Consumption by and agricultural impact of the common hamster, *Cricetus cricetus*, on cultivated fields, *EPPO Bull.*, 423, 1977.
8. **Hamar, M. and Sutova, M.,** Effective control of the hamster (*Cricetus cricetus*) by gassing with polytanol and phostoxin, Report of the international conference on rodents, *EPPO Publ. Ser. A,* 58, 181, 1970.
9. **Hansson, L. and Zejda, J.,** Plant damage by bank voles (*Clethrionomys glareolus*) and related species in Europe, Report of the joint FAO/WHO/EPPO conference on rodents of agricultural and public health concern, *EPPO Bull.*, 7/2 223, 1977.
10. **Kanervo, V. and Myllymäki, A.,** Problems caused by the field vole, *Microtus agrestis* in Scandinavia, Report of the international conference on rodents, *EPPO Publ. Ser. A,* 58, 11, 1970.
11. **Larsson, T. -B.,** Damage caused by small rodents in Sweden, in *Biocontrol of Rodents,* Hansson, L. and Nilsson, B., Eds., Nordic Forestry Research Committee, Lund, Sweden, Ecol. Bull./NFR 19, 47, 1975.
12. **Litjens, B. E. J.,** Report from the Netherlands by the Wildlife division of the Ministry of Agriculture, EPPO working party on the musk rat, Waneningen, Holland, 1981, 28.
13. **Lund, M.,** Vertebrate pest problems in West Europe, in *Proc. Conf. Organisation and Practice of Vertebrate Pest Control,* Dubock, A. C., Ed., ICI Plant Protection, Surrey, England, 1984, 1.
14. **Meylan, A.,** Distribution and damage by *Arvicola terrestris* in Switzerland, First report of the working party on field rodents, *EPPO Publ. Ser. C,* 31, 23, 1974.
15. **Myllymäki, A.,** Outbreaks and damage by the field vole, *Microtus agrestis,* in Finland 1945-1973, First report of the working party on field rodents, *EPPO Publ. Ser. C,* 31, 13, 1974.
16. **Myllymäki, A.,** Outbreaks and damage by field rodents and other harmful small mammals in Finland, in *Biocontrol of Rodents,* Hansson, L. and Nilsson, B., Ed., Nordic Forestry Research Committee, Lund, Sweden, Ecol. Bull./NFR, 19, 17, 1975.
17. **Myllymäki, A.,** Oatbreaks and damage by the field vole, *Microtus agrestis,* since World War II in Europe, Report of the joint FAO/WHO/EPPO conference on rodents of agricultural and public health concern, *EPPO Bull.*, 7/2, 177, 1977.

18. **Nechay, G., Hamar, M., and Grulich, I.,** The common hamster, *Cricetus cricetus*, Report of the joint FAO/WHO/EPPO conference on rodents of agricultural and public health concern, *EPPO Bull.*, 7/2, 255, 1977.
19. **Niklasson, B. and LeDuc, J.,** Isolation of the nephropathia epidemica agent in Sweden, *Lancet*, 1, 1012, 1984.
20. **Osterhaus, A. D. M. E., Spijkers, I., Van Steenis, G., and Van der Groen, G.,** Hantaan virus infection in the Netherlands, *Ned. Tijdschr. voor Geneeskd.*, 128(52), 2461, 1984.
21. **Pelz, H.-J.,** Saatschäden durch Waldmäuse, *Apodemus sylvaticus*, an Zuckerrüben — Ursachen, Prognose und Abwehrmöglichkeiten, *Mitt. Biol. Bundesanstalt Land Forstwirtsch. Berlin-Dahlem*, 223, 275, 1984.
22. **Ryszkowski, L. and Myllymäki, A.,** Outbreaks of *Microtus arvalis* and other microtine rodents in central and eastern Europe, in *Biocontrol of Rodents*, Hansson, L. and Nilsson, B., Eds., Nordic Forestry Research Committee, Lund, Sweden, Ecol. Bull./NFR 19, 57, 1975.
23. **Santini, L.,** *Pitymys savii* in Italy: zones of regular outbreaks and evaluation of the damage done to agriculture in the period 1945-1972, First report of the working party of field rodents, *EPPO Pub. Ser. C*, 31, 25, 1974.
24. **Santini, L.,** European field voles of the genus *Pitymys* and their damage in agriculture, horticulture and forestry, Report on the joint FAO/WHO/EPPO conference on rodents of agricultural and public health concern, *EPPO Bull.*, 243, 1977.
25. **Schindler, U.,** Mäuseschäden und Mäusebekämpfung in Niedersachsen während der Erdmausmassenvermehrung 1951-1953, *Forstw. Cbl.* 73, 7/8, 193, 1954.
26. **Spitz, F.,** Developpement d'un modele de prevision des pullulations du campagnol des champs, Report of the joint FAO/WHO/EPPO conference on rodents of agricultural and public health concern, *EPPO Bull.*, 7/2, 341, 1977.
27. **Tahon, J.,** Damage caused by *Arvicola terrestris* and other subterranean mammals, Report of the international conference on rodents, *EPPO Publ. Ser. A*, 58, 107, 1970.
28. **Tertil, R.,** Impact of the common vole, *Microtus arvalis*, on winter wheat and alfalfa crops, Report of the joint FAO/WHO/EPPO conference on rodents of agricultural and public health concern, *EPO Bull.*, 7/2, 317, 1977.
29. **Tucek, F. J.,** Über Rötelmausschäden in den slowakischen Wäldern im Jahre 1959, *Z. Angew. Zool.*, 47/4, 449, 1960.
30. **Zejda, J.,** Ecology and control of *Clethrionomys glareolus* in Czechoslovakia, Report of the international conference on rodents, *EPPO Publ. Ser. A*, 58, 101, 1970.

Chapter 4

RODENT PROBLEMS IN AFRICA

Lynwood A. Fiedler

TABLE OF CONTENTS

I. INTRODUCTION

After several other major African agricultural pest problems such as locusts and armyworms were reduced, rodents received more attention and are now considered major pests.[183,244] A high degree of social tolerance by Africans to this problem and the tendency of several rodent species to irrupt or dramatically increase their number has resulted in major economic losses.[51] Despite these losses, very little investigation of agricultural rodent problem has occurred, nor is there currently any major existing or proposed effort to deal with rodent pests in Africa. While agricultural problems have received little attention, research on rodent-borne diseases affecting humans has received higher priority.

Taxonomic studies dominated the literature through the mid-1950s, when rodent ecology studies emerged, then expanded and continue to be a primary research topic.[51] Behavioral studies in the laboratory and field increased in the 1970s and have now become an important area of investigation.

Diversity in climate, weather, soil, topography, and vegetation has produced a wide variety of habitats occupied by an equally diverse rodent fauna. Most African countries have 24 to 36 rodent genera,[51] but the majority are not economically important. Rodents have been found at high altitudes and even observed in mines 610 m below the surface.[255] Deserts and semideserts contain mostly cricetids (*Gerbillus, Meriones*, and *Tatera*) and a dipodid (*Jaculus*) which are adapted to extremely arid conditions and frequently damage the dryland crops of subsistence farmers. Wet environments, such as marshes, harbor several rodents (*Dasymys, Otomys, Thryonomys*, and *Rhabdomys*)[144] capable of damaging crops. Grassland and woodland savannas contain *Tatera, Arvicanthis, Praomys*, and *Thryonomys*[51,144] which are responsible for agricultural losses that are normally chronic but can be periodically acute. When new agricultural areas are cultivated, most species increase in population and significant crop damage occurs.[51,81]

Rodent pest outbreaks or "irruptions" occurring at national or regional levels have been reported since at least 1905 (Table 1). The species involved most frequently are *Meriones shawi* in North Africa and *Praomys natalensis* and *Arvicanthis niloticus* in sub-Saharan Africa. These major rodent outbreaks have been somewhat predictable after periods of extended drought followed by normal rainfall. In other cases, excessive precipitation above the norm has sometimes been followed by increased pest rodent populations.[248] The fact that irruptions do not always occur after these unusual weather events indicates that the cause-and-effect relationships are complex and do not presently permit simple explanations.

Factors which may contribute to increased rodent reproduction and survival include diet, vegetative cover (both of which are influenced by rainfall), fewer predators, and decreased mammalian competition (both of which occur after extensive droughts). Seasonal rainfall patterns with distinct wet and dry periods occur in many areas and are associated with annual reproductive periods of many rodent species as well as migration of large mammals.[179,232] The relationship between increased rainfall and greater rodent numbers was first observed in Kenya[209,210] and later noted in *Praomys erythroleucus* in Sierra Leone.[26] Since then, more than 25 species have been reported to have reproductive periods related to rainfall patterns.

A number of agricultural crops are grown during wet seasons. They become susceptible to damage, particularly when rodent populations are increasing and/or establishing themselves within the sheltered environment provided by maturing crops. Major rodent pests prefer habitats of ripening grain.[245] After harvest, however, when conditions become arid and food and shelter are reduced, rodent populations generally decline due to reduced reproduction and increased mortality. The cycle repeats itself during the next wet season. Many areas experience two annual wet seasons — typically March-May (long rains) and September-November (short rains)[178] — resulting in a bimodal rodent reproductive and population pattern.

Table 1
REPORTED NATIONAL OR REGIONAL RODENT OUTBREAKS IN AFRICA

Year	Country	Species	Ref.
1905	Tunisia	*Meriones shawi*	187
1907—09	Tunisia	*Meriones shawi*	187
1920	Uganda	*Arvicanthis niloticus*	144
1925—26	Tanzania	*Praomys natalensis*	98
1929—30	Algeria	*Meriones shawi*	190
1929—31	Tunisia	*Meriones shawi*	187
1930—32	Tanzania	*Praomys natalensis*	98
1932	Morocco	*Meriones shawi*	82
1934	Morocco	*Meriones shawi*	82
1934—36	Algeria	*Meriones shawi*	82
1936	Tanzania	*Praomys natalensis*	144
1939—41	Morocco	*Meriones shawi*	82
1949—50	Morocco	*Meriones shawi*	21, 196
1951—52	Kenya, Tanzania	*Praomys natalensis, Arvicanthis niloticus*	106, 244, 248
1952—54	Morocco	*Meriones shawi*	82
1954	Tunisia	*Meriones shawi*	21
1955—56	Tanzania, Uganda	*Arvicanthis niloticus, Praomys natalensis*	36, 144
1961	South Africa	*Praomys* sp.	153
1962—63	Kenya, Sudan, Tanzania	*Praomys natalensis, Arvicanthis niloticus, Rhabdomys pumilio*	28, 242—244, 248
1963	South Africa	*Gerbillurus* sp.	108
1963—64	Morocco	*Meriones shawi*	82
1966—69	South Africa, Botswana	*Praomys natalensis, Tatera leucogaster*	233, 254
1966	Zambia	*Praomys natalensis*	226
1967	Zimbabwe	*Praomys natalensis*	37, 264
1968	Kenya	*Arvicanthis niloticus*	245, 248
1969	Sudan	*Praomys natalensis, Arvicanthis niloticus*	115
1971	Tanzania	*Praomys natalensis*	144
1970—72	Nigeria	*Praomys natalensis, Arvicanthis niloticus*	28
1975	Zimbabwe	*Praomys natalensis*	37, 70
1975—76[a]	Sudan	*Praomys natalensis, Arvicanthis niloticus*	28
1975—76[a]	Senegal, Mauritania, Mali	*Praomys huberti, Praomys erythroleucus, Arvicanthis niloticus, Taterillus* sp.	28, 66, 67, 118, 197—200
1975—76[a]	Senegal	*Praomys huberti, Arvicanthis niloticus*	66, 67, 198, 199
1975—76[a]	Senegal, Mauritania, Mali	*Praomys huberti, Praomys erythroleucus, Arvicanthis niloticus, Taterillus* sp.	28, 66, 67, 118, 197—200
1975—76[a]	Nigeria, Niger	*Praomys natalensis, Gerbillus*	28
1977—78	Kenya	*Arvicanthis niloticus Praomys natalensis*	4, 42
1978	Somalia	*Praomys natalensis*	14

[a] A major rodent outbreak occurred throughout the Sahel region during this time period.

Extreme population troughs and peaks are accentuated during and after prolonged drought. For example, a drought in Botswana from 1962 to 1965 resulted in reduced breeding by *Praomys natalensis*. After the drought, a massive rodent irruption took place.[233] Outbreaks occurring in Somalia (1978), Senegal (1975-76), and Kenya (1962) were also preceded by prolonged drought. The return of rains initiates a rapid growth of grasses[35,232] which provide food and protective cover for rodents. Since prolonged drought also affects grazing ungulates, which require longer periods to repopulate by reproduction, the rodent has little immediate competition. Raptors, carnivores, and other predators, reduced during drought, repopulate affected areas slowly, favoring rapid rodent population increases.

Seasonal increases in rodent-borne diseases would be expected at or following rodent population peaks. Human plague cases near the Kenyan border of Tanzania, an endemic plague focus area, normally increase in November-December, reaching a peak in January.[159,177] *Praomys natalensis*, a potential carrier of infected fleas, usually reach peak numbers at this time and are commonly found in households and adjacent farmland. Plague has been seasonal in most African countries[46] and a relationship to rodent population dynamics seems possible.

There are many taxonomic problems within several genera in Africa. Some are currently being addressed through a combination of morphometric, blood protein, and karyotyping studies. To some, this may seem more academic than practical, but important applications have been reported in epidemiological studies of an arenavirus[129] and plague. In the multimammate rat, *Praomys natalensis* species complex of southern Africa, the diploid chromosome number $2N = 36$ is susceptible to plague, but the $2N = 32$ is resistant.[124] Two distinct species were proposed naming the $2N = 36$ chromosome number as *P. coucha* and $2N = 32$ as *P. natalensis*.[89] Other rodent pest genera including *Praomys* of West Africa, *Arvicanthis, Aethomys, Gerbillus, Tatera*, and *Saccostomus* have been investigated to help resolve taxonomic questions.[85,130,134,206] Nomenclature and taxonomy in this text follow those of Honacki et al.[112]

II. AGRICULTURAL PROBLEMS

Despite decades of rodent research on biology, taxonomy, ecology, and disease, research on agricultural rodent pest problems in Africa has lagged. Periodic outbreaks involving *Praomys natalensis* or *Arvicanthis niloticus* and the resultant serious crop losses raise concerns, but during the interim years, insidious crop damage due to rodents receives very little attention. The emergency measures taken during these outbreaks are usually ineffective, temporary, acute actions which have not led to long-term solutions.[229]

Africa's climatic, geographic, and habitat diversity has resulted in 12 rodent families with about 240 species[53] of which 77 caused some economic damage (Table 2). Most countries (Figure 1), however, contain only about four to six species which are involved in significant damage to one or more agricultural crops.[71] The genera most frequently involved in field crop damage include *Praomys, Arvicanthis, Meriones, Tatera, Rhabdomys*, and *Thryonomys*, while primarily *Rattus, Praomys*, and *Mus* damage stored foods. Recognition of specific African rodents has been facilitated by an identification manual developed for field use.[162]

Agricultural damage ranges from traditional cereals to plantation, root, forestry, vegetable, and forage crops, as well as postharvest damage to stored foods. When available, cereals are preferred by major rodent pests possibly because of higher water, protein, and carbohydrate content.[171] Important cereal crops significantly damaged by African rodents include wheat, maize, sorghum, rice, millet, and barley. Root crops susceptible to severe damage include groundnut, cassava, and sweet potato, while plantation crops include oil palm, coconut, rubber trees, and cacao. Coniferous tree plantations (including cyprus) and other shrubs and trees receive locally severe damage from several rodent species. Cotton and sugar

Table 2
AFRICAN RODENTS REPORTED TO HAVE DAMAGED AGRICULTURAL CROPS

Family and Genus	Genus Ref.	Species	Ref.
Sciuridae			
Atlantoxerus		*getulus*	263
Funisciurus		*anerythrus*	18, 51, 185, 241, 246
Heliosciurus			
		gambianus	18, 51, 185, 246
		rufobrachium	61, 213
Protoxerus		*stangeri*	61
Sciurus		*carolinensis*	160, 235
Xerus		*erythropus*	18, 76, 78, 115
		inauris	270
		rutilus	70, 172, 173
Pedetidae			
Pedetes		*capensis*	31, 144
Cricetidae			
Cricetomys		*gambianus*	18, 25, 74, 97, 115, 144, 181, 235, 241, 246, 257
Gerbillurus	108	(probably *paeba*)	
Gerbillus	194	*campestris*	263
		gerbillus	5, 263,
		nanus	263
		pyramidum	5, 115, 263
Meriones	194, 247	*crassus*	5, 21
		libycus	21, 113
		shawi	20, 21, 183, 190
Otomys	188	*angoniensis*	70, 249
		denti	144
		irroratus	58, 100, 101, 115, 144
		laminatus	100, 144
		typus	40, 70
Psammomys		*obesus*	5, 194
Steatomys		*caurinus*	25
		pratensis	25, 257
Tatera	18, 38, 144, 247	*brantsii*	115
		robusta	70, 115, 215, 223
		valida	25, 57, 76, 115
Taterillus[a]	25, 28, 66, 67, 115, 118	*gracilis* *pygargus*	
Spalacidae			
Spalax	260	*leucodon*, (= *ehrenbergi*)	5, 113, 194
Rhyzomyidae			
Tachyoryctes		*macrocephalus*	70
		splendens	70, 115, 127, 180, 238
Muridae			
Acomys	260	*cahirinus*	70, 115, 144, 194
		wilsoni	260
Aethomys		*chrysophilus*	234, 235
		kaiseri	70
Arvicanthis		*niloticus*	5, 6, 18, 66, 67, 70, 76, 115, 135, 144, 181, 182,

Table 2 (continued)
AFRICAN RODENTS REPORTED TO HAVE DAMAGED AGRICULTURAL CROPS

Family and Genus	Genus Ref.	Species	Ref.
			194, 198, 215, 242—244, 246, 247
Dasymys		*incomtus*	18, 76, 78, 115, 244
Grammomys (= *Thamnomys*)		*rutilans*	51
Lemniscomys	18	*barbarus*	263
		striatus	33, 70, 76, 81, 90, 115, 144
Lophuromys		*flavopunctatus*	70, 90, 115, 144
		sikapusi	64, 73, 115
Mus (Leggada)		*minutoides*	76, 235
		musculus	21, 70, 115, 182, 194
Nesokia		*indica*	5, 115, 194
Oenomys		*hypoxanthus*	18, 115, 144
Pelomys		*fallax*	81, 97, 144
Praomys		*albipes*	70
		erythroleucus	18, 118
		huberti	198
		morio	25
		natalensis	14, 18, 69, 70, 74, 75, 76, 78, 81, 87, 97, 98, 115, 181, 198, 215, 233, 234, 242—244, 246, 247
		tullbergi	18, 51
Rattus		*norvegicus*	144, 182, 194, 213, 235, 247
		rattus	14, 18, 70, 74, 75, 115, 182, 194, 228, 247
Rhabdomys		*pumilio*	70, 100, 101, 115, 144, 188, 244, 246, 247, 249
Stochomys		*longicaudatus*	51
Uranomys		*ruddi*	17, 76, 115
Gliridae			
Graphiurus	117	*hueti*	64, 74, 241
		murinus	235, 257
Dipodidae			
Allactaga		*tetradactyla*	5, 113, 184, 194
Jaculus		*orientalis*	5, 113, 184, 194
Hystricidae			
Atherurus		*africanus*	25, 52, 144, 213
Hystrix	115, 144, 172, 173, 246, 247	*africaeaustralis*	108, 115, 128, 235
		cristata	25, 52, 70, 213
Thryonomyidae			
Thryonomys		*gregorianus*	144
		swinderianus	10, 18, 25, 52, 65, 69, 70, 76, 78, 144, 150, 235, 246, 247, 252

Table 2 (continued)
AFRICAN RODENTS REPORTED TO HAVE DAMAGED AGRICULTURAL CROPS

Family and Genus	Genus Ref.	Species	Ref.
Bathyergidae			
Bathyergus		*suillus*	49, 50, 108, 207, 235
Cryptomys		*hottentotus*	49, 60, 230, 235
		ochraceocinereus	25, 260
Georychus		*capensis*	207, 235
Heliophobius		*argenteocinereus*	97
Heterocephalus		*glaber*	70, 110, 115, 238

[a] Since external appearance of *T. gracilis* and *T. pygargus* are similar, references to damage are listed by genus only.

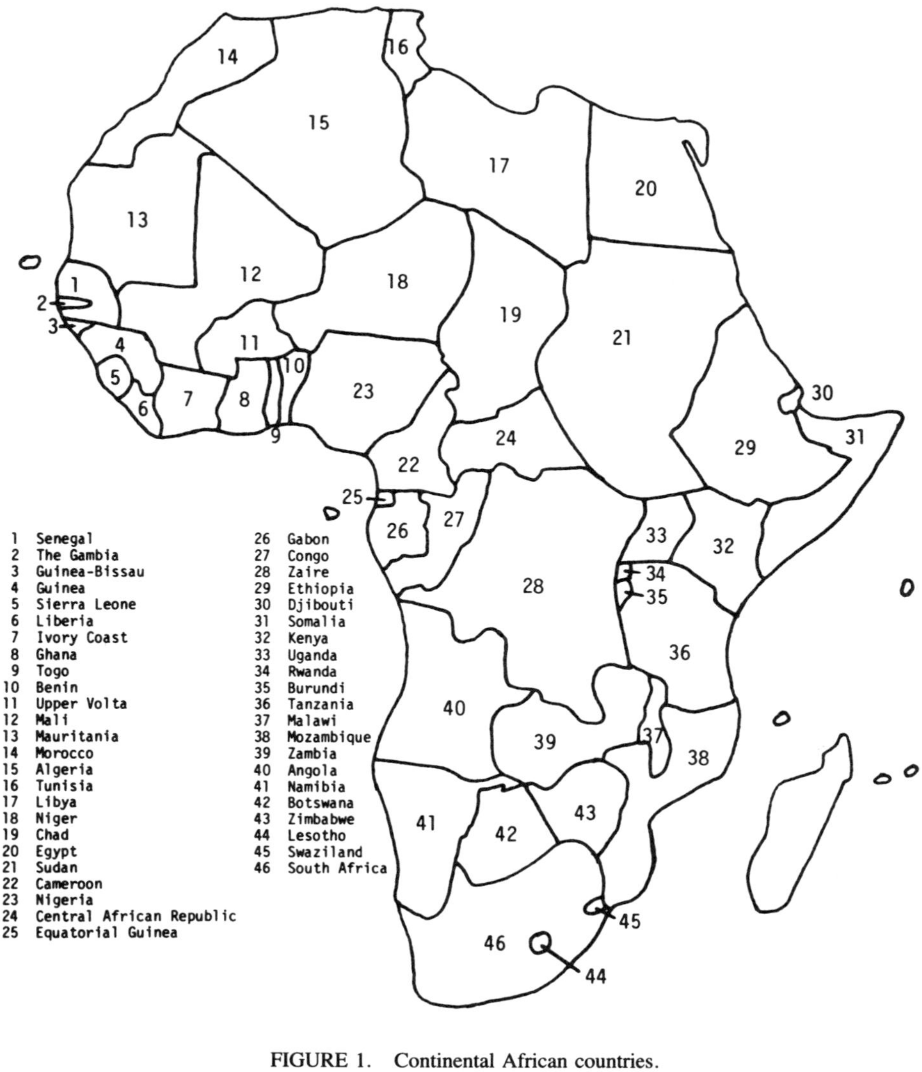

FIGURE 1. Continental African countries.

cane are particularly susceptible during rodent outbreaks. Vegetable and forage crops are not immune to rodent damage. Stored grain and other commodities along with processed food are consumed by several species that enter buildings in large urban areas as well as small village dwellings.[115,125] Estimates of these agricultural losses by rodents have, at best, been good guesses.[126]

Several ecological factors influence pest rodent populations and include climate, weather, soil, vegetation, and topography. When conditions become favorable and susceptible crops are present, agricultural crop damage may be great. Current rodent control measures are insufficient to deal with these periodic irruptions and the result is economic hardship to subsistence as well as cash crop farmers. When ecological factors become only moderately favorable, such as during or after seasonal rains, pest rodent populations usually increase during susceptible crop maturation often resulting in locally severe damage before harvest. The distinct dry and wet seasons result in somewhat predictable patterns of rodent population increases and crop damage,[199,248] but little applied field work has been done to confirm the feasibility of using this knowledge to reduce losses.

A. Major Rodent Pests

The major rodent pests in Africa based on published accounts of economic damage, wide distribution, ability to ''irrupt'', and importance and diversity of the crops damaged include *Praomys natalensis, Arvicanthis niloticus, Meriones shawi, Rhabdomys pumilio, Rattus rattus,* and *Thryonomys swinderianus*. Although 77 rodent species within 45 genera and 11 families have caused economic damage, the six species identified here have been most frequently involved in significantly damaging several crops, have broad distribution ranges, and most have the tendency to irrupt periodically under favorable conditions.

1. Multimammate Rat, Praomys natalensis

Praomys natalensis, the multimammate rat, is the most economically important of all pest rodent species in Africa. Being almost ubiquitous, it ranges from western, central, and eastern Africa to the southern tip. Many field crops at various stages of development are damaged by this small (50 g) rodent, and it readily enters dwellings to feed on stored food. Since it is often found in association with man — in homes, nearby crops, or where villages once were — it is considered an indigenous commensal. However, the introduced roof rat, *Rattus rattus*, is replacing *P. natalensis* in many parts of Africa as the dominant commensal rodent.[144,222,260]

Reproductive potential is high due to the large number of mammae, up to 12 pair, which support very large litters — up to 24[234] — during favorable periods. However, an average litter size *in utero* of 11 is more typical.[226,244] This rodent is mobile,[225] and the combined effect of reproductive productivity and mobility can result in very large numbers occupying agricultural lands over a relatively short time period (3 to 6 months). Major outbreaks, however, probably develop over periods of 1 year or more, but are usually not reported until significant crop damage begins.

Damage to cultivated crops was evident since at least 1931.[97] Wheat, sorghum, maize, mealie (Indian corn), and rice are cereal grains particularly susceptible to damage. During a major outbreak in Kenya, this species fed primarily on seeds and grains in an agricultural area planted with wheat and maize.[244] Damage to seeded maize fields required replanting, while damage to maturing ears, though significant, was less severe. The ability to climb maize stalks was documented in both eastern[244] and West[63] Africa. An average of 30% damage to sorghum was estimated in a 30,000 ha area hear Lake Chad in Nigeria during a major outbreak in 1976/1977.[28]

Rice and corn were heavily damaged in Burundi when a newly opened agricultural area was developed.[81] *Praomys* was primarily responsible, even in the flooded rice fields. Its presence in and damage to maturing rice has also been recorded in Somalia,[14] Liberia,[213] and Nigeria where it was also responsible for significant damage to seeded rice fields.[76,78] In West Africa, *Praomys*, along with *Arvicanthis*, are present around flooded rice fields, but only *Praomys* within the rice field itself.[198] Damage, which occurred during a major outbreak, from both species was 80 to 100%.

Cotton damage by this species was "widely known" in the 1930s in Tanzania where all crop stages were susceptible.[98] *Praomys* continues to be a major vertebrate pest in cotton growing areas.[70,213,215,246]

Root crops, particularly groundnuts (peanut), including the vegetative portion, are heavily damaged.[14,18,50,215,233,247] Other root crops damaged include cassava and yams.[18]

Stored grains and foods are frequently consumed, contaminated, or damaged by this commensal rat.[74,87,115] During a major rat outbreak, a loss of more than 2300 bags of stored maize and wheat was claimed by 44 farmers at Kitale, Kenya; farmers also lost 6352 bags gnawed beyond repair.[244] Storage problems have increased after field crops were harvested[98] indicating that reduced food and shelter brought about by harvest influenced movement to village dwellings. Damage to stored rice occurred in Nigerian milling stations, stores, and domestic pantries; although *Rattus rattus* was considered the main storage pest, *P. natalensis* may also have been responsible for the gnawing of jute and polyethylene bags containing rice resulting in spillage losses of 2 to 5%.[78] Similar damage was reported in poultry houses in Nigeria.[75]

Sugar cane,[18,87] tree saplings,[244] sorghum,[14,98,215] millet,[215] sesame,[14] and mealies[234] have also been reported to be damaged by *P. natalensis*. Being a true omnivore, *Praomys* will also feed on vegetables, vegetation, bark, insects, or whatever may be available. This opportunistic species may even feed on adult chickens in Nigeria.[75] The list of agricultural crops damaged by this species will undoubtedly grow as research efforts better document field agricultural problems.

2. *Unstriped Grass Rat, Arvicanthis niloticus*

Where *Arvicanthis niloticus* is present, it is a serious field pest. Its dominant range is found in eastern Africa only, but the complete range includes Senegal to the west, Egypt to the north, and Zambia to the south.[112,144] In Senegal, it has been a major pest only during outbreaks.[118,198,200] This medium-sized (80 g) rodent is primarily herbivorous, feeding on grasses, and unlike the majority of rodents is diurnal.[54,186,221,257]

The preferred habitat is grassland[221,257] with sufficient cover to protect it from daytime raptors and other predators. In Sudan it was more frequently found near permanent water supplies.[215] Being gregarious, several individuals may be found in burrows which often are modified natural crevices.[144] It may cache food in these burrows[198,215,239] and will occasionally share its burrow with other species — *Lemniscomys striatus*, *Lophuromys flavopunctatus*,[161] and *Praomys natalensis*.[244] In Ethiopia, home ranges for adult males were 2750 m^2 in the wet season and 1400 m^2 in the dry season with adult females and juveniles occupying much smaller ranges.[170]

Field crops damaged include most cereals, cotton, sugar cane, groundnut, and several other root crops. Significant damage to wheat by *Arvicanthis* in Egypt,[194] Sudan,[215] Kenya,[244,246] and West Africa[18,198] has occurred. The potential exists for this rodent to completely destroy tillering wheat fields during outbreaks. Cereal crop damage in Kenya was concentrated around field edges, but in weedy crops, which most likely provided additional cover, damage was throughout fields.[244] Rice seedlings and young stems in Nigeria were damaged primarily by this species.[76] In Senegal, 80 to 100% of growing rice tillers were cut by *Arvicanthis* and *P. huberti*, and seeds 5 to 10 m from field edges were also eaten.[198] Maize in the earlier crop stages is susceptible,[215,244,246] but apparently the developing ears (cobs) are not damaged due to the poor climbing ability of *Arvicanthis*. Other grain crops damaged include millet,[215] barley,[18] sorghum,[67,215] and pennisetum (pearl millet).[215]

Root crops, primarily groundnut,[18,67,215] but also cassava and sweet potato,[18,144,260] are damaged. In some areas, sugar cane and cotton yields may be seriously affected.[18,194,247]

Although not normally a storage pest, *Arvicanthis* occasionally enters grass huts and grain stores.[144,244] Damage to stored food was noted in Egyptian agricultural settlements,[5,135] particularly when food was scarce in fields.[6] Elsewhere, damage to stored grain in eastern Africa[144,244] and at groundnut storage sites in Nigeria[1] indicates that this field pest should be considered at least a minor storage pest.

3. Shaws Jird, Meriones shawi

Meriones shawi, a medium-sized (130 to 200 mm) jird, found only in northern Africa, has been responsible for serious damage to cereals, primarily barley and millet, and forage crops.[21,190] Large, daily food requirements (45% of body weight),[21] continuous food caching, and a habitat preference for moist cultivated areas lend this pest very destructive potentials. Periodic population increases, similar to those of *P. natalensis*, have occurred in Morroco, Tunisia, and Algeria. Vegetable crops, as well as olive groves (low fruit and bark) and almond and pistachio nut orchards (fallen nuts and bark), are also affected. Burrow excavation, which may cover a 15 × 7 m area, has seriously reduced forage crop yields.[20] Two other species of *Meriones, M. libycus*, and *M. crassus*, also damaged cereals, typically in 2- to 3-m clumps in marginal or subsistence crops of North Africa,[5,113] and *M. libycus* has damaged dikes on terraced mountains.[21]

4. Four-striped Grass Mouse, Rhabdomys pumilio

Rhabdomys pumilio, primarily a nonburrowing, diurnally active, small rodent (35 g), was one of three major pests during a serious rodent outbreak in Kenya.[244] It is readily identified by four dorsal black stripes which account for its common name. The range of this species, though not wide, covers eastern and southern Africa from about Kenya south to Angola. Cereals, primarily wheat,[246,247,249] are seriously damaged during high population densities. Although common in maize fields, they are not known to climb stalks to damage maturing cobs.[50,234] In South Africa, 12 to 39% damage by *R. pumilio* and *Otomys* sp. to *Pinus patula* and *P. caribea* trees (many dead or dying) occurred, and 2- to 3-year-old trees were girdled at ground level.[188] A blue dye applied to bark was found in 16 of 76 trapped *Rhabdomys pumilio,* considered a primary pest in coniferous tree plantations in South Africa.[100]

5. Marsh Cane-Rat, Thryonomys swinderianus

Thryonomys swinderianus (in semiaquatic habitats) and *T. gregorianus* (in dryland habitats) are widely distributed and capable of localized severe damage. Commonly referred to as the grasscutter or marsh cane-rat, the larger (4.5 to 9.9 kg) *T. swinderianus* lives in swampy or moist valleys and along margins of rivers and lakes.[144] Its range includes much of western, central, eastern, and southern Africa.

A wide variety of crops are damaged by cane-rats. *T. swinderianus* is a principal mammalian pest of rice in West Africa[18,69,78] where tillering, reproductive, and maturing stages were routinely damaged.[76,150] Also, in West Africa, oil palm seedlings (23.3% damage) and young rubber plants,[18,252] as well as cassava and eggplant[10,150] were consumed by *T. swinderianus*, common in coastal areas. Elsewhere, damage to rice, maize, sweet potato, groundnut, elephant grass, and sugar cane,[52,65,246,247,260] consumption of nuts and fallen fruit, and debarking of young trees in orchards has occurred.[50,144] In southern Africa, cereals (including maize cobs, millet, sorghum, and wheat), sugar cane, pineapple, pumpkin, sweet potato, and groundnut were damaged.[50,233,235]

6. Roof Rat, Rattus rattus

The roof rat, *Rattus rattus*, introduced to many African shipping ports in the late 1800s,[59,213] has normally been a commensal pest damaging stored foods of towns, villages, and remote farms. It was considered the main rodent pest in Egypt, accounting for at least 4 to 5% in total agricultural losses, including poultry operations.[228] It was a primary storage pest in Uganda,[260] Nigeria[78] (including poultry and rabbit houses),[74,75,77] and in villages of western Kenya.[147] With man's assistance, *R. rattus* continues to extend its African range.[213,222]

In some areas, however, it has entered fields and damaged several growing crops, such as irrigated citrus groves in Egypt[247] where it has also damaged rice, sugar cane, and possibly maize.[5,194] This pest has been found with *P. natalensis* in rice fields of Somalia[14] and was one of four rodents responsible for damage to seeded rice in Nigeria.[76] Coconuts growing in West[18] or eastern Africa[97] may also be damaged by the gnawing and penetration of green husks. Another introduced commensal, *R. norvegicus*, normally restricted to large port cities, has reportedly damaged irrigated cereal and vegetable crops in Egypt.[194,247]

B. Other Rodent Pests

In addition to the six primary rodent pests described above, there are at least 71 others that have damaged one or more crops. All 77 rodent pests are summarized and referenced in Table 2. A brief description of damage by most of the remaining species follows.

1. Sciuridae

Squirrels are frequently involved in damage problems such as in Ghana cacao plantations where, over 1 year, 5.2% of the pods were discarded due to damage.[185] *Funisciurus anerythrus* damages West African cacao and due to its selective feeding behavior may destroy a dozen pods before eating one.[18,51,246] This same selective feeding behavior was exhibited by *Heliosciurus gambianus*,[18,51] a serious pest in Ghana and Nigeria.[246] *F. anerythrus* has also consumed oil palm, fruits, and cassava.[52,241]

Xerus erythropus damages root crops including cassava, potato, and yam.[18,52] Beans, maize cobs, cotton bolls, groundnut, melon, and large fruits are also susceptible food items for this species.[52,78,260] Though not a major pest in Nigerian rice fields, it has consumed seedlings and cut a few stems.[76] Very low damage (0.04%) to maize seedlings and occasionally cobs along field edges by a related species, *X. inauris*, was estimated in South African agricultural lands;[270] in Kenya, *X. rutilus* has reportedly consumed planted seed in dry, marginal farming areas.[70]

One squirrel, *Sciurus carolinensis*, introduced to South Africa about 1900,[22,45,50] has damaged fruit (peaches and grapes), almond, and maize.[45,235] Almost 33,000 bounties were paid from 1918 to 1958 due to damage to fruit, grapes in particular, vegetables, and minimal damage to young trees.[22,160]

2. Pedetidae

The springhare, *Pedetes capensis*, found in sandy, open flats of eastern Africa and parts of southern Africa, occasionally damages sweet potato, pumpkin, planted maize seeds, and growing wheat, as well as barley, oats, maize, sorghum, bean, and groundnut.[31,50,144,180] This large (3 to 4 kg) kangaroo-like rodent, unique to Africa, is hunted both as a pest (10 to 15% estimated damage in Botswana field crops) and as an important food source for bushmen.[31,50,233]

3. Cricetidae

Cricetomys gambianus is widely distributed from Senegal to Sudan, south to Mozambique and the northern edges of South Africa. It damaged cacao in West Africa by feeding on pods within 2 m of the ground.[18,241] In 1934, damage to sprouting cereals was recorded in eastern Africa, and though its numbers were not large, it damaged almost any food crop.[97] Root crops, including cassava and yam,[18,246,257] bulbs, tubers, and groundnut;[50,144,235] fruit, including avocado, pear, and mango;[117,235,257] and vegetables, including beans, peas, and pumpkin,[117,144,235,260] were damaged. Among cereals in southern Africa, sorghum and maize (cobs) are eaten.[50,235] It will climb trees, if necessary, and in gardens can be a problem.[117] In Harare, Zimbabwe, 42 animals were removed from a 0.5 ha area containing young peas, lettuce, and cabbage plants.[235] In urban areas of West Africa, it has even invaded sewers.[144]

Gerbillus spp. are potential pests in northern Africa where barley and millet farms are at risk.[194] *G. gerbillus*, in millet and groundnut fields of Niger and Nigeria, *G. pyramidum*, found in association with man and domestic animals in Egypt, *G. campestris*, and *G. nanus* have all been referred to as pests,[5,21,263] but documentation is apparently lacking.

Otomys spp. are particularly damaging to conifers (*Pinus patula* and *P. caribea*) in South Africa where (with *Rhabdomys pumilio*) they accounted for 11.9% (5.0% damaged, 6.9% slightly damaged) to 39.0% (17% dead or dying, 22% slightly damaged) damage by girdling 2- to 3-year-old trees at ground level.[188] In Kenya, small quantities of wheat and the bark of shrubs and trees were taken in dry periods by *O. angoniensis*, but otherwise no significant damage was found.[90,249] In high mountains of Kenya, a large number of *O. typus* was responsible for 30 to 40% removal of vegetation.[40] Both *O. irroratus* and *O. laminatus* were found as secondary contributors to coniferous (*P. patula* and *P. caribea*) forest plantation debarking in which *R. pumilio* was the primary pest.[100] *O. irroratus* damaged forest nurseries and plantations (up to 65% damage) in Zaire and South Africa,[50,58,143,174,191] while in Kenya and Uganda, *O. denti* removed bark and cambium from cyprus trees.[144]

Psammomys obesus, the colonial fat sand rat of North Africa, can be very destructive to grains such as barley and millet;[194] up to 500 barley heads were cached in a single burrow.[99] Adapted for arid environments, it is common near brackish water.[5]

Steatomys pratensis in Tanzania and Zambia brought groundnut and finger millet to their burrow to eat.[257] Commonly referred to as fat mice, they have the ability to form large fat deposits which facilitate estivation during unfavorable periods,[213] thus making them very desirable as food by some natives.[180]

Tatera robusta and *T. valida* have rarely been involved in severe damage[247] but are widely distributed in Africa and affect several crops. Cassava, groundnut, and groundnut vegetation were readily eaten in cultivated areas as well as planted seeds of several crops.[38,144,223,247] *T. valida*, however, was a significant pest on rice seeds and seedlings in Nigeria,[76] and in Zaire, it seriously undermined an airport facility by constructing elaborate burrow systems typical of this genus.[57] Cotton has been damaged by *T. robusta* in Sudan.[215]

Taterillus pygargus was implicated in damage to groundnut in Senegal during a major rodent outbreak.[66,67] Increased rainfall during that time resulted in a population density that rose from 2 to 6 individuals/ha up to 180/ha.[197] *T. gracilis*, differing from *T. pygargus* only in chromosome number,[181] also increased during this outbreak and may have contributed to losses in millet and groundnut in Senegal.[28,118]

4. *Spalacidae*

Spalax leucodon (= *ehrenbergi*) storage chambers sometimes contain barley.[113] This mole-rat chews on plant roots in the littoral semi-desert areas of Egypt, where it is usually less abundant than *Nesokia indica*, but nevertheless damages vegetables and tuberous plants.[5,194] Outside Africa, root crops and rhizomes of sorghum are consumed, and burrows with stored barley, wheat, potato, and sugar beet have been found.[99]

5. *Rhyzomyidae*

Tachyoryctes splendens, an eastern African mole-rat, was reported to be in such high numbers that it hampered grazing in Kenya.[127] With *Heterocephalus glaber* it damaged pyrethrum and other cultivated crops[238] that may include sweet potato, corn, beans, peas, and groundnut.[180] *T. macrocephalus*, found only in Ethiopia,[267] has affected grazing lands and root crops in the higher elevations.[70]

6. Muridae

Acomys spp., spiny mice with characteristic dorsal spines, damaged field crops in Egypt.[194] *A. cahirinus*, common throughout Egypt, has also been a commensal pest.[180,214] *A. wilsoni*, common around sorghum fields, was accused of damaging maize cobs in Uganda.[260]

Aethomys chrysophilus, the bush rat of eastern and southern Africa, damaged maize, sorghum, and legumes (peas and beans) and was a storage pest in southern Africa where it occasionally entered houses.[234,235] *A. kaiseri* has reportedly been a pest of fodder grasses in Kenya.[70]

Dasymys incomtus, the shaggy swamp rat, is found from Senegal to Ethiopia, south to South Africa. It damaged palm plantations in West Africa, more so when a cover crop was exhausted, by eating or injuring the palm heart.[18] Rice seeds, seedlings, or young stems were consumed by this species in Nigeria.[76,78]

Grammomys (=Thamnomys) rutilans, the thicket rat, damaged cacao in Ghana and Nigeria.[51] It ranges from portions of West Africa east to Uganda.

Lemniscomys striatus, the striped grass mouse, ranges from Sierra Leone and parts of eastern Africa south to Zambia and Malawi where it is more abundant.[144] It has damaged oil palm, sweet potato, cassava, groundnut, fallen maize cobs, sorghum, and cotton.[18,33,52,81,260] Though capable of affecting most cultivated crops,[144] only minor damage to Kenyan rice seeds occurs.[90] In the southern range, where it is more abundant, it may be a more significant pest.

Lophuromys flavopunctatus, a brush-furred mouse found primarily in eastern Africa, does not cause significant crop damage.[90] It has, however, been involved in conifer damage that may have been due to *Otomys* spp.,[144] and minor damage to wheat, barley, and sorghum in Ethiopia, and fodder grasses in Kenya.[70] *L. sikapusi*, the rufous-bellied rat, may be a minor pest in oil palm and some other crops in Nigeria.[64,73]

Mus (=Leggada) minutoides (=musculoides), widely distributed south of the Sahara, was a major pest to seeded rice in Nigeria.[76] This small (7 g) mouse may occasionally enter homes, and population irruptions have occurred.[235] *Mus musculus*, the introduced house mouse, has become an important commensal pest whose range includes much of Africa.

Nesokia indica, an Asian mole-rat whose range now includes the Suez Canal Zone and Nile delta of Egypt,[99,112] damaged tuber crops, sugar cane, and the roots of wheat and rice.[194] Corn, barley, and vegetables received enough damage that Egyptian farmers flooded burrows to limit losses;[113] the undermining of water channels also occurs.[5]

Oenomys hypoxanthus, a relatively good climber, damages palm bunches in West Africa[18] as well as millet and rice.[144] Commonly named the rufous-nosed rat, it ranges from Sierra Leone to Ethiopia and into much of central Africa.[112]

Pelomys fallax, the creek rat of eastern and southern Africa, was found in fallow fields, corn, sorghum, and cotton fields of eastern Africa but no damage was indicated.[81] Though not usually considered much of a pest, damage to ripening grain in the coastal plain area of Tanzania, consumption of sugar cane and elephant grass (a perennial grass resembling sugar cane grown for forage), and the ability to fell maize stems does occur.[97,144]

Praomys albipes has reportedly damaged maize and sorghum in Ethiopia[70] and *P. tullbergi* has been responsible for losses to cacao and young rubber plants in West Africa.[18,51] Sugar cane, groundnut vegetation, and root crops (cassava and yam) are eaten by *P. erythroleucus* in West Africa.[18]

Stochomys longicaudatus, the target rat, named for its long guard hairs resembling arrows in a target, has damaged cacao in Ghana and Nigeria.[51] This medium-sized (80 g) murid ranges eastward to Uganda and inhabits mostly forests and plantations.[180]

Uranomys ruddi (=foxi), the Rudd's brush-furred rat, was reported to eat planted rice seed in Nigeria[76] and readily colonizes oil palm plantations.[17] Though widely scattered from Senegal to Kenya south to Mozambique, and occasionally found in abundance, it is generally considered rare.[180]

7. Gliridae

Graphiurus murinus, a small (25 g) African dormouse found throughout the sub-Saharan region, was considered a minor domestic pest to stored food in houses of Tanzania and Zambia[257] and responsible for causing shorts in electrical circuits in southern Africa.[235] Another dormouse, *G. hueti*, damaged cacao pods and beans[74] in West Africa where 4.6%[241] and 2.2 to 11.8%[64] damage estimates were made. Dormice in Tanzania occupied beehives and were considered a nuisance by beekeepers.[117]

8. Dipodidae

Allactaga tetradactyla, the four-toed jerboa of Libya and Egypt, may be involved in damage to several arid zone crops including barley in valleys.[5,113,194] This species is expected to decline when coastal valleys are cultivated and its preferred habitat changes.[5,184]

Jaculus orientalis, another North African dipodid, damages sprouting barley and ripe grain, sometimes severely, in the drier areas of Egypt, especially on desert slopes near the Mediterranean Sea.[5,184,194] This large jerboa (130 g) was common in and around Bedouin fields where barley and dates were found stored in burrows.[113] Bedouins commonly capture and eat this rat.[184]

9. Hystricidae

Atherurus africanus, the African brush-tailed porcupine, is found in western and central Africa where this small porcupine fed on sweet potato and cassava near forest edges.[52,144,213] Fruits such as pawpaw, banana, and guava were consumed in Uganda.[52] Newly opened agricultural areas of forests are very susceptible to damage.

Hystrix spp., large porcupines of Africa, are capable of severe local damage to many crops including sugar cane,[246,247] root crops, groundnut, pumpkin, maize cobs (by severing the stalk), and young cotton.[144] *H. africaeaustralis* damaged maize, potato, and groundnut in Zimbabwe where they were controlled by attraction into baited, fenced enclosures.[128] In southern Africa, it may also affect beet, carrot, onion, maize cobs, pumpkin, melon, and plants of young cotton, beans, and peas.[50,233,235] Almost 360 bounties were paid by South Africa in 1956.[108] *H. cristata*, with a similar appearance and behavior, occurs within the northern half of Africa. It damaged potato and maize in Kenya[70] and cassava, maize cobs, sweet potato, and groundnut in Uganda.[52,260] In some areas, populations of *H. cristata* have declined rapidly from human depredation due to their pest status and tasty flesh.[213]

10. Thryonomyidae

The savanna cane-rat, *Thryonomys gregorianus*, coexists with the marsh cane-rat, *T. swinderianus*, in many areas, and due to a similar appearance, is not readily distinguished. *T. gregorianus* has been a pest of pumpkin, groundnut, sweet potato, and stems of cassava and maize.[50,144] Both species are commonly hunted for food,[50] and attempts at domestication were made in West Africa.[2,3]

11. Bathyergidae

Five species within the family Bathyergidae also two within Rhyzomyidae, one murid (*N. indica*), and one spalacid (*Spalax leucodon*), are commonly referred to as mole-rats. All have been economically important rodent pests in Africa. *Heliophobius argenteocinereus* (=*emini*), the silvery mole-rat, damaged potato and groundnut in Tanzania.[97] Another bathyergid, *Cryptomys hottentotus* damaged cassava in the northern half of Zambia[230] and potato and vegetable gardens of South Africa.[49,50] It also was a nuisance in irrigation schemes[60] and on lawns and golf courses by creating mounds while burrowing.[235] *C. ochraceocinereus* has dug up newly sprouted maize and has damaged sweet potato and groundnut.[25,260] *Georychus capensis*, the Cape mole-rat, also consumed or spoiled vegetable, root, and bulb crops in South Africa.[50,207,235] *Bathyergus suillus*, a dune mole-rat of South Africa, excavated burrows which were a nuisance on golf courses, bowling greens, and tennis courts; damaged reaping machines in wheat fields; contributed to soil erosion in the sandy coastal belt; and undermined railway track beds.[49,50,108,207,235] *B. suillus* has also been a pest of vegetable gardens.[49,50] *Heterocephalus glaber*, the naked mole-rat of eastern Africa, was responsible for losses in pyrethrum and some other cultivated crops including gardens.[110,238] Irrigation dikes in Somalian orchards were apparently affected by *H. glaber*,[70] and Ethiopian citrus tree roots, field crops, and lawns have been damaged.[125]

III. DISEASE PROBLEMS

Tropical diseases abound in Africa, and rodents are responsible for several which affect man or livestock including plague, leptospirosis, murine typhus, and Lassa fever.[86,87] A number of arthropod-borne viruses, many bearing the African name of geographic discovery, are thought to involve rodents.[19,137,139,157,261] One such virus, Rift Valley fever, affecting lambs and calves as well as man, is transmitted by a mosquito, but the vertebrate host, rodent or otherwise, has been difficult to confirm.[43,262] Rodent variants of the rabies virus exist in Africa in several species including *Xerus inauris*.[7,19,138,144] They mimic rabies virus in diagnostic tests and clinical symptoms. Small mammal involvement may yet prove to be significant.[137] Diseases in which African rodents play a part are summarized in Table 3.

A. Bacterial

1. Plague

Plague has been a serious disease in northern Africa since an epidemic in Libya in 50 B.C.,[111] and other parts of Africa since the pandemics in 542 A.D.[86,169] and in the 14th Century.[195] Many epidemics followed known introductions of *R. rattus* and/or their infected fleas in the late 1800s. Some authors, however, have suggested that plague has been present in Africa long before written records indicate.[48,208]

In the early 1900s, plague was discovered in fleas of indigenous rodents — gerbils, *Tatera* spp., and the multimammate rat, *Praomys natalensis*.[234] In some areas of eastern Africa, sylvatic plague was present prior to the known arrival of *R. rattus* and involved *P. natalensis, A. niloticus, R. pumilio*, and *O. angoniensis*.[92,260] *P. natalensis*, being both commensal and a field pest, frequently contacts *R. rattus* in dwellings and *Tatera* spp. in the field, thereby serving as an intermediary to spread the disease.

Plague foci are found in four areas including one each in southern, eastern, western, and northern Africa.[7] These foci, as in many other plague areas of the world, harbor rodents which are resistant and periodically spread plague to susceptible species resulting in epidemics involving humans. For example, in southern Africa, *T. brantsii* has been a principal wild plague reservoir[44,47,122] spreading the disease to susceptible *P. natalensis* populations which transmit infected fleas directly to humans or indirectly by involving a third species, *R. rattus*. Research in southern Africa has demonstrated two "forms" of *P. natalensis*,[84,88,89,95,151] one being plague-resistant (*P. natalensis,* species A) and one susceptible (*P. coucha*, species B).[124] A similar resistant/susceptible cycle may be operating within this species complex influencing plague outbreaks.[250,251]

The urban phase introduced to South Africa in 1899[163] has not been prevalent there since 1938 nor in Mombassa (port of Kenya) since 1941.[44,189] Outbreaks are now confined to rural areas (sylvatic phase) where the conditions for man to become involved in the gerbil/multimammate rat/roof rat transmission cycle are greatest. A 5- to 6-year periodicity was common.[44]

Plague in North Africa is largely restricted to coastal areas, and the indigenous jerboas which are resistant have not readily spread the disease inland to susceptible commensal

Table 3
SOME AFRICAN RODENT-BORNE DISEASES AFFECTING MAN

Disease	Agent	Rodent reservoir[a]	Vector	Ref.
Bacterial				
Brucellosis	*Brucella suis*	*Arvicanthis niloticus*, Commensals, *Praomys natalensis*	Direct contact	19, 105
Leptospirosis	*Leptospira interrogans*	*Arvicanthis niloticus Cricetomys gambianus, Mus musculus, Rattus norvegicus, Rattus rattus*	Urine of infected rodents	12, 15, 16, 19, 50, 86, 156, 227, 247, 256
Plague	*Yersinia pestis*	*Acomys, Arvicanthis niloticus, Aethomys chrysophilus, Aethomys kaiseri, Lemniscomys striatus, Meriones, Mus, Otomys angoniensis, Otomys irrorattus, Otomys unisulcatus, Praomys natalensis, Rattus rattus, Rhabdomys pumilio, Tatera brantsii, Tatera leucogaster, Tatera robusta, Xerus inauris*	Fleas	4, 8, 44, 48, 50, 51, 84, 87, 88, 89, 92, 93, 104, 106, 117, 122, 123, 124, 163, 168, 189, 204, 210, 211, 234, 240, 247, 250, 251
Rat-bite fever	*Spirillum minor, Streptobacillus monoliformis*	Commensals *Meriones shawi, Paraxerus cepapi, Xerus erythropus*	Oral secretion from bite	19, 24, 86, 102, 116, 152, 261
Relapsing fever (tick-borne)	*Borrelia recurrentis*	*Praomys natalensis*	Ticks	19, 86, 139, 166, 224, 269
Salmonellosis	*Salmonella* spp.	Commensals	Contaminated food	19, 50, 86
Tularemia	*Francisella tularensis*	Commensals, Field rodents	Direct contact, blood-sucking arthropods	86
Rickettsial				
Boutonneuse fever (African tick typhus)	*Rickettsia conorii*	*Acomys cahirinus, Aethomys kaiseri, Arvicanthis niloticus, Lemniscomys striatus, Lophuromys, Lophuromys*	Ticks	19, 30, 41, 80, 86, 99, 104, 133, 155, 261

		flavopunctatus, Mus musculus, Otomys angoniensis, Otomys irroratus, Praomys natalensis, Rattus rattus, Rhabdomys pumilio		
Murine Typhus	*Rickettsia typhi*	*Acomys cahirinus, Arvicanthis niloticus, Cricetomys gambianus, Mus musculus, Otomys angoniensis, Rattus norvegicus, Rattus rattus Tatera nigricauda*	Fleas	19, 86, 99, 104, 131, 132, 253, 261
Q-fever	*Coxiella burnetii*	*Acomys cahirinus, Aethomys kaiseri, Apodemus sylvaticus, Arvicanthis niloticus,* Commensals, *Cricetomys gambianus, Lemniscomys barbarus, Lemniscomys striatus Lophuromys, Otomys angoniensis, Otomys irroratus, Praomys natalensis, Rhabdomys pumilio, Tatera nigricauda*	Airborne, direct contact	19, 23, 86, 99, 104, 131, 261
Viral				
Crimean-Congo hemorrhagic fever	Congo virus	*Praomys natalensis*	Ticks	261
Lassa fever	Machupo-like arenavirus	*Mus minutoides, Praomys natalensis, Rattus rattus*	Urine of infected rodents	7, 8, 62, 86, 88, 122, 136, 164, 165, 247, 265
Rift Valley fever	Rift Valley virus (sandfly fever group)	*Arvicanthis niloticus*	Mosquito	7, 19, 43, 86, 89, 144, 157, 220, 262
Protozoan				
African sleeping sickness (Trypanosomiasis)	*Trypanosoma spp.*	*Hystrix cristata, Xerus inauris*	Tsetse fly	9, 144, 201, 213
Babesiosis	*Babesia* spp.	Porcupines,	Ticks	19, 146
Leishmaniasis	*Leishmania donovani L. tropica*	*Acomys cahirinus, Arvicanthis niloticus, Rattus rattus, Tatera nigricauda, Tatera robusta, Xerus rutilus*	Sandflies	11, 13, 19, 27, 34, 86, 91, 94, 103, 104, 107, 109, 114
Toxoplasmosis	*Toxoplasma gondii*	*Rattus norvegicus, Rattus rattus*	Direct contact, water	19, 56, 86, 203, 205

Table 3 (continued)
SOME AFRICAN RODENT-BORNE DISEASES AFFECTING MAN

Disease	Agent	Rodent reservoir[a]	Vector	Ref.
Helminthic				
Angiostrongyliasis	*Angiostrongylus cantonensis*	Several	Third-stage larvae in rat lung, mollusks	19
Hepatic capillariasis	*Capillaria hepatica*	*Otomys irroratus, Praomys albipes, Praomys natalensis, Rattus rattus*	Contaminated food, water, or soil	49, 68, 219
Schistosomiasis	*Schistosoma* spp.	*Arvicanthis niloticus, Dasymys incomtus, Gerbillus pyramidum, Lophuromys flavopunctatus, Oenomys hypoxanthus, Otomys, Pelomys fallax, Praomys jacksoni, Praomys natalensis, Rattus rattus, Tatera, Thallomys*	Water	51, 86, 96, 139, 149, 154, 167, 176, 192, 193, 216, 217, 218, 219, 237
Trichinosis	*Trichinella spiralis (T. nelsoni* proposed for tropics)	*Praomys natalensis, Rattus norvegicus*	Infected meat	19, 50, 268

[a] Includes both true reservoirs and potential reservoirs implicated by serological surveys or parasite presence.

species. In Egypt, rodent control on river boats was apparently successful in stopping the spread of plague from the Suez Canal and northern ports to lower Egypt.[120] Ports along the Mediterranean are particularly susceptible through introduction of infected fleas and/or rats from shipping operations. The fact that *A. niloticus* is now somewhat commensal in Egypt[204] could play a significant role in future plague outbreaks.

Since the first recorded outbreak in eastern Africa in 1697, the Kenya/Tanzania border has continued to be an endemic plague focus.[208,210,240] Through serological surveys, *A. niloticus, R. rattus,* and *P. natalensis* have been identified as important reservoir hosts.[48] Nomadic movements of people, periodic rodent irruptions, and the presence of susceptible rodent species have led to frequent epidemics.[168] It is quite possible that rodent control operations have contributed to plague outbreaks when fleas were not also controlled.

The relationships between rainfall, rodent breeding seasons, and flea indices in eastern Africa were examined early as causative parameters of plague.[209,210,240] Seasonal human case fluctuations have been related to rodent population peaks when high natural mortality begins, thereby increasing the chances for infected fleas of rodents to contact humans. Monitoring the rat/flea index in these endemic areas has helped to identify potential plague areas and time periods.[142] The identification of fleas and rodents has explained why some areas in South Africa have no plague while others do.[122]

Plague was in epidemic proportions in 1929, killing thousands of people in West African coastal areas such as Lagos, but is now relatively rare.[213] *R. rattus* has largely been replaced by *R. norvegicus* as the dominant commensal in much of the coastal area,[32,148,258] but away from the port areas *R. rattus* is "not uncommon in towns and villages throughout the forest belt."[213] The replacement of *R. rattus* by *R. norvegicus* occurred in Egypt also and was believed to be one of several factors responsible for a decline in plague severity.[120]

Certainly the use of drugs, when delivered in a timely manner, has limited the severity of plague outbreaks in Africa today. Within the endemic areas, however, plague remains serious. From 1971 to 1980, the World Health Organization (WHO) annually recorded about 22 deaths and many more cases.[86] Of 30 countries reporting plague from 1958 to 1979, 16 were in Africa.[4]

2. *Salmonellosis*

Another bacterial disease, salmonellosis, is transmitted to humans by food contaminated with rodent feces containing *Salmonella*, of which there are some 2000 serotypes known.[19] Although more commonly reported in temperate countries, the presence of several commensal rodents and unsanitary conditions suggests that salmonellosis may not be uncommon in Africa.

3. *Leptospirosis*

Leptospirosis, a spirochaete transmitted to man in urine of infected commensal and field rodents, occurs worldwide, including Africa.[19,247] Fatality rates are generally low but may increase with age or in the more susceptible human populations of Africa.

4. *Others*

Rat-bite fever, brucellosis, relapsing fever, and tularemia are bacterial diseases affecting Africans and involving reservoirs that include several rodents present in Africa. The causative agent, rodent reservoir, and vector or method of transmission are summarized in Table 3.

B. Rickettsial

Murine typhus, a rickettsial disease commonly transmitted from commensal rodents to man by the flea *Xenopsylla cheopis* has been hosted by several field and commensal rodents in Africa. Boutonneuse fever (African tick typhus), carried by *Arvicanthis niloticus* and other rodents, is transmitted to man by ticks. A third rickettsial disease, Q-fever, spread by the airborne or direct contact route, has also been found in several commensal and field rodents in Africa.

C. Viral

1. Lassa Fever

The virus causing Lassa fever has become well known since its discovery in West Africa.[29,165,265] It is a fatal arenavirus infection related to machupo virus which causes Bolivian haemorrhagic fever in South America. The mode of transmission is not well understood, but the urine of infected rodents, *Praomy natalensis*, can transmit the virus through direct contact or contaminated food; person-to-person infections are also possible during acute stages of illness. A large number of ectoparasites have been found on *P. natalensis*, and involvement of one or more may yet be demonstrated.[122]

The epidemiology of Lassa fever was studied in households of an endemic site in Sierra Leone.[136] Rodent infection was focal with 39% viremic rodents found in human case houses compared to only 3.7% in control houses. Trapping removed an estimated 31 to 82% of the *Praomys* population but did not significantly reduce human seroconversion rates over a 4-week period. Preventing or interrupting epidemics of Lassa fever will require additional study and may prove to be very difficult.

The antibodies to a Lassa-related arenavirus named Mozambique were found in *P. natalensis* (20% seropositive) and *Aethomys chrysophilus* (7.7% seropositive) in Zimbabwe.[129] All of the positive *P. natalensis* were of the diploid number $2N = 32$ while none were of the chromosomal type $2N = 36$. If the Mozambique virus is a naturally attenuated form of Lassa fever, and if the immunity response of Mozambique-exposed Rhesus monkeys to Lassa fever[140] is present in humans, a vaccine against Lassa fever could be developed.[129] There is yet no evidence of human clinical symptoms from this virus although antibodies have been found. Other Lassa-related arenaviruses have been found and may continue to be found in the *Praomys* species complex[83,266] or possibly other rodents.[8]

2. Others

Crimean-Congo hemorrhagic fever and Rift Valley fever have been found in rodents, but their role as reservoirs needs further study.

D. Parasitic

1. Helminthic Infections

Parasitic diseases of humans involving rodents in Africa are present but epidemiologically not well described. *Shistosoma* spp., blood flukes whose larvae develop in freshwater snails, emerging to penetrate the skin of humans in freshwater bodies, have been found in several African rodents. Although irrigated agricultural lands have been associated with a high human incidence, the role of naturally infected rodents is debatable.[167] High parasite incidence rates have been recorded in those species that occupy moist environments, i.e., *Dasymys* sp., *Pelomys* sp., and *Lophuromys flavopunctatus*, which may be involved in maintaining the life cycle of shistosomiasis in Africa.[51,216,217,219] Other genera which have harbored the parasite but which may not be important agents in the life cycle due to drier habitat preferences include *Praomys, Rattus, Otomys*, and *Thallomys*. The fact that some rodents readily eat snails, i.e., *Lophuromys flavopunctatus,*[96] could influence favorably the prevalence of this disease.

Trichinosis, caused by an intestinal roundworm larvae, is found worldwide. When raw or undercooked infested meat is consumed, the encysted larvae mature in the intestine. Adults mate and female worms release larvae which pass through the intestinal wall and travel via the bloodstream to skeletal muscle throughout the host. It is usually mild, but when large numbers of trichinae occur, it can be fatal. Fecal transmission by commensal rodents could result in human infection.[212] Since several rodent species are eaten by Africans, the incidence of human infection by this route could be significant. Trichinae have been found in several predators of rodents (serval, hyena, jackal),[175] and bushpigs, another rodent predator, have caused outbreaks of human trichinosis.

Capillaria hepatica, parasitic worms which principally infect rodents can fatally infect man. Though reported cases in Africa are few, the zoonotic involvement may be more than realized. The infected liver of rodents, when consumed by man or another predator, contain eggs of the adult nematode which develop in the gut and dropped feces. When a second host ingests these eggs, larvae are released, migrate to the liver, and mature as adults. During the first stage, ''spurious infection'' occurs with no clinical infection occurring until the second stage when larvae are produced. Infected *R. rattus* and *P. albipes* were found in commensal habitats of Ethiopia.[68] *P. natalensis* in South Africa had infection rates of 48%, and infected gerbils, mice, and humans were found.[39,145,231] In Zaire, characteristic egg spots were found by visually inspecting the liver in several rodent species.[219] Prevention would appear likely through eliminating human consumption of contaminated food, soil, and possibly water.

2. Protozoan Infections

Leishmaniasis (*Leishmania tropica*, cutaneous type) occurs in most of Africa with *L. donovani* (visceral type) found only in southern Sudan and Kenya.[19] It is transmitted to man by sandflies which have fed on infected rodents or other mammalian hosts. Untreated, it can be fatal. It is more common in rural areas (both types) and in children (visceral type). Dogs appear to be an important reservoir, but *Arvicanthis niloticus* hosted *Leishmania* in Senegal[202] and Ethiopia.[94] In Kenya, the parasite was found in a ground squirrel (*Xerus rutilus*) and gerbils (*Tatera* sp.).[94,107]

In Africa, other protozoan diseases with rodent reservoirs include African sleeping sickness (trypanosomiasis), babesiosis, and toxoplasmosis. These are summarized in Table 3. Malarial species occur in anomalurids (flying squirrels),[141] *Atherurus africanus* (a porcupine), and *Grammomys (= Thamnomys) rutilans*.[79,144,146,213] As in monkeys, however, which carry several malaria species potentially affecting man, natural transmission is probably uncommon.

IV. DISCUSSION

The Sahel region has once again suffered a drought, and when the rains do come again, a repeat of the 1975-76 rodent outbreak could very likely happen. Walsh[259] predicted that economic problems will continue even when rains return since human population increases have canceled modest gains in food production. A severe rodent outbreak now could have consequences worse than the 1975-76 occurrence.*

Monitoring of certain pest rodents to anticipate population increases in selected critical areas would be extremely helpful in Africa. Enough lead time necessary to alert farmers and mobilize control efforts would result in increased effectiveness of limited manpower and materials. Such forecasting tools using rodent activity indices or density estimators in agricultural areas have recently been proposed.[119,199,248]

Procedures have largely been (1) waiting for reported crop damage, (2) confirmation of these reports by on-site visitations, then (3) instituting control methods. Such emergency campaigns have proved costly and difficult to organize.[245] The time lag is too long for effectively stopping or reducing crop losses from short- or long-term rodent outbreaks; nor

* Such an outbreak did occur in 1986 to 1987.

does this approach prevent the chronic, insidious damage that is occurring each crop season resulting in significant losses.

Reducing crop losses due to rodents requires continual effort related more to annual chronic rat damage rather than emergency control efforts. A simulation model for tropical rodent populations was compared to actual field data from Sierra Leone to evaluate the effects of different population characteristics and environmental conditions on the rate of growth of a *Praomys huberti* population.[72] Such models increase the understanding of pest rodent population dynamics and within certain limitations[236] help in designing appropriate monitoring methods to forecast rodent problems in advance.[229]

Applied research on agricultural rodent problems in Africa should be increased. Sampling procedures for crop damage assessment, appropriate for African cropping systems affected by rodents, is lacking. Small-scale field trials are also needed for determining efficacy of current practices compared to new methods. Laboratory testing of candidate chemicals on major rodent pests and food preference tests are needed to develop better bait delivery systems. Nonchemical control methods need to be evaluated and incorporated into an integrated, cost-effective rodent pest management system unique for African rodent problems. Many African pest rodents are used for food[55] and this must be considered in any control program.

Africa contains 31 ''crisis countries'' designated for special attention by international agencies.[121] With so many economic problems needing attention, rodent control has been neglected short of a catastrophy. In fact, African agriculture, in general, has been biased by low government investments in research, training, and extension services.[158] Formally trained rodent pest management specialists are few, training and extension materials and programs for farmers are lacking, and outside donor support to assist in institutionalizing rodent control has been limited. At a time when increased food production is so vitally needed, research and development of cost-effective rodent control techniques specific for Africa need to be encouraged.

Africa is ripe for the development of cost-effective rodent pest management techniques to increase food production and reduce disease incidence. The problem is severe, while the potential gains are large. The passive acceptance by farmers and government officials of damage caused by rodents combined with the lack of a concerted effort by donor agencies supporting agricultural development has prolonged the rodent problem. Until efforts reach the magnitude of what has been done on other major African pests or on rodent damage pests in other parts of the world, rodent damage in Africa will continue to be a major problem.

REFERENCES

1. **Adesuyi, S. A.,** Control of rodents in the groundnut storage site in Kano (Nigeria), Annual Report, Nigerian Stored Products Research Institute, Lagos, 1966, 25.
2. **Ajayi, S. S.,** Wildlife as a source of protein in Nigeria: some priorities for development, *Niger. Field,* 36, 115, 1971.
3. **Ajayi, S. S.,** Giant rats for meat — and some taboos, *Oryx,* 12, 379, 1974.
4. **Akiev, A. K.,** Epidemiology and incidence of plague in the world, 1958-79, *Bull. W.H.O.,* 60, 165, 1982.
5. **Maher Ali, A.,** The changing rodent pest fauna in Egypt, in *Proc. Vertebr. Pest Conf.,* Howard, W. E., Ed., 8, 28, 1978.
6. **Maher Ali, A., and Hafez, H. A.,** Wildlife and vertebrate pests in Egypt in *Proc. Vertebr. Pest Conf.,* Siebe, C., Ed., 7, 276, 1976.
7. **Arata, A. A.,** The importance of small mammals in public health, in *Small Mammals: Their Productivity and Population Dynamics,* Golley, F. B., Petrusewicz, K., and Ryszkowski, L., Eds., International Biological Program 5, Cambridge University Press, Cambridge, 1975, 349.

8. **Arata, A. A. and Gratz, N. G.,** The structure of rodent faunas associated with arenaviral infections, *Bull. W. H. O.*, 52, 621, 1975.
9. **Ashcroft, M. T.,** The importance of African wild animals as reservoirs of trypanosomiasis, *East Afr. Med. J.*, 36, 289, 1959.
10. **Asibey, E. O. A.,** The grasscutter, *Thryonomys swinderianus,* in Ghana, *Symp. Zool. Soc. London,* 34, 161, 1974.
11. **Azab, M. E., Rifaat, M. A., Schnur, L. F., Makhlouf, S. A., El-Sherif, E., and Salem, A. M.,** Canine and rodent leishmanial isolates from Egypt, *Trans. R. Soc. Trop. Med. Hyg.*, 78, 263, 1984.
12. **Ball, M. G.,** Animal hosts of leptospires in Kenya and Uganda, *Am. J. Trop. Med. Hyg.*, 15, 523, 1966.
13. **Barnett, H. C.,** *Sandfly and Sandfly-Borne Diseases in Maramorosch: Biological Transmission of Disease Agents,* Academic Press, New York, 1962, 192.
14. **Barre, M. H.,** *Control Campaign Against a Rat Outbreak in Somalia with Special Reference to the Shebelli and Juba Valley Agricultural Areas,* Department of Plant Protection, Ministry of Agriculture, Magadishu, Somalia, 1978, 9.
15. **Barsoum, I. S., Moch, R. W., Borros, B. A. M., and Kaiser, M. W.,** Leptospires isolated from wild mammals in Egypt, *Trop. Geogr. Med.*, 25, 362, 1973.
16. **Baylet, R. and van Riel, M.,** *Cricetomys gambianus,* a new host involved in the maintenance of leptospires, in French, *Med. Afr. Noire,* 18, 803, 1970.
17. **Bellier, L.,** Evolution of the rodent populations in oil plantations, in French, *Oleagineaux,* 12, 735, 1965.
18. **Bellier, L.,** Rodent pests in West Africa, *Span,* 16, 28, 1973.
19. **Benenson, A. S.,** *Control of Communicable Diseases in Man,* 13th ed., American Public Health Association, Washington, D.C., 1981, 443.
20. **Bernard, J.,** The mammals of Tunisia and neighboring regions, in French, *Bull. Fac. Agron. Tunis,* 24-25, 41, 1969.
21. **Bernard, J.,** Damage caused by the rodents Gerbillidae to agriculture in North Africa and countries of the Middle East, *EPPO Bull.*, 7, 283, 1977.
22. **Bigalke, R.,** The naturalisation of animals, with special reference to South Africa, *South Afr. J. Sci.*, 33, 46, 1937.
23. **Blanc, G. and Bruneau, J.,** Isolation of the virus of Q-fever from two species of rodents in the Nefifik Forest Morocco, in French, *Bull. Soc. Pathol. Exot.*, 49, 431, 1956.
24. **Blanc, G. and Noury, M.,** Infection of *Meriones shawi* Lataste by *Spirillum minus* Carter, *Bull. Soc. Pathol. Exot.*, 29, 383, 1936.
25. **Booth, A. H.,** *Small Mammals of West Africa,* Longmans, Green and Co., London, 1960, 68.
26. **Brambell, F. W. R. and Davis, D. H. S.,** Reproduction of the multimammate mouse (*Mastomys erythroleucus* Temm.) of Sierra Leone, in *Proc. Zool. Soc. London,* 111, 1, 1941.
27. **Bray, R. S.,** Zoonoses in leishmaniasis, in *Parasitic Zoonoses: Clinical and Experimental Studies,* Soulsby, E. J. L., Ed., Academic Press, New York, 65, 1974.
28. **Brei, W.,** South of Sahara: Kenya, Niger, Nigeria, Senegal and Sudan, in *Rodent Pests and Their Control,* Weis, N., Ed., German Agency for Technical Cooperation, Eschborn, West Germany, V So-l, 1981.
29. **Buckley, S. M., Casals, J., and Downs, W. G.,** Isolation and antigenic characterization of Lassa virus, *Nature (London),* 227, 174, 1970.
30. **Burgdorfer, W.,** The spotted fever group diseases, in *CRC Handbook, Series in Zoonoses,* Sect. A. II. Bacterial, Rickettsial, Myotic Diseases, Steele, J. M., Ed., CRC Press, Boca Raton, Fla., 1980, 568.
31. **Butynski, T. M.,** Life history and economic value of the springhare (*Pedetes capensis* Forster) in Botswana, *Botswana Notes and Records,* 5, 209, 1973.
32. **Buxton, P. A.,** Breeding rate of domestic rats trapped in Lagos, Nigeria and certain other countries, *J. Anim. Ecol.*, 5, 53, 1936.
33. **Buyckx, E. J. E.,** Birds and mammals harmful to seed, in French, in *Diseases and Pests of Cultivated Plants in the Congo, Rwanda and Burundi,* Public Institute for Natural Agricultural Studies, The Congo, 1962, 283.
34. **Camerlynck, P., Ranque, P., and Quilici, M.,** Importance of systematic cultures and subcultures in research of natural areas for skin leishmaniasis. Concerning the isolation of 5 strains of leishmania in *Arvicanthis niloticus*, in French, *Med. Trop.*, 27, 89, 1967.
35. **Cassady, J. T.,** The effect of rainfall, soil moisture and harvesting on grass production on two rangeland sites in Kenya, *E. Afr. Agric. For. J.*, 39, 26, 1973.
36. **Chapman, B. M., Chapman, R. F., and Robertson, I. A. D.,** The growth and breeding of the multimammate rat, *Rattus (Mastomys) natalensis* (Smith) in Tanganyika Territory, in *Proc. Zool. Soc. London,* 133, 1, 1959.
37. **Choate, T. S.,** Plagues of rats and rat plagues. I. Outbreaks of rodents and their consequences, *Rhodesian Farmer,* March 21, 17, 1975.

38. **Christiansen, J. T.,** Trainers Guide to Improved Rat Control, U.S. Agency for International Development Mission to Tanzania (AID/afr-c-1480), Publication Number 117, West Virginia University, Division of International Agriculture and Forestry, 1984, 29.
39. **Cochrane, J. C., Sagorin, L., and Wilcocks, M. G.,** *Capillaria hepatica* infection in man, *South Afr. Med. J.*, 31, 751, 1957.
40. **Coe, M. J.,** *The Ecology of the Alpine Zone of Mount Kenya,* W. Junk, The Hague, 1967, 133.
41. **Craddock, A. L.,** Some features of the epidemiology of tick typhus in Kenya, *East Afr. Med. J.*, 38, 227, 1961.
42. **Darlington, J.,** Rodents in Kajiado, *E. Afr. Nat. Hist. Soc. Bull.*, July/Aug., 73, 1984.
43. **Davies, F. G.,** Observations on the epidemiology of Rift Valley fever in Kenya, *J. Hyg.*, 75, 219, 1975.
44. **Davis, D. H. S.,** Sylvatic plague in South Africa: history of plague in man, 1919-1943, *Annu. Trop. Med. Parasitol.*, 42, 202, 1948.
45. **Davis, D. H. S.,** Notes on the status of the American grey squirrel (*Sciurus carolinensis*) in the South Western Cape (South Africa), *Proc. Zool. Soc. London,* 120, part 2, 265, 1950.
46. **Davis, D. H. S.,** Plague in Africa from 1935 to 1949, *Bull. W.H.O.*, 9, 665, 1953.
47. **Davis, D. H. S.,** Plague in South Africa. A study of the epizootic cycle in gerbils (*Tatera brantsi*) in the Northern Orange Free State, *J. Hyg.*, 51, 427, 1953.
48. **Davis, D. H. S., Heisch, R. B., McNeill, D., and Meyer, K. F.,** Serological survey of plague in rodents and other small mammals in Kenya, *Trans. R. Soc. Trop. Med. Hyg.*, 62, 838, 1968.
49. **de Graaff, G.,** On the Mole-rats (Rodentia, Bathyergidae) of South Africa: Thoughts on Possible Economic Implications and Means of Eradication, Paper presented at Entomological Symp. Zoological Pests, Heyns, J., Convenor, September 1971, Pretoria, South Africa, 1971, 11.
50. **de Graaff, G.,** *The Rodents of Southern Africa,* Butterworths, Pretoria, 1981, 267.
51. **Delany, M. J.,** The ecology of small rodents in tropical Africa, *Mamm. Rev.*, 2, 1, 1972.
52. **Delany, M. J.,** *The Rodents of Uganda,* British Museum of Natural History, George Press, Kettering Northamptonshire, England, 1975, 165.
53. **Delany, M. J. and Happold, D. C. D.,** *Ecology of African Mammals,* Tropical Ecology Series, Longman, New York, 1979, 434.
54. **Delany, M. J. and Kansiimeruhanga, W. D. K.,** Observations on the ecology of rodents from a small arable plot near Kampala, Uganda, *Rev. Zool. Bot. Afr.*, 81, 417, 1970.
55. **den Hartog, A. P. and de Vos, A.,** The use of rodents as food in tropical Africa, *Nutr. Newsl. (FAO),* 11, 1, 1973.
56. **De Roever-Bonnet, H.,** Toxoplasmosis in tropical Africa, *Trop. Geogr. Med.*, 24, 7, 1972.
57. **Devers, R., Ruzette, M. A., and Salvi, S.,** Control of gerbils *Tatera valida* Bocage, on the Ndjili Plain, Leopoldville, in French, *Ann. Soc. Belg. Med. Trop.*, 39, 145, 1959.
58. **Dieterlen, F.,** Economic importance relative to agriculture of some rodents in the region of Lake Kivu, in French, *Chron. IRSAC,* 1, 1, 1966.
59. **Dieterlen, F.,** History of the spread of the domestic rat *Rattus rattus* in East Africa, in German, *Z. Angew. Zool.*, 66, 173, 1979.
60. **Eloff, G.,** The Free State mole *(Cryptomys), Farming in South Africa,* 28, 203, 1953.
61. **Emmons, L. H.,** Ecology and resource partitioning among nine species of African rain forest squirrels, *Ecol. Monogr.*, 50, 31, 1980.
62. **Emond, R. T. D.,** Lassa fever, *R. Soc. Health J.*, 100, 48, 1980.
63. **Everard, C. O. R.,** Report on Some Aspects of Rodent Damage to Maize in the Western Region of Nigeria, Ministry of Agriculture and Natural Resources, Research Division, Moor Plantation, Ibadan, Nigeria, 1966, 24.
64. **Everard, C. O. R.,** A Report on the Rodent and Other Vertebrate Pests of Cocoa in Western Nigeria, Ministry of Agriculture and Natural Resources, Research Division, Moor Plantation, Ibadan, Nigeria, 1968, 123.
65. **Ewer, R. F.,** Form and function of the grasscutter *Thryonomys swinderianus* Temm. (Rodentia, Thryonomyidae), *Ghana J. Sci.*, 9, 131, 1969.
66. **Fall, M. W.,** *Sahelian Region Rodent Infestation,* Event 21-76, Center for Short-lived Phenomena, Cambridge, Mass., 1976, 1.
67. **Fall, M. W.,** Rodent Control in Senegal: Present Problems, Future Needs, Trip Report, Denver Wildlife Research Center, Denver, Colo., 1976, 16.
68. **Farhang-Azad, A. and Schlitter, D. A.,** *Capillaria hepatica* in small mammals collected from Shoa Province Ethiopia, *J. Wildl. Dis.*, 14, 358, 1978.
69. **Feakin, S. D.,** Pest control in rice, *PANS Manual,* No. 3, Centre for Overseas Pest Research, London, 1970, 295.
70. **Fiedler, L. A.,** The Status of Rodent Control in Five East African Countries, Unpublished FAO Consultancy Report, Rome, Italy, 1985, 49.

71. **Food and Agriculture Organization of the United Nations,** Cereal Crop Pests in Africa With Particular Reference to Birds, Part 1: Synthesis of Replies to a Questionnaire, FAO, Unpublished Internal Report, Rome, Italy, 1980, 250.
72. **French, N. R.,** Test of model with field data for a population of Mastomys huberti, *Acta Zool. Fenn.*, 173, 119, 1985.
73. **Funmilayo, O.,** The ecology and economic importance of the rufous-bellied rat, *Lophuromys sikapusi sikapusi* in Ibadan, Nigeria, *Nigerian J. Plant Prot.*, 1, 17, 1975.
74. **Funmilayo, O.,** Mammals and birds affecting food production and storage in Nigeria, in *Proc. Vertebr. Pest Conf.*, Clark, J. P. and Marsh, R. E., Eds., 9, 97, 1980.
75. **Funmilayo, O.,** Commensal rats: a threat to poultry production in Nigeria, in *Proc. Vertebr. Pest Conf.*, Marsh, R. E., Ed., 9, 107, 1982.
76. **Funmilayo, O.,** Assessment of rodent damage in field rice and rodent control procedures, in *Integrated Pest Management in Rice in West Africa*, Akinsola, E. A., Ouayogode, B., and Akintayo, I., Eds., West Africa Rice Development Association Regional Training Center, Fendall, Liberia, 1982, 505.
77. **Funmilayo, O. and Akande, M.,** The black/grey rat *Rattus rattus rattus* as a pest in a rabbitry in Ibadan, Western State, Nigeria, *Nigerian J. For.*, 4, 24, 1974.
78. **Funmilayo, O. and Akande, M.,** Vertebrate pests of rice in southwestern Nigeria, *PANS*, 23, 38, 1977.
79. **Garnham, P. C. C.,** Recent research on malaria in mammals excluding man, *Adv. Parasitol.*, 11, 603, 1973.
80. **Gear, J.,** The rickettsial diseases of southern Africa, *South Afr. J. Clin. Sci.*, 5, 158, 1954.
81. **Giban, I. J.,** Control of the multimammate rat, *Mastomys natalensis* (A. Smith) in the irrigated fields of the Republic of Burundi, in *Proc. Vertebr. Pest Conf.*, Howard, W. E., Ed., 8, 40, 1978.
82. **Giban, J. and Haltebourg, H.,** The problem of the *Meriones* of Shaw in Morocco, in French, *Congr. Protect. Cultures Tropic*, Chambre de Commerce et d'Industrie, Marseilles, France, 587, 1965.
83. **Gonzalez, J. P., McCormick, J. B., Saluzzo, J. F., Herve, J. P., Georges, A. J., and Johnson, K. M.,** An arenavirus isolated from wild-caught rodents (*Praomys* species) in the Central African Republic, *Intervirology*, 19, 105, 1983.
84. **Gordon, D. H.,** Distribution of sibling species of the *Praomys (Mastomys) natalensis* group in Rhodesia (Mammalia: Rodentia), *J. Zool. London*, 186, 397, 1978.
85. **Gordon, D. H. and Rautenbach, I. L.,** Species complexes in medically important rodents: chromosome studies of *Aethomys, Tatera*, and *Saccostomus* (Rodentia: Muridae, Cricetidae), *South Afr. J. Sci.*, 76, 559, 1980.
86. **Gratz, N. G.,** The global public health importance of rodents, in *Proc. Conf. on the Organization and Practice of Vertebrate Pest Control*, ICI Plant Prot. Div., Surrey, England, 413, 1984.
87. **Gratz, N. G. and Arata, A. A.,** Problems associated with the control of rodents in tropical Africa, *Bull. W.H.O.*, 52, 697, 1975.
88. **Green, C. A., Gordon, D. H., and Lyons, N. F.,** Biological species in *Pramys (Mastomys) natalensis* (Smith), a rodent carrier of Lassa virus and bubonic plague in Africa, *Am. J. Trop. Med. Hyg.*, 27, 627, 1978.
89. **Green, C. A., Keogh, H., Gordon, D. H., Pinto, M., and Hartwig, E. K.,** The distribution, identification, and naming of the *Mastomys natalensis* (Smith, 1834) species complex in southern Africa (Rodentia: Muridae), *J. Zool. London*, 192, 17, 1980.
90. **Green, M. G. and Taylor, K. D.,** Preliminary experiments in habitat alteration as a means of controlling field rodents in Kenya, *Ecol. Bull.*, 19, 175, 1975.
91. **Grove, S. S.,** Cutaneous leishmaniasis in South West Africa, *South Afr. Med. J.*, 44, 206, 1970.
92. **Guggisberg, C. A. W.,** Translation of "An epidemic of plague on Kilimanjaro in 1912 by R. Lurz", *East Afr. Med. J.*, 44, 215, 1967.
93. **Guggisberg, C. A. W.,** Plague research in East Africa, in German, *Naturhist. Mus. Bern Jahrb.*, 7, 97, 1978/80.
94. **Haile, T. T. and Lemma, A.,** Isolation of *Leishmania* parasites from *Arvicanthis* in Ethiopia, *Trans. R. Soc. Trop. Med. Hyg.*, 71, 180, 1977.
95. **Hallet, J. M.,** Chromosome polymorphism in *Praomys (Mastomys) natalensis* in southern Africa: diploid studies, *South Afr. J. Sci.*, 75, 413, 1979.
96. **Hanney, P.,** The harsh-furred rat in Nyasaland, *J. Mammal.*, 45, 345, 1964.
97. **Harris, W. V.,** Rats and mice, Dept. of Agriculture Leaflet, No. 12, (Revised), Tanganyika Territory, Africa, 1, 1934.
98. **Harris, W. V.,** The grey field mouse, *East Afr. Agric. J.*, 2, 315, 1937.
99. **Harrison, D. L.,** *The Mammals of Arabia*, Vol. 3, Ernst Benn, London, 1972, 670.
100. **Hechter-Schulz, K.,** Rodent control in Sabie forests, *South Afr. J. Sci.*, 58, 161, 1962.
101. **Hechter-Schulz, K.,** Trapping and poisoning experiments to show reaction of rodents in the field, in French, in *Congr. Protect. Cultures Tropic*, Chambre de Commerce et d'Industrie, Marseille, France, 1965, 617.

102. **Heisch, R. B.,** A case of rat-bite fever in Kenya Colony and the discovery of rodents naturally infected with *Spirillum minus* (Carter), *J. Trop. Med. Hyg.*, 53, 33, 1950.
103. **Heisch, R. B.,** The isolation of *Leishmania* from a ground squirrel in Kenya, *East Afr. Med. J.*, 34, 183, 1957.
104. **Heisch, R. B.,** Rodents as reservoirs of arthropod-borne diseases in Kenya, *East Afr. Med. J.*, 38, 256, 1961.
105. **Heisch, R. B., Cooke, E. R. N., Harvey, A. E. C., and de Souza, F.,** The isolation of *Brucella suis* from rodents in Kenya, *East Afr. Med. J.*, 40, 132, 1963.
106. **Heisch, R. B., Grainger, W. E., and d'Souza, J. St. A. M.,** Results of a plague investigation in Kenya, *Trans. R. Soc. Trop. Med. Hyg.*, 47, 503, 1953.
107. **Heisch, R. B., Grainger, W. E., and Harvey, A. E. C.,** The isolation of *Leishmania* from gerbils in Kenya, *J. Trop. Med. Hyg.*, 62, 158, 1959.
108. **Hey, D.,** The control of vertebrate problem animals in the Province of the Cape of Good Hope, Republic of South Africa, in *Proc. Vertebr. Pest Conf.*, 2, 57, 1964.
109. **Heyneman, D.,** Leishmaniasis in the Sudan Republic, *East Afr. Med. J.*, 38, 196, 1961.
110. **Hill, W. C. O., Porter, A., Bloom, R. T., Seago, J., and Southwick, M. D.,** Field and laboratory studies of the naked mole rat, *Heterocephalus glaber*, in *Proc. Zool. Soc. London*, 128, 455, 1955.
111. **Hirst, L. F.,** *The Conquest of Plaque*, 1st ed., Oxford University Press, London, 1953, 478.
112. **Honacki, J. H., Kinman, K. E., and Koeppl, J. W., Eds.,** *Mammal Species of the World: A Taxonomic and Geographic Reference*, Allen Press, and the Association of Systematics Collections, Lawrence, Kansas, 1982, 694.
113. **Hoogstraal, H.,** A brief review of the contemporary land mammals of Egypt (including Sinai). II. Lagomorpha and Rodentia, *J. Egypt. Public Health Assoc.*, 38, 1, 1963.
114. **Hoogstraal, H. and Heyneman, D.,** Leishmaniasis in the Sudan Repbulic. XXX. Final epidemiological report, *Am. J. Trop. Med. Hyg.*, 18, 1091, 1969.
115. **Hopf, H. S., Moreley, G. E. J., and Humphries, J. R. O.,** *Rodent Damage to Growing Crops and to Farm and Village Storage in Tropical and Subtropical Regions*, Center for Overseas Pest Research, Tropical Prod. Institute, London, 1976, 115.
116. **Horrenberger, R.,** A contribution to the study of "sodoku" in North Africa, in French, *Arch. Inst. Pasteur Alger.*, 31, 304, 1953.
117. **Hubbard, C. A.,** Observations on the life histories and behaviour of some small rodents from Tanzania, *Zool. Afr.*, 7, 419, 1972/73.
118. **Hubert, B. and Adam, F.,** Outbreaks of *Mastomys erythroleucus* and *Taterillus gracilis* in the Sahelo-Sudanian zone in Senegal, *Acta Zool. Fenn.*, 173, 113, 1985.
119. **Hubert, B., Adam, F., and Poulet, A. R.,** Modeling of the population cycles of two rodents in Senegal, *Bull. Carnegie Mus. Nat. Hist.*, 6, 88, 1978.
120. **Hussein, A. G.,** Changes in the epidemiology of plague in Egypt, 1899-1951, *Bull. W. H. O.*, 13, 27, 1955.
121. International Agricultural Development Service, Agricultural Development Indicators — Statistical Comparisons of 139 Developing Countries, IADS, New York, 1981, 17.
122. **Isaacson, M.,** The ecology of *Praomys (Mastomys) natalensis* in southern Africa, *Bull. W. H. O.*, 52, 629, 1975.
123. **Isaacson, M.,** Plague, *Med. Educ. Int.*, 127, 1981.
124. **Isaacson, M., Arntzen, L., and Taylor, P.,** Susceptibility of members of the *Mastomys natalensis* species complex to experimental infection with *Yersinia pestis*, *J. Infect. Dis.*, 144, 80, 1981.
125. **Jackson, W. B.,** An Investigational Framework to Assist Research and Management of Rodent Populations in Ethiopia, Institute of Agricultural Research, Addis Ababa, Ethiopia, 1973, 33.
126. **Jackson, W. B.,** Evaluation of rodent depredations to crops and stored products, *EPPO Bull.*, 7, 439, 1977.
127. **Jarvis, J. U. M.,** The structure of a population of mole-rats, *Tachyoryctes splendens*, (Rodentia: Rhyzomyidae), *J. Zool. London*, 171, 1, 1973.
128. **Jarvis, M. J. F. and LaGrange, M.,** Problem vertebrate management in Zimbabwe, in *Proc. Vertebr. Pest Conf.*, Marsh, R. E., Ed., 10, 95, 1982.
129. **Johnson, K. M., Taylor, P., Elliot, L. H., and Tomori, O.,** Recovery of a Lassa-related arenavirus in Zimbabwe, *Am. J. Trop. Med. Hyg.*, 30, 1291, 1981.
130. **Jordan, R. G., Davis, B. L., and Baccar, H.,** Karyotypic and morphometric studies of Tunisian *Gerbillus*, *Mammalia*, 38, 667, 1974.
131. **Juminer, B.,** A study of anti-*Rickettsia* agglutinins in *Cricetomys gambianus* Waterhouse in Dakar, in French, *Arch. Inst. Pasteur Tunis*, 49, 235, 1972.
132. **Juminer, B. and Baylet, R. J.,** Natural infection of *Cricetomys gambianus* Waterhouse by *Rickettsia mooseri* in Senegal, in French, *Arch. Inst. Pasteur Tunis*, 49, 243, 1972.

133. **Kaiser, M. N. and Hoogstraal, H.,** Noteworthy recent tick records from Egypt. II. Hosts, distribution and ecology of *Rhipicephalus simus* Koch, *J. Egypt. Public Health Assoc.*, 42, 231, 1967.

134. **Kaminski, M., Rousseau, M., and Petter, F.,** Electrophoretic studies on blood protein of *Arvicanthis niloticus, Biochem. Syst. Ecol.*, 12, 215, 1984.

135. **Kassab, A., Badran, A. A. F., and Kamelo, A. H.,** Warfarin bait in the control of the field rat, *Arvicanthis niloticus,* (Des.) in Egyptian shounas, Ministry of Agriculture, Plant Protection Department, United Arab Republic (Egypt), Cairo, 1963, 10.

136. **Keenlyside, R. A., McCormick, J. B., Webb, P. A., Smith, E., Elliot, L., and Johnson, K. M.,** Case-control study of *Mastomys natalensis* and humans in Lassa virus-infected households in Sierra Leone, *Am. J. Trop. Med. Hyg.*, 32, 829, 1983.

137. **Kemp, G. E.,** Isolation of viruses from wild mammals in West Africa 1966-1970, *J. Wildl. Dis.*, 10, 279, 1974.

138. **Kemp, G. E., Causey, O. R., Moore, D. L., Odelola, A., and Fabiyi, A.,** Mokola virus: further studies on IbAn 27377, a new rabies-related etiologic agent of zoonosis in Nigeria, *Am. J. Trop. Med. Hyg.*, 21, 356, 1972.

139. **Keogh, H. and Isaacson, M.,** Wild rodents as laboratory models and their part in the study of diseases, *J. South. Afr. Vet. Assoc.*, 49, 229, 1978.

140. **Kiley, M. P., Lange, J. V., and Johnson, K. M.,** Protection of rhesus monkeys from Lassa virus by immunisation with closely related arenavirus, *Lancet,* 2, 738, 1979.

141. **Killick-Kendrick, R.,** Parasitic protozoa of the blood of rodents. III. Two new malaria parasites of anomalurine flying squirrels of the Ivory Coast, *Ann. Parasitol. Hum. Comp.*, 48, 639, 1973.

142. **Kilonzo, B. S., Patel, N. R., and Mtoi, R. S.,** Studies of the seasonal fluctuations of rodents and their fleas in northeastern Tanzania, *Tanzania Vet. Bull.*, 3, 3, 1981.

143. **King, N. L.,** Tree planting in South Africa, *J. South. Afr. For. Assoc.*, 21, 1, 1951.

144. **Kingdon, J.,** *East African Mammals: Hares and Rodents,* Vol. IIB, Academic Press, New York, 1974, 704.

145. **Lämmler, G., Zahner, R., Vollerthun, H., and Rudolph, R.,** Egg production and host reaction in *Capillaria hepatica* infection of *Mastomys natalensis*, in *Parasitic Zoonoses,* Soulsby, E. J. L., Ed., Academic Press, New York, 327, 1974.

146. **Languillon, J.,** A note on two blood parasites of mammals in the Cameroons, in French, *Bull. Soc. Pathol. Exot.*, 50, 362, 1957.

147. **LaVoie, G. K.,** Postharvest Rodent Pests of Maize in Maseno, Kenya, Trip Report, Denver Wildlife Research Center, Denver, Colo., 1985, 12.

148. **Leslie, P. H. and Davis, D. H. S.,** An attempt to determine the absolute number of rats on a given area, *J. Anim. Ecol.*, 8, 94, 1939.

149. **Lo, C. T. and Lemma, A.,** Studies on *Schistoma bovis* in Ethiopia, *Ann. Trop. Med. Parasitol.*, 69, 375, 1975.

150. **Lowes, R. H. G.,** Destruction in Sierra Leone, *Oryx,* 10, 309, 1970.

151. **Lyons, N. F., Green, C. A., Gordon, D. H., and Walters, C. R.,** G-Banding chromosome analysis of *Praomys natalensis* (Smith) (Rodentia: Muridae) from Rhodesia, I. 36 Chromosome population, *Heredity,* 38, 197, 1977.

152. **Mailloux, M.,** Microbiological investigation of the rats of Algier. II. *Borellia hispanica* and *Spirillum mins* (Carter, 1887), in French, *Arch. Inst. Pasteur Alger.*, 40, 344, 1962.

153. **Malherbe, A. P.,** Notes on birds of prey and some others at Boshoek, north of Rustenburg during a rodent plague, *Ostrich,* 34, 95, 1963.

154. **Mansour, N. S.,** *Shistosoma mansoni* and *S. haematobium* found as a natural double infection in the Nile rat, *Arvicanthis niloticus niloticus*, from a human endemic area in Egypt, *J. Parasitol.*, 59, 424, 1973.

155. **Mayoux, A.,** Investigation on rickettsial infection in Madagascar. II. *Rickettsia conori*: tick-bite fever, in French, *Arch. Inst. Pasteur Madagascar,* 41, 49, 1972.

156. **McGuire, C. D. and Meyers, D. M.,** Leptospirosis in Egypt, *Am. J. Trop. Med. Hyg.*, 6, 532, 1957.

157. **McIntosch, B. M., Dickinson, D. B., Meenehan, G. M., and Dos Santos, I. S. L.,** *Culex (Eumenalomyia) rubinotus* Theobold as vector of *Banzi* Germiston and Witwatersrand viruses, II. Infections in sentinel hamsters and wild rodents, *J. Med. Entomol.*, 12, 641, 1976.

158. **McNamara, R. S.,** The Challenges for Sub-Saharan Africa, Sir John Crawford Memorial Lecture, Washington, D.C., Consultative Group on International Agricultural Research, Washington, D.C., November 1, 1985, 49.

159. **Mgeni, A. Y.,** personal communication.

160. **Millar, J. C. G.,** Aspects of the Ecology of the American Grey Squirrel *Sciurus carolinensis* Gmelin in South Africa, M.S. thesis, University of Stellerbosch, South Africa, 1980.

161. **Misonne, X.,** Exploration of Albert National Park (the Congo), the rodents of Ruwenzori and neighbouring regions, in French, *Inst. Parcs Nat. Congo Rwanda Fasc.*, 14, 1963, 165.

162. **Misonne, X.,** Order Rodentia, Part 6, in *The Mammals of Africa — An Identification Manual,* Mester, J. and Setzer, H. W., Eds., Smithsonian Institution Press, Washington, D.C., 1971.
163. **Mitchell, J. A.,** Plague in South Africa: historical summary (up to June 1926), *Publ. S. Afr. Inst. Med. Res.,* 3, 89, 1927.
164. **Monath, T. P.,** Lassa fever: review of epidemiology and epizootiology, *Bull. W.H.O.,* 52, 577, 1975.
165. **Monath, T. P., Newhouse, V. F., Kemp, G. E., Setzer, H. W., and Cacciapouti, A.,** Lassa virus isolation from *Mastomys natalensis* rodents during an epidemic in Sierra Leone, *Science* 185, 263, 1974.
166. **Morel, P. C.,** Tick-borne diseases of man in Africa south of the Sahara, in *Proc. Seminar on the Ecology, Biology and Control of Ticks and Mites of Public Health Importance,* World Health Organization WHO/VBC/68.57, Geneva, Switzerland, 1968, 5.
167. **Mousa, A. H.,** Bilharziasis as a national health problem in the United Arab Republic, in *Bilharziasis,* Wostenholme, G. E. W. and O'Connor, M., Eds., Little, Brown, Boston, Mass., 1962, 1.
168. **Msangi, A. S.,** Entomological observations after the 1968 plague outbreak in Mbulu district, Tanzania, *East Afr. Med. J.,* 46, 465, 1969.
169. **Msangi, A. S.,** The surveillance of rodent populations in East Africa in relation to plague endemicity, *Univ. Sci. J., Univ. Dar-es-Salaam,* 1, 8, 1975.
170. **Müller, J. P.,** Population ecology of *Arvicanthis abyssinicus* in the steppe of the Semien Mountains National Park, Ethiopia, in German, *Z. Saeugetierkd.,* 42, 145, 1977.
171. **Nandwa, S. M.,** Feeding ecology of rodents in Kenya, *East Afr. Wildl. J.,* 11, 407, 1973.
172. **Nandwa, S. M.,** Vertebrate pests in Kenya, in *Proc. 1st East Afr. Conf. Entomol. Pest Contr.,* 1, 111, 1976.
173. **Nandwa, S. M.,** A Survey of Vertebrate Pests in Kenya, Paper presented at 1st Afro-Asian Vertebrate Pests Conference, Cairo, Egypt, November 8 to 11, 1976, 18.
174. **Nänni, U. W.,** Forest hydrological research at Cathedral Peak Research Station, *J. South. Afr. For. Assoc.,* 27, 2, 1956.
175. **Nelson, G. S., Guggisberg, C. W. A., and Mukundi, J.,** Animal hosts of *Trichinella spiralis* in East Africa, *Ann. Trop. Med. Parasitol.,* 57, 332, 1963.
176. **Nelson, G. S., Teesdale, C., and Highton, R. B.,** The role of animals as reservoirs of bilharziasis in Africa, in *Bilharziasis,* Wolstenholme, G. E. W. and O'Connor, M., Eds., Little, Brown, Boston, Mass., 1962, 127.
177. **Ngiweshami, J. Z. L., Shoa, J., and Chizenga, S. H. D.,** Report of Ministry of Health Team which visited Lushoto District to review measures against plague epidemic in the district, Tanzania Ministry of Health, Dar-es-Salaam 1984, 14.
178. **Nieuwolt, S.,** Seasonal droughts in East Africa, *East Afr. Agric. For. J.,* 43, 208, 1978.
179. **Norton-Griffiths, M., Herlocker, D., and Pennycuick, L.,** The patterns of rainfall in the Serengeti Ecosystem, Tanzania, *East Afr. Wildl. J.,* 13, 347, 1975.
180. **Nowak, R. M. and Paradiso, J. L.,** *Walker's Mammals of the World,* 4th ed., Vols. 1 and 2, Johns Hopkins University Press, Baltimore, Md., 1983, 1362.
181. **Organization for Overseas Research, Science and Technology,** The agricultural rodents of Senegal: means of control and principles of the control program, in French, Dakar Center, Laboratory of Applied Zoology, Dakar, 1976, 21.
182. **Tantawy Omar, M.,** The national rat control campaign in Egypt, in *Proc. Conf. The Organization and Practice of Vertebrate Pest Control,* ICI Plant Protection Division, Surrey, England, 1984, 443.
183. **Organization for Economic Co-operation and Development, Improving Rodent Control in Developing Countries,** OECD/FAO/WHO Expert Consultation on Rodent Problems, Control and Research, May 2-5, 1978, OECD, Paris, 1978, 25.
184. **Osborn, D. J. and Helmy, I.,** The contemporary land mammals of Egypt (including Sinai), *Fieldiana Zool.,* 5, 1, 1980.
185. **Owen, H.,** Cacao pod diseases in West Africa, *Ann. Appl. Biol.,* 38, 715, 1951.
186. **Packer, C.,** Demographic changes in a colony of Nile grassrats (*Arvicanthis niloticus*) in Tanzania, *J. Mammal.,* 64, 159, 1983.
187. **Pagliano, M.,** The rodents and their damage in Tunisia, in French, *Bull. Direction Générale de l'Agriculture du Commerce et de la Colonisation,* J. Aloccio, Tunis, 1931, 32.
188. **Parry, M. S.,** The softwood plantations of the East Transvaal and Swaziland, *East Afr. Agric. J.,* 17, 188, 1952.
189. **Patel, N. R.,** The risks of an accidental introduction and/or transfer of plague I. Conditions prevalent in the East African ports of Tanga and Mombassa, *Tanzania Vet. Bull.,* 1, 2, 1979.
190. **Perret, M.,** Vertebrate pests of North Africa, in French, *Déf. Vég.,* 88, 41, 1961.
191. **Pirlot, P. L.,** Rodent pests of plantations in the neighbourhood of Lake Kivu, in French, *Bull. Agric. Congo Belge,* 48, 703, 1957.
192. **Pitchford, R. J.,** Natural schistosome infections in South African rodents, *Trans. R. Soc. Trop. Med. Hyg.,* 53, 213, 1959.

193. **Pitchford, R. J. and Visser, P.,** The role of naturally infected wild rodents in the epidemiology of schistosomiasis in the Eastern Transvaal, *Trans. R. Soc. Trop. Med. Hyg.*, 56, 126, 1962.
194. **Poché, R. M.,** Vertebrate Pest Control in Egypt: Status and Recommendations, Trip Report, Denver Wildlife Research Center, Denver, Colorado, 1981, 47.
195. **Pollitzer, R.,** Plague, *W.H.O. Monogr. Ser.*, 22, 1, 1954.
196. **Portes, M.,** The struggle against arvicolid rodents in Tunisia, in French, Journées d'Etudes Agricoles Nord-Africaines, Rabat, Morocco, 1949, 17.
197. **Poulet, A. R.,** Evolution of the rodent population of a dry bush savanna in the Senegalese Sahel from 1969-1977, *Bull. Carnegie Mus. Nat. Hist.*, 6, 113, 1978.
198. **Poulet, A. R.,** The 1975-76 rodent outbreak in a northern Senegal irrigated farmland, in *Proc. Symp. on Small Mammals: Problems and Control*, Sanchez, F. F., Ed., BIOTROP Spec. Pub., 12, Bogor, Indonesia, 1980, 248.
199. **Poulet, A. R.,** The ecological basis of forecasting rodent outbreaks in a Sahelian agrosystem, *Acta Zool. Fenn.*, 173, 107, 1985.
200. **Poulet, A. R. and Poupon, H.,** The invasion of *Arvicanthis niloticus* (Rodentia, Muridae) within the natural sahelian zone in 1975—1976 and its impact on woody vegetation, in French, *Terre Vie*, 32, 161, 1978.
201. **Ranque, P.,** Study on trypanosomes of Sciuridae from Senegal: *Trypanosoma (Megatrypanum) heliosciuri* sp. n. and *Trypanosoma (Herpetosoma) xeri* Leger and Baury, 1922, *Ann. Trop. Med. Parasitol.*, 71, 255, 1977.
202. **Ranque, P., Quilici, M., and Camerlynck, P.,** *Arvicanthis niloticus* (Rongeur, Muridae), a primary reservoir of leishmaniasis in Senegal, in French, *Bull. Soc. Pathol. Exot.*, 67, 167, 1974.
203. **Rifaat, M. A., Mahdi, A. H., Arafa, M. S., Nasr, N. T., and Sadek, M. S. M.,** Isolation of *Toxoplasma* from *Rattus norvegicus* in Egypt, *Trans. R. Soc. Trop. Med. Hyg.*, 65, 788, 1971.
204. **Rifaat, M. A., Mohammed, A. H. H., Shawarby, A. A., and Arafa, M. S.,** Incidence and distribution of domestic and commensal rodents in UAR, *J. Egypt. Public Health Assoc.*, 44, 127, 1969.
205. **Rifaat, M. A., Nasr, N. T., and Sadek, M. S. M.,** The role of the domestic rat *Rattus alexandrinus* as a reservoir host of *Toxoplasma gondii* in Egypt, *J. Trop. Med. Hyg.*, 76, 257, 1973.
206. **Robbins, C. B., Krebs, J. W., Jr., and Johnson, K. M.,** *Mastomys* (Rodentia: Muridae) species distinguished by hemoglobin pattern differences, *Am. J. Trop. Med. Hyg.*, 32, 624, 1983.
207. **Roberts, A.,** *The Mammals of South Africa*, Trustees of The Mammals of South Africa Book Fund, Johannesburg, 1951, 700.
208. **Roberts, J. I.,** The endemicity of plague in East Africa, *East Afr. Med. J.*, 12, 200, 1935.
209. **Roberts, J. I.,** Rats and plague in Tanganyika, *East Afr. Med. J.*, 12, 220, 1935.
210. **Roberts, J. I.,** The relationship of field rodents to plague in Kenya, *J. Hyg. Cambridge*, 39, 334, 1939.
211. **Roberts, J. I.,** The transmission of plague in Kenya, *J. Trop. Med. Hyg.*, 53, 80, 1950.
212. **Robinson, H. A. and Olsen, O. W.,** The role of rats and mice in the transmission of the porkworm, *Trichinella spiralis* (Owens, 1835) Raillett, 1895, *J. Parasitol.*, 46, 589, 1960.
213. **Rosevear, D. R.,** *The Rodents of West Africa*, British Museum of Natural History, London, 1969, 604.
214. **Rowe, F. P.,** Control of rodents in stored products and urban environments, in *Small Mammals: Their Productivity and Population Dynamics*, Golley, F. B., Petrusewicz, K., and Ryszkowski, L., Eds., International Biological Programme 5, Cambridge University Press, Cambridge 339, 1975.
215. **Schmutterer, H.,** *Pests of Crops in Northeast and Central Africa*, Gustav Fischer Verlag, Stuttgart, 1969, 296.
216. **Schwetz, J.,** On a new schistosome of wild rodents found in the Belgian Congo, *Schistosoma mansoni* var. *rodentorum* var. nov, *Ann. Trop. Med. Parasitol.*, 47, 183, 1953.
217. **Schwetz, J.,** On two schistosomes of wild rodents of the Belgian Congo: *Schistosoma rodhaini* Brumpt, 1931; and *Schistosoma mansoni* var. *rodentorum* Schwetz, 1953; and their relationships to *S. mansoni* of man, *Trans. R. Soc. Trop. Med. Hyg.*, 48, 89, 1954.
218. **Schwetz, J.,** Natural infection of domestic rats (*Rattus rattus*) by *Shistosoma mansoni* in Central Africa, *Bull. Soc. Pathol. Exot.*, 48, 182, 1955.
219. **Schwetz, J.,** Role of wild rats and domestic rats (*Rattus rattus*) in schistosomiasis of man, *Trans. R. Soc. Trop. Med. Hyg.*, 50, 275, 1956.
220. **Scott, G. R. and Heisch, R. B.,** Rift Valley fever and Rift Valley fever rodents, *East Afr. Med. J.*, 36, 665, 1959.
221. **Senzota, R. B. M.,** The habitat and habits of the grass rats (*Arvicanthis niloticus*) in the Serengeti National Park, Tanzania, *Afr. J. Ecol.*, 20, 241, 1982.
222. **Senzota, R. B. M.,** The house rat enters the Serengeti, *Afr. J. Ecol.*, 20, 211, 1982.
223. **Senzota, R. B. M.,** The habitat, abundance and burrowing habits of the gerbil *Tatera robusta* in the Serengeti National Park, Tanzania *Mammalia*, 48, 185, 1984.
224. **Sezi, C. L.,** Relapsing fever at Masaka hospital, *East Afr. Med. J.*, 47, 176, 1970.

225. **Sheppe, W.,** The annual cycle of small mammal populations on a Zambian floodplain, *J. Mammal.*, 53, 445, 1972.
226. **Sheppe, W. A.,** Notes on Zambian rodents and shrews, *The Puku*, 7, 167, 1973.
227. **Shigidi, M. T. A.,** Animal leptospirosis in the Sudan, *Br. Vet. J.*, 130, 528, 1974.
228. **Shuyler, H. R.,** Control of Rodents and Grain-eating Birds, Report to the Government of United Arab Republic, Report No. AGP:RP 7/70, Food and Agriculture Organization, Rome, 1970, 15.
229. **Shuyler, H. R.,** FAO's need for rodent ecology, population dynamics and forecasting data, *EPPO Bull.*, 7, 297, 1977.
230. **Sidorowicz, J.,** Rodents feeding on cassava *Manihot esculenta* Granz. in Zambia, *Mammalia*, 38, 344, 1974.
231. **Silverman, N. H., Katz, J. S., and Levin, S. E.,** *Capillaria hepatica* infection in a child, *South Afr. Med. J.*, 47, 219, 1973.
232. **Sinclair, A. R. E.,** The resource limitation of tropic levels in tropical grassland ecosystems, *J. Anim. Ecol.*, 44, 497, 1975.
233. **Smithers, R. H. N.,** *The Mammals of Botswana*, Museum Memoir No. 4, The trustees of the National Museums of Rhodesia, Salisbury (now Harare, Zimbabwe), 1971, 340.
234. **Smithers, R. H. N.,** *Guide to the Rats and Mice of Rhodesia*, National Museums and Monuments, Salisbury, Rhodesia (now Harare, Zimbabwe), 1975, 50.
235. **Smithers, R. H. N.,** *The Mammals of the Southern African Subregion*, University of Pretoria, Pretoria, South Africa, 1983, 736.
236. **Stenseth, N. C.,** Forecasting of rodent outbreaks: models and the real world, *EPPO Bull.*, 7, 303, 1977.
237. **Stijins, J.,** Natural rodent hosts of *Schistosoma rodhaini*, *Ann. Parasitol. Hum. Comp.*, 27, 385, 1952.
238. **Stockley, C. N.,** Milky eagle owl, *Zoo Life*, 7, 33, 1952.
239. **Suliman, S. M., Shumake, S. A., and Jackson, W. B.,** Food preference in the Nile rat *Arvicanthis niloticus*, *Trop. Pest Manage.*, 30, 151, 1984.
240. **Symes, C. B.,** Notes on Rats, Fleas and Plague in Kenya (II), Kenya Medical Department, Records of the Medical Research Laboratory (No. 3), Government Printer, Nairobi, 1932, 28.
241. **Taylor, K. D.,** An Investigation of Damage to West African Cocoa by Vertebrate Pests, Nigerian Ministry of Agriculture and Natural Resources, Ibadan, Nigeria, 1961, 35.
242. **Taylor, K. D.,** Report on a two-day visit to the Northern Region of Tanganyika, Unpublished Report to the Tanganyika Ministry of Agriculture, Dar-es-Salaan, 1963, 3.
243. **Taylor, K. D.,** Report on rat control in Kenya, Unpublished Report to the Kenya Ministry of Agriculture, Nairobi, 1963, 30.
244. **Taylor, K. D.,** An outbreak of rats in agricultural areas of Kenya in 1962, *East Afr. Agric. For. J.*, 34, 66, 1968.
245. **Taylor, K. D.,** Rodent Research, in 1970 Annual Report, Research Division, Ministry of Agriculture, Nairobi, Kenya, 1972, 22.
246. **Taylor, K. D.,** Rodent problems in tropical agriculture, *PANS* , 18, 81, 1972.
247. **Taylor, K. D.,** Verebrate pest problems in Africa, in *Proc. Conf. Organ. Practice Vertebr. Pest Control*, ICI Plant Protection Division, Surrey, England, 1984, 21.
248. **Taylor, K. D. and Green, M. G.,** An Ecological Study of the Rodent Pests in the Kenya Highlands, Internal Report, Ministry of Agriculture, Kenya, 1974, 53.
249. **Taylor, K. D. and Green, M. G.,** The influence of rainfall on diet and reproduction in four African rodent species, *J. Zool. London*, 180, 367, 1976.
250. **Taylor, P., Gordon, D. H., and Isaacson, M.,** The status of plague in Zimbabwe, *Ann. Trop. Med. Parasitol.*, 75, 165, 1981.
251. **Taylor, P. and Pugh, A.,** Plague in Zimbabwe — a review of the situation in 1982, *Cent. Afr. J. Med.*, 28, 249, 1982.
252. **Toovey, R. W.,** Field planting of oil palms, *J. West Afr. Inst. Oil Palm Res.*, 1, 68, 1953.
253. **Traub, R. L., Wisseman, C. L., Jr., Farhang-Azad, A.,** The ecology of murine typhus. A critical review, *Trop. Dis. Bull.*, 75, 237, 1978.
254. **van der Merwe, C. R. and Keogh, H.,** Cited in de Graaf, G., *The Rodents of Southern Africa*, Butterworths, Pretoria, 1981, 267.
255. **van Eyssen, J. L.,** Animals in mines, *Afr. Wildl.*, 10, 331, 1956.
256. **van Riel, J., Baylet, R., and van Riel, M.,** A microbiological, serological, and epidemiological survey on leptospirosis in Senegal, in French, *Med. Afr. Noire*, 16, 165, 1969.
257. **Vesey-Fitzgerald, D. F.,** The habits and habitats of small rodents in the Congo River catchment region of Zambia and Tanzania *Zool. Afr.*, 2, 111, 1966.
258. **Voelckel, J. and Varieras, G.,** Distribution of species of *Rattus norvegicus* and *Rattus rattus* at Douala, French Equatorial Africa, in French, *Med. Trop.*, 19, 456, 1959.
259. **Walsh, J.,** Sahel will suffer even if rains come, *Science*, 224, 467, 1984.
260. **Watson, J. M.,** The wild mammals of Teso and Karamoja, *Uganda J.*, 14, 53, 1950.

261. **Weber, W. J.,** *Diseases Transmitted by Rats and Mice*, Thompson Publications, Fresno, Calif., 1982, 182.
262. **Weinbren, M. P. and Mason, P. J.,** Rift Valley fever in a wild field rat (*Arvicanthis abyssinicus*), a possible natural host, *South Afr. Med. J.*, 31, 427, 1957.
263. **Weis, N., Ed.,** Rodent Pests and Their Control, German Agency for Technical Cooperation, Eschborn, West Germany, 1981, 189.
264. **Wilson, V. J.,** Notes on the breeding and feeding habits of a pair of barn owls, *Tyto alba* (Scopoli), in Rhodesia, *Arnoldia Rhod.*, 4, 1, 1970.
265. **Wulff, H., Habiyi, A., and Monath, T.,** Recent isolations of Lassa fever from Nigerian rodents, *Bull. W.H.O.*, 52, 609, 1975.
266. **Wulff, H., McIntosh, B. M., Hamner, D. B., and Johnson, K. M.,** Isolation of an arenavirus closely related to Lassa virus from *Mastomys natalensis* in southeast Africa, *Bull. W.H.O.*, 55, 441, 1977.
267. **Yalden, D. W.,** *Tachyoryctes macrocephalus*, *Mamm. Species*, 237, 1, 1985.
268. **Young, E. and Whyte, I. J.,** Trichinosis, *Trichinella spiralis*, infestations in wild animals of the Kruger National Park, South Africa, *J. S. Afr. Vet. Assoc.*, 46, 233, 1975.
269. **Zumpt, I. F.,** Is the multimammate rat a natural reservoir of *Borrelia duttoni?*, *Nature*, 184, 793, 1959.
270. **Zumpt, I. F.,** The ground squirrel, *Afr. Wildl.*, 24, 115, 1970.

Chapter 5

RODENT PROBLEMS IN ASIA

Ishwar Prakash and R. P. Mathur

TABLE OF CONTENTS

I. INTRODUCTION

The largest continent, Asia, which harbored some of the world's oldest civilizations, constitutes one-third of the earths land area. With its massive expanse, it exhibits a spectacular diversity pertaining to climatic and biological features. The arctic zone in the north culminates into tropical environment in the south. The physiography of central Asia is dominated by the mightiest ranges: the Himalayas, Hindukush, Kunlun, and Tien Shan. The great river systems, Indo-Gangetic, Brahmaputra, Tigris-Euphrates, Amur, Yangtze, and Yellow river, among many more, crisscross the vast plains. Asia holds 5 of the 11 deserts of the world — both cold and hot arid zones.[29] As a consequence of the climatic and habitat diversity, the vegetation types exhibit a wide variation; the Siberian coniferous taiga and the wet evergreen and deciduous monsoon forests are separated by belts of steppes and desert grasslands. The mangrove stands of coastal estuaries and backwaters, and the island forests present typical vegetation types. Considering the vast expanse of Asia, surprisingly, less than 15% of the land is under cultivation. Rice and wheat are the most important food crops, and millets constitute the major dryland crop. Oil crops and sugar cane are grown in monsoonal Asia. Jute in India and Pakistan, cotton in these two countries and in Soviet Central Asia, silk in China, and tea and forest products are some of the major nonfood crops. Since the economy of the Asian people is based on agriculture, plant protection assumes special significance in enhancing the production of food, fiber, and fuel through their conservation. Rodents, among all the pests, play an important but a negative role in agricultural economy due to their devastating activities resulting in food losses — both to standing crops and stored foodgrains, and damage to forest produce, soil conservation works, irrigation systems, etc. We have purposely not discussed the role of rodents affecting human health since this aspect is dealt with in great detail in Chapter 7.

II. RODENT PESTS

The distribution and abundance of rodents in Asia indicate a broad pattern which, in a way, corresponds to the terrain physiography and bioclimate of the region. The deserts spread from Syria and Saudi Arabia, to central U.S.S.R., up to Gobi in northeast, and east to the Thar desert. Gerbils and jerboas are the predominant xeric rodent species, and probably the genus *Meriones* outnumbers all other taxa.[95] In the nondesert zones of Soviet Union and Chinese Republic, the voles, *Microtus, Arvicola*, and squirrels, *Citellus, Cricetullus*, appear to be the major rodent pests in the field. In the Indian subcontinent, however, the rodent fauna is slightly different and includes the Oriental elements, *Bandicota*, field *Rattus* spp. *Mus* spp., and *Funambulus* being the more important ones from the food losses point of view. However, *Rattus rattus, R. norvegicus*, and *Mus musculus* constitute the predominant commensal population of rodents in the whole of Asia, as elsewhere in the world. An attempt has been made to list the rodent species of economic importance (Table 1) in various parts of Asia.

Information about the density structure of only a few rodent species is available: *Citellus c. daurica*, 7.68/ha;[27] *Microtus brandti*, 3000 to 4000 burrows/ha (or even 10,000 or more);[27] *Meriones unguiculatus*, 200 to 330/ha in China and 32 burrows with 760 entrances in the U.S.S.R.;[74] *Meriones tamariscinus*, 8 to 10/m^2 in oasis[74] and 10 to 60/ha in the field; *Meriones hurrianae*, 34 to 510 (Av. 290)/ha;[98] *Bandicota bengalensis*, 108/ha in the field[24] and 0.78/m^2 in foodgrain godowns;[116] *Rattus rattus*, 5.7 to 15.6 (Av. 9.8)/house;[66] and *Ochotona curzoniae*, 200/ha.[27] In Korean agricultural fields 33 to 120 harmful rodents (Av. 54)/ha were recorded, dominated by *Apodemus agrarius* (96%).[8] The population of gerbils (*Rhombomys opimus, Tatera indica*), voles, and mole rats has been reported to be fairly high in their respective habitats.[5,11,17,22,30,74-76,131]

Table 1
IMPORTANT RODENT PESTS OF ASIA

Common name	Generic name	Economically significant for	Ref.
Sousliks	*Citellus pygmaeus*	Cereals; grasslands	115
	Citellus suslicus	Cereals; grasslands	115
	Citellus fulvus	Cereals; grasslands	132
	Citellus c. dauricus	Grasslands	27
Northern palm squirrel	*Funambulus pennanti*	Fruits; orchards	11
Indian palm squirrel	*Funambulus palmarum*	Fruits; orchards; lac industry	11, 67
Jungle striped squirrel	*Funambulus tristriatus*	Fruits; orchards; cacao	14, 15
Himalayan marmots	*Marmota bobak sibirica*	Farm machinery	128
Long-clawed ground squirrel	*Spermophilopsis leptodactylus*	Soil erosion	59
Siberian chipmunk	*Tamias sibiricus*	Fruit trees	27
European beaver	*Castor fiber*	Forest produce; water ways	32, 79
Indian crested porcupine	*Hystrix indica*	Root crops, vegetables; trees	11, 21, 49, 52
Fat dormouse	*Glis glis*	Olives; nuts; fruits	132
Migratory hamster	*Cricetulus migratorius*	Stored material	53
Striped hamster	*Cricetulus barabensis*	Field crops	27
Long tailed hamster	*Cricetulus triton*	Field crops	27
Water vole	*Arvicola terrestris*	Field crops	1, 22
Social vole	*Microtus socialis*	Field crops, grasslands; fruits	132
Brandt's vole	*Microtus brandti*	Grasslands; crops	27, 33
Levant vole	*Microtus guentheri*	Cereals; fodder; vegetables	17, 30, 133
Common vole	*Microtus arvalis*	Fields crops; grasslands	22, 45
Hairy footed gerbil	*Gerbillus gleadowi*	Grasslands; crop	11, 88, 90
Indian gerbil	*Tatera indica*	Crops; grasslands; afforestation; vegetables; commensal	2, 3, 11, 121
Indian desert gerbil	*Meriones hurrianae*	Crops; grasslands; afforestation; soil conservation; vegetables	2, 3, 11, 110
Persian gerbil	*Meriiones persicus*	Cereals; grasslands	12, 74, 132
Turkish jird	*Meriones tristrami*	Field crops; grasslands	12
Clawed jird	*Meriones unguiculatus*	Grasslands; homesteads	27, 74
Mid-day gerbil	*Meriones meridianus*	Grasslands; afforestation; tree nurseries	74
King jird	*Meriones rex*	Field crops; fodder; irrigation dams	52
Tamarisk gerbil	*Meriones tamariscinus*	Cereals; pasture; vegetables	74

Table 1 (continued)
IMPORTANT RODENT PESTS OF ASIA

Common name	Generic name	Economically significant for	Ref.
Minor Asian gerbil	*Meriones blackleri*	Cereals; cotton	74
Vinogradov's gerbil	*Meriones vinogradovi*	Cereals; cotton; grasslands	74
Red tailed gerbil	*Meriones libycus*	Grasslands; cereals; young fruit trees; soil conservation	74
Great gerbil	*Rhombomys opimus*	Grasslands; crops; soil conservation	74, 93
Striped field mouse	*Apodemus agrarius*	Field crops	8, 27
Long tailed field mouse	*Apodemus sylvaticus*	Field crops	53
Soft furred field rat	*Rattus meltada*	Crops; grasslands; vegetables; threshing floors	9, 11, 16
House or Roof rat	*Rattus rattus*	Fruits; orchards; coconut; oil palm; field crops; stored crops	7, 11, 21, 44, 65, 66, 109, 120
Norway rat	*Rattus norvegicus*	Storage; crops; ports	8, 10, 47
Buff breasted rat	*Rattus flavipectus*	Houses; storage; ships; trains	27
Turkestan rat	*Rattus turkestanicus*	Fruit fields and storage	5
House mouse	*Mus musculus*	Crops; storage; houses	18, 73, 130
Indian field mouse	*Mus booduga*	Field crops	11, 53
Lesser bandicoot rat	*Bandicota bengalensis*	Field crops; storage; structures	9, 11, 16, 21, 40—43, 116
Larger bandicoot rat	*Bandicota indica*	Storage; crops; kitchen gardens; household structures	11
Short tailed mole rat	*Nesokia indica*	Crops; vegetables; dates; irrigation canals	53, 132
Spiny mouse	*Acomys cahirinus*	Field crops	104
Mole rat	*Spalax leucodon*	Field crops; vegetables; fruits; trees	104
Palestine mole rat	*Spalax ehrenbergi*	Field crops; vegetables; fruits; trees	76
Lagomorpha			
Collard pika	*Ochotona rufescens*	Fruit trees	64
Plateau pika	*Ochotona curzoniae*	Grasslands	27, 130, 134
Altai pika	*Ochotona alpina*	Forest produce	58

Note: Nomenclature according to Reference 36.

Periodic as well as sporadic outbreaks in the rodent populations enhance their numbers menacingly. Such outbreaks can be caused either by failure of monsoon, by favorable reproductive conditions after a fairly long spell of climates (drought or excessive rainfall), or can be associated with flowering of bamboo in forests. In any case, during the upsurge in rodent density losses to the crops increase manifolds. Rodent species prone to outbreaks of their population are *Mus*,[27,33] *Microtus*,[27] *Rattus*,[44] *Arvicola*,[22] *Rhizomys, Cannomys*,[129] *Bandicota, Gerbillus, Rattus (Millardia) meltada*, and *Tatera*.[89,96,97] In China, *Mus musculus* outbreaks resulted in crop losses totaling 150 million kg.[27,130,134] Eruptions in *Mus* population have also been observed in Russia.[125] Over-crowding of *Microtus brandti* in ten foci in the damage area has been reported in Mongolia.[33] Israel faces about 30% crop damage during outbreaks.[53] Explosion in the number of *Gerbillus gleadowi* created famine–like conditions in the Rajasthan desert of India.[89] In the same country about 50% of rain-fed crops were lost to rodent outbreaks.[99] In northeastern India, where *Rattus r. brunneusculus* outbreaks coincide with the time of bamboo flowering, once in 25/50 years, the rats invade the *jhum* (shifting cultivation of mixed crops on hills after slashing and burning the forest patches) fields and ravage the crops in totality.[26,44] Similar outbreaks of *Rhizomys* and *Cannomys* have been reported from Burma,[129] but the phenomenon is not well studied. Cyclonic rainfall can also increase the number of rodents (*Bandicota, Tatera*, and *Mus*) by providing more food and harborage due to lodging of standing crops. Results of surveys prior to and after a cyclone revealed not only an increase in rodent numbers but also a considerable enhancement in their litter size in Andhra Pradesh, India.[100]

III. ECONOMIC LOSSES

Early man probably started growing cereals in Asia more than 5000 years ago so it is quite likely that man first encountered rodent pests on this continent.[132] This has been confirmed by the early pottery rat traps from Pakistan and Iran, and the warning against rodent pests of "our" granaries and barley contained in the old Hindu (Indian) scriptures, which were written about 3000 B.C. Since then, the necessity for increased food production, the conversion of forests, unculturable wastelands, sand dunes, steppes, and grasslands into agricultural fields, the incoming of irrigation canals, the development of oases in the heart of deserts, and many such developmental activities have tended to increase the rodent problem manifolds. Firstly, rodents living away from man in the wild came in contact with man and his agriculture. Consequently they shifted from forests and other habitats to crop fields in the vicinity of man,[90] and to urban areas due to the extension of towns over the rodent habitats.[94] Secondly, certain rodents occupy more than one habitat. *Rattus flavipectus* is the most important species in dwelling houses in south and southwest China, constituting 87.6% of the rodent population, but in tropical regions of China they are found living in the wild.[135] Another man-induced change is the incoming of irrigation systems in arid regions. As a result the true desert elements (gerbils and jerboas) have been replaced by mesic forms like *B. bengalensis, Nesokia indica, R. meltada*, and *M. musculus*.[4,44,90,93,94,96]

In spite of the fact that rodents inflict economic damage to almost all field crops, very little data are available on quantitative assessment of the losses due to field rodents. However, the available information has been gathered from literature for various economic food and other crops grown in different Asian countries (Table 2).

A. Cereal Crops

1. Wheat

Wheat is the major cereal crop grown in Asia, being produced mainly in the Indian subcontinent, the U.S.S.R., and parts of Central Asia. Substantial damage has been reported from India, Pakistan, and Nepal, but only 1% from Korea. Localized damage on individual

Table 2
RODENT DAMAGE TO GROWING AGRICULTURAL CROPS IN NORTHERN ASIA

Crop	Country	Damage (%)	Ref.
Wheat	Nepal	8	37
	Nepal	6—45	21
	India	16.3—21.3	3
	India	21 kg/acre	16
	Pakistan	7.5	9
	Pakistan	6.1	42
	Pakistan	5.66	63
Wheat and barley	Korea	1	8
	India	11	57
Rice	Korea	4	8
	India (Central)	7.1—21.5	57
	India (South)	5.2—65.3	57
	India (South)	Tillering 9.27	102
		Booting 15.48	102
		Ripening 12.38	102
	India (North-East)	15—100	44, 86
	India	72—261 kg/ha	24, 117
	Pakistan	10—50	127
	Pakistan	6	48
	Pakistan	19 (2—43)	41
Maize	Nepal	10	53
	Nepal	1—2	21
	India	14	118
	Iraq	68 (ears)	18
Sorghum	India	5.85—6.28	119
Millet	India	107.7 kg/ha	16
Cereal plants	Yemen	25—30	53
	Israel	30	53
	China	150 million kg	130
Growing crops	Bahrain	1	132
	Iran	0.5	132
	Iraq	1—3	132
	Qatar	1	132
	Syrian Arab Republic	10—20	132
Soyabean	Korea	1.6—11.6	9
	India	24.2	6
Gram	India	1.6	6
Sugar cane	India (Central)	23	6
	India (South)	7.85	6
	India (North)	200 kg/ha	16
	Iraq	1.6	18
Groundnut	India (South)	Seedling 6—9	118
		Growth 18.2—25.8	118
		Maturity 4.1—7.6	118
	India (Central)	17.7	6
Coconut	India (South)	15.5	103
	India (Lakshadweep Islands)	35—50	65
Cotton	India	57	78
Sweet potato	Korea	2.9	8
Cassava	India	9.56	68
Vegetables	India	8.7	2
	Yemen	1—5	53
Apple	Pakistan	20	64
Lime Trees	Arabian Gulf States	10—20	53
Cacao	India	8.0—57.3	14, 15
Oil palm	India	10—57.3	120
Tea	India	50	124

Table 2 (continued)
RODENT DAMAGE TO GROWING AGRICULTURAL CROPS IN NORTHERN ASIA

Crop	Country	Damage (%)	Ref.
Arecanut	India	20	123
Pasture vegetation	Mongolia	70—100	33
	China	50	130
	India	1040 kg/ha	87

farms can be as high as 60%. Principle rodent genera inflicting losses are *Tatera, Meriones, Rattus, Mus, Bandicota, Apodemus, Nesokia, Microtus, Spalax*, and *Arvicola*.

Rodent damage, caused primarily by *Bandicota bengalensis*, in Bangaladesh were estimated to be 12.1% during 1979,[19] 2.5% in 1980,[85] and 8.4% during 1982,[19] amounting to 73,000 tons valued at $15.7 million.[20] A damage assessment carried out at six locations in India[118] revealed that about 11 to 12% of the wheat produce is lost to rodents, mostly due to *B. bengalensis, Rattus meltada, Tatera indica*, and *Mus* spp. In the Thar desert, however, the desert gerbil, *Meriones hurrianae* was found to be the major wheat pest causing losses of 16 to 21%.[3] Other species of *Meriones* are also associated with damages to wheat crops: *M. libycus* in Iran,[12] *M. l. erythrourus* in central Asia (both in arid and irrigated areas),[125] *M. persicus* in Iran[46] where they also store a large quantity of wheat grains (up to 5 kg per burrow), *M. tristrami* in Turkey,[12] and *M. vinogradovi* whose grain stores carry an average of 3.5 kg wheat,[51,125] and *M. shawi* in northern Africa.[12] The damage caused by the gerbils may not be very serious, but is considered socially substantial since it influences the marginal farmer to whom a meager crop yield is essential for survival in the desert environment.

Rain-fed as well as winter wheat crops are vulnerable to rodent attack in Pakistan. *Bandicota bengalensis* is the most abundant species, followed by *Rattus meltada* and *Nesokia indica*.[63] The damage estimates to wheat produce vary from 5 to 7.5%[9,42] *Meriones hurrianae* was found to be a serious pest of wheat in the Northwest Frontier and Sind, whereas *M. libycus* and collared pika, *Ochotona rufescens* were found in Baluchistan.[105]

In Iran, a loss of 30,000 tons of wheat was attributed to the dormice, *Glis glis* and the ground squirrel, *Citellus fulvus*, particularly in Qazin and Rasht regions.[60] In Azarbaijan, *Microtus socialis* seems to be well distributed and attacks 400,000-hectare wheat fields annually.[60] A number of susliks, *Citellus* spp., cause damage to wheat, rye and, barley in U.S.S.R. at the beginning of the sprouting period.[77]

2. Rice

Rice constitutes one of the major crops in several countries, and damage figures are available from Korea, India, Pakistan, Bangaladesh, and Burma and vary from 0 to 100% (Table 2). Such a wide variation in damage assessment may be attributed to different assessment methods applied at various stages of crop growth. The rodent genera involved are: *Bandicota, Apodemus, Nesokia, Microtus, Mus, Rattus*, and *Tatera*.

In Korea, damage to rice (14.2%) is reported to be mainly by the predominant field rodent, the striped field mouse, *Apodemus agrarius*;[8] but, in the coastal areas *Rattus norvegicus* constitutes 60 to 80% of the rodent fauna in paddy fields[111] and the seasonal damage to standing rice crop was estimated to be only 4%.[8] In China, about 18 rodent species are considered to be of economic importance.[130,135] *Apodemus agrarius* is found near agricultural fields in central China, whereas the hamsters, *Cricetulus barabensis* and *C. triton*, are found in northern and central China.[91] Quantified information on actual damage caused by rodents to agricultural crops in China is not available; however, 10% loss in field crops is reported.[35]

Damage assessment to rice crop has been carried out to some extent in the Indian subcontinent. In Bangladesh, 5 to 10% rodent damage in deep water rice fields was observed from 1977 to 1980;[23,83] the species involved were *Bandicota bengalensis* and *B. indica* which are also known to cause extensive damage to rice crops in Burma.[129] The estimated rice crop losses due to rodents in India vary from place to place. In North India the crop yield was reduced by 7.1 to 21.5%, whereas in South India the estimates were 4.5 to 17.7%,[102] but the losses even reached 65.3%.[57] In *jhum* fields, in the northeastern hill region, the damage to paddy may reach 25 to 30%.[86] In most of the Indian region *Bandicota bengalensis* is the major pest in irrigated crop fields, but a considerable proportion of the rodent population is constituted by *Rattus meltada* and sometimes *Tatera indica* in nonirrigated, rain-fed crops.[11,38,89,97] However, in the northeastern hill region the major rodent species inflicting damage to paddy crops is *Rattus rattus brunneusculus*.[26]

The earliest well documented damage assessment data for rice crops are available from lower Sind in Pakistan.[126,127] More recent work has revealed that 2.9% of the tillers of the plants are cut, resulting in yield reductions of 6%.[48] However, 19% yield reduction, with a range of 2 to 43, has also been recorded in Pakistan.[41] In Sri Lanka, *Bandicota bengalensis* was regarded as the most destructive among all field rodents.[82]

3. Maize

Maize is grown in most of the countries of the region and damage estimates from Nepal,[21] India,[118,119] and Iraq[18] indicate that losses due to rodents range from 1 to 14%. In Iraq 68% of the maize ears are attacked and the principle species involved are *Mus, Tatera, Meriones, R. meltada, Hystrix, Bandicota*, and *Rattus*.[18,53,118] The king jird, *Meriones rex*, damages the maize crops in Yemen[52] at the time of germination and early growing stages; and in Nepal *Bandicota bengalensis* damages 2% of the maize crops.[112]

4. Sorghum and Millet

Sorghum and millet are largely rain-fed crops. Damage is caused mainly by *Tatera indica, Meriones* spp., *Nesokia indica*, *Gerbillus* spp., *Bandicota bengalensis, Rattus*, and *Mus*. In the Rajasthan desert (India) the hairy-footed gerbil, *Gerbillus gleadowi*, consumed millet (*Pennisetum typhoides*) seeds during three sowings during 1970, and this being the only staple crop, caused almost famine-like condition.[89]

B. Sugar Cane

The most important rodent species inflicting economic losses to sugar cane in India, Pakistan, Nepal, Iran, and Iraq are *Nesokia, Bandicota, Mus, R. meltada*, and *Rattus*. In Punjab, rodents cause considerable reduction in the yield of molasses which can be 200 kg/ha or 0.11 to 0.42% of the total yield.[16] In Uttar Pradesh, $6 to 8 per hectare are lost in cane fields as a result of ravages by field rodents.[50] In Pakistan, an estimated damage of 25.9% of stalk was considered equivalent to 10.8% loss in sugar production per hectare.[43] The species involved near the Indus delta were found to be *Bandicota bengalensis, R. meltada*, and *Nesokia indica*.[114]

C. Oil Seed Crops

Mustard crop in localized patches of Northern India and Burma is reported to be reduced up to one-third due to *Bandicota, R. meltada, Mus*, and *Tatera*.[6,129]

Groundnut is sown in many Asian countries but quantified damage information is available only from India, though damage is also reported from Burma.[129] In Northern Gujerat, the crop is severely damaged by *R. meltada*[89] and by *B. bengalensis* in Nepal.[21] Damage assessment carried out on an all-India level indicated that losses to groundnut crops ranges from 4 to 26%[118] In central India 12.6% of the pods were hoarded by rodents. Principle rodent species associated with the crop are *Rattus meltada, Bandicota bengalensis, Tatera indica*, and the *Mus* spp.[6]

Soyabean damage in Korea is of two distinct types; in upland fields, the damage is 1.6%, but on the levees of rice field it is 11.6%[8] and in India it is 24.2%[6] *Apodemus, Microtus, Rattus, Mus, R. meltada, Tatera*, and *Bandicota* are the main species involved with damage to the crop.

Coconuts are mainly damaged by *Rattus*, reducing the yield up to 50% in Lakshadweep,[65,104] severely affecting the economy of the people since it is the only crop on the islands. On the mainland, in South India, the losses to coconut were estimated to be 11 to 15%.[103] The loss of nuts and the number of trees attacked were maximum (17.1%) during monsoon and winter. Damage to coconut saplings due to *Bandicota bengalensis* was observed to be 6 to 8% when the rodent damaged the collar part of the young saplings.[121] Plantations in the vicinity of Karachi, Pakistan also suffer losses — up to 30.2 nuts/palm/year.[61] Damage to oil palm, recently introduced in the Andaman and Nicobar group of islands was reported to be considerable, mainly by the *Rattus* spp.[120] About 47 and 57% of the saplings which were 26 and 113 months old were found to be damaged.

D. Fiber Crops

Cotton is the only fiber crop reported to be damaged. The damage is caused by *Tatera indica, R. meltada,* the *Mus* spp., and *Nesokia indica.*[18,78,125]

E. Root Crops

Root crop loss is mainly to potatoes, sweet potatoes (1%),[54] onions, cassava tubes, and sugar beet.[68] Rodent species associated are *Hystrix indica*, *Bandicota bengalensis, Rattus* spp., *Mus* spp., *Apodemus, Microtus, Tatera, Meriones, Spalax*, and *Nesokia*. The porcupine, *Hystrix indica* is reported to hoard around 10 to 20 kg potatoes per burrow in Nepal[21] and inflicts serious losses to potato crops in Yemen.[52]

F. Vegetable Crops

Vegetables and watermelon are produced throughout Asia and a number of rodent species inflict losses to them: viz., *Tatera, Meriones, Rattus, Mus, Hystrix, Bandicota, Microtus, Spalax*, and *Nesokia*. Damage to watermelon in desert countries is as high as 60%.[53] In Syria, watermelon are destroyed by *Spalax leucodon*,[104] by *Hystrix indica* in Yemen,[52] and by a number of gerbillids in the U.S.S.R.[77] In the Indian desert an estimate[2] revealed that losses to about 12 vegetable crops range from 4 to 20% (Av. 8.7%). The cost benefit ratio was found to be 1:900.

G. Fruit Crops and Orchards

Fruit crops such as mango, banana, guava, pineapple, apple, citrus, dates, and avocados are attacked by rats, squirrels, porcupines, and pikas (Lagomorpha: Ochotonidae). Pineapple are damaged by *Hystrix* and *Funambulus*, and mango by *Funambulus* in Nepal[21] and Bangladesh.[84] In Baluchistan, Pakistan, about 20% of the apple crop is attacked by pikas, *Ochotona rufescens*.[62,64] *Rattus* causes a 10 to 20% loss of lime in the Arabian Gulf States,[17] and fruits such as avocado, mango, pomegranate, guava, and banana in Israel.[7] *Nesokia indica* damages dates in the Lower Jordan valley and Israel, and also attacks orchards in Afghanistan and other countries.[53] *Microtus, Rattus, Mus,* and *Apodemus* are serious pests of fruit orchards and fruit storehouses in the U.S.S.R.[5,28] *Tamias* is also reported to damage fruit trees while searching for food in China.[27,130] *Meriones* attack orchards, particularly intercrops like curcurbits.[12] *Spalax* is a serious pest of fruit trees in Syria and Jordon.[104] *Rattus turkestanicus* has been reported to inflict considerable damage in fruit gardens in Pamir, Altai, and the Turkistan mountains, east of Karakum and around Samarkand.[125] Hoarding of 14 kg of walnuts and 7 kg of apples has been reported in the burrows of the Turkestan rat. It was estimated that walnut produce is reduced by 25% due to ravages of this rat. About 30 tons of olives per year are lost to the fat dormouse *Glis glis* and *Citellus* in Iran.[53] In the Caucasian region of the U.S.S.R., damage to grapes due to *G. glis* was estimated to be 12.5 to 41.7%. Likewise, the dormouse gnaws up to 55 pears (33 kg) per day during its active period when its density is 30 per hectare, thus reducing the mountain and foothill crops by 18%.[77]

H. Plantation Crops

Plantation crops include cocao, arecanut, cardamon, tea, and coffee. Cocao and cardamom are attacked by *Funambulus* and *Rattus*,[14,15] whereas tea bushes are damaged up to 50% by *Bandicota*.[124] In Burma, the bamboo rats, *Rhizomys* and *Cannomys*, are reported to damage coffee and tea plantations.[129]

I. Pasture and Fodder Crops

Rodent damage to grasslands and fodder crops is very severe, especially in the arid tracts. In Mongolia, *Microtus* is responsible for 70 to 100% damage to pasture vegetation,[33] whereas in Israel, *M. socialis* damages forage crops substantially.[17,133] In China *Ochotona* causes about 50% damage to grasses.[130] *Meriones, Tatera*, and *Gerbillus* are attributed to inflict serious losses to grasslands in the Indian desert.[87] Annual forage requirement of *Meriones hurrianae* at a high density has been estimated to be around 1040 kg/ha compared with an annual forage production of 1210 kg/ha, hence they are serious competitors of livestock.[87] *Citellus* has also been named to adversely effect the vegetal produce in the semi desert of the U.S.S.R., along with a number of gerbillids, particularly *Meriones meridianus*. Marmots in the Altai region also destroy about 25% of fresh grass.[77] In Turkey, leguminous plants such as alfalfa, clover, and broad beans are attacked by *Microtus, Spalax*, and *Meriones*.[104] The lucerne crop is also damaged by *Meriones libycus* in Turkey[12] and in Arabia.[51] As a result of an unprecedented upsurge of voles (*Microtus socialis* and *Meriones* spp.) during 1957 in Israel, heavy damages occurred to the lucerne crop and the control operations cost $70 to 80 per hectare.[12,133]

J. Forest Trees

Hystrix indica damages trees from Israel to India. In Pakistan it damaged up to 52.5% of *Melia azadarach* and about 24.3% of *Morus alba* trees by girdling activity.[49] Debarking, mainly by *Meriones* and *Tatera*, damage tree saplings of *Albizzia lebbek, Prosopis cineraria*, and *Acacia tortilis*.[88] The beaver, *Castor*, damages forest vegetation and tree species in the U.S.S.R.[32] *Ochotona* cuts herbaceous plants at the root collar level for making hay for winter feeding.[58] Various murid rodents damage acorn and oak sowings in forests,[81,113] whereas pine trees are damaged by voles.[72] Sometimes *Marmota* inflict serious damage by eating planted seed and young forest plantings.[27] *Mus* and *Rattus* ring bark the trunks of nut trees.[53] *Microtus* gnaw bark of young trees and roots of saplings.[17,133] Other rodents implicated in attacking trees and forest plantations are *Apodemus, Clethrionomys*, and *Nesokia*, and *Spalax* which cause serious damage to root systems by their subterranean activity.[55,104,105] The greater woodmouse, *A. speciosus*, eats tree seeds in northeastern China to the detriment of forest regeneration.[55]

K. Soil Conservation

Burrowing rodents cause soil erosion by excavating soil to such an extent that it threatens soil conservation practices. The main species involved are *Meriones, Microtus, Rhombomys*, and a score of gerbils and jerboas. In the U.S.S.R. the number of rodent burrows in Oasis can be as high as 1500 to 1600/ha or even more.[93] In China *Microtus brandti* burrow openings can number 3000 to 4000/ha, or even over 10,000/ha in high density areas, thus causing serious problems to soil conservation programs.[27] The number of burrow openings of the great gerbil, *Rhombomys opimus*, is as many as 468/ ha in the Karakum desert.[59] The intensive burrowing activity of the rodents plays an important role in loosening the sand in the desert. Likewise, no less a part in the sand dynamics is played by the long-toothed ground squirrels, *Spermophilopsis leptodactylus*, that dig up the sand to procure the bulbs of *Poa bulbosa*, their main food source. The number of burrow openings is reported to be 13,135/ha.[59] All the depressions between the sand hills in the area of fixed sand are strewn with these holes. Where there is no *Poa bulbosa*, the ground squirrel eats the roots of the wire-grass, *Aristida pennata*. This loosens the sand, which begins to move due to strong wind action.[59] In the Thar desert, the desert gerbil, *Meriones hurrianae*, is also one of the major factors of desertification since, due to its fossorial activity, it loosens and excavates 61,500 kg fixed sand/day/km^2 in cultivated fields and 1,043,800 kg/day/km^2 in fallow lands.[110]

L. Irrigation Canals

Irrigation canals are mostly damaged by *Nesokia indica*, but other genera, *Meriones, Tatera, Mycogaster, Bandicota, Castor*, and *Rattus (Millardia)* are also involved in causing damage to mud dams and water channels.[12,32,53,89,129]

M. Other Damages

Other damages, very localized and less defined, include damage to the lac industry in India by squirrel, *Funambulus pennanti*[67]; to oasis cultivation in Arabian Gulf States by *Rattus* and gerbils;[53] to fish ponds by nutria, *Myocaster*;[53] and to farm machinery by *Marmota*.[128] Cable damage by *Rattus* and *Mus* can sometimes cause serious concern.[18,70] Around 8 to 10% of poultry farms and adjacent godowns are reported to be infested with *Rattus norvegicus, R. rattus*, and *Mus musculus*, damaging 5% of stored animal food. The damage to eggs and young chicks is also considerable.[13]

IV. HOARDING AND WINTER STORAGE

Rodents possess very high hoarding propensity. In the temperate climate, plant material is stored for the winter; mainly cereals, different types of seeds, and green plants are stored by gerbils, bandicoots, hamsters, mice, and occasionally by *R. meltada*. *Bandicota* is a serious hoarder of paddy crop, 1.75 to 3.32 kg/burrow[24,101] or 92.5 to 972 kg/ha,[40,48,117] 5.7 to 10% of the total yield.[40,106,107] *Bandicota*, *R. meltada, Tatera*, and *Mus* store 1.01 to 13.92 kg/ha wheat in their burrows,[6] whereas *Meriones blackleri* and *M. unguiculatus* hoard 0.7 to 1.4 kg of wheat, the latter may store up to 20.5 kg buck-wheat in their burrows.[74,80] In China, *Cricetulus barabensis* has been reported to carry 40 kg or more cereals and beans with the help of their cheek pouches.[27] *R. meltada, Bandicota bengalensis, Tatera indica*, and *Mus* sp. hoard 12.6% of groundnut pods.[6] *Meriones (meridianus, vinogradova, libycus, tamariscinus)* can store 0.8 to 30 kg of different seeds and green plants, whereas in the colony of great gerbil, *Rhombomys opimus*, the amount of plant material can be as high as 64 kg or more.[74] The great gerbil stores the sedge, *Carex physodes*, in its burrows, drying it beside the openings. The animals put up the hay in the same way that man does and, to prevent its blowing away, sticks 5 cm long pegs made of *Astragalus confirmans* into the sand. The dried sedge collected in one place near two burrows weighed about 2.5 kg.[59] *Microtus arvalis* stores 2 to 3.5 kg of bulbs of *Phleum*.[31,45] In winter, *Mus musculus* accummulates large food reserves in burrows or mounds made in fields.[73] It is not clear whether the hoarding behavior has evolved among rodents as a contingency program ensuring food availability for the adverse times or for their progeny. Most of the time these caches are not utilized by rodents.

Table 3
POSTHARVEST FOOD LOSSES IN NORTHERN ASIA

Country	Loss (%)	Rodent species	Ref.
Burma	$5.12 million per year (only in Rangoon)	*R. norvegicus, R. rattus, M. musculus*	129
China	15—20	*R. norvegicus, R. flavipectus*	27, 35
India	1.69—10	*R. rattus, M. musculus*	
	1.36—3.59 tons/village	*B. bengalensis, B. indica, F. pennanti, T. indica*	53, 82, 119
		R. rattus, M. musculus	66
Iran	0.2	—	132
Iraq	3	—	132
Israel	5 (Animal feed)	*R. rattus, M. musculus, R. norvegicus*	53
Korea	20	*R. norvegicus, R. rattus, M. musculus*	53
Nepal	3—10	*R. rattus, M. musculus* *R. norvegicus, B. indica*	21, 39, 53
Pakistan	5—10	*R. rattus, M. musculus, T. indica, R. norvegicus*	53, 69
Qatar	1	—	132
Syria	5—25	*Microtus guentheri*	132
U.A.E.	5—10	—	132
Yemen	Wide spread but not quantified	*R. rattus, M. musculus*	53, 132

V. POSTHARVEST FOOD LOSS

The postharvest food loss by rodents in Asian countries is no different than other continents. The damage starts from the threshing floor itself[17] where rodent burrows are usually more in number than in the crop fields,[3] and it continues during storage and transportation.[53,132] Damage is more pronounced to stored cereals, pulses, groundnut, vegetables, fruits, coconut, and animal feed. Due to lack of adequate methodology to assess postharvest food losses, or for other reasons, information is not available from many countries. The "guesstimates" of postharvest losses in a few countries are summarized in Table 3. The most important rodent pests are *Rattus rattus* and *Mus musculus*, but *Bandicota bengalensis, B. indica, Rattus norvegicus, Funambulus* spp., *Tatera indica, Meriones hurrianae, Rattus turkestanicus, Microtus socialis*, and *Cricetulus migratorius* also damage stored commodities. *Rattus meltada* is a serious pest of threshing floors in India.[89] The implication of true field rodents in attacking stored crops at a farm level is mainly due to the construction of huts and storage structures in the vicinity of crop fields, and by no means suggest the extension of the range of field rodents into houses or into godowns. However, once these species are established they gradually replace the existing species; *Rattus norvegicus* has almost replaced *R. rattus* in Israel[47] and in Aman City.[71] *B. bengalensis* has expanded its territory in many cities of India at the expense of mainly *R. rattus*.[34,108] The Indian gerbil, *T. indica* has also firmly established its colonies in several desert towns.[96,97]

Yet another type of loss due to rodent pests in the store houses is damage to bags or containers, which sometimes exceed the damage to crop itself.[11,56,70] Damage to grain sacks by *M. musculus* in Burma is reported to be 0.5 to 2.0%.[129] Hundreds of thousands of such bags are cut every day, hence, the loss is immense. Food contamination by urine, feces, or hair of rodents[122] is also a nonapparent yet serious loss in respect to human health, but it is very difficult to calculate it in economic terms as it is not regarded as a total loss.

VI. DISCUSSION

The rodent problems generally fall in two main categories, the first comprising of the group of three cosmopolitan commensal rodents, *Rattus rattus, R. norvegicus* and *Mus musculus*, pests in residential premises and stored foodgrains; and second includes a number of indigenous (e.g., *Gerbillus gleadowi* in the Thar desert, *Rattus r. brunneusculus* in northeastern India, and *Spalax* in Israel) and widely distributed species (e.g., *Microtus, Meriones,* and *Bandicota*), mostly causing damage in crop fields, grasslands, and forest trees. There is a fairly large number of rodent species of which pest status is well documented but there are several others about which very little is known. In recent years some knowledge has been generated about damage assessment but it is more or less restricted to stored commodities. There is meager information on the extent of damage to grasslands, fodder crops, and forests, and its quantitative evaluation is necessary before any control programs can be organized.

The Asian rodents exhibit a spectacular diversity in consonance with the changing topography, climate, and vegetation, and exhaustive work on several species (*Rhombomys,*[74] *Meriones,*[74,92] *Bandicota,*[116] and others) is available, yet there is a need to create more knowledge on the ecology and biology of many more rodent pest species which are of economic importance. Their population dynamics, relative abundances, threshold level, and outbreaks, if known, would greatly assist in formulation of their management strategies. In addition, the knowledge about their predators, parasites, and naturally occurring diseases would also be helpful in achieving the desired objectives. It is hoped that the several foreign-aided programs in many Asian countries, and the national efforts will achieve fruitful results, keeping in view the urgent need of training the commoners, administrators, and policy makers to take up rodent control based on ecological principles. Unless rodent control is integrated as an essential agricultural practice and the masses, up to the village level,[25] are involved in it, only localized success in minimizing the losses due to rodents can be visualized.

REFERENCES

1. **Adamovich, V. L.,** Natural tularemia foci structure in the floodlands of small and medium rivers. II. Epizootological nature of elementary foci, in Russian, *Zh. Mikrobiol. Epidemiol. Immunobiol.*, 48, 108, 1971.
2. **Advani, R. and Mathur, R. P.,** Experimental reduction of rodent damage to vegetable crops in Indian villages, *Agro-Ecosystems,* 8, 39, 1982.
3. **Advani, R., Prakash, I., and Mathur, R. P.,** Assessment of damage by rodents to standing crop of three varieties of wheat, *Z. Angew. Zool.*, 69, 257, 1982.
4. **Ali, A. M.,** Impact of changing irrigation on agricultural pests and wildlife in Egypt, in *Aridland Irrigation in Developing Countries, Environmental Problems and Effects,* Worthington, E. B., Ed., Pergamon Press, Oxford, 1977, 331.
5. **Allabergnov, K.,** Distribution and abundance of the Turkestan rat (*Rattus turkestanicus*) in the cultural landscape of the Fergana Valley (Uzbeck SSR), in Russian, *Uzb. Biol. Zh.*, 18, 138, 1974.
6. **Anonymous,** Proceedings of the Fourth ICAR Workshop on Rodent Research and Training, Lucknow, India, Mimeo 1985, 33.
7. **Avidov, Z., and Benhaim, N.,** Observations on pests of sub-tropical fruit trees in Israel, *Ktavim* 1, 55, 1950.
8. **Bai, D. H., Lee, S. C., Shin, Y. M., Park, G. T., Yook, M. R., Park, I. S., and Son, J. S.,** Studies on the ecology and control of rodents, Institute of Agricultural Sciences, Office of Rural Development, Suweon, South Korea, (Mimeo) 1971, 83.
9. **Baig, M. A. and Khan, A. A.,** Rodent damage to wheat crop in Faisalabad, *Pak. J. Agric. Sci.*, 38, 26, 1977.

10. **Barnett, S. A.,** Damage to wheat by enclosed populations of *Rattus norvegicus.*, *J. Hyg. Cambridge*, 49, 22, 1951.
11. **Barnett, S. A. and Prakash, I.,** *Rodents of Economic Importance in India*, Arnold-Heinemann, New Delhi, India, 1975, 175.
12. **Bernard, J.,** Damage caused by the rodents Gerbillidae to agriculture in north Africa and countries of the middle-east, *EPPO Bull.*, 7, 283, 1977.
13. **Bhardwaj, D. and Prakash, I.,** Rodent pests in poultry farms of Jodhpur, *Rodent Newsl.*, 7, 11, 1983.
14. **Bhat, S. K.,** Rodent damage to cocoa in South India, *Rodent Newsl.*, 2, 91, 1978.
15. **Bhat, S. K., Nair, C. P. R., and Mathew, D. N.,** Mammalian pests of cocoa in South India, *Trop. Pest Manage.*, 27, 297, 1981.
16. **Bindra, O. S. and Sagar, P.,** A study on the losses of wheat, groundnut and sugarcane crops by the field rats in Punjab, *Proc. Int. Symp. Bionomics and Control of Rodents,* Science and Technology Society, Kanpur, 1968, 28.
17. **Bodenheimer, F. S.,** Problems of vole populations in the middle-east, A report on the population dynamics of the levant vole *(Microtus guentheri), Israel Sci. Res. Council,* Jerusalem, 1949, 77.
18. **Brooks, J. E.,** Assignment Report—Rodent Control in Iraq, EM/VBC/3/WHO, Geneva, 1975 (Mimeo).
19. **Brooks, J. E., Sultana, P., and Poche, R. M.,** Damage by *Bandicota bengalensis* to growing wheat in Bangladesh and methods of control, 5, 1982 (unpublished).
20. **Brooks, J. E., Sultana, P., and Siddique, M. A. B.,** Rat damage to the 1982 wheat crop in Bangladesh, Technical Report, Bangladesh Agricultural Research Institute, Vertebrate Pest Division, Dakha, 1982, 16 (Mimeo).
21. **Bruggers, R. L. and Brooks, J. E.,** Vertebrate pest management in agriculture in Nepal: background and technical assistance options, 42, 1984 (unpublished).
22. **Bykovski, V. A.,** Ecology and toxicological bases of direct control of *Arvicola terrestris* (L) and *Microtus arvalis* (Pallas) in Agricultural crops, *EPPO Bull.*, 7, 509, 1977.
23. **Catlin, H. D. and Yasin, M.,** Rat damage to deepwater rice in Bangladesh, *Bangladesh J. Zool.*, 9, 27, 1981.
24. **Chakraborty, S.,** Field observation on the biology and ecology of the lesser bandicoot rat, *Bandicota bengalensis* (Gray) in West Bengal, *Proc. All India Rodent Seminar,* Rodent Control Project, Sidhpur, 1975, 102.
25. **Chaturvedi, G. C.,** Organisation of rodent control in rural areas, *Proc. All India Rodent Seminar,* Rodent Control Project, Sidhpur, 1975, 132.
26. **Chauhan, N. S. and Saxena, R. N.,** Monthly fluctuations in the population of the common field rat, *Rattus rattus brunneusculus* in jhums of Mizoram during the period of flowering of bamboo, *Bambusa tulda* (1976-79), *Rodent Newsl.*, 6, 26, 1982.
27. **Chengzin, W. and Zhi, D.,** Rodent control in China, in *Proc. Organisation and Practice of Vertebrate Pest Control,* Dubock, A. C., Ed., Elvetham Hall, U.K., 1984, 512.
28. **Chichikin, Yu. N.,** The distribution of marine rodents in the nut and fruit forest of the Chatkal range, in Russian, in *Vrednye Grysuny Kirgizii Frunse,* Academy of Science, Kirgizia, Frunse, 1966, 51.
29. **Cloudsley-Thompson, J. L.,** The desert as a habitat, in *Rodents in Desert Environments,* Prakash, I. and Ghosh, P. K., Eds, W. Junk, The Hague, 1975, 1.
30. **Cohen-Shlagman, L., Yom-Tov. Y., and Hellwing, S.,** The biology of the Levant vole, *Microtus guentheri* in Israel. I. Population dynamics in the field, *Z. Saugt.*, 49, 135, 1984.
31. **Corbet, G. B.,** *The Mammals of the Palaeartic Region — A Taxomornic Review,* British Museum (Natural History), London, 1978, 314.
32. **Danilov, P. I.,** Effect of beaver on forest vegetation, in Russian, *Les Khoz,* 5, 76, 1967.
33. **Davaa, N.,** Distribution and harmful effects of *Microtus brandti* Radde 1861 in the Mongolian People's Republic, *Proc. 2nd Int. Theriol. Cong. Brno,* International Theriological Congress Brno, 5, 1978 (Abstr.).
34. **Deoras, P. J.,** Significance of probable change of rat population in Bombay, *Proc. Indian Rodent Symp.*, USAID, New Delhi, 1966, 58.
35. **Dubock, A. C., Ed.,** *Proc. Conf. The Organisation and Practice of Vertebrate Pest Control,* Elvetham Hall, U. K., 1984, 662.
36. **Ellerman, J. R. and Morrison-Scott, T. C. S.,** *Checklist of Palaearctic and Indian Mammals,* British Museum (Natural History), London, 1951, 810.
37. **Fall, M.,** Vertebrate pests: Problems and needs in Nepal, Unpublished report, Denver Wildlife Research Center, Denver, Co., 1975.
38. **Fitzwater, W. D. and Prakash, I.,** *Handbook of Vertebrate Pest Control,* 3rd ed., Indian Council of Agricultural Research, New Delhi (Revised by I. Prakash), 1978, 95.
39. **Frantz, S. C. and Comings, J. P.,** Evaluation of urban rodent infestations — an approach in Nepal, *Proc. 6th Vertebrate Pest Conf.*, Johnson, W. V., Ed., University of California, Davis, 1976, 279.

40. **Fulk, G. W.,** Food hoarding by *Bandicota bengalensis* in a rice field, *Mammalia,* 41, 539, 1977.
41. **Fulk, G. W. and Akhtar, M. T.,** An investigation of rodent damage and yield reduction in rice, *Trop. Pest Manage.,* 27, 116, 1981.
42. **Fulk, G. W., Akhtar, M. T., Salam, S. A., and Smiet, A. C.,** Rodent damage to the 1978 and 1979 wheat crop in Pakistan, *Pak. J. Agric. Res.,* 1, 19, 1980.
43. **Fulk, G. W., Salam, S. A., and Akhtar, M. T.,** Rat damage to sugarcane in Pakistan, *Pak. J. Agric. Res.,* 1, 131, 1980.
44. **Ghosh, A. K., Ed.,** *Studies on Rodents and Their Control,* The Meghalaya Science Society, Shillong, India, 1980, 136.
45. **Gladkina, T. S. and Chentsova, N. Y.,** Food storing by *Microtus arvalis* Pall., in Russian, *Vestn. Zool.,* 5, 17, 1971.
46. **Golvan, Y. J. and Rioux, J. A.,** Ecologie des Merione du Kurdistan iranien. Relations avec l'epidemiologie de la peste rurale, *Ann. Parasit. Humaine Comp.,* 36, 450, 1961.
47. **Gratz, N. G.,** Urban-rodent borne disease and rodent distribution in Israel and neighbouring countries, *Israel. J. Med. Sci.,* 9, 969, 1973.
48. **Greaves, J. H. and Khan, A. A.,** A survey to the incidence of rodent attack on standing rice in the Punjab in 1974, Working Paper, Pak/71/554, Food and Agriculture Organization, Rome, 1975, 4.
49. **Greaves, J. H. and Khan, A. A.,** The status and control of procupines, Genus *Hystrix* as forest pests, *Common. For. Rev.,* 57, 25, 1978.
50. **Gupta, K. M., Singh, R. A., and Misra, G. C.,** Economic loss due to rat attack on sugarcane in Uttar Pradesh, *Proc. Int. Symp. Bionomics and Control of Rodents,* Science and Technology Society, Kanpur, 1968, 17.
51. **Harrison, D. L.,** *The Mammals of Arabia, III. Lagomorpha and Rodentia,* Ernest Benn, London, 1972.
52. **Heckel, Jan-Uwe,** Country Report: Yemen (Y.A.R.), in *Rodent Pests and Their Control,* Weis, N., Ed., German Agency for Technical Cooperation (GTZ), Eschborn, 1980, V Ye, 2.
53. **Hopf, H. S., Morley, G. E. I., and Humphries, J. R. O.,** Rodent damage to growing crops and to farm and village storage in tropical and sub tropical regions, Centre for Overseas Pest Research and Tropical Products Institute, London, 1976, 115.
54. **Howard, W. E., Park, J. S., Shin, Y. M., and Cho, W. S.,** Rodent Control in Republic of Korea, Institute of Agricultural Sciences, Office of Rural Development, Suweon, South Korea (Mimeo), 1975, 278.
55. **Hisia, W. P.,** Population dynamics of small rodents in the Dailing forest regions, Little Hingan mountains. I. Description of fluctuations in numbers of rodents, in Chinese, *Acta Zool. Sinica,* 16, 339, 1964.
56. **Jackson, W. B.,** Evaluation of rodent depredations to crops and stored products, *EPPO Bull.,* 7, 439, 1977.
57. **Jotwani, M. G. and Beri, Y. P.,** Economic losses due to rats, *Proc. Int. Symp. Bionomics and Control of Rodents,* Science and Technology Society, Kanpur, 1968, 9.
58. **Jung, E.,** The natural regeneration of *Pinus sibirica* at low altitudes in the Altai near Lake Teletzkoye, *Arch. For.,* 15, 617, 1966.
59. **Kashkarov, D. N. and Kurbotov, V.,** Preliminary ecological survey of the vertebrates of Central Karakum desert in western Turkestan, *Ecology,* 11, 35, 1930.
60. **Kaukenin, D. L. and Rampaud, M.,** A review of brodifacoum efficacy in the U.S. and worldwide. *Proc. 12th Vertebr. Pest. Conf.,* Marsh, R. E., Ed., University of California, Davis, 1986, 92.
61. **Khan, A. A.,** Rat damage to coconut plantation in Karachi, *Pak. J. Zool.,* 10, 292, 1978.
62. **Khan, A. A.,** Field trials of some rodenticides against the collared pika, *Ochotona rufescens* in apple orchard, *Intl. Pest Control,* 23, 12, 1981.
63. **Khan, A. A., Ahmed, S., and Choudhry, M. A.,** Comparative evaluation of brodifacoum and bromadiolone against field rats in wheat and paddy crops, in *Proc. Organization and Practice of Vertebr. Pest Control,* Dubock, A. C., Ed., Imperial Chemical Industries, Haslemere, U. K., 1984, 363.
64. **Khan, A. A. and Smythe, W. R.,** The problem of pika control in Baluchistan, Pakistan, *Proc. 9th Vertebr. Pest Conf.,* J. P., Clark, Ed., University of California, Davis, 1980, 130.
65. **Kidavakoya, P. I. K.,** Rat hunt in Laccadives, *Bull. Indian Cent. Coconut Comm.,* 9, 104, 1955.
66. **Krishnamurthy, K., Uniyal, V., Singh, J., and Pingale, S. V.,** Studies on rodents and their control. I. Studies on rat population and losses to food grains, *Bull. Grain Technol.,* 5, 147, 1967.
67. **Krishnaswamy, S. and Chouhan, N. S.,** A note on insects consumed as food by squirrels and birds at Kundi Forest Palamau district, *J. Bombay Nat. Hist. Soc.,* 54, 457, 1957.
68. **Lal, S. S. and Pillai, K. S.,** Cassava pests and their control in Southern India, *Trop. Pest Manage.,* 27, 480, 1981.
69. **Maqsud, M. N., Saleem, Z. I., and Rasheed, A. K.,** Rats and their control in food grain stores, *Agric. Pak.,* 8, 310, 1957.
70. **Meehan, A. P.,** *Rats and Mice. Their Biology and Control,* The Rentokil Library, East Grinstead, U.K., 1984, 384.

71. **Meyer, A. N.,** Planning information for Amman City rat control programme, WHO/VBC/78.702, World Health Organization, Geneva, 1978, 14.
72. **Mosgovoi, D. P. and Mosgovaia, O. A.,** Damage to Pine trees by voles in Bashkirshi National Park, in Russian, *Theriology (Novosibirsk),* 1, 391, 1972.
73. **Naumov, N. P.,** The ecology of the hillock mouse, translated by Severtzov, A. N., *Inst. Evol. Morph.,* Moscow, 3, 35, 1940.
74. **Naumov, N. P. and Lobachev, V. S.,** Ecology of the desert rodents of the U.S.S.R. (Gerbils and Jerboas), in *Rodents in Desert Environments,* Prakash, I. and Ghosh, P. K., Eds., W. Junk, The Hague, 1975, 465.
75. **Neronov, V. M. and Farhang-Azad, A.,** Distribution and range structure of the great gerbil (*Rhombomys opimus*) in Iran, in Russian, *Zool. Zh.,* 51, 715, 1972.
76. **Nevo, E.,** Observations on Israeli Populations of the mole rat *Spalax ehrenbergi* Nehring 1898, *Mammalia,* 25, 127, 1961.
77. **Ognev, S. I.,** *Mammals of Eastern Europe and Northern Asia,* Vol. 1., Translated from Russian, Sivan Press, Jerusalem, 1962.
78. **Panchabhavi, I. S. and Thimmiah, G.,** Rat damage to green cotton balls at Dharwar, *Curr. Res.,* 4, 13, 1975.
79. **Panov, G. M.,** The European beaver in densely populated regions of the Ukrainan SSR, in Russian, *Vestn. Zool.,* 5, 83, 1974.
80. **Papanyan, S. B.,** Ecology of *Meriones blackleri bogdanovi* Heptner in the Armenian SSR, in Russian, *Biol. Ah. Arm.,* 25, 59, 1972.
81. **Petrov, O. V. and Kransnjuk, A. A.,** Increasing acorn yields, and the intensity of acorn destruction by rodents on seed areas in forest-steppe oak stands, in Russian, *Vestn. Leningrad Univ. Ser. Biol.,* 15, 122, 1960.
82. **Pingale, S. V., Krishnamurthy, K., and Ramasivan, T.,** *Rats,* Food Grain Technologist's Research Association of India, Hapur, 1967, 91.
83. **Poche, R. M., Mian, M. Y., Sultana, P., Haque, M. E., and Siddique, M. A.,** Rodents in Bangladesh deepwater rice, Technical Report, Bangladesh Agricultural Research Institute, Vertebrate Pest Division, Dakha, 1980, 11, 32 (Mimeo).
84. **Posamentier, H.,** Rodent damage and evaluation of rodent control in pineapple, Bangladesh, *Bangladesh J. Zool.,* 9, 41, 1982.
85. **Posamentier, H. and Alam, S.,** Rodent damage and control in wheat in Bangladesh during the 1980 season, *Bangladesh J. Zool.,* 8, 99, 1980.
86. **Prabhakaran, P. K. and Michael, R. G.,** Preliminary note on the rodent outbreak in Mizoram, in *Studies on Rodents and their Control,* Ghosh, A. K., Ed., The Meghalaya Science Society, Shillong, India, 1980, 73.
87. **Prakash, I.,** Eco-toxicology of Indian desert gerbil, *Meriones hurrianae* Jerdon. V. Food preference in the field during monsoon, *J. Bombay Nat. Hist. Soc.,* 65, 581, 1969.
88. **Prakash, I.,** Rodent damage in the field and their control in prevention of waste and loss of food grains, Asian Productivity Organisation, Tokyo, 1974, 199.
89. **Prakash, I.,** *Rodent Pest Management, Principles and Practices,* Central Arid Zone Research Institute, Jodhpur, Monogr. 4, 28, 1976.
90. **Prakash, I.,** Impact of changing land use pattern of the rodent communities in the Indus valley, *Proc. Symp. Land and Water Management Indus Basin, Punjab,* Agricultural University, Ludhiana, India, II, 481, 1978.
91. **Prakash, I.,** Arid Zone Research and development in China, Deputation Report, Central Arid Zone Research Institute, Jodhpur, 59, 1978, (Mimeo).
92. **Prakash, I.,** Ecology of the Indian Desert Gerbil, Meriones hurrianae, Central Arid Zone Research Institute, Jodhpur, Monogr. 10, 87, 1981.
93. **Prakash, I.,** Deputation report on visit to the USSR, Central Arid Zone Research Institute, Jodhpur, 16, 1983 (Mimeo).
94. **Prakash, I.,** Rodents acquire newer territories: strategies for their management, in *Frontiers of Research in Agriculture,* Roy, S. K., Ed., Indian Statistical Institute, Calcutta, 1983, 436.
95. **Prakash, I. and Ghosh, P. K., Eds.,** *Rodents in Desert Envirnments,* W. Junk Publishers, The Hague, 1975, 616.
96. **Prakash, I. and Mathur, R. P.,** *Bandicota bengalensis* in Bikaner town, *Rodent Newsl.,* 3, 12, 1979.
97. **Prakash, I. and Mathur, R. P.,** *Rodent Control,* Indian Council of Agricultural Research, New Delhi, 1986.
98. **Prakash, I., Taneja, G. C., and Purohit, K. G.,** Eco-toxicology and control of Indian desert gerbil, *Meriones hurrianae* (Jordon). VIII. Relative numbers in relation to ecological factors, *J. Bombay Nat. Hist. Soc.,* 68, 86, 1971.
99. **Quadiruddin, M.,** Control of field rats in Hyderabad state, *Plant Prot. Bull.,* 6, 38, 1954.

100. **Rao, A. M. K. M.,** A report on rodent population in cyclone affected area in Andhra Pradesh, *Rodent Newsl.*, 2, 3, 1978.
101. **Rao, A. M. K. M.,** Demography and hoarding among lesser bandicoot rat, *Bandicota bengalensis* in rice fields. *Saugt. Mitteil.*, 28, 312, 1980.
102. **Rao, A. M. K. M. and Singh, C. D.,** Notes on distribution of rodent damage in Tella Hamsa rice around Hyderabad, *Rodent Newsl.*, 7, 9, 1983.
103. **Rao, A. M. K. M. and Subiah, K. S.,** Coconut losses due to rodents in Krishna district, Andhra Pradesh, *Rodent Newsl.*, 3, 17, 1982.
104. **Richards, C. G. J.,** Methods for the control of Mole rats, *Spalax leucodon* in Northern Syria, *Trop. Pest Manage.*, 28, 37, 1982.
105. **Roberts, T. J.,** Country Report, Pakistan, in *Rodent Pests and Their Control,* Weis, N., Ed., German Agency for Technical Cooperation (GTZ), Eschborn, W. Germany, 1980, VPa, 1.
106. **Roy, S. K.,** Preharvest loss of rice due to field rodents, *Econ. Pol. Week.*, 9, 66, 1974.
107. **Roy, S. K.,** The burrows, grain stores, movements and runs of the Indian mole rat, *Bandicota bengalensis,* in *Studies on Rodent and their Control,* Ghosh, A. K., Ed., The Meghalaya Science Society, Shillong, India, 1980, 30.
108. **Seal, S. C. and Banerji, R. N.,** Changing patterns of rodent populations in Calcutta and Howrah, *Proc. Indian Rodent Symp.*, USAID, New Delhi, 1966, 69.
109. **Shah, D. R. and Subiah, K. S.,** Notes on Rodent Control in Lakshadweep Islands with Roadafarin (Warfarin) wax blocks, *Pestology,* 2, 36, 1978.
110. **Sharma, V. N. and Joshi, M. C.,** Soil excavated by desert gerbil, *Meriones hurrianae* (Jerdon) in the Shekhawati region of the Rajasthan desert, *Ann. Arid Zone,* 14, 268, 1975.
111. **Shin, Y. M.,** Influence of rice ripening time and neighbouring habitat on rodent species composition and crop damage, *Korean J. Plant Prot.*, 13, 135, 1974.
112. **Shroff, S.,** The rodent menace in Nepal, *Rising Nepal,* 11, 5, 1974.
113. **Shitl'mark, F. R.,** Protecting oak sowings from Muridae in the Tellerman forests, in Russian, *Soobshch. Lab. Lesoved,* 3, 3, 1961.
114. **Smiet, A. C., Fulk, G. W., and Lathiya, S. B.,** Rodent ecology in sugarcane in lower sind, Pakistan, *Acta Theriol.*, 25, 81, 1980.
115. **Sokur, I. T.,** Measures to protect harvests from harmful rodents in the Ukraine and possibilities of their perfection, in Russian, *Vestn. Zool.*, 6, 1972.
116. **Spillet, J. J.,** *The Ecology of the Lesser Bandicoot Rat in Calcutta,* Bombay Natural History Society and John Hopkins University, Bombay, 1968, 323.
117. **Sridhara, S. and Krishnamoorthy, R. V.,** Grain losses by spoilage by wild rodents under laboratory conditions, *Prot. Ecol.*, 1, 103, 1979.
118. **Srivastava, A. S.,** Project Coordiantor's final report on the "All India Coordinated Scheme for Research on the study of Field rats" (1959-1969), Indian Council of Agricultural Research, New Delhi, Unpublished, 1970, 47.
119. **Srivastava, A. S.,** *Rodent Control for Increased Food Production,* Rotary Club, Kanpur, India, 1970, 152.
120. **Subiah, K. S.,** Rodent problem in oil Palm Plantation in Hut Bay, Little Andaman and suggested control measures, *Indian J. Farm Chem.*, 1, 32, 1983.
121. **Urs, Y. L.,** Damage to coconuts in a nursery plantation by *Bandicota bengalensis, PANS,* 25, 147, 1978.
122. **Urs, Y. L., Krishnakumara, M. K., and Majumder, S. K.,** Parameters for assessing commensal rodent populations in rural ecological conditions, *Proc. Indian Acad. Sci.*, 88B, 163, 1979.
123. **Valsala, P. G.,** Pests of arecanut and their control, *Coco. Bull.*, 12, 13, 1958.
124. **Veeraraghavan, R.,** Weak bushes in teagardens, *Plant Chron.*, 61, 195, 1966.
125. **Vinogradov, B. S. and Argiropulo, A. I.,** *Fauna of U.S.S.R.—Mammals — Key to Rodents,* Zoological Institute of Academic Sciences, Moscow, U.S.S.R., 1941, 230.
126. **Wagle, P. V.,** The rice rats of lower sind and their control, *J. Bombay Nat. Hist. Soc.*, 32, 330, 1927.
127. **Wagle, P. V. and Gul Mohomad, K. B.,** Injurious field rats of lower sind and their extermination, *Dept. Agric. Bombay Poona Bull.*, 138, 34, 1927.
128. **Walker, E. P.,** *Mammals of the World,* John Hopkins University Press, Baltimore, 1975, 1500.
129. **Walton, D. W. and Brooks, J. E.,** Country Report: Burma, in *Rodent Pests and Their Control,* Weis, N., Ed., German Agency for Technical Cooperation (GTZ), Eschborn, W. Germany, 1981, V Bu, 1.
130. **Wang, L.,** Acute rodenticides in the control of rodent pests in China, a review, *Proc. 11th Vertebr. Pest. Conf.*, Clark, D. O., Ed., University of California, Davis, 1984, 54.
131. **Weis, N., Ed.,** *Rodent Pests and their Control,* Germany Agency for Technical Cooperation (GTZ), Eschborn, West Germany, 1981.
132. World Health Organization, Report on Seminar on Public Health and Economic Aspects of Rodent Control, WHO, Alexandria, 1975 (Mimeo).

133. **Wolf, Y.,** The levant vole, *Microtus guentheri* (Danford et Alston, 1882). Economic importance and control. *EPPO Bull.*, 7, 277, 1977.
134. **Wuping, Xia,** A bird's eye view on the progress of rodent control and rodent biology during the past ten years in China, *Acta. Zool. Sinica*, 22, 250, 1976.
135. **Zhi, Deng and Chengxin, W.,** Rodent Control in China. *Proc. 11th Vertebr. Pest Conf.*, Clark, D. O., Ed., University of California, Davis, 1984, 47.

Chapter 6

RODENT PROBLEMS IN SELECTED COUNTRIES IN SOUTHEAST ASIA AND ISLANDS IN THE PACIFIC

M. M. Hoque, F. F. Sanchez, and E. A. Benigno

TABLE OF CONTENTS

I. INTRODUCTION

Serious pest situations involving rodents are common throughout Southeast Asia, the Pacific Islands, and much of the developing world. Rodents are among the major constraints to crop production in these areas. Many times they mean the difference between a good crop and total crop failure.

The severity of rodent problems is already an established fact in Southeast Asian agriculture. However, estimates of agricultural crop losses caused by rodents are few. This difficulty is compounded by a lack of damage assessment methods appropriate for specific crops or by the fact that no damage surveys have been conducted to estimate losses to the crops. This common inability to express rodent damage in economic terms is probably one of the principal reasons why control of rat damage has been given much less attention than that caused by insects, weeds, and plant pathogens. Nevertheless, it has been generally agreed that rodents are serious problems both in agriculture and public health.

The countries in Southeast Asia (SEA) and the Pacific are within the tropic zone, sharing common agroecosystems and also some common rodent pest species. The problem pests evolved with the developments in agriculture. For instance, in the Philippines and neighboring countries, the increased use of irrigation in rice culture and the planting of varieties with shorter growing periods have resulted in highly variable crop schedules. Moreover, the tropical agriculture environments are dynamic and subject to rapid changes in habitat and available food, thus contributing to the creation of rodent pest situations which are sporadic and highly variable. For example, in rice, damage caused by rats occurs in two patterns. One, appearing every year in every area, is highly variable even between adjacent rice fields. Another is an exceptionally heavy damage associated with rapid increases in the number of rats over wide areas resulting in rodent outbreaks with consequent high infestation over a wide area. This pattern is still poorly understood.

Rats also pose danger to public health, a damage that may not be possible to express in numerical terms. As far as the Pacific Islands and Southeast Asian countries are concerned, rat-borne diseases are known in these regions. With the close association between rodents and man, the incidence of rodent-borne diseases transmissible to man becomes unavoidable. Some of these diseases are murine typhus, plague, scrub typhus, leptospirosis, rat-bite fever, salmonellosis, rickettsial pox, hymenolepsia, and angiostrongylosis. Apparently, some of these diseases are acquired through contact with an intermediate host, often times an invertebrate living in the bodies of rodents or in the vicinity of rodent burrows and harborages.[28] Although these diseases may not be reported in any country in Southeast Asia and the Pacific regions their proximity to known endemic foci such as Mainland China, the Indian subcontinent, and Java should be enough reason for surveillance. Moreover, with the presence of a potential rodent reservoir, arthropod vectors, and a susceptible human host, in combination with an increase in transportation, a possible epidemic of these diseases could take a country off guard.

Although this paper may not give a detailed report on rodent problems in public health, it does not necessarily mean that the problem is not important. Comprehensive reports on research regarding ths problem are dealt with in World Health Organizations (WHO) reports. This paper provides an overview of rodent problems in Southeast Asia and the Pacific Islands, with an emphasis on agriculture in places where such information is available.

II. RODENT PROBLEMS

There are at least 29 species of rodents that are considered pests of economic importance in the two regions (Table 1). These species are either pests to agriculture crops, stored products, vectors of rat-borne diseases, or cause all of these nuisances to man. Of the 29

Table 1
RODENT PESTS OF ECONOMIC IMPORTANCE IN SELECTED COUNTRIES IN SOUTHEAST ASIA AND ISLANDS IN THE PACIFIC

	Locality														
Species	Philippines	Malaysia	Indonesia	Thailand	Singapore	Vietnam	Micronesia	Mariana	Palau	Caroline	Hawaii	West Samoa	Fiji	Marshall	Tokelau
Rattus rattus			X	X			X	X	X	X	X	X	X	X	
R. r. rattus							X	X	X	X					
R. R. diardii		X	X		X										
R. r. mindanensis	X														
R. r. mansorius										X					
R. norvegicus	X	X	X	X	X					X	X	X			
R. argentiventer	X	X	X	X											
R. tiomanicus		X	X	X											
R. losea				X											
R. muelleri		X	X												
R. exulans	X	X	X	X	X	X	X	X	X	X	X	X	X	X	X
R. e. micronesiensis							X	X		X					
R. annandalei					X										
R. flavipectus						X									
Bandicota spp.		X	X												
B. savilei				X											
B. indica		X	X	X											
B. bengalensis		X	X			X									
Chrotomys whiteheadii			X												
Mus musculus	X	X		X	X		X	X	X	X			X		
M. m. castaneus			X	X											
M. m. moniyamai										X					
M. cervicolor				X											
M. caroli				X											
M. caroli ouwensi			X												
Callosciurus notatus		X													
C. prevosti		X													
C. finlaysoni				X											
Manetes berdmorei				X											

Table 2
MAJOR RODENT PESTS OF AGRICULTURAL CROPS IN SELECTED COUNTRIES IN SOUTHEAST ASIA AND ISLANDS IN THE PACIFIC

	Crops damaged							
Species	**Rice**	**Maize**	**Sugar cane**	**Coconut**	**Oil palm**	**Cocoa**	**Root crops**	**Pineapple**
Rattus rattus	X	X	X	X	X	X		X
R. argentiventer	X	X	X		X			
R. exulans	X	X	X	X		X	X	X
R. norvegicus			X					
R. tiomanicus	X				X	X	X	
R. losea	X							
R. muelleri	X					X		
R. flavipectus	X							
Bandicota spp.	X		X				X	
Mus spp.	X							X
Callosciurus spp.					X			

species, *Rattus exulans* is the most widely distributed, occurring in 15 localities, followed by *R. rattus* in 10 localities, *R. norvegicus* in 8 localities, *Mus musculus* in 9 localities, *R. argentiventer* and *R. r. rattus* in 4 localities each, and the rest are distributed in less than 4 locations. In SEA and the Pacific, the genus *Rattus*, as a pest, occurs 66.3%, followed by *Mus* (17.4%), Bandicota (10.5%), *Callosicurus* (3.4%), and *C. whiteheadii* and *M. berdmorei* (1.2% each). The importance of one species over another varies from one locality to another. For example, while *R. norvegicus* is considered a minor pest in agriculture in the Philipines, it is a major pest of sugar cane in Hawaii.[10] *R. r. diardii*, a less important pest in the Philippines, is both a pest in rice fields and a vector of plague-borne fleas in Indonesia.[31,18] The variation in severity of one species seems to depend on certain factors inherent to a locality even if the areas are within the same geographical location.

The importance of the genus *Rattus* as agricultural pests can be gauged on the wide range of crops damaged by the species, with *R. rattus* topping the list followed by *R. exulans* (Table 2). Eight major rodent species attack rice, five on sugar cane, four on cocoa, three each on oil palm, pineapple, and root crops, and two each on maize, coconut, and vegetable crops. Although not all the species were indicated to inflict damage to these crops, they have the potential to cause considerable losses. The magnitude of damage inflicted by rodent pests depends on myriads of ecological factors in a given location and it is not even a surprise to find variations occurring within a location or country.

The following reports the magnitude of rodent problems in selected Southeast Asian countries and selected islands in the Pacific. Estimates of crop damage selected from the rodent literature are summarized in Table 2. We caution the reader that this tabulation was not intended to be exhaustive and the figures are the best available based on ocular observations and a few systematic surveys of crop loss.

A. Philippines

Rodent collections throughout the country indicate that *R. r. mindanensis*, *R. argentiventer*, *R. exulans*, and *R. norvegicus* are the major species inhabiting rice lands.[1] *R. argentiventer* is widely found in the islands of Mindano and Mindoro. Together with *R. r. mindanensis*, this species was implicated in the rodent outbreaks that occurred in Mindanao in 1952-1953 resulting in total losses of rice, corn, and other crops, and widespread food shortages during those outbreaks.[4] *R. exulans* is the rice-field rat in the island of Palawan and is encountered in a variety of habitats, particularly uplands. *R. norvegicus* is a pest found locally on stored foods and household effects and is a potential problem in rice fields, especially near houses. However, recent findings show that this species has invaded rice fields in Cebu, Panay, and Central Luzon. Mice (*Mus musculus*), another rodent pest species in the islands, are restricted to houses and food storage.

A national survey on damage to rice was initiated in the 1970s by the Rodent Research Center (now the National Crop Protection Center) and the Bureau of Plant Industry in order to estimate the losses due to pests. Losses to standing rice based on percent cut tillers were as follows: 1970, 4.53%; 1971, 1.65 to 3.22%; and in 1973, 2.47 to 4.31% for dry and wet seasons, respectively.[29] Up to 90% of the total fields surveyed during this period had measurable damage while about 7% of the fields had over 10% cut tillers. The national average of less than 5% preharvest damage was valued at $67.3 million in 1975.[29] However, a great reduction to an annual average loss of 0.52% between 1976 and 1980 was achieved through the National Rodent Control Program.[3] With these national damage surveys, the country was able to establish rodent problem areas and areas to watch for possible rodent outbreaks.

Corn, another staple crop of the Filipinos is not exempted from pest damage. Estimates of losses could go as high as 80%.[15] No figures on losses for the country have been reported except from a limited survey from six corn producing provinces, pegging the damage at 5.42% damage to maturing corn earns.[15] It is expected that a national corn damage survey shall soon be conducted with the development of rodent damage assessment methods on corn.

Coconut, one of the major exports of the Philippines, suffers from rodent depredation. Of the four agricultural pest species *R. r. mindanensis* and *R. exulans* have been found to climb trees and damage the nuts on the crown.[25,29] Estimates of losses ranged from 12.4 to 65% of the harvestable nuts. These figures were based on an aggregate of 11 studies with durations ranging from 5 months to 3 years. Based on damaged dropped nuts, loss ranged from .2 to 1.3 nuts/tree/month or an average of 8 nuts/tree/year.[14] In a coconut field adjacent to rice fields, a higher loss of 1600 nuts/ha/year or an average of 11 nuts/tree/year was reported.[12]

The idea that palms might be able to compensate for coconuts damaged by rats was tested in Fiji.[40] It was found that such compensation, by increased copra content of undamaged nuts, played at most a minor (7% or loss) part in the differences in nut weights observed in experimental plots.[26] Furthermore, the lower copra weights (about 10%) of nuts from the undamaged plots were unimportant when compared with the improved yields (about 150%). The big differences in yield between protected and unprotected plots were due mostly to factors other than compensation for rat-damaged nuts.

Surgar cane, another major export product grown in 18 provinces on 6 islands in the Philippines, is affected by rats. Besides causing direct damage to the crops, rats cause secondary losses by bacterial and fungal infection that induce souring with subsequent losses in the sucrose content of the canes. In Negros island, one of the major cane producing provinces, *R. r. mindanensis* was the predominant rat species caught. Of the 189 rats caught there was only one *R. norvegicus*.[7] This shows that the latter is a minor pest of sugar cane in the Philippines, unlike in Hawaii.

Sugar cane yield losses due to rats were seldom reported in the Philippines until 1961 when the difference in yield between rat-infested and noninfested sugar cane varieties was studied. As much as 3.7 piculs less sugar per ton of cane was lost from infested canes compared with the normal yield of uninfested ones. Researchers at the Victorias Milling Company in Negros initiated the measurement of sugar losses due to rats.[24] Their survey on six farms showed that the damaged stalks ranged from 11.1 to 24.3%, or an average of 15.3%. Corresponding sugar losses ranged from 2.2 to 9.6% with an average of 5.1%. The same workers reported that under unirrigated conditions 10% stalk damage corresponded to a 4% sugar loss. In the same province (Negros) 14 mill district surveyed in 1976 to 1977 showed rat damage ranging from 2 to 13% with an average of 6.7% rat damage to cane stalks.[7] The estimated total sugar loss ranged from 0.77 to 7% with an average of 2.3%. Rat damage was highly correlated with dead stalk tonnage reduction, juice deterioration, and total sugar loss. The study also showed that the percent of total sugar loss (TSL) could be estimated by the following equation: $\%\ TSL = 0.778 + 0.556\ RD - 0.014\ RD,^2$ where the percent of canes attacked (RD) could be easily estimated from damage survey. The most practical, reliable, and least expensive sampling method in terms of man-hour per area involves counting the stem with single or multiple rats gnawing on five sampling areas of 4 m × 10 m/ha.

Pineapple is not grown as extensively as sugar cane, coconut, or rice by individual farmers except on plantations. Rodent problems on big pineapple plantations have not been reported. However, pineapple grown as intercrops with coconut, coffee, cocoa, and bananas are affected by the pest. In a study of pineapple intercropped with coconut, the ground-baited plots suffered no loss, whereas the untreated reference plots had a loss of 10.2% of the harvestable fruits.[13] This loss would be equivalent to about 1300 pineapple fruits per hectare.

B. Malaysia

The rodent pest species reported in Malaysia are *R. argentiventer, R. diardii, R. tiomanicus, R. exulans, r. norvegicus, B. indica, Bandicota* spp, *B. bengalensis, M. musculus*, and two squirrels, *C. notatus* and *C. prevosti*.[47] *R. argentiventer* is generally the dominant species found in rice fields and is seldom found in houses. *R. tiomanicus* is the major species found in most oil palm estates although *R. argentiventer* is also present in young palms.[44] *R. r. diardii, R. exulans*, and *M. musculus* infest stored grains, with *R. r. diardii* the predominant pest while *R. exulans* occasionally inhabit rice fields. *B. bengalensis* is restricted to Penang while *B. indica* is restricted to Kedah and Southern Perlis. *R. norvegicus* is found mostly in ports.[47] Squirrels (*C. notatus*) have been observed to damage young oil palms. However, the plants were found to recover 6 weeks after the attack, with only 9% showing no sign of regrowth. Their damage was reported to be less lethal than that of rats.[43]

Various estimates on yield losses during preharvest ranged from 1% to 11%.[33,47] Surveys conducted in 1978/1979 in Pulau Pinang showed seasonal yield losses to rice in the average of 5 to 11%.[2,33] An earlier conservative estimate of 1.5% in 1974[33] placed the value of losses at $3.4 to 16.9 million annually. However, serious infestations of up to 80% have been reported.[47]

In Malaysia, oil palm is one of the major crops to suffer significant damage due to rats and squirrels.[19] The economic losses due to rats in the oil palm have been estimated at 240 kg oil/acre/year.[19] In 1981 this represented a potential loss of $45.6 million. Another estimate of loss was based on a study of the mean consumption of nuts by rats in captivity. This was estimated by multiplying the rat population per hectare by the average daily consumption of the animal, giving a loss of 1.8 to 3.0 ton/ha.[44] This study also gave an estimate of 200 to 600 rats (dominated by *R. tiomanicus*) per hectare in oil palm without control.

If there is one crop that has been extensively studied in Malaysia, it is the oil palm, where research has been continuing now for almost a decade. The results of these studies are worth stating here, especially for a population to which the effective management of the pest has great relevance. Evidently, *R. tiomanicus* is the only rat species regularly found in mature oil palm. In one situation, where no systematic control measures had been taken, a very heavy rat attack affected the fruit of palms dominated by this species.[43,44] *R. argentiventer* estimated to constitute over 90% of the population of rats in the younger mature palms (4 year old) decreasing progressively with palm age to less than 10% at 7 years.[44] The worker postulated that *R. argentiventer*, with burrowing habit, established easily in younger oil palm because of the grassland-like environment of newly planted palm. On the other hand,

R. tiomanicus is a more adept climber, displacing *R. argentiventer* in older palms. It was also found that control by baiting acts as a selection pressure against *R. argentiventer*, thereby facilitating the transition of *R. tiomanicus* to the sole species in the oil palms.[44] Other studies also showed that *R. tiomanicus* gradually replaced *R. argentiventer* in unbaited plots of palms up to about 7 years old; in the plot subjected to regular systematic baiting, *R. argentiventer* rebounded for several years. The study also reported that in an area baited to near elimination of *R. tiomanicus*, three other species, *R. exulans, R. rajah,* and *C. whiteheadii*, (the last two are jungle species) appeared in small numbers and fed on the oil palm fruit, before *R. tiomanicus* re-established itself. The significant result of this control work was the recommendation to carry out a wide control in one continuous campaign at intervals that do not permit the rapid increase phase of population growth to proceed. The control measure should be applied every 6 months.[44]

Cocoa is another economic crop in Malaysia that is attacked by rodents. The damage levels to cocoa pods by rats and squirrels are more variable, ranging from 50 to 60% in monoculture cocoa to as high as 92% in cocoa intercropped with coconuts.[19] Others observed that this crop could be severely attacked by several rat species,[44] although long term losses have not yet been worked out. Short term losses reach 70 to 100% loss. The species found in cocoa are *R. tiomanicus, R. argentiventer*, and *R. muelleri*. The first two may have migrated from oil palms while the third one came from the nearby jungle (in Sabah). Apparently, it is not yet clear if any of the rat species could become "the cocoa rat" in large areas.[44] The worker thinks that cocoa probably does not support an independent population, but supplies food to rats that can continually move in from adjacent environments.

C. Indonesia

Rodent pests in Indonesia have been long considered as a constraint to crop production and storage. Several important species affecting agricultural crops are *R. argentiventer, R. exulans, R. rattus, R. r. diardii, R. tiomanicus, R. exulans*, and secondary pest species, *Chrotomys whiteheadii, M. musculus castaneus, M. caroli ouwensi, Bandicota* spp., *B. indica*, and *B. bengalensis*.[32] The number one rice field pest species, *R. argentiventer*, is widely distributed over the country but is apparently restricted to grassland and rice fields. *R. tiomanicus* is abundant in plantations of coconut and oil palm. *R. exulans* is common in croplands and houses, and it is a potential pest to rice fields in Central Sulawesi and Java. The *Bandicoots* are common in Sumatra and West and Central Java, mainly inhabitating gardens near houses and fallow rice fields. *R. r. diardii* and *R. norvegicus* also damage food in storage.

The importance of rodents as pests in Indonesian agriculture is clearly seen in the magnitude of losses and the occurrence of serious rat infestation to crops in the country. Serious damage was been reported in 1915, and again in 1931 and 1933 when a few thousand hectares of rice suffered heavy losses in Ceribon, West Java.[32] Rat outbreaks in rice occurred again in 1961 and 1963 in Java and Madura with damage estimated at 35% in 1,000,000 ha and 28% in 822,000 ha. In 1972 to 1973, 232,626 ha of rice were badly affected. Estimates of losses prior to a rat control program in Indonesia were quoted as follows: 27% yield loss from 317,212 ha rice affected in 1979 to 1980, and 16% loss from 248,225 ha affected in 1980 to 1981.[16] Although there was a considerable reduction in loss on rice in the 1980 to 1981 crop season, the loss was still higher compared to those in the Philippines and Malaysia. It was noted that even with a 16% loss this already represented a gain of 110,230 MT valued at $31 million/year.[16] The national 5 year rat control program in Indonesia, which started in 1979, has greatly contributed to the reduction of the rodent problem in rice in Indonesia.

Statistics on losses to other crops such as corn, cassava, mungbean, peanuts, soybean, and sweet potato were given in terms of areas affected, with damage ranging from light to moderate in 1976 and 1977. Cassava had the widest area affected by rats followed by soybean, corn, sweet potato, peanuts, and mungbean.[31] Damage to sugar cane was estimated at 30 to 100% loss from 14,000 ha affected by rats.[31]

D. Thailand

There are about 12 economically important rodent pests species in Thailand. *R. argentiventer* is the common rice field rat of the Central Plain and Southern (Peninsular) portion of Thailand, while *R. losea* is predominant in northern Thailand.[35] *R. argentiventer*, considered the number one rodent pest causing damage to rice, is the one often recovered during massive rodent campaigns in rice fields. *R. rattus* is a common house rat but occasionally is found also in rice fields near houses, along with *R. norvegicus* and *R. exulans*. The other two Bandicoot rats are less important, although at times they inhabit crop lands. *B. indica* is found in all rice growing areas of the country except the peninsula. The lesser Bandicoot species, *B. savilei*, often mistaken as *B. bengalensis*, is also widely distributed in rice and sugar cane growing areas surrounding the Central Plains.[35] Two species of mice are common in paddy fields all over the country. *M. caroli* and *M. cervicolor* are often found together in the same field. Besides rats and mice, two species of squirrels, namely *Manetes berdmorei* and *C. finlaysoni*, damage fruits.[46]

Available statistics on crop losses due to rodents, particularly rats and mice, show the importance of these pests in Thai agriculture. From 1971 to 1973 rats and mice infestation of major crops, such as rice, corn, soybean, sugar cane, cotton, watermelon, coconut, fruits, and vegetables, caused damage amounting to an average of $5.5 million. In 1971, the total area affected was 377,152 ha, with a damage estimate of $2.3 million; in 1972, 1,373,980 ha at $8.2 million; and in 1973, 990,214 ha at $5.9 million.[46] Estimates of losses were also reported from experiments by the Department of Agriculture in Thailand. In 1970, the yield of the second rice crop in the rice-producing province of Central Plains was reduced by 36% due to rat damage. In the main season rice production was reduced by about 5 to 15% in the Central Plains, 20 to 60% in the South, 2 to 4% in the North, and less than 2% in the Northeast. It was further estimated that the monetary loss caused by rodents in fields alone is about $7.96 million annually during this period.[33]

The limited reports on individual crop losses could have been due to lack of techniques for damage assessment, except in rice and sugar cane. In rice, the national survey in 1975-76 estimated an average loss of 7% all over the country.[37] The study on sugar cane showed an average loss in yield of 5.3% depending on the age of crop damaged and the position of the damage on the stem.[8]

E. Vietnam

Rodents are reported to cause significant damage to field crops. At the village level, rats migrate from fields to houses after a harvest and back to the fields as the next crop matures.[27] There are at least three major rat pest species identified to cause economic losses in agriculture: *B. bengalensis, R. exulans*, and *R. flavipectus*.[26] These species not only infest standing crops but also cause considerable damage to stored food. However, no systematic surveys have been reported with regard to crop damage and magnitude of loss, except for a very limited survey on rice conducted in 1972 on the southern part of Vietnam.[29] Scientists from the Philippines initiated damage surveys to rice in three provinces of former South Vietnam. The results showed rat damage below 10%, with a range of 0.2 to 9.6%. Mean damage was generally highest among transplanted high yielding varieties and lowest in floating rice. Although the damage was generally low, all of the fields sampled had a measurable damage.

F. Singapore

The rodent problem in Singapore is medical rather than agricultural in nature. Because the country is strategically situated in the center of southeast Asia, it has become the hub of air travel as well as a foci of various communicable diseases. Rodent control is therefore of utmost importance. Regular surveillance is carried out in public places, e.g., shops, eating houses, and bin centers.[20] Areas with high rodent infestation are zoned off for priority control. The government of Singapore has also instituted legal measures (the Environment Public Health Act of 1970) to encourage citizens to do their part for the maintenance of clean premises in order to prevent the build-up of rodent populations. Although Singapore has been declared safe from wide epidemics of plague by WHO there is still the increasing possibility of limited outbreaks, especially in seaports, airports, and coastal areas with history of the plague.

Five rodent species have been identified as vectors of rodent-borne diseases, namely *R. norvegicus, R. exulans, R. r. diardii, R. annandalei, and M. musculus*, and an insectivore, *Suncus murinus*.[21] Collections of rodents in the country describe the habitat and distribution of the pest species. *R. norvegicus* is abundant and mainly restricted to urban areas near the sea. *R. r. diardii* is more abundant in rural areas although it is also present in port and city areas. *R. exulans* is common in every habitat and throughout the country. Likewise, *M. musculus* is distributed over the whole island, but found mostly in areas with human habitation, along with *S. murinus*. All of these species are vectors of the plague flea (*Xenospsylla cheopis*).

As far as the disease is concerned, the last outbreak of plague occurred in 1933. Although there is no current report on its occurrence, the country is still vulnerable because of its high rodent and flea populations. The plague flea index is above the acceptable WHO threshold of one thoughout most of the year. In 1968 to 1985, a total of 29 locally contacted murine and 10 scub typhus cases were reported. However, rodent-borne diseases of epidemic significance have not been recorded.

G. Hawaii

Rodents in Hawaii are considered both a threat to public health and agriculture. The problem of plague in rodents and in man was formerly a major public health significance, but it has diminished since the 1940s.[38] However, even in the 1960s serological evidence still show that plague was present in the rodent populations of the Hanokoa region.[5] However, recent reports indicate that the disease-causing microorganisms are already under control. Three rat pests species, *R. norvegicus, R. exulans, and R. rattus* damage sugar cane, one of the major crops in Hawaii.[9,15] Other crops that are also damaged by the pests are pineapple and macadamia nuts.[6,20]

In Hawaii, sugar cane is a 2-year crop that maybe kept on a 3-year cycle in some plantation. An acre of sugar cane may yield up to 100 tons of cane, which may then be processed to yield an average of 11 tons of sugar.[36] Rats injure cane by gnawing through the hard outer rind and chewing into the soft fibrous center. Damage may occur in 1 or as many as 15 to 20 internodes of a cane stalk. A few inches eaten from a stalk may kill the stalk or severely reduce sugar production by allowing souring microorganisms to enter the wound. These losses are of far greater importance in the low cane producing areas where up to 40% of the stalks maybe affected. The loss of raw sugar in these areas is about 0.6 tons/acre compared to 0.2 tons/acre in drier or irrigated regions.[10] Estimates of losses of cane range from 4 to 40% and damaged stalks were valued at $2 to $5 million/ton in the sugar industry annually, or an average of $4.5 million.[10,11] Studies relating damage to sugar loss showed a loss of about 0.2 tons sugar per acre for each 10% increment in the number of stalks chewed.[36] Sugar cane researchers have reported an average damage of 20 to 30% of cane-stalk in source plantations. Earlier estimates on damaged stalks revealed that with 32% damage there is an equivalent reduction of sugar by 5.9% and with 100% damaged stalks sugar is reduced by 17%.[8] It was generally recognized that an accurate, yet simple, damage assessment to use in sugar cane seems one of the important factors in the estimation of sugar cane losses due to rats. As one scientist has mentioned there is no latest report on losses due to rats even in Hawaii where this crop is grown extensively.[8] For instance, a rapid assessment technique based on the linear distance between damage allows the calculation of the number of damaged nodes per square meter, which has been related for specific varieties and conditions to the loss of sugar. However, this sampling system, if it can be adopted by ordinary laymen, may still need the assistance of the researchers since calculations of the sugar loss is apparently complicated.[8]

Macadamia nuts are one of the important crops that are affected by rodents. In 1968 the industry was selling between $4 and 5 million worth of the processed product, but the prediction was for this crop's sales to reach $25 million in 10 years time.[6] Losses of nuts due to rats is on the average of 15.76% with variation of 13.12% to 18.61% at the field center where no rat control was practiced.[6] This loss was valued at $300,000 million or equivalent to 1.5 million pounds of nuts with an estimated percentage of damaged nuts of 13%. A rat control method used by the growers in Hanokoa, Hawaii, which consisted of thallium sulfate coated macadamia kernels placed in wax paper distributed in the field, held the damage level below 5%. The problem of rodent control in macadamia orchards is the selection for a bait comparable to the taste of macadamia nuts because apparently, macadamia nut kernels are so far the most acceptable bait compared with others.

H. Watern Samoa

Western Samoa, one of the countries in the South Pacific, has a total land area of about 900 mi^2 and relies mainly on its agriculture for foreign exchange.[39] The country derives its major income from the export of copra, cocoa, and bananas, in this order. Rodents have been known to be the limiting factor in the production of these crops in the islands.[39] Not only are these major export crops badly affected by rodents but peanuts, cassava, sweet potato, and other stored products are also affected.

Four rodent species that cause damage in Western Samoa have been identified: *R. rattus, R. norvegicus, R. exulans*, and the common house mouse, *M. musculus*. Damage caused by the pest to crops seems to depend on the upkeep of the individual plantation. Severe damage occurs in small weedy plantations and especially where susceptible crops are grown.

Rat damage to coconuts is estimated to be 10 to 20% of the nuts, although no systematic survey has been conducted on the extent of losses due to the pest in the country. It was also reported that dwarf coconuts were more susceptible to the pest than tall ones.[39]

The estimates on losses to cocoa in Western Samoa ranges from 10 to 15% of the harvestable pods, although losses between 30 and 35% have been recorded by one of the major cocoa producers in the country (The Trust Estate Corporation of Western Samoa). Losses to cocoa pods is not only due to what is eaten by the rats, but also what is lost due to secondary infections. The pests prefer ripe pods, but unripe pods are also eaten when ripe pods are less available. Since cocoa is a valued crop in western Samoa, losses as low as 10% are equivalent to the loss of several thousand dollars worth of cocoa usually exported.

Peanuts and cassava are also damaged by rats in Western Samoa. Although peanut is not as widely cultivated as cocoa and coconut, significant losses to this crop have been reported and have reached as high as 90% in exceptional cases. Cassava is another root crop which has a high potential for wide cultivation in the islands of Samoa, both for human consumption and for use as animal feed. Few cassava growers in the country have reported damage to their crops by rats, but one farmer reported a loss of 100%.

On the whole, the damage inflicted by rodents to agricultural crops in Western Samoa has drawn much interest from the outside. The rodent control program is currently a joint undertaking by the Western Samoan and German governments. The launching of the Western Samoan-German Crop Protection Program in the 1980s brought technical expertise into the country.

I. Fiji

Rats have been considered an agricultural problem in Fiji for many years, particularly in relation to coconuts.[41] Two species of rats, *R. rattus* and *R. exulans* have been established and both inflict damage to the nuts in the crown. It was observed that *R. rattus* forages arboreally far more extensively at greater heights than *R. exulans*. Thus it was expected that the former would inflict more damage to Fiji's predominantly mature plantations. A study was done in Fiji to assess the extent of damage inflicted by the two species. Identification of the species was made through teeth markings left by each species on the damaged nuts. A survey of freshly damaged nuts from an area containing short trees showed that both species damage the nuts, but *R. rattus* is responsible for most of the damage in Fiji and reflects the vertical distribution of the species. This figure is much lower than the 1934 quoted figure of a loss of 23 nuts/palm/year, and the explanation given that the trees were shorter, thus damage higher on the tree was observed.[41] This result is similar to the findings in the Philippines wherein *R. rattus mindanensis* were found in greater proportion than *R. exulans* that were caught in the coconut crowns.

The 4 year coconut research project conducted in Fiji, which started in 1969, has provided much information on the influence of rats on coconut — from damage assessment to ecology and control of the pest. Previous reports on the problem of rodent pests in Fiji have given various estimates on losses; 10%, 6.8% of 3.6 nuts/palm/year.[41] In another detailed assessment the loss was 28% (about 23 nuts/palm/year) of the total crop. In 1969 to 1972 a long-term assessment of damage in Fiji revealed many interesting facts about rodents and coconuts. Short palms (under 10 m) had higher damage 5.5 ± 0.7 nuts/tree/year. The overall loss based on a long-term survey on different sites gave an average loss of 4.5 ± .4 nuts/palm/year. These studies, conducted in Fiji in the 1970s, led to the conclusion that the actual loss due to rats is possibly only one-half of that observed from surveys since the trees are able to compensate for nut loss by partial replacement. The worker found that female flower production increased, that yield remained stable, and that trees could fully replace damage nuts by harvest time when even 50% or more of the nuts were artificially damaged.[40]

J. Tokelau Islands

The species of rat found to be of economic importance with regard to agriculture and public health in the Tokelau islands is *R. exulans*. Mean estimates of rat damaged nuts were 43 to 71% based on 18 and 11 sample quadrats surveyed for rat damage.[45] However, varying damage of 80, 90, and even up to 100% of the nuts fell were reported due to rats in the islands.

Besides the loss caused by rodents on coconuts, reports indicated a correlation between the occurrence of rats and filariasis.[45] Rodent damaged nuts serve as breeding habitats for the mosquito, *Aedes polynesiensis*, the carrier of filariasis (*Wuchereria bancrofti*). The worker found the greatest density of mosquitoes in islets with the largest proportion of rat-damaged coconuts on the forest floor.

K. Gilbert and Ellice Islands

Copra provides the main source of income for the islands. The coconut provides a very important part of the diet of the people there. This crop, however, is severely affected by the rodent pest. The two species of rats known to occur are *R. exulans* and *R. rattus*.[48]

Several estimates of losses to coconuts in the Gilbert and Ellice islands have been reported as follows: 21% and 73%.[42] A damage assessment relating the number of damaged nuts to yield give crop losses ranging from 10 to 63% with an average of 23%.[48] More damaged nuts have been observed on islands that are predominated by *R. rattus*, while with the *R. exulans* damage to coconut seem to have a lesser significance.

L. Other Islands

Three other islands in the Pacific, French Polynesia, New Hebrides, and the Solomon Islands have rodent problems in coconut and cocoa.[42] Statistics on losses to the crops are given in Table 3. It maybe safe to assume that the *Rattus* spp. would be responsible for the damage to these crops.

Table 3
SELECTED LISTS OF MAJOR CROPS DAMAGED BY RODENTS AND ESTIMATES OF LOSSES

Location	Crops	Damage estimates	Comments	Ref.
Philippines	Rice (preharvest)	1970, 4.5% (Wet season); 1971, 1.65% (dry season); 1974, 4.31% (dry), 2.47% (wet); 1973, 2.47% (dry), 4.31% (wet)	National averages	29
		1974, 3.2% (range of 1.65—4.53% cut tillers valued at $67.3 million)	National average, estimated value of loss based on 1975 exchange rate	30
		1975, 4.63% cut tillers; 1976—1980, ave. of 0.52% cut tillers	National survey, reduced loss due to control program	3
	Corn	1983—84, ave. of 5.42% (1.8—80%) damaged ears; equivalent to 69.53 kg (25.2—1120 kg/ha)	Survey from 18 municipalities from 6 corn producing provinces	15
	Coconut	1600 nuts/ha/yr	Counts of damaged fallen nuts	12
		Ave. of 43.5% (12.4—65%) loss from potential harvest	Based from 11 studies; duration of experiment ranged from 5 months to 3 yr	14
	Sugar cane	30—50% damaged cane stalks	Estimated	24
		2.2—9.6% sugar loss with 11.1—24.3% damaged cane stalks	Survey from 6 farms	24
		2—13% damaged stalks	Experiment and survey from 14 mill districts	7
	Pineapple	10.2% of harvestable fruits	Untreated reference from rat control experiment; no damage in treated	13
Malaysia	Rice (preharvest)	5—11% yield loss	Survey	2
		1—5%	Estimated	33
	Oil palm	240 kg oil/acre/yr valued at $45.5 million	Estimated value of loss based on 1981 price	19
		146.5 lb oil/acre; $26.6 million/ha	Questionnaire	17
	Cocoa	50—60% of harvestable pods	Estimated in cocoa monoculture	19
		92% of harvestable pods	Estimated in coconut-cocoa intercropped	19
Indonesia	Rice	27% yield loss from 317,212 ha affected in 1979—80 16% yield loss from 248,225 ha affected in 1980—81	Estimated and from survey	16
	Sugar cane	30—100% loss from 14,000 ha affected by rats	Estimated	31

Table 3 (continued)
SELECTED LISTS OF MAJOR CROPS DAMAGED BY RODENTS AND ESTIMATES OF LOSSES

Location	Crops	Damage estimates	Comments	Ref.
Hawaii	Sugar cane	0.2—0.6 ton/acre or approx. $28—$83/acre; 0.7 ton sugar/acre valued at $4.5 million/yr with 4—40% damaged stalks	Estimated	10, 11 36
		$4.5 m/yr	Estimated loss with 32% damaged stalks	11
		17% reduction in sugar content	Estimated loss with 100% damaged stalks	8
	Macademia nut	15.8% (13—19%) loss on nuts; valued at $300,000	Determined from nut counts	6
Fiji	Coconut	28%	Study	41
		5—13%; 4.5 nuts/palm/yr	Survey from 11 islands	
	Cocoa	9—63%; 9,000 pods/ha	Pod count	17
Tokelau	Coconut	43—71% of harvestable nuts	Survey using quadrats	47
Gilbert and Ellice	Coconut	23% (10—63%) of harvestable nuts	Estimated from a 12 week study	48
Solomon	Cocoa	1—9% of pods; 5,640 pods/acre	Survey from random sample of 20 acres	14 17
French Polynesia	Coconut	25—30% of nuts	Estimated	42
New Hebrides	Coconut	20—50% of potential yield	Counts from 2 plots	42
Indonesia	Corn Cassava Soybean Peanuts Mungbean Sweet Potato	1,149 ha affected 4,676 ha affected 1,302 ha affected 168 ha affected 45 ha affected 505 ha affected	Areas affected in 1976—77 had damage ranging from light to moderate. No estimate of loss was stated.	31
Thailand	Rice (preharvest)	Ave. of 7% cut tillers	Survey in 1975—76	37
	Sugar cane	5.3% loss	Study	8
	Major crops — rice, corn, soybean, coconut, vegetables, fruits, and sugar cane	An average of 913,783 ha affected; value of loss is in the ave. of $5.5 million	Estimated in 1971—1973	46
Vietnam	Rice (preharvest)	0.2—9.6% cut tillers	Survey from 78 hamlets representing 7 provinces	29
Western Samoa	Coconut	10—20% of harvestable nuts	Estimated	39
	Cassava		Farmer's claim	
	Peanuts	Up to 100%	Farmer's claim	
	Cocoa	Up to 90%	Estimated	
		10—15% of harvestable pods	Limited survey	23

REFERENCES

1. **Barbehenn, K. R., Sumangil, J. P., and Libay, J. L.** Rodents of the Philippines croplands, *Phillip. Agric.*, 56, 217, 1973.
2. **Buckle, A. P., Rowe, F. P., and Husin, A. R.** Field trials of warfarin and brodicacoum wax block baits for the control of the ricefield rat, *Rattus argentiventer* in Peninsular Malaysia, in Proc. Int. Conf. Plant Prot. in Tropics, Kuala Lumpur, Malaysia, March 1 to 4, 1982.

3. **Bureau of Plant Industry Annual Reports,** 1976-1980, Manila, Philippines, unpublished.
4. **Crucillo, V., Otanes, F., and Morales, J. L.,** What we're doing to control field rats in Cotabato, Philippine Islands, in *Pest Control,* 22, 10, 1954.
5. **Devick, W. S.,** Rodent plague in Hawaii, in Proc. Rodents as Factors in Diseases and Economic Loss, East West Center, Honolulu, Hawaii, June 17 to 27, 1968.
6. **Erskine, F. C.,** Rat control in the Honokoa Sugar Company macadamia nut orchard, in Proc. Rodents as Factors in Disease and Economic Loss, East West Center, Honolulu, Hawaii, June 17 to 27, 1968.
7. **Estioko, B. R.,** Sugarcane yield losses due to rodents with reference to species distributions, infield stalk damage, varietal preference and movement under Philippine conditions, Ph. D. dissertation, University of Philippines Los Banos, Philippines (mimeo), November, 1978.
8. **Hampson, S. J.,** A review of rodent damage to sugarcane with criteria for the use of rodenticides, in *Proc. Conf. on the Organisation and Practice of Vertebrate Pest Control,* Dubock, A.C., Ed., Elvetham Hall, Hampshire, England, 1984, 227.
9. **Hilton, H. W.,** The rat problem in Hawaiian sugarcane, in Proc. Rodents as Factors in Diseases and Economic Loss, East West Center, Honolulu, Hawaii, June 17 to 27, 1968.
10. **Hood, G. A.,** Rat control research in Hawaiian sugarcane in Proc. Rodents as Factors in Disease and Economic Loss, East West Center, Honolulu, Hawaii, June 17 to 27, 1968.
11. **Hood, G. A., Nass, R. D., and Lindsey, G. D.,** The rat in Hawaiian sugarcane, in Proc. 4th Vert. Pest Conf., West Sacramento, Calif., March 3 to 5, 1970.
12. **Hoque, M. M.,** Notes on rodent pest affecting coconut, *Phillip. Agric.,* 56, 280, 1973.
13. **Hoque, M. M.,** Rat Control in coconut-pineapple intercrops, *Phillip. Agric.,* 66, 82, 1983.
14. **Hoque, M. M. and Aterrado, E. V.,** Development of rat control technology in the Philippines, Paper presented during the 2nd Natl. Coconut Res. Dev. Symp., Los Banos, Laguna, Philippines, 1984.
15. **Hoque, M. M.,** Determination of national losses to corn due to rats, Terminal Report for PCARRD study No. 34-260-21, U.P., Los Banos, College, Laguna (mimeo), 9, 1985.
16. **Indrarto, N.,** Five year rat control program in Indonesia, in *Proc. Conf. Organisation and Practice of Vertebrate Pest Control,* Dubock, A. C., Ed., Elvetham Hall, Hampshire, England, 1984, 475.
17. **Jackson, W. B.,** Evaluation of rodent depredations to crops and stored products, *EPPO Bull.,* 7, 439, 1977.
18. **Kadarsan, S.,** On parasites and diseases of the wild rats in Java, Indonesia, in Proc. Rodents as Factors in Diseases and Economic Loss, East West Center, Honolulu, Hawaii, June 17 to 27, 1968.
19. **Khoo, C. K.,** The practice of rodent control in Malaysian plantation crops, in *Proc. Conf. on the Organisation and Practice of Vertebrate Pest Control,* Dubock, A. C., Ed., Elvetham Hall, Hempshire, England, 1984, 209.
20. **Manke, K. F.,** The economic importance of rodents in the pineapple industry in Hawaii, in Proc. Rodent as Factors in Diseases and Economic Loss, East West Center, Honolulu, Hawaii, June 17 to 27, 1968.
21. **Ong, K. H.,** Status of rodent surveillance and control in Singapore, in *Proc. Symp. on Small Mammal Problems and Control,* Sanchez, F. F., Ed., U.P. Los Banos, College, Laguna, Philippines, 55, 1980.
22. **Partoatmodjo, S.,** Rat problems in foodcrops in Indonesia, in Proc. of the Regional Training Sem. on Field Rat Cont. and Res., Manila, Philippines, March 4 to 15, 1974.
23. **Perez, R. V.,** Technical assistance to West Samoa on rat control, *Plant Prot. News,* 9, 18, 1980.
24. **Porquez, P. H. and Ledesma, F. I.,** Sugar losses from cane damaged by rats, in *Sugar News,* 461, 364, 1970.
25. **Reidinger, R. F. and Libay, J. L.,** Crown-baiting in coconut trees with anticoagulant rodenticides to reduce damage, in *Proc. Symp. on Small Mammal Problems and Control,* Sanchez, F. F., Ed., U.P. Los Banos, College, Laguna, Philippines, 211, 1980.
26. **Reidinger, R. F. and Libay, J. L.,** Size and weight of coconuts damaged by rats, *Phillip. Agric.,* 64, 281, 1981.
27. **Richards, C. G. J.,** Rodent control in Vietnam using "klerat" rodenticide, Plant Protection Division, ICI, Surrey, England, 1984, 2 (unpublished).
28. **Salazar, N.** Rodent-borne diseases and epidemiology of field and commercial rodents in the Philippines, in Proc. 2nd Regional Training Sem. on Field Rat Cont. and Res., 1977, Manila, Philippines, March 16 to 25, 1977.
29. **Sanchez, F. F. et al.** Rodent Research Center Annual Progress Report, 1972—1976, University of the Philippines College, Laguna, Philippines.
30. **Schaefer, J.,** Crop gains—crop losses (views on the field rat situation in Philippine rice crops), *Plant Prot. News,* 4, 12, 1975.
31. **Soekarna, D.,** The ricefield rat and its control in Indonesia, in Proc. Rodents as Factors in Disease and Economic Loss East West Center, Honolulu, Hawaii, June 17 to 27, 1968.
32. **Soekarna, D., Partoatmodjo, S., and Boeadi,** Problems and management of small mammals in Indonesia with special reference to rats, in *Proc. Symp. on Small Mammal Problems and Control.,* Sanchez, F. F., Ed., U.P. Los Banos, College, Laguna, Philippines, 35, 1980.

33. **Soon, C. O.,** Rodent control problems and activities in Malaysia, in *Proc. 2nd Regional Training Sem. on Field Rat Cont. and Res.,* Manila, Philippines, March 16 to 25, 1977.
34. **Strecker, R. L. Marshall, J. T., Jackson, W. B., Barbehenn, and Johnson, D.,** Pacific Island Rat Ecology, Report of a study made on Ponape and adjacent islands, 1955-1958, *Bernice P. Bishop Mus. Bull.,* 225, 274, 1962.
35. **Sudto, P.,** Problems on small mammals in Thailand: taxonomy and distribution, in *Proc. Symp. on Small Mammal Problems and Control,* Sanchez, F. F., Ed., U.P. Los Banos, College, Laguna, Philippines, 17, 1980.
36. **Teshima, A. H.,** Rats and the Hawaiian sugar industry, in *Proc Rodents as Factors in Disease and Economic Loss,* East West Center, Honolulu, Hawaii, June 17 to 27, 1968.
37. **Tongtavee, K.,** Experience with brodifacoum in laboratory and field conditions in Thailand, in *Proc. Conf. on the Organisation and Practice of Vertebrate Pest Control,* Dubock, A. C., Ed., Elvetham Hall, Hampshire, England, 1984, 357.
38. **Tomich, P. Q.,** Mammal ecology and disease control in Hawaii, in Proc. *Rodents as Factors in Disease and Economic Loss, East West Center,* Honolulu, Hawaii, June 17 to 27, 1968.
39. **Tuisuga, S. A.,** Rodent damage in Western Samoa, in Proc. 2nd *Regional Training Sem. on Field Rat Cont. and Res.,* Manila, Philippines, March 16 to 25, 1977.
40. **Williams, J. M.,** The effect of artificial rat damage on coconut yields in Fiji, *PANS,* 20 (3), 275, 1974.
41. **Williams, J. M.,** Rat damage to coconuts in Fiji. I. Assessment of damage, *PANS,* 20, 379, 1974.
42. **Wilson, E. J.,** Rat problem in the Pacific Basin, in *Proc. Rodents as Factors in Diseases and Economic Loss,* East West Center, Honolulu, Hawaii, June 17 to 27, 1968.
43. **Wood, B. J.,** Population studies on the Malaysian wood rat (*Rattus tiomanicus*) in oil palms, demonstrating an effective new control method and assessing some older ones, *The Plantar,* 45, 510, 1969.
44. **Wood, B. J.,** Progress in the control of tropical field rats, *Proc. Int. Conf. Plant Prot. in Tropics,* Kuala Lumpur, Malaysia, March 1 to 4, 1982.
45. **Wodzicki, K.,** Rat ecology and control in the Tokelau Islands, Central Pacific, in *Proc. Rodents as Factors in Diseases and Economic Loss,* East West Center, Honolulu, Hawaii, June 17 to 27, 1968.
46. **Yenbutra, S. and Wuthijindaroj, O.,** Report on the present status of vertebrate pest problem and program in Thailand, in *Proc. Regional Training Sem. on Field Rat Cont. and Res.,* Philippines, March 4 to 15, 1974.
47. **Yong, C. Y.,** Status of rodent problem in Peninsular Malaysia, in *Proc. of the Regional Sem. on Field Rat Cont. and Res.,* Manila, Philippines, March 4 to 15, 1974.
48. **Smith, F. J.,** Rat damage to coconuts in the Gilbert and Ellice Islands, in *Proc. Rodents as Factors in Diseases and Economic Loss,* East West Center, Honolulu, Hawaii, June 17 to 27, 1968.

Chapter 7

RODENTS AND HUMAN DISEASE: A GLOBAL APPRECIATION

Norman G. Gratz

TABLE OF CONTENTS

I. INTRODUCTION

Much has been written about the economic importance of rodents, describing the extent and cost of their predation on food and crops and stored food products. There are, as well, many individual papers describing the involvement of a given rodent species in a given country or locality in the epidemiology of a specific disease of veterinary or human importance. A small number of brief reviews have also been published summarizing in general terms the public health importance of rodents as an order.[9,93] However, no substantial publication exists reviewing and evaluating the public health importance of just those zoonotic diseases in whose epidemiological cycle rodents serve as reservoirs or hosts and presenting in some detail the nature, epidemiology, distribution, and description of their vectors and reservoirs. The importance of rodents as reservoirs of such "classical" human diseases as plague is generally well recognized, but their importance and role as reservoirs for a considerable number of other infectious diseases of man is far less well known even though some of these diseases give rise to substantial morbidity and mortality over vast geographical areas. Furthermore, it is only comparatively recently that the role of rodents as hosts of some newly emergent diseases, such as Lassa fever and Haemorrhagic Fever with Renal Syndrome, has been determined with certainty. The lack of recognition of the magnitude of this problem has certainly created difficulties for those persons responsible for rodent control programs in obtaining adequate funding to support and expand such programs. The economic consequences of failing to effectively control rodent species attacking crops or stored foodstuffs will probably be readily apparent both at the level of the individual farmer and to agricultural authorities; unfortunately the possible consequences of permitting the serious proliferation of commensal rodent populations is all too often little appreciated by public health authorities in ministries of health or municipalities who are responsible for the allocation of resources to disease prevention or control programs.

The following chapter will describe those diseases in whose transmission rodents have been implicated to any important extent and will attempt to give a picture of the public health importance of these diseases, their epidemiology and distribution, their vectors where these exist, and their rodent reservoirs.

It must be stated at the very onset that it is difficult to give even an approximate estimate of the incidence or prevalence in human populations of this group of diseases. While many of the groups present themselves in a dramatic manner in the human patient and are readily recognized and often reported, others are difficult to diagnose clinically and frequently occur under conditions and in geographical areas where their presence remains unrecognized and unreported and where, to the detriment of the patient, appropriate treatment is not provided. Even when in some instances it is suspected on a clinical basis that the infection is due to one of the members of this large group, the laboratory facilities necessary to provide diagnostic confirmation may not exist or the laboratory personnel may not have the necessary equipment, reagents, or even the motivation or training to carry out the usually exacting diagnostic procedures. This situation is not, of course, peculiar to the rodent-borne diseases alone, but this group of many different types of infections suffers perhaps more frequently than others in being unrecognized and, indeed, unsuspected and, as a result, little accurate information is available from many countries on their incidence and prevalence. In some geographical areas the very epidemiology of some of the infections remains obscure.

Nevertheless, with the passage of time and the accumulation of information, through research and the gradual improvement in diagnostic facilities, a clearer picture is emerging of at least the general distribution of these rodent-borne diseases and of their occurrence in particular rodent species and human populations and of their epidemiology. However, a great deal remains to be learned both regarding their rodent reservoirs and the manner in which the diseases are transmitted to man. The better and more complete our understanding of the epidemiology of this group of diseases, the more selective and effective can be the control measures that must be developed to control their transmission.

The fact that rodent-borne diseases are so widespread should give no cause for surprise. Few other animal groups live in closer contact with man than commensal rodent populations and few others have more successfully adapted to such an association. Rodents share men's dwellings and consume or foul part of his food both in the fields before it is harvested and in his granaries and warehouses after it has been processed and stored. In urban areas human populations are expanding at a rate which in many, if not most of the cities of the developing world, far exceeds the resources of municipal services to provide adequate housing or satisfactorily dispose of solid or liquid wastes.[91]

One of the serious consequences of this accumulation of uncollected and untreated wastes is the production of conditions that greatly favor the multiplication of rodent populations and lead to greatly increased contacts between rodents and man. In rural areas, the ever increasing demand to grow more food for urban populations has resulted in a general shift towards single crop culture and more storage of food in villages while waiting for its transport to cities. Such storage is usually under poor conditions that encourage rodent predation; thus in villages in rural areas as well as in cities, conditions favor the increase of commensal rodent populations, and their contact with man also increases. Unfortunately rodent control, where it is carried out at all, is applied in a piecemeal manner and has little real impact on rodent populations. Aside from the increased consumption of his food by rodents, under such conditions man almost inevitably suffers from an increased transmission of rodent-borne diseases. The commitment of virtually all countries to a strategy of "health for all by the year 2000" will, it is hoped, result in a gradual improvement of health care facilities and a growing primary health care network; under such conditions it can be expected that any important or unusual outbreak of disease will be quickly recognized and the necessary resources marshaled to deal with it. This will certainly require a great deal more training, an increased understanding of the possibility of these infections occurring, and an increased ability to deal with them. It is hoped that the present review will help to increase the awareness of the extent of the rodent-borne diseases, the areas in which they occur, the species of rodents that are likely to carry them, and, in the cases in which they are involved, of the arthropods that are their vectors to man. No attempt will be made to review the control measures that can be carried out to control rodent populations; there are an abundance of publications on this subject. The main problem remains to motivate health and agricultural authorities to implement and assist in carrying out the control and the communities to take part in them and support them in whatever manner they can.

Most of the infections which will be reviewed below are zoonoses, i.e., infectious agents with rodents or other animals as their hosts or reservoirs. While the hosts or reservoirs may not suffer from the pathogen, it serves to maintain it in the environment and eventually may serve to transmit it to man or his domestic animals by contact with infected urine or feces, by bite, or by means of an arthropod vector.

II. THE VIRAL ZOONOSES

A. Venezuelan Equine Encephalitis (VEE)

The causative agent of Venezuelan equine encephalitis is a group A togavirus or alphavirus which is transmitted by mosquitoes from small mammal reservoirs to equine hosts or to man.

While the virus has been isolated from about two dozen different species of domestic and wild animals, and many others have no doubt been infected, the disease is only clinically recognized and of economic importance in the horse and other equines.

Following a short incubation period of 1 to 3 days, the disease may manifest itself as a benign febrile disease lasting no longer than 1 or 2 days or, in other animals the infection may lead to the development of an encephalomyelitis. The mortality rate among horses that develop such symptoms is high and may reach 80%. When the infection is transmitted to man by the bite of an infected vector mosquito the incubation period is also brief, from 2 to 5 days. The symptoms in man may vary from a mild influenza-like fever, to serious encephalitis. The high fever that frequently occurs is accompanied by malaise, chills, myalgia, cephalagia, and frequently nausea, vomiting, and diarrhea. The course of the disease may be 1 to 4 days. The symptoms of encephalitis occur much more frequently among children than among adults. Fortunately the mortality rate is low and is estimated to be 0.2 to 1% of the clinical cases in contrast to the much higher mortality in equines.

The distribution of the disease and of the infectious agent is limited to the Western Hemisphere. Human and equine cases have been reported from the west coast of South America in Peru, on the east coast in foci in Venezuela, and then to the north through Colombia and Central America to as far as Texas. A virus closely related to VEE has been isolated from rodents in Brazil. Epidemic outbreaks of considerable size have occurred in Venezuela, Colombia, and Ecuador. In Venezuela 23,283 human cases of the disease were reported and 31,000 from Ecuador during an epidemic which occurred from 1962 to 1964.[174]

In Colombia, in an outbreak in 1967, an estimated 67,000 to 104,000 equine cases occurred with an average mortality of 28%. The appearance of such outbreaks is cyclic with equine and human epidemics occurring every 7 to 10 years. Fortunately, as has been stated above, human mortality is low though involvement of the central nervous system is not infrequent and this may give rise to serious illness requiring hospitalization and may result in central nervous system (CNS) sequelae.

From the above it can be seen that this infection is one of very considerable concern both in the area of animal health and that of human health.

1. The Vectors

Mosquitoes were first incriminated as vectors of VEE during an outbreak of the disease in Trinidad in 1943.[88] The vector species on the island were considered to be *Mansonia titillans* and *Aedes taeniorrhynchus* which, on epidemiological grounds, was thought to have brought the first cases of the disease from Venezuela to Trinidad. During outbreaks in Venezuela, *A. taeniorrhynchus* appeared to be the most frequent carrier of the virus.[200] The virus has been isolated from numerous other species of mosquitoes throughout the geograhical range from which the disease has been reported; these species include one or more representatives from the genera *Aedes, Anopheles, Culex*, (subgenus *Melanoconion*), other *Culex* spp., *Haemagogus, Mansonia*, and *Psorophora*, among others.[203]

2. The Reservoirs

In the couse of epidemic outbreaks of the disease and at a period when an epizootic strain of VEE is circulating in equine populations, transmission may occur from equines to man through the agency of various vector mosquito species. However, during interepidemic periods, the disease is maintained in a sylvan endemic cycle by rodent to rodent transmission also through the intermediary of a vector mosquito species. Much of the detailed epidemiology of the maintenance cycle remains to be clarified.

Jonkers[115] reviewed the extensive studies that had been carried out on VEE in southern Florida, Mexico, Panama, Colombia, and Trinidad; although many different animal species were shown to have become infective during the endemic periods, he concluded that none of them played an essential role as reservoirs of the infection except for rodents. As an example, despite a variety of potential hosts in Trinidad, the only isolations of VEE virus were those made from the cotton rat, *Sigmodon hispidus* in open country and from *Oryzomys laticeps, Heteromys anomalus*, and *Zygodontomys brevicauda* in the forest. In serological surveys in Brazil, *Proechimys guyannensis oris*, the spiny rat, and *Oryzomys capito goeldii*, the rice rat, were recorded as seropositive to VEE. The virus was frequently isolated from *O. capito goeldii* which appears to be the most important reservoir species of rodent in Brazil though, as noted, a number of other species of rodent were also found to be seropositive for the virus. In Venezuela, marsupials, and particularly *Didelphis marsupialis*, have shown high antibody titers to VEE.[233] In Panama, 29 species of indigenous vertebrates have been found naturally infected with VEE virus with infections being particularly frequent in the cotton rat *Sigmodon hispidus* and *P. semispinosus*.[94] Studies were carried out in Guatemala by Scherer[195] et al. as a follow up to an outbreak of VEE in that country. Fully 80% of the opposums and rodents which were tested were seropositive for VEE. The rate of positivity was particularly high in Caues' rice rat, *Oryzomys palustris*, and the cotton rat, *S. hispidus*.

Four species of wild mammals were found infected with VEE virus in Veracruz State, Mexico,[194] these include two species of opposums, *D. marsupialis* and *Philander opossum* and two species of bats. Virus was not recovered but antibody tests were positive in the same area for a number of species of rodents, i.e., *Peromyscus mexicanus, O. palustris, Rattus rattus*, and *S. hispidus*.

In the U.S., VEE antibody has been detected in human populations in Florida and several surveys have been carried out to attempt to determine the probable reservoirs of the disease in that state. Specific antibody for VEE was found in 16 out of 25 cotton mice, *Peromyscus gossypinus,* and 3 out of 16 cotton rats, *S. hispidus,* from the single site sampled.[57] In a later study,[146] small numbers of *O. palustris* were also found to be seropositive.

In the summer of 1971, a major epidemic of VEE occurred in southern Texas. More than 1500 equines died of the disease and 110 human cases were reported but there were no human deaths. The most important mosquito vector was determined to be *Psorophora confinnis*. Serological surveys found only small numbers of rodents positive and the infection apparently failed to establish itself in the local rodent populations. It is considered that horses were the most important reservoirs at the time of the epidemic.[213]

Transmission at the time of an epizootic of VEE appears to have an equine-mosquite-equine cycle but rodents apparently serve as the most important interepizootic reservoirs of this arbovirus. Fortunately, effective vaccines now exist and over 10 million horses have been vaccinated with the attenuated TC-33 strain of VEE virus; this has probably prevented the deaths of tens of thousands of equines and untold numbers of human cases.[156]

B. California Group Viruses

The several different species of viruses that comprise this closely related group of arboviruses have been reported from the western and central U.S. and the southern provinces of Canada as well as from Central and South America, Europe and, Africa. Among the many different viruses reported, 12 are considered as distinct species.[129]

C. California Encephalitis Infections (CE)

This arbovirus was first isolated and identified in 1943,[103] though at the time it was not known to be a virus of man. Infections in man are now known to be caused by several closely related members of the La Crosse virus group. The disease is most frequently diagnosed in children in whom it gives rise to acute central nervous system symptoms with fever, headaches, nausea, or vomiting, and at times convulsions; the disease may lead to death. Several hundred cases have been reported from the U.S. since the disease was first identified as a clinical entity.

1. The Vectors

The original isolates of CE were made from two pools of *Aedes melanimon* and one pool of *Culex tarsalis*. Following more intensive studies, it appears that *A. melanimon* is the most important vector of the disease.[188] There is now evidence that transovarial and transstadial transmission of CE occurs in both *Aedes dorsalis*, which is now thought to be a vector in California, and *A. melanimon.*[226]

2. The Reservoirs

Many different species of small mammals have been found to be seropositive for CE virus. In addition to several species of lagomorphs, the California ground squirrel *Spermophilus beecheyi* and a number of other rodent species have been recorded as seropositive at one time or another, including *Sciurus griseus, Erethizon dorsatum, Neotoma fuscipes, N. lepida, Peromyscus maniculatus,* and *Rattus rattus*.

D. La Crosse Virus (LAC)

Of all the California group of arboviruses, LAC probably represents the most important cause of arboviral encephalitis in the U.S. except for those years in which epidemics of St. Louis encephalitis occur. The infection is endemic in the upper Midwest region, particularly in the states of Wisconsin, Illinois, Indiana, and Ohio.

1. The Vectors

Aedes triseriatus is both the most important vector and also a species in which transovarial transmission occurs; the virus probably overwinters in the eggs of *A. triseriatus* which are usually laid in treeholes.

2. The Reservoirs

The summer amplification cycle of the LAC virus involves small mammals; the most important vertebrate hosts are chipmunks, i.e., the eastern chipmunk, *Tamias striatus* and the gray squirrel, *Sciurus carolinensis*.[124,167]

E. Keystone Virus (KEY)

Keystone virus has been reported from Pennsylvania and Maryland in the northeastern U.S., south to Florida, and as far west as Texas. While there is serological evidence of human involvement with the KEY virus, the infection apparently produces no clinical disease.

1. The Vectors

The likely vectors of KEY virus are the mosquitoes *Aedes tormentor* and *A. atlanticus*; transovarial transmission of the virus is known to occur.

2. The Reservoir

The probable reservoir of this virus in the southeastern part of the U.S. is the gray squirrel, *S. carolinensis*.[215]

F. Colorado Tick Fever (CTF)

This infection is known to occur and cause human disease in the mountainous or highland regions of at least 11 western states of the U.S. and 2 provinces of Canada, Alberta and British Columbia.[69] Infection in man gives rise to an acute, febrile, biphasic illness, usually mild in nature though it may be much more severe in children; death is uncommon. Some one to two hundred cases a year are reported from the U.S. and the number of cases appears to be on the increase as human movement in tick-infested areas increases, especially in mountainous recreational areas.

1. The Vectors

While at least eight species of tick have been found naturally infected with the virus, the only important vector species is *Dermacentor andersoni*.[44]

2. The Reservoirs

In nature CTF virus is maintained in a cycle involving immature stages of its tick vector and some of the host animals. Nymphal stages of the tick carry the virus through the winter and in the spring when feeding on animals they infect young rodents that develop an unapparent infection with a prolonged viremia. Among the rodent species which have been found naturally infected with CTF are the golden-mantled ground squirrel, *Spermophilus lateralis tescorum*, the Columbian ground squirrel, *Citellus c. columbianus*, the pine squirrel, *Tamiasiurus hudsonicus richardsoni*, as well as chipmunks and meadow voles, and various other species. Chipmunks are amplifying hosts for the virus. Other ground squirrels, *Per-*

omyscus spp., *Microtus* spp., and the porcupine may all be important reservoirs in various geographical areas.

G. Powassan Encephalitis

Only a few cases of disease resulting from infection with this virus have been reported from Canada. The virus however, has been isolated from ticks and small mammals both in Canada and the U.S.

1. The Vectors

Transmission of the virus is probably through species of Ixodid ticks; isolations of virus have been made from *Ixodes marxi* and *I. cooki* in Canada and *I. cooki, I. spinipalpus*, and *Dermacentor andersoni* in the U.S.

2. The Reservoirs

Isolations of Powassan virus have been made from the woodchucks, *Marmota manax* and *M. flaviventris*, red squirrels, *Tamiasciurus hudsonicus*, and deermice. It is thought that ground squirrels and woodchucks are of importance in the maintenance cycle.[12]

H. Group C Bunyaviral Fevers

The distribution of all the members of this group of related arboviruses is restricted to the Americas and, for the most part, to South America. The group includes such species as Caraparu, Itaqui, Madrid, Marituba, Murutuca, Oriboca, and Nepuyo viruses. Disease resulting from infection by these viruses is relatively benign. While individual human populations may show a high rate of infection to one or another of this group, they are mainly found among persons working in rural and forest areas. Many vertebrate hosts have been found positive both through isolation of virus and through serological surveys; these include the rodents *Proechimys semispinosus, P. guyannensis, Nectomys squamipes, Zygodontomys brevicauda, Heteromys anomalus, Oryzomys squamipes, O. capito*, and various marsupials.[141]

I. Group B, Tick-Borne Encephalitis (Tick-Borne European Menigoencephalitis)

Tick-borne encephalitis is a very widely spread infection ranging from the far eastern areas of the U.S.S.R., where it is present as a subtype of Russian Spring-Summer Encephalitis, and west to Switzerland and France; in these latter countries the subtype of the disease is present in a somewhat milder form as Tick-Borne Encephalitis (TBE). The far eastern form is often lethal; in some endemic areas, case fatality rates may reach 20%, whereas the mortality rate from TBE in central Europe may be only 1 to 5% of the cases; usually however, TBE gives rise to a benign meningoencephalitis. The persons at greatest risk to infection throughout the endemic geographical range of the disease are forestry workers and hunters and, to an increasing extent, persons entering for recreational purposes into the natural foci, i.e., tick-infested deciduous forests, young forests, areas of underbrush, or forest-fringe areas.[244] Almost 5000 cases of the disease were reported in Austria between 1972 and 1982; after a vaccination campaign in which some 1,900,000 people were vaccinated, the incidence of the disease now appears to be falling in that country. Similar indications of decline in incidence have also occurred in other countries where vulnerable members of the population have been vaccinated.

1. The Vectors

The main vector of the western form of TBE is the tick *Ixodes ricinus*, which is a species related to *I. persulcatus*. West of the Ural mountains, the distribution of the TBE virus and that of *I. ricinus* coincide. Other species of ticks have also been found infected with the virus but none have been shown to have the same vectorial importance as *I. ricinus*. The virus survives best in a habitat with a relatively stable population density of ticks and small mammals such as insectivores and especially rodents and birds. There are many studies that have been carried out on the natural foci of infection in the U.S.S.R. and central and western Europe, and an extensive literature of the subject exists.[123]

In the Far East, *I. persalcatus* is the most important vector. Once infected from feeding on a viremic host, the immature ticks remain infected for life and can transmit the virus to another host as long as 3 years later. The disease can also be transmitted from one generation of ticks to subsequent ones transovarially; the extent to which this can maintain the infection in a given focus in the absence of vertebrate hosts is uncertain and such a mode of transmission may not, in fact, be of epidemiological importance.

2. The Reservoirs

Domestic and wild ungulates may maintain *I. ricinus* populations at some density; however, ecological areas with a dense population of small mammals with short generation times especially favor the development of larval ticks while the presence of bird populations favors the development of nymphs. The most commonly infected rodent hosts of the virus are the field mouse, *Apodemus flavicollis*, and the bank vole, *Clethironomys glareolus*. *A. sylvaticus*, *A. agrarius*, *Microtus arvalis*, and the insectivore *Sorex araneus* have also been found with antibodies to the virus.

J. Crimean-Congo Haemorrhagic Fever (CCHF)

This very serious disease was first recognized in the steppe region of the U.S.S.R. in the 1940s and has since been determined to be antigenically identical to a virus known to be widespread in East and West Africa, Zaire, South Africa, Bulgaria, Yugoslavia, Iraq, and Pakistan. It can therefore be concluded that the virus is enzootic in the Paleartic, Oriental, and Ethiopian faunal regions in steppe, savannah, semidesert, and foothill biotopes where *Hyalomma* ticks are the predominant species parasitizing domestic and wild animals.[106] Seroepidemiological surveys revealed the possible occurrence of unapparent infections in man and animals in Bulgaria, Iran, Egypt, Turkey, Afghanistan, India, Greece, Hungary, Yugoslavia, and Nigeria.[229] The virus has recently been isolated from the blood of patients in southern Xinjiang, China during an outbreak of a haemorrhagic fever in that area.[251]

1. The Vectors

The vectors of CCHF are usually ticks of the genera *Hyalomma, Boophilus, Dermacentor*, and *Rhipicephalus*; *H. marginatum marginatum* is the most common vector in the Ukraine, Crimea, and Astrakhan areas of the U.S.S.R. and elsewhere in Eastern Europe. It has also been isolated from *Ixodes ricinus, D. marginatus, Haemaphysalis punctata*, and *R. bursa* in Europe. In the southeastern part of the Paleartic region, *H. anatolicum anatolicum, H. asiaticum asiaticum, H. marginatum turanicum, H. detritum, R. bursa, R. pumilio, R. sanguineus, R. turanicus, D. marginatus, D. daghestanicus*, and *Boophilus annalatus* are common vectors. In Pakistan, two strains of CCHF virus have been isolated, one from *H. a. anatolicum* and one from *B. microplus*. In Africa CCHF virus has been isolated from a number of different species of *Hyalomma, Amblyomma, Boophilus*, and *Rhipicephalus* species. In China, isolates of virus were made from *H. asiaticum*. In all, virus has been isolated from 16 species of ticks and a 17th species is implicated as a vector on the grounds of antigenic studies.[106]

2. The Reservoirs

The extensive review of the disease by Hoogstraal[106] describes numerous surveys of rodent populations for the presence of CCHF virus or antibodies to it. Generally, *H. m. marginatum* does not appear to parasitize rodents. While reports have appeared in the literature of antibodies being found in *Apodemus sylvaticus, Citellus pygmaeus,* and the long-clawed ground squirrel in Turkmenia, i.e., *Spermophilus l. leptodactylus*, the role of rodents as reservoirs of the disease though widely suspected to be important, must remain as equivocal and not as yet demonstrated. Due to the importance of the disease and the possibility that rodents may be involved, it must, therefore, be mentioned.

K. The Rodent-Borne Haemorrhagic Fevers

Three of the diseases in this group, Junin haemorrhagic fever, Machupo haemorrhagic fever, and Lassa fever, are members of the genus *Arenavirus* while the fourth member, Hantaan haemorrhagic fever with renal snydrome (HFRS), has been proposed for inclusion in the Bunyaviridae.[245] Junin and Machupo viruses are the causative agents, respectively, of Argentinian and Bolivian haemorrhagic fevers in the two countries from which the names are taken, while the presence of Lassa fever has been demonstrated both by clinical disease and serological evidence in many countries of West Africa and, most recently, in Mozambique where Lassa, or a very closely related virus, has been found. HFRS is now known to be widely distributed in many different countries of the world. Another member of the arenaviruses, lymphocytic choriomeningitis (LCM), is not a haemorrhagic fever and will, therefore, be dealt with separately below.

L. Lassa Fever

The disease was first recognized as a clinical entity in Nigeria in 1969 and, as noted above, it has now been reported as being present, either on the basis of clinical reports or by serological surveys, throughout a wide area of Africa including Burkina Faso, the Central African Republic, Gambia, Ghana, Guinea, the Ivory Coast, Liberia, Mali, Nigeria, Senegal, and Sierra Leone. Fluorescent antibody studies in the western part of Sudan have shown the presence if antibodies in 12 to 13% of the human sera tested though there is no evidence of human illness related to Lassa fever. In Sierra Leone between 3 and 11% of the febrile illnesses in two study villages were caused by Lassa fever; in surveys of several rural areas of the country, 6% of the population had complement fixing antibody to Lassa although only 0.2% of these individuals gave a history of illness from the disease.[80] In north-western Liberia, 13% of all febrile cases admitted to a hospital in Zorzor, as well as 17% of all patients tested at the hospital were positive for Lassa fever and the overall mortality from the disease was 13.6%.[164] In a serological and/or virus isolation study on 1909 human sera in the Central African Republic, only 4 human sera or 0.4% were positive for Lassa fever. In contrast, in Sierra Leone, the disease accounts for 10% of all febrile illnesses admitted to hospital and 1.7% of the general death rate.[245]

1. The Reservoirs

There is no known arthropod vector of Lassa fever nor has the virus been isolated from any of a large number of arthropods tested in a number of intensive searches for the virus among various insect and tick species.

Lassa virus has been isolated from the multimammate rat, *Mastomys natalensis* in Sierra Leone,[163] and in Nigeria.[248] Two additional virus isolations have been made in Nigeria, one from a *R. rattus* and one from a *Mus minutoides*; the authors who reported the finding of virus in *R. rattus* felt that further confirmation of the identification of the particular rodent was necessary. More recently an isolation of an arenavirus closely related to Lassa was made from *M. natalensis* in Mozambique,[249] and in 1981 six strains of the same virus, now termed Mozambique virus, were recovered from *M. natalensis* in Zimbabwe.[114] Despite the frequent finding of Lassa virus in as common a commensal species as *M. natalensis*, an attempt to reduce the rate of seroconversion in a human population in a Lassa endemic village in Sierra Leone through a program of intensive rodent trapping failed to achieve any significant reduction in the incidence of infection in trapped houses as compared to untrapped ones.[119] While it appears that the major risk factor in transmission of Lassa fever is from the excreta and urine of rodents infected with Lassa virus, it is apparent that much still remains to be learned about the dynamics of transmission of the disease agent.

M. Argentinian Haemorrhagic Fever

This disease is caused by the Junin virus of the Tacaribe group; it has been clinically recognized since the mid-1940s but the causative virus was only first isolated in 1958. The disease is restricted to Argentina and within that country covers a geographical area of more than 100,000 km^2 in an area of intensive agricultural production in the pampas; over one million inhabitants are resident in the endemic area. There appears to be evidence that the endemic area of the disease is steadily expanding.[151]

Between 1958 and 1984, over 20,000 cases of the disease were reported with a mortality rate of 10 to 15% in cases which received no treatment.[175]

The disease is seasonal in occurrence and natural infections are usually found among agricultural workers in corn and wheat fields, coinciding with the harvest period. Treatment of the disease is by provision of immune plasma within 8 days of onset of the symptoms; a live, attenuated virus vaccine is under development.

1. The Reservoirs

As is the case with Lassa virus, there is no known or suspected vector of the disease from its animal reservoirs to man, transmission probably occurring through the medium of dust or foodstuffs contaminated by rodent excreta or urine carrying the virus. While the precise mechanism of transmission remains unknown, the fact that chronically infected rodents persistently shed virus in their saliva and urine, means that environmental contamination, as has been indicated above, therefore seems the most likely mode of transmission.

The two principal reservoirs or natural hosts of Junin virus are *Calomys callosus* and *C. laucha*. Due to the extensive distribution of these two species in the disease endemic area and their high population densities, effective control of transmission through measures against the rodent reservoirs would probably be difficult and uneconomic to achieve.

N. Bolivian Haemorrhagic Fever

The causative agent of this virus is known as Machupo virus and it is closely related to Junin virus. Reports of an unrecognized febrile illness were received from a sparsely settled area of northeastern Bolivia in 1958. By mid-1962, 470 cases of the diseases had been registered and in 30%, or 142 of these, there was a fatal outcome.[150] As a result of this outbreak, the 600 inhabitants of Orobayaya, in the Province of Iténez, abandoned their community. Between 1960 and 1964, more than 1000 cases of Bolivian haemorrhagic fever with a mortality rate of 22% were recorded in the Provinces of Iténez, Marmoreé, and Yacuma in the Department of Beni. In 1971 an outbreak occurred in a hospital in Cochabamba, a city situated outside the endemic area of the disease. The index cases appeared to have contracted the infection on a ranch located in the community of Fortaleza, Province of Yacuma, Department of Beni. Five of the persons that were in contact with the patient during his stay in the hospital contracted the disease and four of them died. No human cases of Bolivian haemorrhagic fever have been registered since 1974. Neither the disease nor its causative agent have been reported from outside Bolivia.

1. The Reservoir

Serological studies that showed that the virus was closely related to Junin and Tacaribe viruses made a rodent host suspect. This was confirmed by the isolation of the etiologic agent, Machupo virus, from several specimens of *Calomys callosus* within the geographical area of the outbreak.[113] Because of the severity and high mortality rate of the disease, an intensive rodent reservoir control program was started in 1964 using both traps and zinc phosphide baits containing tartar emetic. A total of 3020 rodents were known to have been killed in the disease endemic area of which 2896 were *Calomys callosus*. Two weeks after the initiation of the program, human case rates in the area under rodent control dropped from an average of 1.1 to 0.24 cases per day.[124] Since 1964, the control program against the peridomestic reservoir has been maintained at a cost of some $30,000/year but, as a result, no cases of the disease have been reported since 1974 even though the virus is thought to persist in the endemic area.[160]

O. Haemorrhagic Fever with Renal Syndrome (HFRS)

It is now clear that this disease has been described under a variety of names at different times and from different places in areas extending from Japan to the Baltic. A World Health Organization (WHO) meeting held on the subject of this disease[241] recommended that the name Haemorrhagic Fever with Renal Syndrome (HFRS) be used for all forms of the disease so far described and this usage will be followed below.

A disease with clinical features similar to HFRS was described in the Tula area of the U.S.S.R. in the early 1930s and was given the name of Tula fever. Some 1000 cases of the disease were reported between 1930 and 1934. Later, the same disease was seen to have a even greater distribution and was given the name of Far-Eastern nephroso-nephritis by Russian physicians working in the Amur river valley. Since then several thousand cases have been recognized as occurring annually in the Soviet Union. The disease was recognized as an entity in northeastern China in 1934-35 and some 30,000 cases were reported as occurring annually. In 1984 the Government of China reported that 90,000 cases and almost 3000 deaths had occurred. Mortality from the disease in China has varied from 3.2 to 6.4%. In Sweden the disease was named nephropathia epidemica; during the World War II, an epidemic of more than 1000 cases occurred among German and Finish troops and some 100 cases a year are still reported from Finland. During the Korean conflict, more than 3000 United Nations troops stationed in the demilitarized zone developed an apparently new disease with a mortality rate of 10 to 15%. Cases of the disease later appeared further south on the peninsula and the infection is now known to be widespread. In 1960 the disease was first recognized in Japan, in Osaka City; since that time 119 cases and 2 deaths have been reported from other port cities as well as from rural areas of Japan. In addition, a number of cases have occurred among laboratory workers tending rodent colonies.

The virus causative agent of the HFRS infection was only first isolated in 1976 in Korea from specimens of *Apodemus agrarius*[134] and from humans in 1978. The etiological agent was named Hantaan virus after the Hantaan River which runs near the 38th parallel in Korea.

It is now recognized from seroepidemiological surveys and human clinical case reports, that Hantaan and Hantaan-related viruses are widely distributed throughout the world. This has been demonstrated by the presence of antibodies against the agents in sera from man, urban and rural rodent populations, and laboratory rat colonies.[245] There are at least two distinct forms of the disease, mild and severe, the latter form being more common in Asian countries. In Europe, the majority of the cases of HFRS are mild; several hundred cases of HFRS, similar to the mild forms of the Korean disease are reported annually in Scandinavia including Finland, Sweden, and Norway, in Bulgaria, Czechoslovakia, Hungary, Rumania, and Yugoslavia in eastern Europe, and a few cases have also been reported recently from Belgium, France, and Greece.

Antibodies against Hantaan or Hantaan-related viruses have been detected in human sera in the Americas in Alaska, Brazil, Canada, Columbia, and the U.S., in the Western Pacific on Fiji, in Southeast Asia in Burma, Hong Kong, India, Malaysia, Philippines, Taiwan, and Thailand, and in Africa in the Central African Republic, Egypt, Gabon, Nigeria, and

Uganda. Antibodies against Hantaan or Hantaan-related virus have been detected in urban rat sera in Argentina, Brazil and the U.S., in various cities in Asia and the Pacific area in China, Fiji, Hong Kong, Japan, the Republic of Korea, Malaysia, Philippines and Thailand and in Egypt in the eastern Mediterranean. The records of these serological findings have been summarized in a view of the disease.[78]

1. The Reservoir

From the time at which the earliest epidemics of the disease were recognized as a distinct entity, it has been speculated that the occurrence of cases of HFRS were associated with contact with wild rodents. As has been noted earlier, this was first proven when antigen to the virus was first found in wild caught striped field mice, *Apodemus agrarius*, and virus was later isolated from lung tissues of a laboratory strain of *A. agrarius coreae*.[133] In 1980 the virus was recovered from the lung tissues of *Clethrionomys glareolus*, the bank vole, in Finland.[40] Tkachenko[219] et al. summarized the findings of antigen to HFRS which included *C. glareolus, C. rutilus, Apodemus agrarius, A. sylvaticus*, and *Microtus agrestis* as positive in various European foci of HFRS in the U.S.S.R. in all 14 different animal species, whereas in the far eastern U.S.S.R. foci, HFRS antigen was found in 9 species of animals, viz, *Microtus fortis, Clethrionomys rufocanus, C. rutilus, Cricetulus barabensis, Apodemus peninsulae, A. agrarius, Rattus norvegicus, Mus musculus, and Eutamias sibiricus*. As a reservoir, the dominant role was served, in the author's opinion, by those species found in the highest population densities. In Czechoslovakia, a serological study of wild rodent species showed antigen in lung tissue of *C. glareolus, A. flavicollis*, and *A. agrarius*.[98] An epidemic of the disease involving 114 human cases in the Sarajevo region of Yugoslavia was preceded by a greatly increased density of the *A. flavicollis* population, an increase larger than any which had been seen for 20 years.[84]

As has been remarked earlier, antibodies to Hantaan or Hantaan-related viruses have been widely found in the Americas. In the U.S., antibody to Hantaan or a Hantaan-related virus was found in *Microtus pennsylvanicus* from widely separated areas in Virginia and Alaska.[136] Further tests on *Microtus pennsylvanicus* and *Peromyscus leucopus* trapped in Maryland showed that the virus detected appeared to be of another member of the HFRS group.[135]

The finding of Hantaan or Hantaan-like viruses in domestic rodents as well as in laboratory rat colonies in several parts of the world has important epidemiological implications. The viruses have been reported from *Rattus norvegicus* and *R. rattus* populations in Korea,[132] in Japan in Osaka,[214] Kobe, and Yokahama,[166] in Belgium,[229] and in the U.S. in Baltimore,[58] Philadelphia, and Houston.[130] In South America, *Rattus norvegicus* has also been found positive in Brazil where 56% (30 out of 54 rats) of the individuals tested were positive. Lower percentages of positivity were found in Sao Paulo (14%), and Recife-Olinda (6%). Tests on *R. norvegicus* in Buenos Aires, Argentina showed a positivity rate of 11%.[131]

Although HFRS transmission from infected rodents via arthropods has been postulated, only a single isolation has been made from pools of chigger mites,[159] and in view of the many negative searches that have been made for virus in a large number of arthropod species, this route of transmission seems unlikely.

Large quantities of virus are excreted in the saliva, urine, and feces of infected rodents and virus infection may persist for several months. The main route of infection among *Apodemus* field mice is through the respiratory tract but the infection can also be transmitted via the saliva, urine, or feces. Studies on the route of transmission among domestic rodents have not as yet been done and much remains to be learned about the epidemiology of this world-wide infection under the many different ecological conditions in which it occurs.

P. Lymphocytic Choriomeningitis (LCM)

The virus which is now known to cause LCM, has been recognized since the early 1930s though it was only later that the virus was associated with a disease syndrome in man, i.e., acute benign aseptic meningitis. Still later it was shown that the virus was actually the cause of only a small proportion of the cases of this disease. In 1935, the virus was detected as a chronic infection in a colony of albino laboratory mice.[221] Extensive surveys have now shown that the virus probably is of world-wide distribution but despite this, cases of human disease resulting from infection with the virus are rare, or at least cases diagnosed and reported as such are rare. The disease has generally been likened to influenza and is only rarely fatal. Three small epidemics of the disease involving 239 human cases have occurred in the U.S. since 1960, all of them associated with laboratory colonies of the Syrian hamster.[97] In Germany, a survey of 2000 rural inhabitants showed that only 3.4% were seropositive for LCM.[31]

1. The Reservoir

The LCM virus has been isolated from a number of different animals in nature but the most important of these from the epidemiological viewpoint is the house mouse, *Mus musculus*. Since the time of the first finding of the virus in house mice which were taken from the home of two persons in the U.S. suffering from nonbacterial meningitis,[11] there have been repeated associations between the occurrence of the disease and infected mice both wild-caught and from laboratory colonies. The mouse develops no antibody to the virus and continues to shed virus for a good part of its life after infection. The virus is probably transmitted to man by infected urine, feces, and nasal or oral secretions contaminating food or water. As mentioned earlier, the Syrian hamster *Mesocricetus auratus* has been found to be infected and has been the cause of outbreaks of the illness both in the U.S. and Germany.

III. THE RICKETTSIAL DISEASES

A. Spotted-Fever Group Diseases

Almost all of the rickettsial agents of the diseases of this group, i.e., Rocky Mountain Spotted Fever, North Asian Tick Typhus, Boutonneuse Fever, and Queensland Tick Typhus, are tick-borne. The sole exception so far known is rickettsialpox whose vector is a mite. In the case of probably most, if not all of these diseases, rodents serve as the natural host to one extent or another.

B. Rocky Mountain Spotted Fever

The etiologic agent of the disease in man is *Rickettsia rickettsii*. The disease is only known to occur in North and South America; in the U.S. it has been reported throughout the country, except for states of Alaska, Maine, Vermont, and Hawaii. More than 97% of the cases are reported from the eastern and southeastern states. In Canada the disease has been reported from British Columbia, Alberta, and Saskatchewan. South of the U.S. it has been reported from Costa Rica,[81] Mexico, Panama, Columbia, and Brazil.[46] Some 847 cases were reported in the U.S. in 1984, with the highest incidence being in Oklahoma, though the single highest number of cases, 178, was reported from North Carolina.[55] Most of the patients were under 20 years of age, male, and most of those for whom reports were received had been treated by antibiotics; the overall mortality rate was 3.6% though it was 23% among those who received no treatment. A typical macuopapular rash over most of the body, high fever, severe headache, and chills characterizes the symptoms of the disease. The mortality rate in untreated patients increases with age.

1. The Vector

Five species of ticks are known to transmit *R. rickettsii* in the U.S., i.e., *Dermacentor andersoni, D. variabilis, D. parumapertus, Amblyomma americanum*, and *Haemaphysalis leporispalustris*. *Amblyomma cajennense* and *Rhipicephalus sanguineus* are both vectors in

Mexico.[42] Of the above species, the most important vectors in the U.S. are the wood tick, *D. andersoni* and the dog tick, *D. variabilis*.[45]

Ticks become infected when feeding on rickettsemic host animals, particularly on young rodents. Passage of the disease also occurs transovarially in *D. andersoni, D. variabilis*, and *H. leporispalustris* which were shown to pass the rickettsiae to 100% of their progeny via the eggs.[47] Transmission of the infectious agent to man occurs through the bite of an infected tick when it is feeding. This may occur in the natural foci or from ticks brought into the household on dogs.

2. The Reservoir

A substantial number of different animals including domestic pets, especially dogs and, to a lesser extent, cats, as well as hares and rodents can serve as hosts and reservoirs of the infectious agent, in addition to the role served by ticks through transovarial transmission of the rickettsiae from one generation to another. Most of the vector species of ticks are not particularly host specific and may thus feed on both large and small domestic and wild mammals. Ticks may become infected via two mechanisms, either through transovarial transmission of the infectious agent or by feeding on an infected host, usually a rodent, which itself recently became infected. In susceptible animal species, rickettsiae generally circulate for 5 to 8 days, rarely longer, but this time is adequate to represent a constant source of infection to uninfected ticks.

In the U.S., *R. rickettsiae* has been isolated from meadow voles, *Microtus pennsylvanicus*; a pine vole, *Pitymys pinetarum*; a white footed mouse, *Peromyscus leuopus*; a cotton rat, *Sigmodon hispidus*; cottontail rabbits, *Sylvilagus floridanus*; the opossum, *Didelphis marsupialis virginiana*,[33] as well as from a snowshoe hare, *Lepus americanus*; chipmunks, *Eutamias amoenus*; and golden mantled ground squirrels, *Spermophilus lateralis tescorum*.[48] In California, serological evidence has implicated brush rabbits, *S. bachmani*, the jack rabbit, *L. californicus*, and the kangaroo rat, *Dipodomys californicus* along with *P. maniculatus* and *P. boylii* as actual or potential reservoirs of the infectious agent.[127]

C. North Asian Tick Typhus

This is a relatively benign disease which does not give rise to any mortality. Its distribution is from islands in the Sea of Japan, through the Pacific Far East and Siberia, throughout the steppes of Asiatic Russia, and possibly into Pakistan and Eastern Europe. The causative agent is *Rickettsia siberica* or *R. rickettsii* var. *siberica*.

1. The Vector

The known vectors of the infectious agents are ticks, and it has been isolated from *Dermacentor nuttalli, D. marginatus, D. silvarum, Haemaphysalis punctata*, and *H. concinna*. The disease is transmitted to man by the bite of adult ticks whose larvae and nymphs have fed on infected mammals.

2. The Reservoirs

Wild rodents are the most important hosts in the natural cycle. Isolations of the rickettsial agent have been reported from the following species of rodents, *M. musculus, R. norvegicus, Apodemus gregorius, Microtus* spp., *Cricetulus* spp., *Eutamias* spp., *Spermophilus* spp., *Lagurus* spp., and the European hare, *Lepus europaeus*.[46]

The disease is transmitted to man by the bite of adult ticks whose larvae and nymphs have fed on infected small mammals.

D. Boutonneuse Fever

The causative agent of this infection in man is *Rickettsia conorii*. The infection in man may give rise to a mild to moderately severe illness with a fever of a 1 to 2 week duration in man. It is characterized by a primary lesion, the "tache noire", at the site of the bite by the vector tick which is then followed by the development of a generalized macuopapular rash. The mortality rate is low, less than 3% even in cases which receive no antibiotic treatment. Within its area of distribution the disease is quite common and in some areas it may, in fact, be increasing in incidence due to humans being more frequently bitten by dog ticks as the numbers of pet dogs increases. The infection is widely distributed throughout Africa where it is known as "African Tick Typhus" and around the Mediterranean basin where it is also known as "Marseilles Fever." The disease has been reported from the Black and Caspian sea regions and as far east as India, with doubtful reports from Malaysia and Vietnam.

1. The Vector

The vector tick in the Mediterranean area is the common dog tick, *Rhipicephalus sanguineus*. In South Africa other species of ticks which have been found infected in nature include *Haemaphysalis leachii, Amblyomma herbraeum, R. appendiculatus, Boophilus decoloratus*, and *Hyalomma aegyptium*.[85] In the southern U.S.S.R., the infection is associated only with *R. sanguineus*.[105]

2. The Reservoir

Dogs are usually considered to be the most important reservoir of the infectious agent but considering the frequency with which this species of rickettsia is found in rodents, it is quite likely that they may serve a still more important role as a reservoir than dogs. In South Africa, strains of *R. conorii* have been recovered from *Otomys irroatus* and *R. rattus*, and in Kenya from *Arvicanthis niloticus, R. rattus, Mastomys natalensis, Otomys* sp., *Aethomys kaiseri, Lophuromys flavopunctatus*, and *Lemniscomys striatus*. In Malaysia, a *R. conorri*-like strain of rickettsia has been isolated from specimens of *R. mulleri*, *R. whiteheadi, R. rattus argentiventer*, and *R. rattus diardii*.[153] Generally speaking it may be supposed that rodents are probably the most important hosts of the infectious agent in the wild cycle while dogs and dog ticks probably maintain a secondary, domestic cycle of the disease.

E. Queensland Tick Typhus

The causative agent of this infection in man is *Rickettsia australis* and the resulting disease is mild and generally similar to Boutonneuse Fever even to the extent that an eschar appears at the site of the infected tick bite. The known distribution of the infection is limited to Queensland, Australia. The true incidence of the disease is uncertain.

1. The Vector

Isolations of *R. australis* have been made from the tick *Ixodes holocyclus* and *I. tasmani*.

2. The Reservoir

No isolations of the rickettsial causative agent have been made from any small mammal host; however, the short-nosed bandicoot, *Isaodon obesulus*, the long-nosed bandicoot, *Perameles nausta*, the opossum, *Trichosurus vulpecula*, and other small mammals have been found to be seropositive in geographical areas where the human infection has been reported.

F. Rickettsialpox

The disease was first recognized as an entity in 1946 in New York City when an outbreak of more than 100 cases occurred in a housing development;[95] the causative agent *Rickettsia akari* was isolated from the blood of persons who had fallen ill from the disease as well as from house mice. The disease is characterized in man by the appearance of an initial eschar-like lesion at the site of the point of feeding of the vector mite, followed by lymphadenopathy, fever, and a skin rash that appears briefly.

From the time of the initial outbreak in 1946 to 1949, 500 cases of rickettsialpox were reported from New York City and the disease has now also been reported from several other urban areas of the U.S. as well as from the U.S.S.R., West Africa, South Africa, and Costa Rica. The reported incidence of the disease has fallen in recent years which is possibly related to the generally better level of control of house mouse populations in most urban areas.

1. The Vector

The rickettsial causative agent has been isolated from the development stages of the rodent mite *Allodermanyssus sanguineus* which has recently fed on man and thus transmitted the disease. Although the mites acquire the infectious agent from rickettsemic mice, they can also pass it on to their progeny by transovarial transmission. No other species of mite has been found positive for *R. akari*.

2. The Reservoir

R. akari was isolated from *Mus musculus* trapped in the same housing development where the initial outbreak of the disease was first recognized,[109] and this rodent appears to be the only species of reservoir in the U.S. *R. akari* has also been isolated from a field vole in Korea, *Microtus fortis pelliceus*,[110] and it apparently has a field cycle of transmission in Africa indicating the existence of as yet unidentified wild reservoirs.

G. Scrub Typhus (Tsutsugamushi Disease, Chigger-Borne Typhus)

Scrub typhus is a widespread and often very serious disease of man. The site of the infection is usually marked by a lesion or eschar. The infection may give rise to an acute febrile illness with headaches and lymphadenopathy followed by the appearance of a macropapular rash on the trunk and then on the extremities. While the disease readily responds to antibiotic therapy, the mortality rate in untreated cases may range from 1 to 60% depending upon the strain of the rickettsia and the extent of previous exposure. The causative agent is *Rickettsia tsutsugamushi* (which is also reported in the literature as *R. orientalis*). There is a great antigenic diversity among different strains of the agent which may be reflected in the great differences in the severity of the disease. Few current, accurate reports exist on the incidence of infection in most of the endemic countries, but there is little doubt that it remains a common disease causing much morbidity and some mortality in its natural foci.

1. Distribution

Scrub typhus is known to occur in suitable habitats in the following regions and islands: Japan and neighboring islands; Sakhalin island; Korea; southern China including Tibet; on Taiwan; the Primorye district of the U.S.S.R., including the islands of the Bay of Peter the Great and Tadshikistan; the northern part of the Indian sub-continent and west to the Indus valley; much of Indo-China peninsula including Vietnam, Malaysia, and Burma; the Philippines; New Guinea and adjacent islands including the Solomons and New Hebrides; the coastal portion of North Queensland; the Nicobar Islands; Sri Lanka; Indonesia and Borneo.[223] The infection has also been reported, on the basis of serological evidence, from eastern Iran.[102] The disease is not known from the rest of Asia, Europe, or the New World and a single report from West Africa is considered conjectural. The distribution of the disease is characterized by the presence of one of the vector mite species, the presence of a rodent reservoir host, the presence of the causative agent, and the occurrence of transitional vegetation usually resulting from the felling of primary forests for agricultural or military purposes. The resulting "scrub" clearings or fringe habitats favor the proliferation of rodent hosts of the genus *Rattus* at the expense of less densely distributed wild, forest species of rodents. There is extensive literature describing the vegetational sequence which encourages the development of scrub typhus foci.[14,15,137,168,222,223]

2. The Vector

In every geographical area in which the disease has been reported, the vectors have been found to be mites of the subgenus *Leptotrombidium* or the *L. deliense* group. Schluger and Kudryashova[196] regard the subgenus as the only vectors of the disease to man. Only the larval form of these mites, the "chiggers", are parasitic and they usually feed only once; since transovarial and trans-stadial transmission of rickettsiae from the infected female mite to the next generation has been demonstrated in *Leptotrombidium*, these mites are also considered to be the true hosts of the scrub typhus rickettsiae.[178] The most important vector mite species of the group are *L. deliense, L. akamushi, L. fletcheri, L. arenicola, L. pallidum, L. scutellare*, and *L. pavlovski. L. deliense* is a particularly widespread species and has been found infected with the causative agent in Burma, China, India, Malaysia, New Guinea, Pakistan, the Pescadores and Philippines Islands, Taiwan, Thailand, and Australia. *L. pavlovskyi* and *L. pallidum* are the vectors in the U.S.S.R.[121] and *L. pallidum, L. akamushi*, and *L. scutellare* are the vectors in Japan.

3. The Vertebrate Reservoir

Most warm-blooded vertebrate are susceptible to infection by *R. tsutsugamushi* and are readily fed on by mites in infested foci. Between 56 and 59 species of mammals from 27 to 30 genera have been found naturally infected by the causative agent.[223] Rodents of the genus *Rattus* are the principal hosts of the vector mites in South-east Asia and *R. rattus* and its various subspecies and forms is the most commonly and densely distributed member of the genus. However, in Japan, 16 of the isolates of *R. tsutsugamushi* have been from field voles and most of these (82%) from voles of the genus *Microtus. M. montebelli* is a common host of the vector mite in that country. None of the isolates of the agent have been from commensal rodents and all *R. rattus* found infected have been from wild populations. Other rodent species found infected include the house mouse, *M. musculus*, as well as the insectivore *Suncus* sp., but, again, none of these have been from house-inhabiting populations.

The disease remains of public health importance in most of its originally known endemic areas and persists even in Japan though at a much lower level of incidence than in the past. In areas of South-east Asia where deforestation or transitional agriculture occurs, infection rates in the human population may be surprisingly high. It was concluded[191] that on the basis of the prevalence of antibiodies to scrub typhus, that 3% of the rural population of Malaysia is infected with the disease each year and that the positive antibody rates in some focal areas may range from 6 to 69%. The highest antibody rates were among persons who worked in forested areas, while antibody rates among urban populations were very much lower. It was noted that there was even a disparity between the relatively low reported incidence of cases of scrub typhus among the rural populations of Malaysia and the high prevalence of antibody to *R. tsutsugamushi* among the same populations.[38] They examined a large population of oil-palm workers and found that there were an estimated 400 cases of scrub typhus annually among the population of 10,000 living on one plantation. The disease accounted for 23% of all febrile illnesses in one hospital but it was recognized that its presence was not always correctly diagnosed clinically. Even higher rates were found in a recent seroepidemiological survey of a rural village in northeastern Thailand where there was a 77% prevalance for antibody to *R. tsutsugamushi* in the adult population of the village; the villagers work in the surrounding forests logging and they are exposed to the bites of the vector mites.[112]

H. Murine Typhus

This disease in man is also often referred to as flea-borne typhus or endemic typhus to distinguish it from louse-borne or epidemic typhus. The causative agent is *Rickettsia typhi* (=*R. mooseri* of authors). Commensal rats are the principal reservoir and the oriental rat flea, *Xenopsylla cheopis*, is the most important vector. The course of the disease resembles that of a mild cases of epidemic or louse-borne typhus but the mortality rate is almost always far lower, usually not exceeding 2% in untreated patients; older patients are more susceptible to the disease. The disease is characterized by chills, headache, and fever with a rash appearing about the fifth day of illness; the high fever and rash persist for up to 2 weeks frequently along with respiratory symptoms. The infection readily responds to treatment with tetracycline and other broad spectrum antibiotics such as chloramphenicol or doxycycline.

1. Distribution

In considering the distribution of this infection, one must distinguish between reports of actual human cases of the disease and serological evidence of the presence of antibodies to *R. typhi* in rodent populations and of actual isolations of the causative agent from rodents. Traub et al.[224] provided an extremely complete summary of the distribution of the human disease and of the distribution of the infectious agent in rodent populations as determined by microbiological tests; this review should be consulted by anyone wishing further detail on this subject.

Both the infectious agent and disease arising from infection with it are virtually worldwide in distribution; murine typhus has been reported from the majority of those urban areas where there is a substantial rodent population present. In the Americas it has been reported from the U.S. where the number of cases recorded yearly reached between 2000 to 6000 in the early 1940s; at present no more than 60 to 80 cases are reported annually and the majority of these are from the state of Texas. Human cases have also been reported from Puerto Rico. Cases of the disease have been reported from Mexico, Guatemala, Panama, French Guiana, Venezuela, Brazil, Argentina, Chile, and Ecuador. In Asia, cases have been reported from most parts of India, China, Burma, Vietnam, Indonesia, Malaysia, Philippines, and Australia. In Europe, Albania, France, Greece, Italy, Rumania, some parts of the U.S.S.R., and Yugoslavia have reported cases. In the Middle East, the disease is present in Israel, Jordan, Lebanon, Kuwait, and Egypt; it is present in numerous countries in Africa but little accurate data exist on its distribution or incidence on that continent. Globally speaking, the disease is certainly under reported.[240] As an example, though the number of cases in the U.S. reached its highest annual incidence in 1944 with 5401 cases reported, it has been estimated that the actual number of cases was no less than three to four times this number.[43] The frequent and early use of broad-spectrum antibiotics for the treatment of as yet undiagnosed febrile illnesses no doubt contributes to the lack of accurate reporting of this disease.[186] In an outbreak in Kuwait in 1978, 254 cases of murine typhus occurred between April and August of that year and the authors[2] suggest that the disease is hyperendemic in the country rather than endemic and that incidence may be much higher than is commonly thought in most endemic foci.

2. The Vectors

The most important vector transmitting the infectious agent from its rodent reservoirs to man is the oriental rat flea *Xenophylla cheopis*; while other species of fleas may also transmit the disease, *X. cheopis* is particularly common wherever the disease occurs in man. Transmission to man usually occurs when man is bitten by the rat flea and the bite wound is contaminated by flea feces containing the infectious agent *R. typhi*. *X. cheopis* has frequently been found infected with *R. typhi* in various parts of the world and this species of flea is no doubt the main vector of murine typhus from rat to rat as well as from man to man.

There are some indications that the mouse flea *Leptopsylla segnis* and the cat flea *Ctenocephalides felis* may have a more significant role in the transmission of the disease than was hitherto thought.[74-76]

Farhang-Azad and Traub[74] have recently presented evidence that *X. cheopis* may transmit murine typhus to man by bite as well, rather than only via infective feces or through crushing an infected flea on an open bite wound.

3. The Reservoirs

The causative agent of murine typhus was first isolated from domestic rats in Mexico City in 1931,[165] and shortly thereafter Dyer et al.[68] isolated the rickettsiae from fleas collected from rats in Baltimore. Since these initial studies, there have been many isolations of *R. typhi* from domestic or commensal rats, primarily from *R. norvegicus* and *R. rattus*. Positive antibody rates in rat populations may be very high in some localities; Smith[206] studied serological findings in the southern U.S. from 1946 to 1951 and found that the monthly incidence over that period varied from 4.3% in February to a maximum of 33% in May in *R. norvegicus* populations. The peaks in serological incidence correspond with the peaks in *X. cheopis* indices on the rats. Because of the large number of human cases which were occurring in the southeastern part of the U.S., major DDT dusting campaigns were carried out using a 10% DDT dust applied to rat runs and burrows in Georgia and other endemic states in 1945. *R. norvegicus* and *R. rattus* were the only two species of rats incriminated in the transmission cycle of the disease as reservoirs; numerous publications analyzed the effect of the DDT dustings and reported that the reductions in the *X. cheopis* indices which occurred after these dustings were followed by reductions in the prevalence of *R. typhi* positive rats and, subsequently, a reduction in the incidence of human cases of the disease. A similar sequence of events as a result of a DDT dusting campaign was reported from Israel by Gratz[92] when rat burrows in 12,000 courtyards in Tel Aviv were dusted with a 10% DDT dust in 1953 and 18,000 courtyards were dusted in 1954. All the rats trapped in the surveys connected with the control campaign were *R. norvegicus* and the predominant flea species was *X. cheopis*. After the completion of the insecticidal dustings of the rat burrows, there was a steady decline in the incidence of cases of murine typhus in the human population living or working in the area.

R. typhi has also been isolated from the house mouse but it is not considered as an important reservoir and is usually only found infected when in the same niches as *R. typhi* bearing rats. The extensive surveys which have been carried out in a search for *R. typhi* in wild rodent populations have been described in detail by Traub et al.[224] and the authors concluded that only commensal rats and peridomestic mice species were reservoirs of the infectious agents. In geographical areas in which other rat species are also important commensals of man, e.g., Rangoon, Burma where the lesser bandicoot rat, *Bandicota bengalensis,* is the dominant commensal rat species,[235] it has been found to be very frequently seropositive for *R. typhi* (reports to the WHO).

I. Q Fever

The disease in man causes an acute febrile illness, giving rise to chills, headaches, malaise, and, in many instance, a pneumonitis occurs. Chronic endocarditis, pericarditis, and hepatitis have been reported as associated with the infection. Mortality in untreated cases is low, being no more than 1%. The infectious agent is *Coxiella burnetti*, (*Rickettsia burnetti*). Transmission is generally by foamites but infected cows are not an uncommon source of infection.

1. Distribution

This rickettsial disease has a world-wide distribution but seems to be on the decline in many countries; only a few dozen cases a year are reported from the U.S. The disease occurs mainly in countries where herds of cattle, sheep, and goats are common. Occasionally small epidemic outbreaks may occur among workers in stockyards and meat processing plants.

2. The Vectors

C. burnetti has been isolated from some 50 species of ticks, mites, fleas, and lice infesting infected animals. It would appear that ticks serve as both the vector and reservoir of the infectious agent in various areas. *Dermacentor, Hyalomma, Rhipicephalus, Haemaphysalis, Ixodes, Amblyomma, Argas, Ornithodoros, Boophilus species* have all been found infected as well as *Octobius megnini*.

3. The Reservoirs

Many different species of vertebrates have been found infected with *C. burnetti* in nature. These include a number of rodent species, among them the short-nosed bandicoot in Australia, *Isodon macrourus*; a large number of rodent species in the U.S. including woodrats *Neotoma* spp., ground squirrels, *Spermophilus* spp., field mice *Peromyscus* spp., *Perognathus* spp., *Dipodomys microps* and *D. ordii*, and chipmunks *Eutamias* spp.; in Morocco, the jird, *Meriones shawi*, and the field mouse *Apodemus sylvaticus*; in Czechoslovakia *R. norvegicus, M. musculus*, and the red bank vole *Clethrionomys glareolus*; in Spain the golden dormouse, *Eliomys quercinus*; and in the U.S.S.R. the suslik, *Citellus relictus*, the grey squirrel *Cricetulus migratorius*, the jerboa, *Allocataga elater*, the great gerbil, *Rhombomys opimus*, and the ground squirrel, *S. leptodactylus*, have been found infected.[211]

It appears that the infection is maintained in nature by wild animals and their ectoparasites and transferred to man, mainly from domestic animals when the wild cycle of infection overlaps to domestic animals. Some transovarial transmission occurs in *Hyalomma asiaticum* but it is not an important means of maintaining the infection in nature.[61]

IV. BACTERIAL DISEASES

A. Spirochetal Diseases

1. The Relapsing Fevers

A number of different diseases fall within this group, all of them caused by members of the genus *Borrelia*. Various species of the genus are found in almost every part of the world. One of the better known members of the genus is *Borrelia recurrentis* or louse-borne or epidemic relapsing fever; it is transmitted from man to man by the human body louse when an infected louse is crushed on the site of a louse bite. There is no known animal reservoir and the disease will not be among those considered below.

2. Tick-Borne Relapsing Fevers (TBRF)

This group of *Borrelia* spp. are apparently derived from the louse-borne species and cross-protection occurs between the louse-borne and tick-borne strains or species. There is a great deal of geographical variation among the borreliae with a great deal of tick-*Borrelia* specificity.[90]

All of the tick vectors of the tick-borne relapsing fevers (TBRF), are argasids, i.e., all members of the family *Argasidae* and all belong to the genus *Ornithodoros*. The tick transmits the infection by saliva or by coxal fluid during the time it is feeding on an animal or human host. The epidemiology of the given TBRF is dependent on the tick vector, its degree of infection from the natural host, which are often wild rodents, and on the degree of contact between man and the vector. Silent foci of TBRF can persist for many years; some species of infected ticks can live for many years under suitable conditions and transovarial transmission of the *Borrelia* also takes place.

3. The African Tick-Borne Relapsing Fevers

Three *Borrelia-Ornithodoros* complexes are found in Africa: *O. moubata* and *B. duttoni* in East and Central Africa; *O. erraticus erraticus* and *B. hispanica* in the Maghreb countries of North Africa, Spain, Portugal, Cyprus, and Greece; and *O. erraticus sonrai* and borreliae of the "crocidurae" group which extend from West Africa to North Africa and through the Middle East to Iran.

4. East African Relapsing Fever

In man, the disease caused by East African Relapsing Fever is characterized by a long incubation period of more than 7 days to 2 weeks and a high number of recurrent attacks or relapses of the fever, often as many as 11. Each period of high fever and toxemia lasts for several days and is separated by an afebrile period of about a week; fever may rise to 40°C, vomiting is common, petechial eruptions may occur, and a varying degree of mortality can follow. The actual incidence of cases of the disease is uncertain since in most of the areas where transmission takes place epidemiological surveillance is poor or nonexistant.

The Vector — The vectors are ticks of the *O. moubata* complex involving some four different species, each transmitting a different strain of *B. duttoni*, some of which confer only partial immunity against one another.[192]

The Reservoir — No animal has ever been found infected by *B. duttoni* and it appears that man is the only reservoir in the tick-man-tick cycle.[86]

5. Hispano-African Replasing Fever

This disease has been reported from Spain, Portugal, Morocco, Algeria, Tunisia, Greece and Cyprus. In man the disease is milder than that of East African Relapsing Fever and only 4 to 5 relapses occur. As with the others, the vectors are ticks and its main reservoirs are rodents. As noted above, the causative agent of the disease is *B. hispanica*.

The Vector — Although it has been suggested that *Rhipicephalus sanguineus* was implicated as a vector of the disease,[201] the vector is actually the large form or variety of *O. erraticus*, i.e., *O. erraticus erraticus*. The tick is widely distributed in the Maghreb and found in the burrows of porcupines, foxes, and wild rodents as well as in the burrows of domestic rodents inside houses. Transmission of the infectious agent is more commonly through contamination with infected coxal fluid than by the bite of the tick.

The Reservoir — While *B. hispanica* is sometimes found infecting wild animals, the natural reservoirs are the *Rattus* spp. and the large West Africa rat, *Cricetomys gambianus*. The ticks that feed on these rats will also readily feed on man as well and rodent-tick-man contact is especially frequent when *Rattus* spp. are nesting within houses.

6. Crocidurae Group Relpasing Fevers

The "crocidurae" group of relapsing fevers consists of a number of closely related strains of *Borrelia* with similar biological properties. They are found in tropical Africa ranging from Senegal to Kenya and in North Africa from Morocco to Egypt. They have also been reported from Turkey and Iran. The relapsing fevers of this group are generally benign diseases though there is a considerable geographical variation; most cases probably are not diagnosed. The various strains of the agent are usually associated with different reservoirs from which most of the varietal names are taken.

The Vector — Three relapsing fevers are transmitted by the small form of *O. erraticus*, i.e., *C. erraticus sonrai*. A number of other subspecies, forms, or varieties of the tick have been proposed in the literature, but morpholigically and ecologically they are extremely difficult to separate.

The Reservoirs and Strains — Rodhain[192] has provided the following classification for the various strains of this group of TBRF and their reservoirs:

1. *B. crocidurae*: reported from Senegal, Egypt, and Turkey; its usual hosts are peridomestic *C. gambianus* and *Rattus rattus*. It has also been found in *Arvicanthis* sp. in Senegal.
2. *B. merionesi:* described from southern Morocco by Blatazard[18], it was later found in *Meriones shawi*.
3. *B. microti:* commonly found in wild rodents of the genera *Microtus, Tatera, Cricetulus,* and *Meriones* in Iran.
4. *B. dipodilli*: reported only from Kenya as a parasite of gerbils, *Dipodillus* sp.[104]
5. *Borrelia* sp. found in *Psammomys obesus* in Jordan and in Israel.
6. *Borrelia* sp. isolated from *O. erraticus sonrai* from a *P. obesus* burrow in Mauritania.

7. Asiatic Tick-Borne Relapsing Fevers

As with the "crocidurae" group of *Ornithodoros-Borrelia* in Africa, complexes are found in Asia as well. The most important of these, *B. persica*, is transmitted by *O. tholozani*. The several Asian forms of relapsing fever extend from Egypt to China. The disease caused by each of the various strains varies in severity, as in Africa, from one geographical area to another; the most severe forms can lead to coma and death.

8. Asiatic/African Tick-Borne Relapsing Fever

The disease caused by the infectious agent *B. persica* has an extremely wide distribution. This species, or closely related strains, has been reported from Libya, Egypt, Cyprus, Israel, Lebanon, Syria, Iraq, Iran, the southern part of the U.S.S.R., Pakistan, Afghanistan, and western China. It may also be found in the Kashmir area of India. Within this vast zone the severity of the disease shows considerable variation ranging from a relatively benign illness to an extremely severe one. From two to eight relapses can occur. Cases often occur in groups when an individual house is infested with ticks from domestic rodent hosts or when groups of individuals seek shelter in a tick-infested cave or crevice. This frequently occurs among herdsman and more and more commonly among hikers and tourists seeking shelter from inclement weather in caves.

The Vector — The tick vector of *B. persica* is *Ornithodoros tholozani* (= *O. papillipes*). The distribution of the tick is equivalent to that of the disease. Within the many different ecological zones in which *O. tholozani* is found, it may infest rodent burrows, caves, rock crevices, cellars of ruined or inhabited houses especially if they have soil floors, and mud-walled animal shelters as well as a variety of other similar habitats. Transmission of infection in the *O. tholozani* group is by bite rather than by coxal fluid. Caves may be infested by literally thousands of ticks; third stage nymphs and adults may survive in such a favorable habitat for several years without a blood meal and survival without food in the laboratory for 5 to 10 years has been reported.[184] The long life span of this vector and its ability to survive for long periods of time without a blood meal have obvious epidemiological implications and in part explain the periodic outbreaks of relapsing fever in areas from which no cases may have been reported for a number of years. Recent investigations in the U.S.S.R.[232] have shown that while infection rates from this diseases have dropped sharply since the 1940s, many rural structures in Uzbekistan still remain infested by the tick. In other areas of the world, fewer cases now occur among people who live in mud structures as housing improves, but more cases occur among students and other hikers in recreational pursuits who enter tick-infested caves.

The Reservoirs — While the ticks feed readily on a variety of small and large animals, rodents and other small mammals appear to be the only true reservoirs of the disease in nature. In domestic and peridomestic situations, *Rattus* spp. are the reservoirs, while several species of wild rodents including *Acomys* spp. as well as hedgehogs and procupines have also been found infected with relapsing fever in nature. Inasmuch as *Borrelia* strains lose their pathogenicity through transovarial transmission in the tick, the maintenance of the disease in wild rodent populations is of paramount importance epidemiologically.

9. Caucasian Tick-Borne Relapsing Fever

The causative agent of this form of relapsing fever is *B. caucasica*. It has been reported from the Ukraine, the Caucasus, Soviet Armenia, Azerbaidzhan, Baku, and in the Georgian Republic, in Iraq and in Iran. The human infection gives rise to a severe disease with as many as 10 to 15 relapses over a period of about 3 months.

The Vector — The vector of Caucasian relapsing fever is the tick *O. verrucosus*; this species has been reported from Iraq, Iran, and the central Asian Republics of the U.S.S.R. It is found in caves, rock crevices, and rodent burrows and readily feeds on man, rodents, and other small mammals as well as on foxes and birds.

The Reservoirs — Various species of the genus *Meriones*, the gerbil *Alloctaga jaculus*, voles, and other field mice are the known reservoirs of the infectious agent in its area of distribution.

10. Caucasian Tick-Borne Relapsing Fever, Low Pathogenicity Form

The spirochaetal agent of this mild form of the disease, *B. latyschewii*, was first reported from Iran in *O. tartakowskyi* ticks taken from rodent burrows. It has now been recorded from much of central Asian U.S.S.R. The disease in humans is quite benign, causing a fever of no more than one day's duration or it may even be completely asymptomatic. No human cases have been reported in recent years.

The Vector — The Vector is *O. tartakowskyi*; this species of tick is found in the semidesert steppes of southern and central Asia (e.g., in the U.S.S.R. and Iran). It can be taken from rodent burrows, as well as the burrows of other animals, from stables and even houses. Infection of the vertebrate hosts is through contamination of the tick's bite by coxal fluid. The tick readily adapts to newly cultivated conditions and readily feeds upon both wild and commensal rodent species in newly farmed areas.

The Reservoir — In Iran the hosts of the tick and reservoirs of the spirochete are reported as *Rhombomys opimus* and *Meriones libycus*.[19] In the Turan lowland deserts of the U.S.S.R., the main host is the great gerbil, *Rhambomys opimus*; the tick vector is an extremely common inhabitant of the burrows of this species. Other animals in which spirochete infections were found in the Turan area where the long-tailed Libyan jirds, *M. libycus*, and *Dryomys nitedullus*. Between 3 to 20% of the *R. opimus* populations have been found infected in this area.[67] The tick has also been noted to feed on lizards and toads.

11. American Tick-Borne Relapsing Fevers

There are four species of TBRF found in North America; virtually all cases of these diseases in the U.S. have been reported from the western states with the exception of a single case reported from a child in Ohio who had apparently never left the state.[144] TBRF is found in Mexico, probably elsewhere in Central America, and in Panama. Each of the four species of causative agents will be considered separately below along with the particular tick vector with which each is associated.

a. Borrelia hermsi and Ornithodoros hermsi

This is the most important of the North American TBRF and is responsible for frequent but sporadic cases and, occasionally, small outbreaks of the disease. The largest single outbreak recorded occurred at the Grand Canyon National Park with 16 confirmed and 62 possible cases among visitors to the park and park employees.[32] The cause of this outbreak was apparently due to increased feeding of ticks on humans as a result of a diminution of the local rodent population following a severe winter. Cases of the disease are reported, for the most part, from geographical areas over 1000 m in altitude. Cases have been reported from California, Colorado, Idaho, Nevada, Oregon, Washington, and, in Canada, from British Colombia. The single case of relapsing fever in Ohio referred to above was due to *B. hermsi*. Many cases occur among persons sleeping in tick and rodent-infested mountain cabins. Fortunately the disease in man is not particularly severe; bouts of fever may last for 2 to 9 days with intermediate afebrile periods.

The Vector — The vector of *B. hermsi* is *O. hermsi*. This species of tick is usually found at higher altitudes, from 1000 to 2400 m. *O. hermsi* generally feeds on chipmunks, tree squirrels, and pine squirrels, and can be found in numbers in rodents nest in dead trees or inside of mountain cabins.

The Reservoir — Inasmuch as transovarial transmission in the tick does not appear to be frequent with only 0.3 to 2% of the larvae from infected females being themselves infected, the rodent hosts of the disease are the most important natural reservoir. There is no direct man-to-man transmission. The most important rodent reservoirs are probably *Tamiasciurus douglasii* or *Eutamias* sp.

b. Borrelia parkeri and Ornithodoros parkeri

All the cases of relapsing fever due to this *Borrelia* species have been reported only from the western states of the U.S. As with relapsing fever due to *B. hermsi*, infection by this species does not lead to serious illness. Unlike *B. hermsi*, cases of the disease only occur at lower elevations, a fact related to the ecology of the tick vector and rodent reservoir.

The Vector — The vector of *B. parkeri* is the tick *O. parkeri*; it is a widely distributed species in the western U.S. and Canada and is frequently found in the nests of the burrowing owl, *Speotyto cunicularia*, on which it avidly feeds, as well as in the nests of various species of rodents. Transmission of the disease is by bite of the tick and not through the coxal fluid.

The Reservoir — The known hosts of the tick and probable reservoirs of the disease are *Citellus* spp., including *C. columbianus*, *C. richardsoni*, *Cynomys leucurus*, *Marmota* sp., *Peromyscus* sp., *Lepus* sp., *Sylvilagus* sp., and *Mustela* sp.

c. Borrelia turicatae and Ornithodoros turicata

The distribution of this species of *Borrelia* is considerably wider than that of the two species described above; it has been reported from many areas of the western U.S. and in Mexico as well as from the state of Florida. Disease resulting from infection by this species is also mild with a lower fever and three to eight short relapses may occur.

The Vector — The vector of *B. turicatae* is *O. turicata*; the tick itself has a wider distribution than the reported cases of the disease and has been found in western Canada, California, Arizona, Colorado, Kansas, New Mexico, Utah, Texas, Oklahoma, and in several locations in Florida as well as in Mexico and Central America. Transovarial transmission of the infectious agent is common, and 60 to 100% of the eggs of infected females give birth to infected larvae. The tick has been found in rodent burrows, burrowing owl burrows, terrapin holes, rabbit burrows, in caves, feeding on rattlesnakes, on pigs in pig sties, and frequently within houses where it will readily and avidly attack man. Transmission of the spirochetes can occur even after very brief periods of feeding.

The Reservoir — Although this species of tick has many, varied hosts, rodents are probably the most important, if not the only, reservoir of the disease.

d. Borrelia mazzotti and Ornithodoros talaje

This is also a very wide spread species of *Borrelia*; it has been reported from California, Arizona, Nevada, Kansas, Texas, Florida, and southwards through Guatemala, Panama, and Colombia. Infection in man is not common.

The Vector — The vector of *B. mazzotti*, and probably the vector of other common

Borrelia species as well, is *O. talaje*. The distribution of *O. talaje* is very widespread, ranging from California through Kansas and Florida, and through Mexico and Central America to Argentina. It is usually found in rodent burrows and occasionally in houses where it feeds on domestic rodent species. Transmission of the infection is by tick bite.

The Reservoir — The reservoirs of the disease are almost certainly only rodents. The tick has been found feeding on *Dipodomys* sp., *Neotoma albigula, N. floridana baileyi, Citellus beecheyi, Rattus rattus, R. norvegicus*, and on several occasions on monkeys. *Borrelia* infections have been found in many rodents as well as in the opossum and monkeys.

*e. **Borrelia venezuelensis*** and *Ornithodoros venezuelensis*

This species of *Borrelia* has been reported from Panama, Colombia, Venzeuela, Ecuador, and Paraguay. The disease caused by this agent is severe involving several relapses in the human patient.

The Vector — Transmission is usually by *O. venezuelensis* (=*O. rudis*), although *O. talaje* may also be vector. The distribution of the vector is the same as that of the disease. *O. venezuelensis* is very anthropophilic and is frequently found inside rodent burrows including those inside houses. Transmission is by infected coxal fluid and perhaps by bite as well.

The Reservoirs — According to Marinkelle and Grose,[155] this spirochete has been found in nature infecting man, monkeys, and marsupials, as well as rats and mice.

12. Lyme Disease

Lyme disease is a systemic, tick-borne infection first recognized in the town of Lyme, Connecticut in 1975.[210] The disease is characterized by the appearance of a skin lesion, erythema chronicum migrans (ECM), fever, headache, myalgias, and arthralgias. The clinical condition, ECM, was known in Europe for some years but more recently the name Lyme disease has also come into use on that continent. ECM begins as a red papule that expands to become an annular lesion that may reach up to 70 cm in diameter and multiple skin lesions can occur. Some patients subsequently develop arthritis, neurologic or cardiac complications weeks or even months after the initial lesion. Early evidence[209] suggested an association of cases of the disease with bites by ticks; in 1982 a spirochete was isolated from an *Ixodes dammini* tick,[6] and later from ECM skin lesions on human patients. From the time of the first reports of the disease to the Centers for Disease Control the number of cases has been steadily increasing, perhaps, in part, as a result of better diagnosis; in 1984, 1500 cases of the disease were reported in the U.S.[56] Lyme disease was reported from 11 states in 1980 and 1982, from 18 states in 1983, and 21 states in 1984. Considerable under-reporting of the disease is suspected and the disease is already the most common tick-borne disease in the U.S. and of growing public health significance in that country. In the light of improved knowledge of the parasite and better diagnostic methods, the disease has been shown to occur with considerable frequency in 18 countries on three continents, Europe, North America and, most recently Australia, though in the latter ticks have not been implicated in the transmission.[197]

The Vector — The vectors of the disease are now known to be the ticks *Ixodes dammini* in the eastern U.S. and *I. pacificus* in the western states, i.e., in California, Nevada, Oregon, and Utah. *B. burgdorferi* has also been recovered from *Amblyomma americanum* in New Jersey[198] which is now considered a secondary vector in some areas, as well as from *Dermacentor variabilis*. The vector tick species in Europe is probably *I. ricinus*.

The Reservoirs — Anderson et al.[6] found *B. burgdorferi* positive ticks on western chipmunks, *Tamias striatus,* raccoons, *Procyon later,* white footed mice, *Peromyscus leucopus,* and a red squirrel, *Tamiasciurus hudsonicus* in Connecticut. Spirochetes were isolated from the blood of a raccoon and the white footed mouse. Considering the frequency with which *I. dammini* is found on field mice, Spielman et al.[207] suggested that they serve as the principal reservoir of the Lyme spirochete should it ensue that transovarial transmission is not the principal means of transmission from one generation of the vector to another. In light of a recent study of field collected *I. dammini* nymphs in which only 2 nymphs out of 274 collected from an area endemic for Lyme disease were positive for spirochetes, it would appear that transovarial transmission is only of limited importance in maintaining *B. burgdorferi* in nature,[179] and that rodents are indeed the most important reservoirs. Levine et al.[140] carried out a study on *P. leucopus* populations in coastal Massachusetts and concluded that this species was the main reservoir of Lyme disease in the region surveyed.

Work in Switzerland[52] has shown that spirochaetal infected ticks, *I. ricinus,* associated with *Apodemus* sp. and *Clethrionomys glareolus,* may be the vectors of a disease, tick-borne meningo-radiculitis, which is similar if not very closely related to Lyme disease.

13. Leptospirosis

Leptospirosis in man is a very widespread infection caused by spirochetes belonging to two species. All of the many parasitic and pathogenic serovars (= serotypes) belong to the species *Leptospira interrogans.* Leptospirosis is, in all probability, the most prevalent of the zoonotic diseases. The disease in man has many synonyms among them Weil's disease, canicola fever, haemorrhagic jaundice, mud fever, swineherd's disease, etc. Manifestations of the disease include fever, headache, chills, severe malaise, vomiting, myalgia, conjunctivitis, and occasionally meningitis, jaundice, renal insufficiency, and haemorrhage in the skin and mucous membrane. The influenza-like clinical illness lasts from a few days to 3 weeks but some infections may be quite asymptomatic. The degree of severity of the disease varies with the infecting serovar. While mortality is low, it can reach 20%, especially in older patients when associated with jaundice and kidney damage. Over 120 different serovars have been described which fall into about 20 different serogroups. The disease is transmitted from infected animal reservoirs to man by skin contact, especially abraded skin, with contaminated water, moist soil, or vegetation contaminated with urine of infected animals and occasionally through the ingestion of food or water contaminated with the urine of infected rats. The disease is a special occupational hazard to persons working in a moist environment likely to be contaminated with the infectious agent or individuals who frequently come into close contact with infected animals or their tissues. Cases of the disease are common among rice and sugar cane-field workers, farmers, miners, dairy workers, abattoir employees, fishermen, and troops in the field, and is particularly common among sewer workers. It was found to be the most common cause of acute fevers among American servicemen stationed in suburban and rural Vietnam.[27]

The Reservoirs — Almost every known mammal can be considered a possible carrier of the leptospires and a disseminator within its own species. Aside from man who is an accidental host from whom the largest number of serovars have been isolated (99), 24 serovars have been isolated from cattle, 21 from swine, 19 from dogs, and 15 to 24 have been isolated from various species of rodents.[220]

The review by Torten[220] should be referred to for the extensive list of the serovars identified and the animal species from which they have been isolated. In the following section only the primarily rodent-borne serovars will be considered.

In general, in the tropics, a wider range of rodent species and other small mammals are infected by a greater number of different serovars than in temperate climates.[71] Rodents, especially rats, may infect both farm animals as well as other individuals of their own species. This is a common cycle of infection of cattle and pigs, and man may be infected from either animal source. Rodents can also serve to contaminate water or soil which then becomes a source of infection for pigs, cattle, or sheep which then, in turn, become carriers and excretors infecting other rodents or others of their own species or man. A water buffalo-man cycle is a common one in rice growing areas of the world.

The infection cycles are confined to wild rodents, are self-maintaining and are related to the territorial limits of individuals, families, and animals in their natural habitats. The intrusion of man into these habitats, or alternatively, the intrusion of infected domestic rodents into man's habitats, involves the risk of infection of man with leptospirosis.

In farm and home environments, rats, mice, voles, gerbils, and coypu are the most important source of human infection. The more dense the domestic and peridomestic rodent populations, the greater the risk of infection to man. *L. icterohaemorrhagiae* is the most common of 24 serovars that have been isolated from *R. rattus;* isolations of 22 different serovars have been made from *R. norvegicus,* 9 from *Apodemus agrarius,* 17 from *A. sylvaticus dichurus,* 9 from the water vole, *Clethrionomys glareolus,* 8 from the short-nosed bandicoot *Isodon macrourus,* 14 from the common field vole *Microtus arvalis,* 13 from *Mus musculus,* etc. So extensive is the listing of rodents from which leptospires have been isolated, that reference is again recommended to Torten,[220] who has summarized much of the extensive literature on this subject.

When one considers the high prevalence of infection of leptospirosis recorded in the many surveys which have been carried out for these pathogens on urban rat populations, the high degree of frequency of this disease in human populations is not surprising. As examples of the result of surveys in urban rat populations, Lindenbaum and Eylan,[143] in a survey urban rats in Israel, isolated leptospires from 21.2% of the *R. norvegicus* and 9.9% of the *R. rattus* which they trapped; *L. icterohaemorrhagiae* was the most common serovar found in these rats populations. Thiermann[218] tested 358 rat sera from *R. norvegicus* in Detroit; only *L. icterohaemorrhagiae* was found and the seropositivity rate was 90.3%. In Colorado, a serological survey of a *R. norvegicus* population showed 66.4% to be positive for *L. icterohaemorrhagiae,*[5] while infection rates of up to 46% were found in a series of surveys carried out in Canada;[158] All infections were again *L. icterohaemorrhagiae*. In Taipei City, Tsai et al.[225] found 26.7% of the *R. norvegicus* and 9.3% of the *R. rattus* positive for leptospires; the serovars isolated were *javanica, australis* and *Bataviae*. An interesting find in West Africa was the finding by Baylet and Riel[22] that 11.2% of the giant rat, *Cricetomys gambianus* trapped in Dakar, Senegal were positive for leptospirosis. From the above surveys and many others that have been carried out in urban and periurban areas, the overwhelming role of rodents as reservoirs of certain serovars of this disease is evident.

14. Rat-Bite Fever

Two separate diseases are included under the name of "rat-bite fever"; one of these, streptobacillary fever, is caused by *Streptobacillus moniliformis* and the other by a spirochete, *Spirillum minor*. The two infections are clinically similar to one another but different enough to warrent separate description. There is little information available on the incidence of either type of rat-bite fever inasmuch as the disease is virtually nowhere reportable and most cases probably go undiagnosed or misdiagnosed. Since both forms of the disease readily respond to treatment by antibiotics, it is quite likely that many, if not most, cases go unrecognized as a result of their successful treatment.

15. Streptobacillary Rat-Bite Fever (Haverhill Fever, Epidemic Arthritic Erythema)

As noted above the infectious agent causing this disease in man is *Streptobacillus moniliformis*. There is usually a history of rat-bite within a 10 day period before the appearance of chills and high fever followed by a macropapular or petrichial rash, mostly on the extremities. Some of the joints may become swollen, red, and painful. Bacterial endocarditis and focal abscesses may occur in late untreated cases with a fatality rate of 7 to 10%.

The infectious agent is transmitted only through the secretions of the mouth and nose of an infected rat, usually at the time of a bite. Infection has also occurred in persons living or working in rat-infested buildings or by contamination of milk that has been fed upon by infected rats. Occasionally other animals may be infected by the causative agent.

Cases of this disease have occurred worldwide but are uncommon in North and South America and, for the most part, in Europe. Infection in laboratory rats is quite common in the U.S. and human disease following the bite of a laboratory colony rat is not uncommon.[54]

The reservoirs are laboratory or wild rats, and rarely, other rodent species such as squirrels.

16. Spirillary Fever (Sodoku)

The infectious agent causing this disease in man is *Spirillum minor*. This sporadic rat-bite fever is the most common from of rat-bite fever in Japan and in Asia. Clinically it differs from streptobacillary fever in having no arthritic symptoms and in the occurrence of a distinctive rash or reddish or purplish plaques. The incubation period is from 1 to 3 weeks after the bite of an infected rat and the previously healed bite wound reactivates when symptoms of the disease appear. Mortality can be 10% in untreated cases. All other aspects are as in streptobacillary fever.

B. Plague

Of all the rodent-borne diseases, plague is certainly the best known to the general public. While its relation to rodents was not known or understood, the disease itself was recognized and feared from antiquity as the ''Black Death''. In the past great pandemics of plague literally decimated the human populations of many countries. The first recorded pandemic began in Arabia during the time of Justinian and reached Egypt in 542 AD; from Egypt it spread west through North Africa and east through Palestine and Syria to Europe and then throughout the Roman Empire as far as the British Isles and Ireland. The next, and perhaps the most infamous of the great pandemics of this disease, was that of the ''Black Death'' of Europe which ravaged that continent from the middle of the 14th century to the end of the 16th century. Many historical accounts of the time describing this pandemic are still extent; the pandemic was apparently due to a particularly severe form of the disease. The very high mortality that occurred wherever the plague appeared was also doubtlessly associated, at least in part, with the poor nutritional state of the human population at the time.

Plague is essentially a disease of rodents which is passed from one rodent to another, both in commensal and sylvatic or wild rodent populations, by species of vector fleas which have fed upon infected rodents. When such infected fleas feed on man the disease can be transmitted to humans. The etiologic agent of plague is known as *Yersina pestis (=Pasteurella pestis).*

In man the initial response to infection through the bite of a flea is a local infection which may not be visible or even clinically evident. Within 1 to 5 days the infection spreads via the lymphatics draining the site of the infection to the lymph nodes either in the inguinal, axillary, or cervical areas. The involved nodes are swollen and are usually excruciatingly tender and may suppurate. Fever is very often present. The enlarged nodes are called buboes and this form of the disease is called *bubonic* plague. Milder infections may not present these symptoms. Untreated bubonic plague in its most virulent form may have a mortality of approximately 50% though this can now be reduced to negligible proportions through the timely use of chemotherapy.

When a generalized infection occurs with bacteriemia in the blood, *septicemic* plague has developed, including pulmonary infection, and this generally gives rise to a 100% mortality if the disease is not quickly recognized and given the appropriate treatment. Secondary pneumonic infection or *pneumonic* plague is estimated to occur in only 5% of such cases since death usually takes place before secondary respiratory spread can happen as a result of pulmonary invasion. When an individual with bubonic plague develops a secondary pneumonia and the infection is spread through the respiratory route, a person-to-person chain of infection may then result which is no longer dependent on an animal/flea contact; this may have the gravest of consequences from the public health viewpoint. Persons with such a form of plague must be isolated and all their contacts treated. Untreated pneumonic plague can cause a 95% mortality or higher. Nevertheless it may respond to treatment providing that the disease is recognized and given the correct chemotherapy with the greatest possible dispatch.

Even though plague as an epidemic disease has greatly declined since the early years of the present century, and even more so since the introduction of synthetic organic insecticides after World War II, it still remains endemic in a substantial number of foci and, in fact, is probably still present in most of the natural foci in which it previously existed. It persists in these natural foci within well defined geographical areas where the etiological agent, its vector, and its reservoir hosts form an ecological association allowing infection to circulate for indefinite periods of time. These natural foci may cover very large geographical areas, e.g., most of the western U.S., and may remain quiescent for varying and indefinite periods of time with the appearance of sporadic epizootics. Natural plague foci are completely independent of man and generally have a cyclical pattern of activity with epizootics revealing themselves by the presence of dead rodents in the endemic area and by occasional human cases. Such foci can also be the source of sizeable epidemics when an epizootic gives rise to a temporary plague area with many susceptible hosts and proficient flea vectors are seeded from the permanent focus.[238]

1. Distribution

Most natural foci of plague persist over extremely long periods of time. Table 1 records only those human cases of the disease which have been reported to the WHO. It is likely that a number of silent foci of rodent plague still exist in some countries where both rodent and human plague were once common. Many of these old or once active foci of plague are not adequately or regularly surveyed and the table represents the best available guide to the present distribution of active plague foci based on officially reported information, though the figures may be an under estimation of the actual situation.

2. The Vectors

When a flea feeds upon a plague-infected rodent it will ingest plague bacilli. In a suitable vector species, these multiply in the gut of the flea and may block the lumen; a flea in such a condition is called a "blocked flea". Another host animal may be easily infected by the bite of flea in this condition as the blocked flea will repeatedly attempt to feed to satisfy hunger and *Y. pestis* bacilli are likely to be regurgitated from the mass of the *Y. pestis* bacilli in the block and from the esophagus and proventriculus into the wound caused by the flea bite. While a substantial number of the approximately 1500 known species of fleas have been found positive for *Y. pestis* on some occasion, the most important vector species, particularly in urban and periurban areas and on *Rattus* species in most areas, is the oriental rat flea *Xenopsylla cheopis*. Due to the frequent development of blockage of the digestive system by the relatively rapid growth of *Y. pestis* bacilli, this species is a very efficient vector of plague at ambient temperatures up to 28°C. If the flea does not clear itself of the blockage despite repeated efforts to feed, it will eventually die of starvation. Among the reasons for the importance of *X. cheopis* as a vector is the readiness with which it will bite humans when starved or in the absence of a rodent host, its adaptation to commensal conditions, its global distribution, and the often very high indices on *Rattus* species. A number of closely related flea species are also efficient vectors of plague though, for the most part, their distribution is more limited. Among the other important vectors of plague is *X. braziliensis* which is found in Africa, South America, and India, usually on *Rattus* species and most frequently in rural areas and in villages; *X. astia* on *Rattus* species in South-east Asia, again primarily in rural areas; and *X. vexabilis* which is an ectoparasite of *Rattus* on Pacific Islands. *Pulex irritans,* the human flea, has a global distribution, is often found in very high densities, has a wide range of host species, and though not a good

Table 1
HUMAN PLAGUE: NUMBER OF CASES (AND DEATHS) REPORTED IN THE WORLD, 1977—1984[246]

Country	1977	1978	1979	1980	1981	1982	1983	1984
Africa								
Angola	—(—)	—(—)	—(—)	21(4)	6(—)	—(—)	—(—)	—(—)
Kenya	—(—)	166(9)	227(1)	5(2)	—(—)	—(—)	—(—)	—(—)
Lesotho	—(—)	—(—)	—(—)	—(—)	—(—)	—(—)	—(—)	—(—)
Libya	11(6)	—(—)	—(—)	—(—)	—(—)	—(—)	—(—)	8(—)
Madagascar	58(16)	25(6)	23(13)	11(5)	44(13)	38(19)	24(10)	39(18)
Mozambique	97(14)	12(—)	—(—)	—(—)	—(—)	—(—)	—(—)	—(—)
Namibia	—(—)	—(—)	—(—)	—(—)	—(—)	—(—)	—(—)	—(—)
South Africa	—(—)	—(—)	—(—)	—(—)	—(—)	19(1)	—(—)	—(—)
Tanzania	2(2)	—(—)	—(—)	44(9)	3(—)	36(6)	226(32)	132(—)
Uganda	—(—)	—(—)	—(—)	—(—)	—(—)	153(3)	—(—)	—(—)
Zaire	4(3)	—(—)	1(1)	—(—)	—(—)	1(—)	—(—)	—(—)
Zimbabwe	—(—)	—(—)	—(—)	—(—)	—(—)	3(2)	1(—)	—(—)
Total	172(41)	203(15)	251(15)	81(20)	53(13)	250(31)	251(42)	179(18)
Americas								
Bolivia	29(9)	68(2)	10(—)	26(2)	21(1)	1(—)	21(4)	12(2)
Brazil	1(—)	11(—)	—(—)	98(—)	59(—)	151(1)	82(—)	37(2)
Ecuador	—(—)	—(—)	—(—)	—(—)	8(—)	—(—)	65(—)	7(1)
Peru	—(—)	6(—)	—(—)	—(—)	27(7)	11(—)	17(2)	413(31)
USA	18(2)	12(2)	13(2)	18(5)	13(4)	41(6)	40(6)	31(6)
Total	48(11)	97(5)	23(2)	142(7)	128(12)	182(4)	225(12)	500(42)
Asia								
Burma	591(26)	171(6)	73(2)	73(4)	1(—)	165(1)	96(3)	—(—)
China	. . .	. . .	8(6)	30(20)	1(—)	—(—)	25(15)	—(—)
Kampuchea	. . .	. . .	. . .	. . .	—(—)	—(—)	—(—)	—(—)
Viet Nam	667(. . . .)	314 (8)	306(8)	180(5)	11(—)	116(—)	118(—)	229(6)
Total	1258(26)	485(14)	387(14)	283(29)	13(—)	281(1)	239(18)	229(6)
World Total	1478(78)	785(34)	661(33)	506(56)	194(25)	713(36)	715(72)	908(66)

transmitter of plague under laboratory conditions, is thought by a number of workers to occasionally transmit plague among human populations. The northern rat flea, *Nosopsyllus fasciatus* probably has a more important role in maintaining plague among rodent populations than as a vector of disease to man.

Among sylvatic or wild rodent populations, there is a great diversity of flea vectors varying both with the geographical area, ecologically, and with the different species of animal hosts. For a detailed review of this subject, the review by Brown[37] should be consulted.

3. The Reservoirs

There is much confusion in the literature resulting from attempts to separate the rodent species responsible for the maintenance of plague in the wild cycle or natural foci of the disease from those species and populations involved in epizootics and possible sources of plague to man. The reason for this is that many different species of rodents and other small mammals are susceptible to infection by *Y. pestis;* Pollitzer[182] listed more than 200 rodent and lagomorphs that were known to be implicated in the epidemiological cycle of plague in one geographical area or another.

Poland and Barnes[181] have classified the animal hosts of plague into enzootic (maintenance) hosts and epizootic (amplification) hosts. The first group includes rodents from genera that are relatively resistant to plague and consequently mortality from plague infection among these species is low even though antibody surveys will generally show a 100% positivity rate among the rodents trapped from such populations. For various reasons the organisms may overlap into a more susceptible small mammal population and an epizootic, sometimes of very considerable proportions, may occur. Such an epizootic takes place in those rodent species that are highly susceptible to *Y. pestis* infection and the epizootic will cause a high rate of mortality among their populations. The results of such a substantial mortality may cause a dangerous situation in which the number of surviving host animals available to the flea population is greatly reduced and the fleas concentrating on these survivors greatly enhance their chances of being infected and dying; this may result in large numbers of free-living fleas avidly seeking a blood meal from available alternative hosts and the risk of involvement of humans in the area becomes very great. Obviously, should such a situation occur in an urban area in the presence of a dense population of a plague susceptible rodent species, then the risk of human involvement or an epidemic becomes great.

It is difficult to place the many different species of rodents, lagomorphs, and other small mammals into groups which will fit the classification proposed above; the susceptibility of a given species to plague infection may vary within the geographical limits of the foci and may also vary temporally with variations in the density of the host populations, their flea ectoparasites, and possible changes in the virulence of the strain of plague involved. There are many foci where, despite their general susceptibility to the disease, plague appears to be maintained over long periods of time within *Rattus* populations, generally *R. rattus* or one of its subspecies or closely related species. Nevertheless, enough information is already available to identify the main reservoir species in most of the better known endemic foci of plague.

Plague Reservoirs in Southern Africa — The main natural reservoir in the existing foci in this area is the gerbil, *Tatera brantsi,*[227] from this species the infection often spreads to *Mastomys natalensis* and then on to *R. rattus.*[64] After the initial outbreaks of plague which occurred in Zimbabwe in 1974, an extensive survey was carried out[217] on the antibodies in the indigenous rodent species. The wild rodent species found positive were *Aethomys chrysophilus, A. namaquensis, Tatera leucogaster,* and *Pedetes capensis.* The authors concluded that *M. natalensis* linked the wild rodent populations with the human cases. In Lesotho, *T. brantsi* is regarded as the main reservoir of plague during outbreaks of the disease in the 1960s, while it appears that *T. brantsi* is the main reservoir of infection in Namibia as well.

Plague Reservoirs in East Africa — The ecology of plague in Madagascar, Kenya, and Tanzania has been extensively investigated though many areas within these large countries have not as yet been studied. Davis et al.[65] found high titers of HA antibody in *M. natalensis, R. rattus,* and *Acomys* sp. with lower titers in a number of other rodent species. Some time later another focus of plague was detected in the Tana area of Kenya where *Tatera robusta, Arvicanthis niloticus, Lemniscomys striatus,* and *Petrodromus tetrodactylus* all showed high titers to *Y. pestis.*[205] In Tanzania, *M. natalensis* is doubtless the main reservoir with *Arvicanthis abyssinicus* also playing some role. Unlike the situation on the mainland, it is thought that only *R. rattus* is the reservoir of plague in Madagascar.[41]

Plague Reservoirs in Central Africa — For a considerable period of time, two foci of plague have been known to exist in Zaire; one of these is to the west of Lake Edward and the other to the west of Lake Albert extending up to the Uganda border. Misonne[161] has published an extensive review of the status of these foci at the time and the conditions have probably not changed greatly since then. Only two species of commensal rodents are found in the foci, *R. rattus and M. natalensis,* both of which frequently reach very high population densities in infested villages. *A. abyssinicus* is commonly found in the immediate environs of the villages. The sylvatic species which are probably involved in the maintenance of plague in the foci were considered as *L. aquilus,* along with *Otomys tropicalis* and *Dasymys* spp., with *A. abyssinicus* serving in the intermediate role or bridge between the purely sylvatic species and the domestic rats.

Plague Reservoirs in North Africa and the Middle East — The only focus of plague in North Africa which has recently been reported to be still active is that in Libya. From surveys made in the Nofilia focus of that country, it was found that *Rattus* species are absent in the focus and that plague is probably transmitted directly from the wild rodent reservoirs to man by fleas. Misonne[162] thought that the probable reservoir and the rodent species found in the greatest density was the gerbil, *Gerbillus gerbillus.* However, later investigations failed to find any seropositive gerbils while seropositive *Meriones libycus* were found in the Tobruk area; it was then concluded that this species is the likely wild reservoir.[254]

Various unconfirmed reports have been received about cases of plague which have appeared in Egypt over the last decade but no information is available as to their number, distribution vector, or reservoirs.

A focus of plague is known to exist on the Yemen-Saudi Arabian border; other than the fact that gerbils, including *Meriones rex,* and *R. rattus* were found in the area where the most recent outbreak occurred, nothing else has been published on the probable reservoirs.

An extensive endemic focus exists in Iran where the wild reservoirs are gerbils, mainly *M. persicus, M. shawi,* and *T. indica,* with *M. tristrami* and *M. vinogradovi* at times being involved in the cycle of transmission.

Plague Reservoirs in Asia — A large natural focus of plague remains active in the Asian part of the U.S.S.R. In the preCaspian region, the main reservoir in the steppes is *Citrellus pygmaeus,* and in the sandy stretches *Meriones meridanus* and possibly *M. tamoriscinus* are also reservoirs. In the central Asian focus, the main reservoir in the desert lowland is *Rhombomys opimus* and *Meriones erythrourus* and in the high mountain areas of the focus, *Marmota baibacina* and *M. caudata.* In the transcaucasia area, the gerbils, including *M. libycus* among others, are the main reservoir while *Marmota siberica* and *Citellus dauricus* are involved in the Transbaikalian plague focus.[183]

Plague is endemic over a considerable area of Burma with cases being reported from isolated hamlets, villages, and cities. The species showing the highest rate of antibody to plague in Rangoon and vicinity among 1620 animals tested in 1976 was *Bandicota bengalensis* which is also the most common species of rodent in the city of Rangoon; its rate of positivity was 15.4% while *R. norvegicus* showed 11.1% positive, *R. exulans* 7.6%, and the insectivore *Suncus murinus* 3.35%.[36] Although there have been extensive studies of the ecology of plague in the cities and villages of Burma, little is known of the epidemiology or maintenance cycle in the sylvatic areas.

The most important reservoirs of plague in Vietnam are *R. norvegicus, R. rattus,* and *Suncus murinus.*[231] *Bandicota indica* has been found seropositive to plague with increasing frequency and, taking into consideration its relative resistance to plague, it is likely that this species has an urban or periurban role in the plague cycle in the country. A forest rat species, *R. nitidus* has also been found infected along with *Dremomys rufigensis, Mus castaneus,* and *R. exulans.*[230]

A plague focus existed and was, at one time, active in the Boyolali area of central Java, Indonesia. No evidence of any activity, either serological or bacteriological, has been found in this focus since 1974. The rodent species from which *Y. pestis* was detected prior to 1974 were *Rattus rattus diardii* and *R. exulans ephippium.*

Although it is considered a source of at least one of the plague pandemics, only relatively few cases of plague are reported from China today. Little is known of the current epidemiology of the disease in the country but the Siberian marmot, *Marmota bobak,* was considered an important wild reservoir in the past.

Plague Reservoirs in North America — In the U.S., wild rodent plague is endemic throughout much of the western states of the country including Washington, Oregon, and California on the Pacific coast and east to Montana, Wyoming, Colorado, New Mexico, and Texas. The majority of human cases also have occurred in the Rocky Mountain region. From 1983 to 1984, 71 human cases occurred in the U.S., 42 of them in New Mexico, 12 in Arizona, 7 in California, 4 in Colorado, 3 in Utah, and 1 each in Orgeon, Texas, and Washington. The 40 cases which occurred in 1983 were the highest annual number since 1920. During the 1970 to 1980 period, evidence of plague infection was found in 75 species of 4 mammalian orders, Rodentia, Lagomorpha, Insectivora, and Artiodactyla, aside from man. The animal species associated epidemiologically with plague in 1974 to 1980 and the percentage of individuals found positive were as follows: Sciurid rodents-*Spermophilus variegatus,* 41.9%; *S. beecheyi,* 10.5%; *Cynomys gunnisoni,* 3.8%; *Etutamias* spp., 0.9%; unidentified sciurids, 2.9%; Cricetid rodents-*Neotoma cinerea,* 1.9%; unidentified rodents, 10.5%; Lagomorphs, 6.5%; carnivores-*Canis latrans,* 0.9%; and *Felis catus,* 3.8%.[20] Plague epizootics are sporadic and unpredictable within the enzootic area.

While sylvatic plague is reported from Canada, no human cases have occurred. Plague has been found in *Citellus richardsoni* in Alberta and Saskatchewan. *Y. pestis* infected fleas have been taken from *Marmota flaviventris* ssp.[173]

No human cases of plague have occurred in Mexico since 1963 though prairie dogs, *Cynomys mexicanus* were found naturally infected with plague in northern Mexico in 1969.

Plague Reservoirs in South America — Two widely separated foci of plague occur in Bolivia, one in the north-west of the country around La Paz and the other in the south central-part. *Y. pestis* infections have been found in *R. rattus* and, on occasion in *M. musculus,* though this was probably a chance infection. Wild rodents and lagomorphs found naturally infected in Bolivia are as follows: *Dasyprocta variegata boliviae, Galea musteloides, Graomys griseoflaus, Hesperomys fecundus, H. venustus, Oryzomys flavescens, Oxymycterus paramensis, Phyllotis wolffsohni, Rhipidomys leucodatylus,* and *Sylvilagus braziliensis gibsoni.*[149] The relative importance of each of these species from the epizootic and human epidemic viewpoint is not entirely clear and requires further investigation.

Plague is endemic in northeastern Brazil where the number of cases reached a peak of 496 human cases in 1975; the figures are certainly not accurate as many patients received antibiotics before any bacteriological confirmation could be carried out, and furthermore, facilities for antibody tests were not available.[4] Numerous surveys have been carried out in Brazil on the distribution and relative population density of both urban and wild rodent species, but very few bacteriological and/or serological studies have been carried out either in the main focus of the north-east or in other rural or urban areas of the country where sporadic cases have appeared, or have been recognized. De la Borrea[173] believes that *R. rattus* is the most important of the commensal species involved in the epidemiology of plague in the country, while among the wild rodents *Zygodontomys pixuna, Oryzomys subflavus, Galea spixii,* and *Cercomys cunicularius laurentius* are the most commonly involved species to the extent that this is known from the few microbiological studies that have been done on the native rodent fauna. From the work that De la Borrea and his colleagues[4] carried out De la Borrea confirmed that *Z. pixuma* was the most important species from both bacteriological and serological findings. *Y. pestis* was also isolated from *O. subflavus, R. rattus,* and *O. eliurus.*

Aside from an outbreak in 1983, the number of cases of plague in Ecuador has generally been declining in recent years. *Rattus* species have little role in the epidemiology of plague in the country. On one occasion, 12 plague-infected *Mus musculus* were found in a house in a village but were, in all probability, themselves dead-end victims of an epizootic of wild-rodent plague that had reached the village. Plague-infected *Akodon mollis* and *Oryzomys xanthaeolus,* which adapt themselves to human habitations, have been found inside houses, as have *O. longicaudatus, O. flavescens,* and *Phyllotis frusticicolus.* Both *Sigmodon peruanus* and *S. puna* have been found naturally infected with plague in Ecuador and the squirrel, *Sciurus stramineus nebouxi,* which is comparatively resistant to plague, is considered the source of acute epizootics that occur in the highly plague susceptible *A. mollis* and *O. xanthaeolus* populations.

There was a large outbreak of plague in Peru in 1984 which affected the Departments of Cajamarca and Piura and seems to have originated from an intensive epizootic of wild rodent plague. Earlier outbreaks of plague in Peru were associated with *R. rattus* as well as with guinea pigs infected with plague which had been kept in houses.[187] Macchiavello[148] observed that in the Department of La Libertad, epizootics of plague in populations of *S. peruanus* were followed by epizootics in the closely associated populations of *R. rattus;* this occurred especially during the warm season when *X. cheopis* infestations increased and was often followed by human cases of the disease. In the Department of La Piura, *Sciurus stramineus nebouxi* is the species of wild rodent most frequently found infected with plague, while *O. xanthaeolus xanthaeolus* has also been found infected on a few occasions.

De la Barrera[173] concluded that in all the plague affected areas of Peru, *A. mollis* and *O. xanthaeolus* are the rodents most frequently found as well as the species most frequently found infected with plague. He was of the opinion that *S. peruanus* was less important than the previous two species. It is also apparent that, in Peru, the epidemiology of plague transmission has many aspects that remain to be clarified, particularly those related to the wild rodent reservoirs.

C. Yersiniosis

An acute enteric disease in man which is manifested by diarrhea of varying degrees of severity, entercolitis, acute mesenteric lymphadenitis occasionally mimicking appendicitis, low grade fever, headache, pharyngitis, anorexia, vomiting, erythema nodosum (in adults, especially in women), arthritis, cutaneous ulceration, abscesses, and, rarely, septicemia. The causative agents may be *Yersina enterocolitica* and *Y. pseudotuberculosis. Y. enterocolitica* infections more commonly present with gastro-enterocolitis syndrome and *Y. pseudotuberculosis* with abdominal pain.

Fecal-oral transmission takes place by contact with infected persons or animals or by eating or drinking fecal contaminated food or water containing one of the infectious agents. The principal reservoir for these two species of *Yersina* are animals. Some theories have been put forward that the original source of at least *Y. pseudotuberculosis* is through mutation from the infectious agent of medieval plague.

Most records of the occurrence of *Y. pseudotuberculosis* have been from northern and central Europe but human cases have also been reported from the far eastern regions of the U.S.S.R., Japan, Australia, and New Zealand. Cases in the Mediterranean littoral are rare and only one human case has been reported from Italy and one from Greece. With the exception of Algeria and Morocco, the African continent is free from *Y. pseudotuberculosis* and it has never been reported from India. Little more than 15 cases have been reported from the U.S. *Y. enterocolitica* has a much wider geographical distribution and is found in Europe including France, Belgium, Sweden, Norway, and Finland and more than 5000 cases of the disease have been reported from that continent. It has also been reported from the Americas, the Congo, and South Africa.

1. The Reservoirs

Y. pseudotuberculosis has been recovered from a wide variety of animals and birds including farm mammals, household pets, and wild animals; infections are extremely widespread in domestic and wild birds and these along with rodents are the principal reservoirs of the infectious agent. In the U.S., *Y. pseudotuberculosis* has been reported from 19 mammal species including 6 species of rodents, 5 ruminants, 4 lagomorphs, 2 carnivores, a marsupial, a swine, and 5 avian species, as well as from man.[108] In California it has been recovered from Norway rats.[100] In a survey carried out in Sapporo, Japan in 1976, *Y. pseudotuberculosis* was isolated from 8 of 259 *R. norvegicus* trapped in barns, slaughter houses, and a zoo. No isolations were made from 11 *R. rattus* taken from the same area.[117]

Y. enterocolitica has been isolated from *Microtus arvalis* and *Apodemus sylvaticus* in Czechoslovakia;[3] in Norway, Sweden, and Finland 24 strains of this species were isolated from 551 small rodents over a 1 year period including from *Clethrionomys glaareolus, C. rufocanus, Apodemus sylvaticus, A. flavicollis, Microtus oeconomus,* and *M. agrestis*. An extensive listing of all reports of epidemiological records in man and animals is given in a comprehensive review on pseudotuburcular yersiniosis.[212]

D. Tularemia (Rabbit Fever)

The infectious agent causing the disease tularemia is *Francisella tularensis* (=*Pasteurella tularensis*). There are two types of the organism, i.e., Jellison type A and Jellison type B, which may be differentiated epidemiologically. Jellison type A organisms are common in rabbits, are restricted in distribution to North America, are frequently transmitted by ticks, and are more virulent than Jellison type B. The latter type originates from sources other than rabbits in North America and this type, along with other strains from elsewhere in the northern hemisphere, are less virulent than type A.

In man, the clinical manifestations of the disease are related to the route of infection and the virulence of the strain. The disease is characterized by chills and fever and generalized prostration at the onset. A localized ulcer frequently appears at the site of the original infection. Lymph nodes draining the area of the ulcer become swollen, tender, and suppurating, sometimes leading to a clinical misdiagnosis of plague in those geographical areas where both the diseases are found. Fatality in untreated cases is about 5% but the infection quickly responds to treatment by a number of antibiotics. The disease has a complex epidemiology which may involve a variety of wild rodents and lagomorphs; it may be transmitted to man by ticks, several species of diptera, water contaminated by the urine of infected animals, or even dust or grain contaminated with mouse feces or by the handling of infected rodents.

The distribution of tularemia is strictly holartic. It is found throughout North America, many parts of continental Europe, the U.S.S.R., China, and Japan. It has also been reported in animals in Iran.[10] The number of cases in the U.S. has averaged about 260 a year for the period 1980 through 1984.

1. The Vector

Transmission of tularemia to man can take place either through the bite of an infected tick or biting fly or through the medium of contaminated water, direct contact with an infected animal or consuming poorly cooked meat from an infected animal. Transmission by Ixodid ticks occurs both from animal to animal and, by bite of the tick to man though, as has been observed above, this is not necessarily the only mode of transmission. While transovarial transmission has been stated to occur in ticks, this has not been clearly demonstrated in all species but does certainly occur in *Amblyomma americanum*. Type A epizootics are those in which *F. tularensis* is transmitted to sheep by *Dermacentor andersoni*, to cottontail rabbits (*Sylvilagus* spp.) by *Haemaphysalid leporispalustris, A. americanum*, or *D. variabilis*, and to man either by tick bite or contact with infected animals.[26] Tick-borne infection is of minimal direct importance to man in Japan, but infected *Ixodes* and *Haemaphysalis* species are found on hares in that country. Tularemia infections of man through tick bites are considered unusual in the U.S.S.R. In the U.S., the deerfly, *Chrysops discalis*, is an important vector of the disease to man while tabanid species are vectors in the U.S.S.R.

2. The Reservoir

Bell[26] has presented an extensive listing of the mammals known to be infected with *F. tularensis* with comments on the role of each group in the epidemiology of tularemia and this is reproduced in Table 2.

One is struck by the number of different species of animals which have been found infected, but clearly not all of these serve as reservoirs of the disease. Hopla[107] emphasized that to isolate tularemia organisms from an animal does not prove it to be a reservoir and to be a true reservoir the animal must be shown to be involved in the maintenance of the infectious agent, to be highly susceptible to infection and capable of sustaining a long-term bacteriemia, and low sensitivity (unapparent infection). He therefore proposed that the tick and the principal vertebrate host upon which it feeds forms a more valid concept of a reservoir for tularemia than does the tick itself. This concept is readily applicable to the combination of *Dermacentor parumapertus* and *Lepus californicus* as well as, to a lesser extent, *D. andersoni* and rodent populations. However, in addition to the cycle described above, cricetid rodents also have a role in the transmission of tularemia through the shedding of *F. tularensis* into water which is later consumed by man and other animals. Hopla believes that the evolution of *Francisella* has been mainly in association with rodents.

E. Salmonellosis

A very large number of serotypes of the genus *Salmonella* are pathogenic for both man and animals. The prevalence of the different serotypes which have been isolated varies greatly from one country to another. In all more than 2000 different serotypes are known of which *Salmonella typhimurium* is the most commonly reported. Salmonellosis, the disease resulting from infection from one of the serotypes, has an equally wide range of clinical features; it is generally characterized by acute gastroenteritis, abdominal pain, diarrhea, nausea, and vomiting. Dehydration may be severe and fever is nearly always present. Death generally occurs only among the very young, the very old, or persons in a generally debilitated state. Only a small proportion of cases are recognized clinically and fewer are serotyped. There are probably upwards of two million cases of salmonellosis a year in the U.S. and the number of cases which occur every year on a global basis must be a substantial multiple of this figure.

Transmission to man is usually through the ingestion of the organism in food or water which has been contaminated by feces from an infected person or animal, or it may result from the consumption of infected commercially produced meat products such as inadequately

Table 2
MAMMALS KNOWN TO BE INFECTED WITH *FRANCISELLA TULARENSIS*

Family	Common names	Comments
Soricidae	Shrews	Several species naturally infected in Old and New World
Talpidae	Moles	Occasionally found infected
Callithricidae	Marmosets	One record in a domiciled squirrel monkey
Hominidae	Humans	About 1% mortality in Old World (*Francisella tularensis* B) and about 5% in New World (*Francisella tularensis* A and B) before antibiotics
Leporidae	Rabbits, hares	*Lepus* spp. commonly infected in Old and New World and *Sylvilagus* spp. in New World; susceptibility variable among genera; *Oryctolagus cuniculi* native of Europe, not important in epidemiology in Old World; domestic form rather resistant to *Francisella tularensis* B
Erethizontidae	New World porcupines	Isolation of *Francisella tularensis* reported
Didelphidae	Opossums	Seropositive and a rare source of human infection
Canidae	Dogs, wolves, coyotes, foxes	Occasional evidence of infection; some with mortality (*Urocyon* sp.), but adult animals in good condition usually suffer little from experimental infection
Mustelidae	Weasels, polecats, mink, martens, badgers	Sparse evidence of infection in wild, but epizootics have occurred on fur farms (mink)
Felidae	Cats, lynx	Irregularly susceptible; fatal infection recorded after large oral doses
Suidae	Pigs	Not fatally susceptible in limited trials but a source of human infection
Procyonidae	Raccoons	Seropositive and a rare source of human infection
Sciuridae	Squirrels, ground squirrels, prairie dogs, chipmunks, marmots, susliks	Ground squirrels (*Citellus[Otospermophilus]beecheyi)* were the first source of cultures by McCoy and Chapin; not prominently reported since the studies of McCoy and Chapin[157]
Heteromyidae	Pocket mice, kangaroo rats	Occasional infection
Castoridae	Beavers	Epizootics in wild and on fur farms
Cricetidae	New World rats and mice, hamsters, voles, lemmings, gerbils	Various species probably the most commonly infected of all animals in Old and New World, especially *Microtus* spp. and *Arvicola* sp.
Dipodidae	Jerboas	Old World
Muridae	Old World rats and mice	*Mus musculus* very susceptible; suffers epizootics; *Rattus norvegicus* and *R. rattus* relatively resistant
Zapoidae	Jumping mouse	Isolation
Cervidae	Deer	White tail and mule deer (*Odocoileus* spp.) occasionally source of human infection when animals are dressed
Antilocapridae	Pronghorn antelope	Serologic evidence of infection
Bovidae	Cattle, buffalo, goats, sheep	Evidence (serology) of nonlethal infection is common in all species, especially in sheep; high mortality has been documented in sheep on range, but folded, well-fed sheep are resistant

Table 2 (continued)
MAMMALS KNOWN TO BE INFECTED WITH *FRANCISELLA TULARENSIS*

Family	Common names	Comments
Equidae	Horses	Foals heavily infested with ticks have succumbed to tularemia
Camelidae	Camels	Culture isolated

(From Stoenner, H., Kaplan, W., and Torten, M., *Handbook Series in Zoonoses, A, Bacterial, Rickettsial, and Mycotic Diseases*, Vol. II, CRC Press, Boca Raton, Fla., 1980, 176.)

cooked poultry, raw or lightly cooked eggs, unpasteurized milk, or other dairy products. In general, the epidemiology of infection represents a complex cycle between man, reservoir mammals, reptiles and birds, as well as feces contaminated food and water. Even some arthropods, especially house flies and cockroaches, are considered to be vectors through mechanical transmission of the *Salmonella*.

Although, as noted above, there is considerable variation in the prevalence of the different serotypes from one geographical area to another, the distribution of most of the serotypes is global. Diagnosis, and consequently reporting, is usually more accurate in Europe and North America where most of the outbreaks of the disease are usually of food-borne transmission though they may arise from a number of different sources including animal ones. The real incidence of infection with any of the serotypes is probably everywhere unknown; in the U.S., Steele[208] estimated that the 20,000 cases which were reported in that country in 1968 represented no more than 1 to 2% of the true incidence of the disease. In most tropical countries, the disease is neither reportable nor, when recognized, do facilities necessarily exist to determine the particular serotype responsible of any given outbreak or individual human or animal case. The true public health importance of this infection can be estimated from the results of the relatively small number of specific surveys that have been carried out with support of good laboratory facilities. As an example, Anderson et al.[7] carried out bacteriological and serological examinations of 741 patients who had been admitted to a hospital in Jakarta, Indonesia because of febrile illness. Of these, evidence of *Salmonella* infection was found in 317 or 43% of the patients.

1. Rodent Reservoirs of Salmonellosis

In veiw of the plethora of animal species that may serve as reservoirs of the infectious agents, rodents cannot be considered as the main reservoir. There is, nevertheless, ample evidence to show that their importance is considerble both in transmitting the disease organism to man by contamination of man's foodstuffs or water, or by transmitting the infection to other animal species who in turn may transmit it to man. It is of particular note that most of the common serotypes which are found in rodents are also those which are the most common in the environment and are those which most frequently give rise to outbreaks of salmonellosis of man in any given area.

In fact, in Europe *Salmonella* strains were used as a method of controlling rodent populations until comparatively recently in some countries. Loeffler[145] developed a mouse-typhoid culture (later known as *S. typhimurium*), that he used against an eruption of voles in wheat fields in Thessaly, Greece with some measure of success. Danysz (in Wodzicki)[247] examined voles which were found dying during an epizootic in France and from them cultured a microorganism (later named *S. enteritidis* var. *danysz*), which he determined also killed rats in his laboratory. He later used this strain, first in the control of voles,[63] and later in the field against commensal rats, i.e., *Rattus rattus* and *R. norvegicus*. Bahr[17] recommended the combined use of a bacterial preparation which he had earlier developed and named

''Ratin'' *(S. eteritidis)* and ''Ratinin'' (=red squill). He claimed that Ratin was deadly to rats but not to man. While the use of various *Salmonella* cultures became quite popular in several countries in Europe, it was soon suspected that the bacteria being used in these preparations were not as species specific as had originally been thought and claimed. In 1942, Leslie[139] examined the six principal ''viruses'' (as they were then termed) which were commercially available and in use for the control of rodents in England at that time, i.e., ''Liverpool'', ''Danysz'', ''London'', ''Ready Rat Relief'', ''Institute Pasteur'', and ''Ratin''. All of these cultures were found to be identical with *S. enteritidis,* both of the *danysz* subgroup and of the classic *S. enteritidis* type. The author noted that both of these strains are pathogenic to man, domestic animals, and some poultry and reviewed the considerable evidence that had accumulated by that time from Denmark, Germany, the U.S., and the U.K. which suggested that these cultures of *Salmonella,* when used for rodent control, had been responsible both for small outbreaks of salmonellosis and many individual human cases, including some fatal ones. Taylor[216] pointed out that in the U.K., between 1944 and June, 1953, 1267 cases of salmonellosis due to *S. enteritidis* var. *jena,* 21 of them fatal, had been reported and that, in addition, there were another 413 cases reported which were due to *S. enteritidis* var. *danysz,* of which 2 were fatal. Most of these cases were associated with the use of rat baits using preparations containing one or two of the above serotypes and the author advised that the use and sale of all of these be stopped. As a result of such reports, the use of *Salmonella* containing preparations has virtually ceased. The third report of the WHO/FAO Expert Committee on Zoonoses[237] concluded:

> '' . . . that salmonellas should under no circumstances be used as rodenticides. Rodents rapidly develop resistance to *Salmonella* serotypes; thus, this method has little practical value. Moreover, it has been shown in different countries that such practices are a public health hazard because the serotypes used are dangerous to man.''

A few examples of recent investigations on the prevalence of different *Salmonella* serotypes in rodent populations are of interest. A survey carried out in the Netherlands[99] in the early 1960s found that out of 735 *R. norvegicus* trapped in slaughter houses, farms, and mink farms, salmonellae were isolated from 73 (30.8%) of the 237 rats caught in slaughter houses, 17 (4.0%) of 429 rats trapped on randomly selected farms, and 12 (17.4%) of 69 rats caught on mink farms. *S. typhimurium* was found in 31.8% of the positive rats. On the island of Luzon in the Philippines, 11.8% of the rats (species and serotypes not listed) were found with salmonellae in their gastro-intestinal tracts.[8]

On the basis of 3401 strains isolated from animals or with an animal related source, Gerichter[87] reported on the animal sources of *Salmonella* in Israel. Of the strains isolated, 114 strains of 13 serotypes were isolated from rats; *S. typhimurium* and *S. enteritidis* were the most common, 47.4% and 15.8%, respectively; but, in addition, *S. sofia, S. blackley, S. emek,* and *S. newport* were also frequently detected.

Between July, 1973 and January, 1974, 300 rodents and shrews comprised of 149 *R. norvegicus,* 94 *R. exulans,* and 57 *S. murinus* were trapped in Saigon and another settled town in Vietnam; 31 strains of *Salmonella* were isolated, i.e., from 10.3% of the total. *S. weltevreden* was the most common (14/31) of the 10 serologic types that were encountered in all 3 species of mammals trapped.[170]

Similar surveys have been carried out in many countries with results more or less consistent with those presented above. The variation from one place to another depends upon the *Salmonella* serotypes present in the environment, the rodent species, their densities, and their access to sources of *Salmonella.* Most of the surveys show that one serotype or another of *Salmonella* is present in commensal rodent populations where the survey was carried out and further highlight the role of rodents in the transmission of the infectious agent of the disease.

V. THE PARASITIC ZOONOSES

This group of parasitic infections are particularly prevalent in the human populations of tropical and subtropical countries where they give rise to a very high level of morbidity and, in the cases of certain members of this group, a considerable mortality as well. As human populations expand in rural areas and especially into lands not previously farmed, they come into increasing contact and involvement with the natural foci of many infections which have wild rodents as their reservoirs. Commensal rodents in urban areas may, in some cases, also serve as suitable reservoirs for these infectious agents and the diseases are readily transferred to the cities. Where sanitation is poor and rodent populations dense, foci of some of the parasitic zoonoses have indeed been established and are increasing in incidence in urban areas.

As with the reviews of viral, rickettsial, and bacterial diseases, the following section will consider only those human infections where there is presumptive evidence that rodents serve to one extent or another as important reservoirs of the infection and will not touch upon reports in the literature of what must be considered as no more than an unusual finding of a given parasitic disease in a rodent host.

A. Protozoal Infections

1. Toxoplasmosis

The causative agent of toxoplasmosis is the intracellular coccidian protozoan, *Toxoplasma gondii*. The domestic cat is the definitive host though occasionally wild cats may also serve as a definitive host; a very wide variety of birds and mammals have been found infected. It is only in the cat, however, that the sexual stages of the parasite occur. After the consumption of the infective forms of *Toxoplasma* in the flesh of prey animals, the infective forms invade the intestine and grow into trophozoites; the latter multiply by endodyogeny and later by scizogony resulting in the production of merozoites which are released into the lumen of the host cell and in turn invade other epithelial cells. Several such generations of propagation are followed by the development of microgametocytes and macrogametocytes. After fertilization, zygotes are formed and shed into the external environment with the feces where, after a period of incubation, sporoblasts are formed each containing two sporocysts with in turn four sporozoites. The ingestion of the sporulated oocysts by a new host is followed by the appearance of the asexual stages of the parasite which invades and proliferates in many types of tissues. *T. gondii* will invade all orders of mammals and a wide variety of species of birds, reptiles, and amphibia.[111]

The disease in man is generally subclinical. Symptomatic toxoplasmosis can be congenital or acquired postnatally. Intrauterine infection is the most serious. Infection of the fetus usually takes place when the mother acquires a primary infection in the second trimester of pregnancy. Infection of the mother in the first trimester of pregnancy causes few cases of fetal infection but when they do occur, they are extremely serious. Congenital infection normally occurs only when a woman has a primary exposure during pregnancy. Clinical manifestations in an infected mother are rare.

Postnatal transmission of *Toxoplasmosa* to man, i.e., acquired infection, is almost exclusively by the oral route through the ingestion of raw or poorly cooked meat containing cysts or trophozoites, or of meat which is contaminated by oocysts excreted by a cat.

The infectious agent is world-wide in distribution in animals, birds, and man; and it has been estimated that one third of the human population of the world possesses antibodies for the parasite. In the U.S. alone, it is estimated that some 3000 infants are born each year with congenital toxoplasmosis.[1] The infection is absent from those few parts of the world, such as a small number of islands in the Pacific where there are no cats. In the U.S. the prevalence of *T. gondii* in domestic cats ranges from 15 to 64% in different surveys.[49]

a. Rodent Reservoirs

As has already been noted, domestic cats are the definitive hosts of the parasite although occasionally wild cats of the genus *Felis* and *Lynx* may also serve as a definitive hosts. Cats may become infected by eating infected raw meat, birds, mice, and other small rodents including rats. In surveys in Tennessee, *T. gondii* antibodies were found in 20% of the *R. norvegicus* tested.[70] In a serum survey in Florida, 13% of the *R. rattus* and 2 out of 8 *R. norvegicus* examined were positive for antibodies to the parasite along with 5% of the *Sigmodon hispidus,* 5% of the rice rats, *Oryzomys palustris,* 1% of the cotton mice, *Peromyscus gossypinus,* and 1% of the gray squirrel, *Sciurus carolinensis;* the highest prevalence was 19% in the armadillo, *Dasypus novemcinctus.*[50] In California, *M. musculus, Peromyscus maniculatus,* and the ground squirrel *Otospermophilus beecheyi* have all been found positive with antibody levels of 3 to 4%. A survey was carried out in Costa Rica[193] to determine the infection rates in potential intermediate hosts of *T. gondii* which were known to be prey for domestic cats, the most important of which were mice, rats, and sparrows, all of whom are common in and around houses as are cats. From among 202 *M. musculus, T. gondii* was isolated from 3.5%, from 12.5% of 120 *R. norvegicus,* and 16% of the sparrow, *Zonotrichia capensis.* The authors emphasized that transmission from cat to cat is not as efficiently accomplished with oocysts as transmission via cysts in the tissue of the intermediate hosts which highlights the role and importance of rodents as intermediate hosts of this infectious agent though generally speaking, several infected prey species are available to cats as intermediate hosts. The overall antibody rates to *T. gondii* in the human populations of a number of different areas of Costa Rica was 61.4%.[79]

2. Leishmaniasis

The disease known as leishmaniasis has many local names, among them ''Oriental sore'', ''Aleppo'', and ''Baghdad boil'' in the Old World; and ''chiclero ulcer'', ''espundia'', ''pian-bois'', and ''uta'' in the New World. The causative agent is a flagellate protozoan of the genus *Leishmania* which occurs in vertebrates including man in the aflagellate form or amastigote form, and in the vector insects, i.e., phlebotomine sandflies, in the flagellate or promastigote form. Speciation of the leishmaniae has been based on the type of clinical manifestations caused by a given form, but this distinction is not absolute; visceral forms may produce cutaneous lesions and cutaneous forms may visceralize. Recent developments in the typing of strains by isoenzyme and DNA buoyant density techniques have greatly assisted in gaining an understanding of the epidemiology of the different species of leishmaniasis in many different geographical areas. The majority of leishmaniases are zoonoses involving wild or domestic animals (rodents, marsupials, edentates, and canines) as the reservoir hosts. As will be seen below, among the most important of these reservoir hosts are the rodents. Despite the clinical uncertainties referred to above, the following consideration of the disease will, nevertheless, follow the various broad clinical forms commonly used.

a. Visceral leishmaniasis (VL)

Visceral leishmaniasis is caused by *Leishmania donovani* and its subspecies and may be endemic, sporadic, or epidemic. The clinical features of the disease tend to differ between the three situations.

Endemic visceral leishmaniasis — This form of the disease especially affects children. In the Mediterranean area, South-east Asia, China, and Latin America, the most common age group affected is that of 1 to 4 years. In East Africa and India the peak age of infection is 5 to 9 years, though the disease may occur among teenagers as well. Males are affected at twice the rate of females. The common symptoms are fever, malaise, weight loss, anorexia, and discomfort in the left hypochondrium. The common clinical forms are marked nontender splenomegaly, moderate hepatomegaly and lymphadenopathy, and wasting and pallor of the mucous membranes. Signs of malnutrition including edema and skin and hair changes develop. Intercurrent pneumonia, dysentery, and pulmonary tuberculosis are common and may be the cause of death.

Sporadic visceral leishmaniasis — Nonindigenous people of any age, entering an endemic area may contract VL. The onset of fever is usually abrupt, beginning 2 to 3 weeks after exposure. The disease may progress acutely with chills, high undulating fever, often with two daily peaks, drenching sweats, rapid weight-loss, and profound malaise. Such patients are more likely to develop the rare complications of severe acute haemolytic anemia, acute renal damage, and severe mucosal haemorrhage.

Epidemic visceral leishmaniasis — All age groups are susceptible except those persons old enough to have been infected in a previous epidemic. Acute forms are rare. Subclinical cases are thought to be very common and indeed outnumber the frank clinical cases.

Post-kala azar dermal leishmaniasis — Caused by *L. donovani,* this form is mostly restricted to the Indian subcontinent and may occasionally occur in East Africa. It commences one to several years after the apparent cure of visceral leishmaniasis. The chronic lesions consist of multiple nodular infiltrations of the skin, usually without ulceration. Macules on any part of the body may later become nodular, especially on the face.

b. Cutaneous Leishmaniasis (CL)

The clinical features of cutaneous leishmaniasis (CL) tend to differ between and within regions and reflect the different species of parasite or the type of zoonotic cycle involved. Usually a lesion starts as a nodule at the site of the inoculation; a crust develops centrally which may fall away exposing an ulcer which heals gradually leaving a depressed scar with an altered pigment. Nodules around the edges of the lesion and typical. There are a wide variety of clinical forms which will be briefly listed below:

Cutaneous leishmaniasis of the Old World — This form is commonly caused by three species of *Leishmania, L. tropica, L. major, and L. aethiopica.* Cutaneous leisions caused by *L. donovani* have been reported from Africa and from the Mediterranean basin.

Anthroponotic or urban cutaneous leishmaniasis — This form of the disease is caused by *L. tropica* and produces a painless ulceration of the skin which often leads to disfiguring scars. These dry ulcers usually heal spontaneously in a year or sometimes longer.

Leishmania recidivans — This lupoid or tuberculoid form of leishmaniasis may last many years and does not easily respond to treatment. The slowly progressing lesion, which usually appears on the face, is characterized by a scar with peripheral activity. If untreated, the disease is destructive and disfiguring.

Zoonotic or rural cutaneous leishmaniasis — The causative agent of this form is *L. major;* it produces painless lesions that are often severely inflamed and ulcerated and heal in 2 to 8 months. Frequently the lesions are multiple, becoming confluent and secondarily infected, especially when they occur in nonimmune immigrants to a disease endemic area. Such lesions are slow to heal and may leave large, disfiguring scars.

Cutaneous leishmaniasis due to *L. aethiopica* — This form of the disease may give rise to three types of lesions, oriental sore, mucocutaneous leishmaniasis, and diffuse cutaneous leishmaniasis; ulceration is late or absent and healing may take 1 to 3 years or even longer.

Cutaneous leishmaniasis of the New World — This condition may be caused by numerous species and subspecies of *Leishmania.* The clinical features are similar to those of Old World cutaneous leishmaniasis but the lesions tend to be more severe and chronic. Each of the species giving rise to this condition is listed below:

1. *L. braziliensis braziliensis:* In the primary cutaneous stage, this form produces single or multiple ulcers which seldom heal spontaneously and may become very large.

Lymphatic involvement is very common in his form of the disease. Up to 80% of the untreated primary lesions are followed by the development of mucocutaneous leishmaniasis.

2. *L. braziliensis guyanensis:* This subspecies is the cause of "pian bois", frequently producing painless, single, dry lesions that are often tumorous and tend to persist. Metastatic spread along the lymphatic system and ulcers all over the body are not uncommon.
3. *L. braziliensis panamensis:* This subspecies causes lesions that are normally ulcers, which do not heal spontaneously; lymphatic involvement is usually a feature of the disease.
4. *L. mexicana mexicana:* This is the cause of "chiclero ulcer" or "bay sore". The painless lesion often heals spontaneously within a few months, but sometimes chronic progressive lesions develop and destruction of part of the ear is not uncommon. Most lesions are single and 60% are located on the ear.
5. *L. mexicana amazonensis:* This subspecies causes single or multiple skin lesions, which seldom heal spontaneously. Though this infection is common among forest rodents, the disease is rare in man. About 30% of patients with lesions attributed to this parasite have diffuse cutaneous leishmaniasis.
6. *L. mexicana venezuelensis:* This subspecies causes indolent nodular lesions, which heal spontaneously in 6 months.
7. *L. mexicana garnhami:* This subspecies causes single or multiple lesions which usually heal spontaneously in 6 months.
8. *L. peruviana:* This is the cause of "uta", which usually mainly affects children. The single or few painless lesions usually heal spontaneously in about 4 months.

Diffuse cutaneous leishmaniasis (DCL) — This condition is caused by various species and subspecies of *Leishmania* and is characterized by widely disseminated thickening of the skin in plaques, papules, or multiple nodules, especially on the face and exterior surfaces of the limbs, sometimes resembling lepromatous leprosy. There is no real ulceration or mucosal involvement. The disease does not heal spontaneously, and there tends to be a relapse after treatment.

Leishmania mexicana pifanoi — This is known only from cases of diffuse cutaneous leishmaniasis in Venezuela. In the focus of cutaneous leishmaniasis in the Dominican Republic, all patients have DCL, but the parasite responsible has not yet been identified. *L. m. amazonensis* is the parasite most commonly associated with DCL in the New World. In the Old World, DCL, caused by *L. aethiopica,* has been reported in Ethiopia and Kenya, and DCL due to an unidentified parasite has been reported in Namibia and the United Republic of Tanzania.

Mucocutaneous leishmaniasis (MCL) — Mucocutaneous leishmaniasis of "espundia" is caused by *L. braziliensis* and possibly by *L. b. panamensis* and *L. b. guyanensis.* The primary lesions are similar to other types of cutaneous leishmaniasis. Metastatic spread to the oronasal/pharyngeal mucosa may occur during the presence of the primary lesion or up to 30 years later. Ulceration and erosion progressively destroy the soft tissue and cartilage of the oronasal/pharyngeal cavity and swelling of the nose and lips may cause "tapir nose". The condition may be painful or painless. Secondary infection is common. In contrast to cutaneous leishmaniasis, the lesions do not heal spontaneously. Suffering and mutilation are severe and death occurs as a result of bronchopneumonia or malnutrition. A small number of cases of mucocutaneous leishmaniasis have been reported from the Old World, most of them in adult males from the Sudan. The Sudnese form is characterized by the slow evolution of ulceration of the buccal mucosa; the causal agent is probably *L. donovani* s.l. In Ethiopia, the primary lesion following infection with *L. aethiopica* manifests itself as mucocutaneous leishmaniasis.

c. Distribution of the Leishmaniases

It is difficult to present what might be considered a truly accurate picture of the distribution or prevalence of the leishmaniases in man. In many countries the disease is not reportable or is present in areas from which reports are extremely difficult to obtain and in which diagnostic facilities may be virtually absent. Furthermore, there may be considerable variation in the incidence of the disease over a period of time. There is now an increasing awareness of the effect of environmental and ecological changes on the extent of transmission of the disease.

A recent WHO publication[242] has summarized in considerable detail the individual reports from various countries including the species and subspecies of the causative agents reported in the surveys and studies. The following section will present a general picture of the distribution of the leishmaniases, but the WHO summary should be consulted in the original for more detailed information.

Visceral leishmaniasis — Visceral leishmaniasis as caused by a variety of subspecific forms of *L. donovani* has been reported in the New World from a limited area of Argentina as well as in Bolivia, Brazil, Colombia, Ecuador, Paraguay, and Venezuela in South America, from Guatemala, Honduras, Mexico, El Salvador, and Guadeloupe in Central America, and the Caribbean. In the Old World, the disease has a very broad geographical distribution; in Africa it has been recorded from Morocco, Algeria, Tunisia, and Libya (as *L. donovani infantum* for the most part), as well as the Central African Republic, Chad, and the Sudan (as Sudanese cutaneous, ''mucocutaneous'' and visceral leishmaniasis),[125] Gabon, Kenya, and possibly southern Ethiopia and Somalia, (in a form presenting both cutaneous and visceral leishmaniasis), Nigeria, Malawi, Niger, Burkina Faso, Zaire, and Zambia. In Europe, it is found in Portugal, Spain, southern France including Corsica, Italy including Sardinia and Sicily, Hungary, Rumania, Yugoslavia, Greece including Crete, and Turkey. In the Middle East, the disease is present in Iraq, Iran, Israel, Lebanon, Saudi Arabia, Syria, North Yemen, and South Yemen (as *''infantum''*), and in Central Asia and the Indian subcontenent in foci in the central Asian republics of the U.S.S.R. and in Afghanistan, also as *''infantum''*. *L. donovani donovani,* known as visceral leishmaniasis, kala azar, Dum-Dum fever, and a variety of other local names, has been reported from India, Bangladesh, Nepal, and China.

Cutaneous Leishmaniasis of the Old World — This disease is caused either by a variety of subspecies of the *L. tropica* complex or by *L. major*. Although there is some question as to the exact identification of the causative agent, cutaneous leishmaniasis caused by *L. tropica tropica* is found in most countries of the Mediterranean basin and some countries neighboring upon them. In Asia it is reported from Israel and Jordan and east to foci in Pakistan and India. It is also found in several of the southern republics of the U.S.S.R. *L. tropica aethiopica* is known in Africa in the Ethiopian highlands as simple cutaneous leishmaniasis and a focus also exists in Kenya; a very similar disease condition has been reported from Namibia. *L. major* (or *L. tropica major*) and closely related members of the complex are widely spread, being reported from Algeria, Burkina Faso, Djibouti, Egypt, Guinea, Libya, Mali, Mauritania, Morocco, Senegal, and Sudan in Africa; from Afghanistan, Yemen, India, Iran, Iraq, Israel, Jordan, Pakistan, Saudi Arabia, Syria, and several of the southern republics of the U.S.S.R. in the Middle East and western Asia, as well as from China.

Cutaneous Leishmaniasis of the New World — As stated earlier, cutaneous leishmaniasis of the New World is due to members of the *L. mexicana* and *L. braziliensis* complexes. Cutaneous leishmaniasis in the New World occurs from southern Mexico to northern Argentina. Members of the *L. mexicana* complex have been reported from Mexico, Belize, Honduras, Guatemala, Panama, Trinidad, in the Amazon basin of Brazil, and in Venezuela. *L. braziliensis* is widely spread from Costa Rica through the northern tier of South America to Paraguay, Bolivia, and northern Argentina and on the slopes of Peruvian Andes.

d. The Public Health Importance of Leishmaniasis

From the above it can be seen that leishmaniasis in its various forms is widespread, but it is not always easy to determine the relative importance of the disease due to the poor reporting in many of the countries in which it is known to occur. It is certain, however, that leishmaniasis is a very considerable public health problem in many countries and that in some places where it had earlier declined, a resurgence of the disease is occurring. In Africa, new foci of visceral leishmaniasis have been described in recent years but there is very little transmission in West Africa and none in southern Africa. Generally speaking, the extent of the disease and the morbidity and mortality due to it, are little known in Africa, though more frequent outbreaks appear to be occurring, often associated with nonimmune individuals moving into endemic areas where development projects are taking place. Since the disease is fatal when untreated and probably very many cases go unrecognized, it must be considered that visceral leishmaniasis is a serious public health problem in some parts of the Africa Region. Devastating epidemics of visceral leishmaniasis have frequently occurred in Sudan in the past and there do not appear to be any ecological factors which would prevent them from doing so again, especially in the southern part of the country; however, because of the difficulties in obtaining information from that area there are few accurate reports on the current incidence of the disease.

Some of the newly discovered foci of visceral leishmaniasis in Africa may indeed actually be new foci of transmission and the disease may well be on the increase in older, known foci.[16,77,96]

In any event it is clear that there is a serious need for more detailed epidemiological studies to better define the problem and the threat that it poses and enable more selective and effective control measures to be undertaken.

Cutaneous leishmaniasis must be considered far less of a public health problem since it is self-curing within several months or years. However, when it occurs in epidemic form and especially when it extends to new areas it can represent a serious problem for health authorities, and for the populations involved. Belazzoug[23] and Belazzoug et al.[25] reported on an outbreak of cutaneous leishmaniasis in Algeria, south of the City of Algiers, from an area in which it had not been previously reported; the number of cases totaled almost 900 and may indicate that a serious spread of the disease can occur in similar ecological areas of the country.

In the Americas visceral leishmaniasis is only a serious public health problem in Brazil, with the highest prevalence in the States of Ceara and Bahia and in scattered foci elsewhere. The apparent increase may however, represent a greater interest in the disease in this country; prior to 1952 only 33 cases of visceral leishmaniasis had been reported from Brazil but by 1977 more than 2000 cases were reported in Brazilian medical journals.[236] Cutaneous leishmaniasis is, as all other leishmaniases in the region, principally a zoonoses and infection in man is usually associated with forested areas and persons entering them for work. In Costa Rica morbidity rates show that the infection is a serious problem with a rate of 100 cases per 10,000 population. The rate may very well be higher in other countries.[242] Because of the very serious nature of the disease, mucocutaneous leishmaniasis must be considered as a important problem wherever it occurs in any numbers.

In Southeast Asia, the gradual decline in indoor residual insecticide coverage due to reductions in the scale of malaria eradication and control programs has resulted in an increase in sandfly vector densities which has happened in many countries. The epidemic outbreak of 70,000 to 100,000 cases of kala azar in North Bihar, India in 1977 occurred in an area where insecticide spraying for malaria vector control was stopped in 1946; when insecticide spraying against the *Phlebotomus* vectors was begun again in an effort to contain the epidemic, the number of reported cases gradually declined to 12,500 in 1980. Transmission of cutaneous leishmaniasis is also reappearing in a good many areas of the region from which it had earlier disappeared, also mainly as a result of the cessation of indoor residual insecticide applications for malaria vector control. Visceral leishmaniasis, arthroponotic cutaneous leishmaniasis, and zoonotic cutaneous leishmaniasis occur in almost all of the countries of the Eastern Mediterranean Region and in many of them the incidence has increased in recent years; again it would appear that the cessation of antimalarial sprayings are an important factor in the recrudescence of the disease in man and in canines. Cutaneous leishmaniasis is the most important of the forms of leishmaniasis in the Region and serious epidemics of it have occurred in the last 10 years in Iran, Israel, Libya, Pakistan, Saudi Arabia, and Sudan, many of them associated with development projects or other major population relocations and movements.

In the Western Pacific Region visceral leishmaniasis was until the early 1950s a very serious public health problem, particularly in China. As a result of an intensive campaign against the disease, its vectors and reservoirs, no more than a few dozen cases are now reported yearly in that country.

The Vector — In all cases, the vectors of leishmaniasis belong to the subfamily Phlebotominae which contains about 600 species and subspecies of sandflies; of these some 70 are proven or suspected vectors of leishmaniasis. The vectors in the Old World belong to the genus *Phlebotomus* and those in the New World to the genus *Lutzomyia*.

Vector species of sandflies must, of course be anthropophilic and support the full development of the life-cycle of the parasite which culminates in the invasion of the pharynx and mouth parts of the female sandfly by the parasite. The various vector species are often specific for the transmission of a single species of parasite.

Table 3 lists the proven vectors of leishmaniasis in the Old and New World:[242]

Small Mammal Reservoirs — Those vertebrate species in which the parasite population is maintained indefinitely are the reservoirs of *Leishmania*. Usually there is only one reservoir host for each *Leishmania* species in a given biotope though other animals may become incidentally infected. Most *Leishmania* species in the natural foci of infection are maintained by wild animals who are, for the most part, rodents. The dog is the most important domestic reservoir of *L. donovani* and *L. peruviana* and is occasionally infected with other species, but commensal rats, i.e., *R. norvegicus* and *R. rattus,* have now been shown to harbor at least three *Leishmania* species and probably play a maintenance role for *L. donovani* in Italy.

Reservoirs of leishmaniasis in the New World — There are many records of *Leishmania* infections in a large variety of mammals in the New World; however, many of these must be regarded as incidental infections. Table 4 lists those species which are known or thought to be the main reservoirs of each of the *Leishmania* parasites.

Reservoirs of leishmaniasis in the Old World — At one time it appeared that the main host for visceral leishmaniasis was the dog and this animal still remains perhaps the most important domestic host. Further studies however, have now shown an increasing variety of proven and potential reservoirs of both visceral and cutaneous leishmaniasis in their various forms. Table 5 lists the most important reservoir hosts of leishmaniasis in the Old World including Europe, Africa, and Asia.

3. American Trypanosomiasis or Chagas' Disease

American Trypanosomiasis or, as it is more commonly called, Chagas' Disease, is caused by a flagellated protozoan with a well-developed undulatory membrane, *Trypanosoma cruzi;* the parasite has a complex developmental cycle and undergoes several transformations in the vertebrate host and in the triatomid vector.

Infection of man occurs with penetrations of metacyclic trypomastigotes (i.e., the last phase of the parasite development in the insect vector) into the macrophages of the connective tissue of the dermis or into the macrophages of the subcutaneous tissue where they become

Table 3
PROVEN VECTORS[a] OF HUMAN LEISHMANIASIS[242]

Old World: genus *Phlebotomus*

Subgenus of *Phelebotomus*	Species	Parasite transmitted[b]	Main areas of endemic foci
Phlebotomus	*duboscqi*	*L. major* (ZCL)	Senegal and parts of West Africa
	papatasi	*L. major* (ZCL)	South-west Asia, North Africa, and USSR
	saheli	*L. major* (ZCL)	India (Rajasthan)
Paraphlebotomus	*sergenti*	*L. tropica* (ACL)	Afghanistan, India, Islamic Republic of Iran, Iraq, Pakistan, and U.S.S.R.
Larroussius	*ariasi*	*L. donovani* (VL)	France
	longicuspis	*L. donovani* (VL)	North Africa
	major syriacus	*L. donovani* (VL)	South-east Asia
	orientalis	*L. donovani* (VL)	Sudan
	perniciosus	*L. donovani* (VL)	Italy and North Africa
	smirnovi	*L. donovani* (VL)	U.S.S.R.
	tobbi	*L. donovani* (VL)	Cyprus
	longipes	*L. aethiopica* (ZCL)	Ethiopia
	pedifer	*L. aethiopica* (ZCL)	Ethiopia and Kenya
Synphlebotomus	*martini*	*L. donovani* (VL)	Kenya
Adlerius	*chinensis*	*L. donovani* (VL)	China
	longiductus	*L. donovani* (VL)	U.S.S.R.
Euphlebotomus	*argentipes*	*L. donovani* (VL)	India

New World: genus *Lutzomyia*

Nyssomyia	*flaviscutellata*	*L. mexicana amazonensis* (CL)	Amazon basin in Brazil
	olmeca	*L. mexicana mexicana* (CL)	Southern Mexico and Central America
	trapidoi	*L. braziliensis panamensis* (CL)	Panama
	umbratilis	*L. braziliensis guyanensis* (CL)	Northern Brazil and French Guiana
Psychodopygus	*wellcomei*	*L. braziliensis braziliensis* (CL)	Brazil

[a] Proven vector: anthropophilic and most animal reservoirs, found naturally infected wtih a parasite indistinguishable from that found in man and reservoirs. Other sandflies including species in the *Phlebotomus (Paraphlebotomus) caucasicus* group and *Lutzomyia longipalpis* group are suspected of being vectors, but final incrimination awaits further study.

[b] ZCL, zoonotic cutaneous leishmaniasis; ACL, anthroponotic cutaneous leishmaniasis; VL, visceral leishmaniasis; CL, cutaneous leishmaniasis.

amastigotes (leishmanial forms). This intracellular form, without flagellum or membrane, multiplies by binary fission for 4 or 5 days. The host cell is eventually destroyed and the parasites invade other macrophages. Some parasites of the primary focus reach the blood stream where they become trypomastigotes (flagellated forms, with a membrane supporting the flagellum) and spread throughout the host invading cells in many different viscera including the heart, where they again take on the leishmanial or amastigote form and multiply.

The second developmental cycle occurs in the triatomid vector. These hematophagous insects, when feeding on a mammal, ingest the trypomastigotes circulating in the blood. These stages are abundant in the circulatory system of lower animals and tend to appear during febrile periods in man. The insect vector may also ingest intracellular amastigote forms which undergo a new series of transformations and multiplication phases in the insect. The parasites multiply in the form of crithidia and after 15 to 30 days the metacyclic

Table 4
MAIN RESERVOIR HOSTS OF HUMAN LEISHMANIASIS IN THE AMERICAS[242]

Parasite	Probable main reservoir host	Suspected reservoir hosts
L. donovani	Dog	Foxes, *Cerdocyon thous* and *Lycalopex vetulus*
L. braziliensis braziliensis	*Oryzomys capito, O. nigripes, Akodon arviculoides*	—
L. braziliensis guyanensis	The anteater *Tamandua tetradactyla*	—
L. braziliensis panamensis	The sloth *Choloepus hoffmanii*	—
L. Mexicana mexicana	Tree-rat *Ototylomys phyllotis*, spiny pocket mouse *Heteromys desmarestianus*, vesper-rat *Nectomys sumichrasti*	—
L. mexicana amazonensis	Proechimys guyannensis, rice-rat *Oryzomys capito*	Marsupials, foxes
L. mexicana venezuelensis	*Heteromys anomalus*	—
L. mexicana garnhami	Unknown	Marsupials
L. peruviana	Dog	—

Table 5
RESERVOIR HOSTS OF LEISHMANIASIS IN THE OLD WORLD[34]

Country	*Leishmania* species	Reservoir	Ref. for rodents
Europe			
Cyprus	*L. d. infantum*	Dog	
France	*L. d. infantum*	Dog, fox	
Greece	*L. d. infantum*	Dog	
Italy	*L. donovani*	Dog, fox, *Rattus rattus*	28, 185
Malta	*L. d. infantum*	Dog	
Portugal	*L. d. infantum*	Dog	
Spain	*L. d. infantum*	Dog	
USSR	*L. d. infantum*	Dog, jackal, fox	
Yugoslavia	*L. d. infantum*	Dog, *Rattus rattus*	177
Africa			
Algeria	*L. d. infantum*	Dog, fox	24
	L. major	*Psammomys obesus*	
Ethiopia	*L. aethiopica*	Hyrax spp.	101
	L. major	*Arvicanthis niloticus*	
Gambia	*L. d. infantum*	Dog	
Kenya	*L. donovani*	Dog, hyrax spp.	
	L. aethiopica	*Cricetomys* sp.	169
Libya	*L. d. infantum*	Dog	
	L. major	*Psammomys obesus, Meriones lybicus*	13
Morocco	*L. d. infantum*	Dog, *Xerus getulus*	30
	L. major	*Meriones shawi*	189
Senegal	*L. major*	*Arvicanthis niloticus, Mastomys erythrolecus, Tatera gambiana*	128 66
Tunisia	*L. major*	*Meriones shawi shawi*	190

trypanosomes form in the rectum. These forms of the parasite are eliminated in the feces when the triatomid feeds on man or another vetrebrate host. The parasite invades the host through the bite in the skin or through abrasions in the skin or mucous membranes and thus begins the cycle again.

The clinical manifestations of *T. cruzi* infection in man may vary with the intraspecific variations of the parasite and the degree of immune response of the host. The prepatent period is about 5 to 10 days. The infection usually evolves into acute and chronic stages of the disease.

The acute stage usually occurs in children and can last for 1 to 3 months. It is characterized by continuous moderate fever, malaise, lymphadenopathy, and hepatosplenomegaly. An inflammatory response at the site of the infection (''chagoma'') may last up to 8 weeks. Ophthalmoglandular syndrome, or Romana's sign, occurs in a significant percentage of acute cases. Life threatening or fatal manifestations in acute disease include myocarditis and meningoencephalitis.

The chronic stage of Chagas' disease may manifest itself in 10 to 15 years after infection has occurred through slowly evolving heart damage which includes gross cardiac enlargement. In Argentina it is estimated that approximately 20% of the Chagas' disease patients suffer from myocarditis.[1]

There is an asymptomatic or latent period between the acute and chronic stages with only transient parasitemias. The asymptomatic phase may also last indefinitely. Digestive manifestations of *T. cruzi* infections, mainly megacolon and megaesophagus, have been reported mainly from central Brazil.

Besides transmission of the disease through a vector, many cases are the result of transmission by blood transfusion. In addition, transplacental transmission of the disease also occurs.

Chagas' disease is restricted entirely to the Americas. It has been reported in an area extending from the southern states of the U.S. as well as from California to Patagonia, Argentina in the south. It has been reported from every country of the Americas within this geographical range except Guyana and Surinam. It has been estimated that 7 to 12 million people are infected with *T. cruzi* and that 35 million people are exposed to the risk of infection.[35,239] Zeledon and Rabinovich[253] reported that the number of infected persons may total 4.5 million in Brazil, 3 million in Argentina, 1.2 million in Venezuela, 1.7 million in Colombia, 0.7 million in Bolivia, and 0.3 million in Chili, and extrapolated that the total number of people infected was 13 to 14 million.

Serological studies in rural areas of Latin America have shown positivity rates of 40% or more. The disease is also found in urban areas as well, particularly near parks or other ecological areas that favor the vector's survival. Many of the human clinical cases that are seen in urban areas are often among immigrants from rural areas that are suffering from chronic Chagas' disease, a great many of whom suffer from chronic myocarditis. All in all Chagas' disease must be considered to be among the major public health problems in Latin America. A detailed review of the status of the disease in the Americas on a country by country basis has appeared in the WHO Weekly Epidemiological Record for 8 February, 1985.[246]

1. The Vector

The only vectors of the disease are Hemiptera of the family Reduviidae, subfamily Triatominae. Some 92 species of triatomids have been described from the New World,[252] of these about 66 species have been found naturally infected with *T. cruzi* but not all of these can be considered as important vectors.

About 40 of these species have been found in artificial ectotopes that are close to man. Probably half of these species have become adapted to one degree or another to human dwellings and this close degree of association encourages transmission of the disease. Of the total perhaps a dozen are of considerable importance as vectors from the epidemiological standpoint.

The most important of the domestic vectors of Chagas' disease is *Triatoma infestans* and it is the principal vector in large areas of Chili, Argentina, Uruguay, Bolivia, Paraguay,

Brazil, and southern Peru. This species shows a high preference for the blood of man and several domestic animals.

Panstrongylus magistus is another important species of vector but with a narrower range of distribution. In Brazil it is found living under wild, peridomestic and domestic conditions and in the city of Salvador it has become urban.[21] Piesman et al.[180] found a very high rate of correlation between children seroreactive to *T. cruzi* and infection rates in *P. megistus* infected households in Bahia State, Brazil; in 63 infested houses which he studied, no less than 54% of bugs, collected from houses with seroreactive children, were infected. This species is also found in homes in small areas of Paraguay and Argentina.

Other species that can be found breeding in households in Brazil in more restricted areas are *T. sordida, T. braziliensis, T. pseudomaculata,* and *T. rubrofasciata. T. sordida* has been found to readily move into houses when *T. infestans* has been eliminated.[204] *T. sordida* is also found as a domestic vector in Argentina and Bolivia. *T. rubrofasciata* is an urban species found in the coastal cities of Recife, Salvador, and Rio de Janeiro.[21,60]

T. maculata is present and a vector in parts of Colombia, Venezuela, and some of the West Indies and appears to be adapting to house dwelling more frequently than in the past. In Venezuela it moves into houses when the more domestic triatomine species *Rhodnius prolixus* has been eliminated by insecticides.[83]

In southern Peru the vector of Chagas' disease is *T. infestans,* while in the north-west the vectors are *P. herreri* and *R. ecuadoriensis,* and in the eastern mountainous regions the vector is *R. robustus.*[147]

T. dimidiata is the most important vector in the area of Guayaquil, Ecuador, is distributed as far north as Mexico, and is an important domestic vector in certain areas of Colombia, Panama, and the Central American countries.[253]

The most important vector of Chagas' disease in the northern part of South America is *Rhodnius prolixus;* it is very common in the poorly constructed mud or mud and thatch huts of rural and periruban areas, often in densities high enough to be both a vector and a serious pest of man. In it's range in Venezuela and Colombia, this species is also commonly found in extradomiciliary situations as well, frequently infected with *T. cruzi.* D'Alessandro et al.[62] have found it as the most common species in palm trees in the oriental plains of Colombia.

R. pallescens is the vector of Chagas' disease in Panama, while in Mexico, *T. dimidiata, R. prolixus, T. phyllosoma, T. pallidepennis, T. mazzottii,* and *T. barberi* play roles in the transmission of the disease in different parts of the country.

2. *Small Mammal Reservoirs*

Natural infections with *T. cruzi* have been recorded for a very large number of species of domestic and wild animals though many of the strains may not necessarily correspond to *T. cruzi.* Different animal species may serve as reservoirs of the infectious agent under different ecological conditions. Table 6 lists the mammals that have been found to be naturally infected with *T. cruzi.*

B. The Cestode Infections

The cestodes or tapeworms include several species of public health importance in man. They may be maintained in nature in a number of animals and fish, especially those which require an intermediate host to complete their life cycle. All tapeworms have a mechanism for attaching their scolex or head to the intestinal wall of their host. The body beyond the scolex consists of a series of segments or proglottids containing the eggs. When the eggs or larvae that emerge from them are ingested by a suitable host, the development cycle commences varying with the particular group and species.

Table 6
MAMMALS REPORTED TO BE NATURALLY INFECTED WITH *TRYPANOSOMA CRUZI*

Marsupialia

Didelphis azare(= paraguayensis)
Didelphis marsupialis
Didelphis marsupialis aurita
Didelphis marsupialis etensis
Didelphis marsupialis mesamericus
Didelphis virginiana
Lutreolina crassicaudata crassicaudata
Lutriolina crassicaudata paranalis
Marmosa agilis agilis
Marmosa cinerea
Marmosa mitis casta
Marmosa pallidior
Metachirus nudicaudatus
Monodelphis domestica
Caluromys philander

Chiroptera

Antrozous pallidus pacificus
Artibeus fallax lituratus
Artibeus jamaicensis jamaicensis
Artibeus jamaicensis lituratus
Carollia perspicillatus
Carollia perspicillatus azteca
Choeronyteris minor
Desmodus rotundus murinus
Eptesicus brasiliensis argentinus
Eptesicus furnalis
Eptesicus fuscus fuscus
Eumops abrasus
Eumops bonariensis beckeri
Glossophaga soricina leachi
Hemiderma perpicillatum
Histiotus macrotus laephotis
Histiotus montanus
Lonchophylla mordax
Macrotus mexicanus
Micronycteris megalotis
Molossus major major
Molossus obscurus
Myotis nigricans
Myotis occultus
Myotis velifer velifer
Noctilio labialis albiventer
Peropterix macrotis macrotis
Phyllostomus elongatum
Phyllostomus hastatus
Phyllostomus hastatus panamensis
Pipistrellus hesperus maximus
Pteronatus davyi fulvus
Saccopteryx bilineata
Tadaria brasiliensis
Tadaria mollossys
Uroderma bilobatum

Primates

Alouatta seniculus
Cebus apella
Cebus capucinus
Cebus sp.
Hapale jacchus
Homo sapiens
Hylobates pileatus
Macaca irus
Macaca mulatta
Nyciticebus coucang
Saimiri boliviensis
Saimiri sciurea

Edentata

Cabassous tatouya
Cabassous unicinctus
Chaetophractus vellerosus pannosus
Chaetophractus vellerosus vellerosus
Chaetophractus villosus
Dasypus hybridus
Dasypus novemcinctus
Dasypus novemcinctus fenestratus
Dasypus novemcinctus mexicanus
Dasypus novemcinctus texanus
Dasypus paraguayensis
Dasypus pentadactylus
Euphractus sexcintus
Tamandua tetradactyla
Tolypeutes matacus
Zaedyus pichi caurinus

Lagomorpha

Orycytolagus cuniculus
Sylvilagus sp.

Table 6 (continued)
MAMMALS REPORTED TO BE NATURALLY INFECTED WITH *TRYPANOSOMA CRUZI*

Rodentia	
Akodon arviculoides cursor	*Neotoma alleni*
Cavia porcellus	*Neotoma fuscipes macrotis*
Cercomys cunnicularius laurentius	*Neotoma lepida lepida*
Citellus leucurus cinnamoneus	*Neotoma micropus canescens*
Coelogenus subniger	*Neotoma micropus micropus*
Coendous mexicanus laenatus	*Octodon degus degus*
Coendous prehensilis	*Peramys domesticus*
Cuniculus paca paca	*Peromyscus boylii rowlei*
Dasyprocta agouti	*Peromyscus truei gilberti*
Dasyprocta rubrata	*Peromyscus truei montipinoris*
Galea spixii spixii	*Rattus norvegicus*
Guerlinguetus gilirigularis	*Rattus rattus alexandrinus*
Leptosciurus argentinus	*Rattus rattus frugivorus*
Mus musculus	*Rattus rattus rattus*
Mus musculus brevirostris	*Sciurus gerrardi morulus*
Nectomys squamipes amazonicus	*Sciurus* sp.
Neotoma albigula albigula	*Zygodontomys pixuna*
Carnivora	
Canis familiaris	*Grison vittatus*
Cerdocyon thous	*Mephitis mephitis nigra*
Dusicyon culpaeus culpaeus	*Nasua narica*
Dusicyon gracilis gracilis	*Procyon lotor*
Dusicyon griseus graislis	*Tayra barbara*
Felix domesticus	*Urocyon cinereoargenteus*
Grison cuja	
Artiodactyla	
Sus scrofa domesticus	*Capra hircus*

(From Jacobs, L. and Arambulo, P., *Handbook Series in Zoonoses, C, Parasitic Zoonoses*, Vol. I, CRC Press, Boca Raton, Fla., 1982, 110.)

1. Hymenolepsis

This infestation may be caused by two species of the genus *Hymenolepsis, H. nana* and *H. diminuta*. The most common of the two is *H. nana*. Man serves as both the definitive and intermediate host in the majority of the infestations due to *H. nana*. The gravid proglottid of this small species of tapeworm disintegrates in the human intestine and the eggs pass with the feces to the outside environment; when another human host ingests an embryonated egg the larva hatches and develops into an adult to complete the cycle. In strains of the parasite that are adapted to rodents the cycle can also be completed within the same animal species.

The highest rates of infection of *H. nana* are found in infant and child populations of southern Europe, North Africa, parts of the Middle East, and Latin America, and are usually more common in urban than in rural areas. Where the infection is present, rates of infection in children may range from less than 1% to more than 50% in the population. Biswas et al.[29] studied the epidemiology of *H. nana* in a village in Rajasthan, India; out of 757 individuals examined 6% were infected by this tapeworm including 10.1% of the children and 3.9% of the adults. Of the persons infected, 35% showed symptoms ascribed to the infection. Depending on the worm load, infections may be asymptomatic or, when worm loads are heavy, give rise to gastrointestinal disorders such as nausea, vomiting, abdominal pains, diarrhea, and pruritus. Autoinfection occurs in the host but the infection will eventually die out if the patient is not reinfected.

Rodent Hosts — Despite the fact that the infection can be transmitted directly from one human host to another, high rates of infection can be found in rodent populations and there is some divergence of opinion as to the exact importance of rodents in the transmission of the disease. At the present stage of knowledge the importance of rodents in maintaining the organism in nature is uncertain.

2. *Echinococcosis*

Otherwise known as hydatid disease, the disease in man is produced by cysts which are the larval stage of tapeworms of the genus *Echinococcus*. There are four species which are regarded as valid: (1) *E. granulosus* involves the dog as final host and livestock as intermediate hosts with man as an accidental but not infrequent host; (2) *E. multilocularis* has as its final hosts foxes, coyotes, cats, and dogs while the intermediate hosts are mainly microtime rodents and occasionally domestic mice; (3) *E. oligarthrus* occurs in Central and South America and large cats such as the puma, the jaguar, and others are the definitive hosts while agoutis and possibly other rodents are intermediate hosts; and (4) *E. vogeli* uses dogs as its final hosts and the paca, possibly the agouti, and the spiny rat as intermediate hosts in Central and South America.

a. Echinococcus multilocularis

The disease in man is caused by alveolar larval cysts of *E. multilocularis*. They are usually found in the liver and more rarely may be found in the lungs and brain. Growth of the cysts, unlike those due to *E. granulosus,* is not restricted by a thick cyst wall and so they continually expand by external proliferation throughout the liver and contiguous organs producing chronic large lesions with a grave prognosis.

Man is infected through the ingestion of infective eggs passed in the feces of infected canines or felines which may soil the animal's hair or other objects handled by man.

While it is relatively wide spread in the northern hemisphere, human infection with *E. multilocularis* is only sporadic and, where it occurs, far less frequent than infection with *E. granulosus*.

The Reservoir — Leiby et al.[138] reported that this species of tapeworm was widespread in the north-central U.S. in North Dakota, South Dakota, Minnesota, Iowa, and eastern Montana; in these areas the deer mouse, meadow vole, and house mouse serve as intermediate hosts. Kritsky et al.[120] found it infecting the bushy-tailed woodrat, *Neotoma cinerea rupicola,* in Wyoming, thus considerably extending the range of the species to the west and also adding a new host to the records.

b. Echinococcus oligarthrus

This species has been reported from Panama and Colombia. Infection in man is rare and little information is available on its rodent hosts, other than it has been reported from the agouti.

c. Echinococcus vogeli

Cases of this species in man have been reported from Colombia, Ecuador, Panama, and Venezuela. The cysts of the larval tapeworms may occur in the liver, lungs, and other organs. The symptoms vary according to the cyst size and location.

The Reservoir — The principal definitive hosts are the bush dog and domestic dogs. The main intermediate host is the paca, *Cuniiculus paca,* and the spiny rat.

3. Raillietiniasis

The disease in man is caused by one of two species of cestodes, *Raillietina demerariensis* or *R. celebensis*. The life cycle of these two species is not known, but the intermediate host is assumed to be an arthropod as in other species of the genus while the definitive hosts are mainly rodents. Man is probably infected by ingestion of food contaminated by cercoid-infected arthropods.

a. Raillietina demerariensis

This is a neotropical species and human infections have been found in Ecuador, Cuba, Guyana, and Honduras. The largest endemic focus is one which is found near Quito, Ecuador, and elsewhere the disease is rare. In man it has been reported that infection with the parasite may cause digestive disorders, headaches, convulsions, circulatory disorders, weight loss, and retarded growth in children.

The Reservoir — Various rodent species exist, but little information is available.

b. R. celebensis

This species has been reported from several countries in eastern Asia including Japan, Taiwan, and India. The symptoms are probably the same as in *R. demerariensis* though it has been reported that in the Philippines the infection may be asymptomatic.

The Reservoir — Niphadkar and Rao[171] found *R. celebensis* in 5% of the *Rattus rattus* and 7% of the *Bandicota bengalensis* population they surveyed in Bombay, including 200 specimens of each species. Higher rates of infection have been reported in rodents in other parts of Asia.

There are a number of other cestodes with rodents as their definitive or intermediate host; among these are *Multiceps (Taenia) multiceps* and *Inermicapsifer madagarascariensis*. The number of human cases is very small and they are of very little public health significance.

C. Trematode Infections

The only human trematode infection in which a rodent is involved in the life cycle is schistosomiasis or bilharziasis. In most areas the disease is caused by one of three species of blood flukes and each of these will be considered separately below. Two additional species which infect man are only of local importance.

1. Schistosomiasis

Schistosomiasis is one of the most widespread of the parasitic diseases and in socioeconomic and public health importance is considered second only to malaria due to its debilitating effect on populations throughout large geographic areas. The disease is endemic in 74 countries and it is estimated that more than 200 million persons residing in rural and agricultural areas are infected with it. In recent years many water development projects and dams have served to spread the disease and intensify its prevalence.

a. Schistosoma mansoni

S. mansoni is found primarily in the mesenteric veins which drain the large intestine. The eggs are evacuated mainly with fecal matter and when they reach water they hatch and release a miricidium or larva which swims in search of a suitable intermediate host — a snail of the genus *Biomphalaria*. Asexual multiplication occurs in the snail and in approximately a month's time cercariae emerge from the snail into the water to seek a definitive host. A single miricidium can give rise to more than 100,000 cercariae. The cercariae pierces the epidermis of a suitable host, loses its tail and becomes a schistosomule which enters the blood stream and, passing through the lungs, migrates to the liver where it develops to sexual maturity. The male then couples with the female worm and carries it to the mesenteric veins where egg laying begins. The time of development, from penetration of the skin until oviposition, in *S. mansoni* occupies some 40 to 60 days.

The main sequelae of human infection with this species is liver and spleen enlargement; the frequency with which symptoms occur is often related to the intensity of infection and mild infections may be asymptomatic. The main lesions caused by *S. mansoni* are found in the intestinal wall and with time they spread to the liver, producing fibrosis and portal hypertension, ascites, and splenomegaly. In advanced states there may be pulmonary lesions and respiratory symptoms.

S. mansoni is the only species of schistosome known to occur in the Americas, where it occurs in large areas of Brazil, in northern and central Venezuela, on the coast of Suriname, and in several of the Caribbean islands. It occurs over a large portion of Africa, in many of the countries of the Eastern Mediterranean, especially Egypt, Yemen, and Saudi Arabia, and in China and the Philippines. In all this species occurs in 53 countries.[243]

The Reservoir — Man is the main definitive host of *S. mansoni* and on his own can maintain the infection under favorable ecological conditions. In the endemic areas of northern Brazil, several naturally infected species of wild rodents have been found, a large proportion of which eliminate viable eggs in their feces. Because of their population densities and frequent contact with waterways, they can contribute to the spread of this parasite. In Pernambuco, of 27 *Rattus rattus frugivorus* examined 16 or 59% were infected with *S. mansoni*. In Minas Gerais, five species of rodents and one marsupial were found to be naturally infected. *Nectomys squamipes* is of special interest because of its affinity to water and the high percentage of fertile eggs it excretes — 68.4%.[255] Rodents have been found naturally infected with *S. mansoni* in a number of other countries, including *R. norvegicus* and *R. rattus* in Guadeloupe,[59,116,125] *R. rattus* in Venezuela,[89] *Pelomys* sp. in Kenya,[118] and various other species in Zaire, Uganda, and South Africa. There is still some uncertainty as to whether rodents can maintain *S. mansoni* infections in nature in the absence of man; however, the ability of certain species of rodents to pass viable eggs in their feces does suggest that they play a role in the transmission, and the entire question is certainly deserving of further investigation. In Africa, some primates, particularly baboons, may have a role in the maintenance of *S. mansoni* in some foci.

b. Schistosoma haematobium

This species of schistosome is very wide spread in man in Africa and in the Eastern Mediterranean regions in a total of 52 different countries. Most of the individuals infected are in the 5- to 15-year age group.

The adult worms are widely distributed in the pelvic and mesenteric venous plexuses and oviposition takes place mainly in the lower urinary tract and distal gut.

Wild animals have rarely been found infected with *S. haematobium* and where they have been found, it appears that they are no more than incidental infections in small numbers and are probably of little importance in the epidemiology of the disease.

c. Schistosoma japonicum

In the past, *S. japonicum* occurred in six countries, but today it is found only in China, Indonesia, and the Philippines. A related parasite, *S. mekongi,* has been reported from Kampuchea and Laos. The disease caused by *S. japonicum* is similar in its symptomatology to that caused by *S. mansoni,* but in general it causes more serious illness. The early lesions following infection are located primarily in the small intestine, and intestinal and hepatic fibrosis develops more rapidly than with *S. mansoni* as *S. japonicum* lays more eggs.

The intermediate host of *S. japonicum* is the amphibious snail, *Oncomelania,* which can survive several months out of water and maintain the larval stages of the parasite during such a period.

The Reservoirs — As with the species above, man is the main reservoir of *S. japonicum,* but domestic and wild animals play an important role as reservoirs of the infection. The disease can, in fact, be considered as common to man and animals and the parasite can move freely between the species through the intermediate hosts. Dogs, cats, bovines, equines, swine, sheep, and goats as well as rats, mice, felines, mustelids, and monkeys have been found infected. In China, 29 species of mammals have been found to be infected and, in some areas, micotine rodents are found with a very high number of parasites. In other areas great importance is attached to rats in maintaining the infection, especially when the snail population is low. Pesigan et al.[176] examined 203 *Rattus rattus* at Palo, Leyte in the Philippines and found that 86 or 22.6% were positive for *Schistosoma* ova in their feces; they did not consider that these rats played an important role in the dissemination of the disease as they believed that the mean daily egg output and the hatchability were too low. Cabrera[51] also carried out a study on the prevalence of *S. japonicum* in *R. rattus* in Leyte; the presence of eggs in liver sections of 557 rats from 3 localities was 76.1%, 73.0%, and 65.1%. Of the stools, 48 or 12.5% were positive for ova. Cabrera considered that the high density of the rats, their feeding in rice fields, and their habit of defecating several times in 24 hr indicated that rats could play a role in maintaining the infection in nature even in the absence of human transmission. In 1973, Carney et al.[53] carried out a survey of 286 rodents from a newly discovered focus of *S. japonicum* in the Napu valley of central Sulawesi, Indonesia. *S. japonicum* was found in the mesenteric veins and liver of *R. exulans,* trapped in two locations, 3 or 44 rats (7%) in one village and 3 of 37 (8%) in another were positive.

D. Nematode Infections

1. Angiostrongyliasis

Angiostrongyliasis or eosinophilic meningoencephalitis in man is a disease of the central nervous system due to infection by the third stage infective larvae of the nematode *Angiostrongylus cantonensis*. The meninges are predominantly involved. While the infection may be asymptomatic or mild, more commonly it is characterized by severe headaches, vomiting, stiffness of the neck and back, and temporary facial paralysis occurs in 5% of the patients. There is usually a prolonged fever. Eosinophilic meningitis has been recorded in a number of patients. Illness may last for a few days to several months. Death from the infection is rare but can occur.

The disease is widespread and is endemic in Hawaii, Tahiti, many other Pacific islands, Vietnam, Thailand, Malaysia, Indonesia, Taiwan, the Philippines, and Cuba, while the parasite itself is also found in Australia, Madagascar, and Egypt. It has recently been reported from Papua, New Guinea[199] and Bombay, India.[202]

Transmission is through the ingestion of raw or insufficiently cooked land or water snails, or slugs, which are the intermediate transport hosts, harboring infective larvae. Prawns, fish, and land crabs that have ingested snails or slugs may also transport the infective larvae. Lettuce and other similar leaf vegetables contaminated by molluscs may also serve as a source of infection. The molluscs are infected by first-stage larvae excreted by an infected rodent; when the third-stage larvae have developed, rodents (and man) can be infected when they eat the mollusc. In the rat the larvae migrate to the brain and mature to the adult stage; the adults then migrate to the surface of the brain and through the venous system to reach the final site in the pulmonary arteries. After the worms mate, the eggs hatch in the branches of the pulmonary arteries, the first-stage larvae enter the bronchial system and pass up the trachea, are swallowed and passed in the feces.

Rodent Reservoirs — *A. cantonensis* has been found in at least ten different species of the genus *Rattus,* in *Bandicota indica,* in *Melomys littoralis* and, on one occasion, in the shrew *Suncus murinus riukiuanus* on Okinawa.[172] Generally speaking, rats of the genus *Rattus* are the most important reservoirs of the infection. Infection rates in wild-caught rodents may be very high; Wallace and Rosen[234] examined 1127 *R. norvegicus, R. rattus,* and *R. exulans* on the Hawaiian and Society Islands and found the overall percentage positive for *A. cantonensis* to be 45%, with rates of infection usually being highest in *R. norvegicus.* The rates of prevalence in rats on those Pacific Islands found positive was usually higher than on the mainland of Asia. On Taiwan, the highest percentage of infection in rodents was in *Bandicota indica nemorivaga,* in which infestation rates ranged from 25 to 50%, while in Jakarta and its suburbs 2.4% of *R. rattus dirdii,* 21.4% of *R. argentiventer,* and 18.1% of *R. norvegicus* were infected.[154] In Malaysia, infection rates in those wild-caught rodents which were found positive, ranged from 2% in a single *R. annandalei* out of 47 examined to 41% or 219 out of 526 *R. jalorensis* examined.[142] Results from many other surveys carried out on rats in various areas within the range of *A. cantonensis* show similar findings to those already listed.

a. Abdominal Angiostrongyliasis

This condition may result from infection by *Angiostrongylus costaricensis.* It is predominantly found among children under the age of 13 and causes symptoms similar to those of appendicitis with abdominal pain, fever, anorexia, vomiting, abdominal rigidity, a tumor-like mass in the right lower quadrant, and pain of rectal examination. In the rodent reservoir, the adult worm lives in the mesenteric arteries in the cecal area and its eggs are carried into the intestinal wall. The first-stage larvae migrate into the lumen, are excreted in the feces, and then may be ingested by a slug. The infective larvae are found in the slug's slime left on the soil or vegetable surfaces; when ingested by man, the infective larvae penetrate the gut wall, maturing in the lymphatic nodes and vessels. The adult worms migrate to the mesenteric arterioles of the ileocecal region where oviposition occurs.

Human infections of *A. costaricensis* are known from Costa Rica, Honduras, El Salvador, Mexico, and Brazil and suspected cases have been reported from Nicaragua and Venezuela. More recent reports have recorded the parasite in Colombia[152] and the U.S., in Texas.[228]

Rodent Reservoirs — In Costa Rica, the principal host is either the cotton rat, *Sigmodon hispidus,* or *R. rattus.* In Colombia, the rat *Oryzomys caliginosus* is the only species so far found positive. The worm is present in *Rattus rattus* in the southern U.S. In Panama the highest prevalence of infection is found in *S. hispidus,* which is also one of the most abundant rodent species in the areas which were studied, and lives close to human dwellings. *Zygodontomys microtinus, Liomys adspersus,* and *Oryzomys fulvescens* have also been found infected by the worm.

2. Capillariasis

There are three types of this disease, each due to a different species of parasite. *Capillaria philippinesis* is the cause of intestinal capillariasis and its intermediate hosts are fish. Adult worms have only been found in humans and the natural host is still unknown; experimental infections of the gerbil, *Meriones unguiculatus* and *Mastomys natalensis* have been successful but no infection has ever been found in nature in rodents. *Capillaria aerophila* is a parasite of dogs and cats and other carnivorous animals and humans may be become infected by ingesting infective eggs in soil or soil-contaminated food. The number of recorded human infections is however, very small.

Capillaria hepatica is found on all continents among synanthropic and wild rodents, but disease in man resulting from infection from this species is rare and probably no more than a few dozen cases have been recorded. When it occurs however, it is serious and frequently fatal. Clinical cases are due to a massive invasion of *C. hepatica* in the liver where they reach maturity and begin egg production in a manner similar to that which occurs in the rodent host. Symptoms are hepatomegaly, fever, nausea and vomiting, diarrhea or constipation, abdominal distension, etc. The infection is transmitted by the ingestion of embryonated eggs picked up from the soil where they have been disseminated by carnivores; in the case of human infections, the disseminating hosts may be cats and dogs which have hunted and eaten rodents infected with the worms.

Rodent Reservoirs — In some areas of the world 50 to 80% of the domestic rats, *Rattus norvegicus,* have been found infected with *C. hepatica*. Farhang-Azad[72] studied the dynamics of infection among Norway rats taken from the zoo in Baltimore, Maryland; 845 rats were examined and 75% of them were found to be infected with *C. hepatica*. Galvao[82] examined 138 *R. norvegicus* in Salvador Bahia in Brazil; of these 78 or 57% were positive for the parasite. Over a 5 year period, 986 house shrews *Suncus murinus* were trapped[39] in various parts of Indonesia and examined for parasites; 12 shrews from 6 different localities in West Java were found to be infected with *C. hepatica,* which was the first record of infection in this shrew species by the worm parasite. Farhang-Azad and Schlitter[73] examined 1110 small mammals for *C. hepatica* infection in Shoa Province, Ethiopia. Of these, 19 out of 308 (6.2%) *R. rattus* and 1 out of 212 (0.5%) *Praomys albipes* were infected. Elsewhere in Africa, 48% of a *Mastomys natalensis* population in Johannesburg, South Africa were infected.

3. Trichinosis

Also known as trichiniasis and trichinellosis, the disease is caused by infection with a small filiform nematode, *Trichinella spiralis*. In the stomach of a carnivore or omnivore which has ingested meat containing the encapsulated infective larva, the capsule is digested and the larvae of the parasite are released to lodge in the villi of the mucosa and glandular crypts of the small intestine where it continues to develop, reaching the adult stage in 2 to 3 days. A short time after copulation, viviparous females begin to deposit larvae. Larvae appear 4 to 7 days after the ingestion of infected meat and may continue to be laid for several weeks. The larvae penetrate the intestinal wall to enter the lymphatic vessels and mesenteric venules reaching the posterior vena cava, then they move through the heart and lungs via the arterial system to all the organs and tissues where they may stay for a limited period. After entering the circulatory system again they migrate to and encapsulate in the striated muscles. The larvae that remain in the other organs and tissues, including the smooth muscles, die in a short time. The encapsulated larvae can survive for years in the muscle tissue of the host. The entire cycle develops within a single host but in order to initiate a new cycle it is necessary for the larva encapsulated within the tissue to be ingested by a new host whether of the same or another animal species.

Clinical disease in man is highly variable and can range from unapparent infections to a fulminating, fatal disease depending on the number of larvae that have been ingested. Among the symptoms of infection are the sudden appearance of muscle soreness and pain and edema of the upper eyelids is a common early and characteristic sign. Other ocular signs may occur followed by thirst, profuse sweating, chills, weakness, prostration, and rapidly increasing eosinophila. Gastrointestinal symptoms such as diarrhea due to the intraintestinal activity of the adult worms may occur. Remittent fever is usual, sometimes as high as 40°C; the fever terminates after 1 to 3 weeks. Respiratory and neurological complications may appear in the 3rd to 6th week following infection and death, when it occurs, can take place in the 1st or 2nd week and between the 4th and 8th weeks.

The disease is worldwide in distribution but very variable in incidence and is related to the practices of eating and preparing pork or wild animal meat such as bear meat in parts of North America and wild boar in Europe. Cases and outbreaks are usually sporadic and localized and are often associated with the homemade preparation of sausages and other meat products. In most cases the studies that have been made on the incidence of the disease have been related to the domestic cycle in swine, rats and man, and the infection can persist in wild animal species without cases being recorded in commensal animals or man. No effort will be made to review the handful of cases recorded in most countries since in any event the failure of accurate diagnosis in many instances probably masks the actual distribution and incidence of the disease. In general, while the distribution of the disease is widespread, it is not now a serious public health problem in any given area.

Rodent Reservoirs — *T. spiralis* has a wide range of animal hosts among both domestic and wild animals. Pigs however, constitute the main source of infection for man. The infection rate in pigs is greatest for those that are fed on raw household wastes, raw garbage, or forage in the city streets of some countries. Infection rates in pigs in the U.S. and Europe are now from 0.1 to 0.5%.

Rats are part of the commensal animal cycle of trichinosis. In the U.S. rats in rural areas are generally not infected, but those found living in garbage dumps may have a high rate of infection. A survey of 1268 rats from a garbage dump in the U.S. showed 5.3% positive and in the U.S.S.R. 1.5% of 8037 rats were infected. Very high infection rates have been found in Lebanon, 36% of the rats examined. In rats captured in refuse dumps of Santiago and Antofagasta 8% and 28.6%, respectively, were infected and in the city abattoir of Santiago, 25% of the *R. norvegicus* were positive in 1965.[1]

Pigs are infected when fed on refuse containing infected muscle fibers of swine origin and cats, dogs, and rats are infected in the same manner. Pigs are easily infected through eating dead rats on refuse dumps whose bodies are infected by *T. spiralis* and this is a frequent source of infection where pigs are free to consume refuse from the streets of villages or towns or are able to forage on garbage dumps where dead rats are found.

VI. CONCLUSIONS

In considering the totality of the rodent-borne diseases which have been reviewed above, it is clear that there are few geographical areas free from one or another of these diseases and that together they account for a very great deal of human morbidity and mortality. As has been indicated earlier, the true extent of the illness and death which is actually due to these diseases is difficult to judge as many of their most important foci are in countries from which the least accurate reporting is received. In many countries, there are no requirements to report these diseases nor as has been stated at the outset, could many of the endemic countries do so, as they are among those which have the least facilities for the diagnosis and laboratory confirmation of these and many other communicable diseases of public health importance in their territories. Without this information it is difficult to establish an order of priorities for how the disease problems of the given country should be dealt with and, indeed, all too often, little is, or can be, done to control their transmission.

To the health problems caused by the rodent-borne diseases must be added the very considerable economic damage that is incurred as a result of rodent depredations both in urban and rural areas. In some farming areas rodent consumption and destruction of food crops is so great that it can not but adversely affect the nutritional status of the human populations in the same areas. In many towns and cities, stored food products are the subject of extensive attack by rodents who consume large quantities of foodstuffs and foul and destroy at least the same quantities. Unchecked, these rodent populations can then become the reservoirs of the diseases which have been discussed above.

It is hoped that an understanding of even the approximate magnitude of the rodent-borne disease problem may assist in convincing health authorities of the importance of undertaking technically sound and well planned rodent control programs both to reduce the economic burden resulting from the dense urban and rural rodent populations in some of the countries that can least afford to support them.

REFERENCES

1. **Acha, P. and Szyfres, D.,** Toxoplasmosis, in *Zoonoses and Communicable Diseases Common to Man and Animals,* Pan American Health Organization Sci. Publ. 354, 1980, 402.
2. **Al-Awadi, A. R., Al-Kazemi, N., Ezzat, G., Saah, A. J., Shepard, C., Zaghloul, T., and Gherdian, B.,** Murine typhus in Kuwait in 1978, *Bull. W.H.O.,* 60, 283, 1982.
3. **Aldova, E. and Lim, D.,** *Yersina enterocolitica* in small rodents. Communication 1: pilot study in two wildlife areas, *Zentralbl. Bakteriol. Abt. I Orig.,* A226, 491, 1974.
4. **Almeida, C. R., Aleida, A. R., Vieira, J. B., Guida, U., and Butler, T.,** Plague in Brazil during two years of bacteriological and serological surveillance, *Bull. W.H.O.,* 59, 591, 1981.
5. **Al Saadi, M. and Post, G.,** Rodent leptospirosis in Colorado, *J. Wildl. Dis.,* 12, 315, 1976.
6. **Anderson, J. F., Magnarelli, L. A., Burgdorfer, W., and Barbour, A. G.,** Spirochetes in *Ixodes dammini* and mammals from Connecticut, *Am. J. Trop. Med. Hyg.,* 32, 818, 1983.
7. **Anderson, K. E., Joseph, S. W., Nasution, R., Sunoto, Butler, T., Van Peenen, P. F. D., Irving, G. S., Soroso, J. S., and Watten, R. H.,** Febrile illnesses resulting in hospital admission: a bacteriological and serological study in Jakarta, Indonesia, *Am. J. Trop. Med. Hyg.,* 25, 116, 1976.
8. **Arambulo, P. V.,** Salmonellosis in the Philippines, a major zoonosis, *J. Philipp. Vet. Med. Assoc.,* 2, 104, 1970.
9. **Arata, A.,** The importance of small mammals in public health, in *Small Mammals: Their Productivity and Population Dynamics,* International Biological Programme, Cambridge, 5, 349, 1975.
10. **Arata, A., Chamsa, M., Farhang-Azad, A., Mescerjakova, I., Neronov, V., and Saidi, S.,** First detection of tularemia in domestic and wild mammals in Iran, *Bull. W.H.O.,* 49, 597, 1973.
11. **Armstrong, C. and Sweet, L. K.,** Lymphocytic choriomeningitis: report of two cases with recovery of virus from gray mice *(Mus musculus)* trapped in the two infected households, *Public Health Rep.,* 54, 673, 1939.
12. **Artsob, H.,** Powassen encephalitis, in *CRC Handbook Series in Zoonoses, B: Viral Zoonoses,* Vol. I, Beran, G. W., Ed., Boca Raton, Fla., 1981, 156.
13. **Ashford, R. W., Schnur, L. F., Chance, M. L., Samaan, S. A., and Ahmed, H. N.,** Cutaneous leishmaniasis in the Libyan Arab Republic: preliminary ecological findings, *Ann. Trop. Med. Parasitol.,* 71, 265, 1977.
14. **Audy, J. R.,** Scrub typhus as a study in ecology, *Nature,* 159, 295, 1947.
15. **Audy, J. R.,** The localization of disease with special reference to the zoonoses, *Trans. R. Soc. Trop. Med. Hyg.,* 52, 308, 1958.
16. **Ayele, T. and Ali, A.,** The distribution of visceral leishmaniasis in Ethiopia, *Am. J. Trop. Med. Hyg.,* 33, 548, 1984.
17. **Bahr, L.,** Einige experimentelle Untersuchungen Spezielf an Wanderratten *(Epimys norvegicus),* mit "Ratin" und "Ratinin" (1907-1938), *Z. Infekt. Kr.,* 59, 44, 1938.
18. **Baltazard, M.,** Presence au Maroc d'un spirochète type *duttoni* transmis dans le nature par *Ornithodoros erraticus, Bull. Soc. Pathol. Exot.,* 30, 551, 1937.
19. **Baltazard, M., Bahmanyar, M., Pournak, R., and Mofidi, C.,** *Ornithodoros tartakovskyi* Olenev 1931 et *Borellia (Spirochaeta) latychevi* Sofiev 1941 (Note préliminaire), *Ann. Parasit. Hum. Comp.,* 27, 311, 1952.
20. **Barnes, A. M.,** Surveillance and control of bubonic plague in the United States, *Symp. Zool. Soc. London,* 50, 237, 1982.
21. **Barreto, M. P.,** Reservatorios e vetores do *Trypanosoma cruzi* no Brasil, *Arq. Hig. Saude Publica,* 28, 43, 1963.
22. **Baylet, R. and Riel, V.,** *Cricetomys gambianus,* nouvel hote de conservation de leptospires, *Med. Afrique Noire,* 18, 803, 1971.
23. **Belazzoug, S.,** Une épidéemie de leishmaniose cutaneé dans la region de M'sila (Algérie), *Bull. Soc. Pathol. Exot.,* 75, 497, 1982.
24. **Belazzoug, S.,** Le nouveau foyer de leishmaniose cutanée de M'sila (Algérie). Infestation naturelle de *Psammomys obesus* (Rongeur, Gerbillidae), *Bull. Soc. Pathol. Exot.,* 76, 146, 1983.
25. **Belazzoug, S., Ammar-Khodja, A., Belkaid, W., and Tabet-Derraz, O.,** La leishmanoses cutanée du Nord de Algérie, *Bull. Soc. Pathol. Exot.,* 78, 615, 1985.
26. **Bell, J. F.,** Tularemia, in *CRC Handbook Series in Zoonoses, A, Bacterial, Rickettsial and Mycotic Diseases, Vol. I,* Stoenner, H., Kaplan, W., and Torten, M., Eds., CRC Press, Boca Raton, Fla., 1980, 161.
27. **Berman, S. J., Irving, G. S., Kundin, W. D., Gunning, J.-J., and Watten, R. H.,** Epidemiology of the acute fevers of unknown origin in South Vietnam: effect of laboratory support upon clinical diagnosis, *Am. J. Trop. Med. Hyg.,* 22, 796, 1973.
28. **Bettini, S., Pozio, E., and Gradoni, L.,** Leishmaniasis in Tuscany (Italy). II. *Leishmania* from wild rodentia and carnivora in a human and canine leishmaniasis focus, *Trans. R. Soc. Trop. Med. Hyg.,* 74, 77, 1980.

29. **Biswas, H., Arora, R. R., and Saraljit Sehgal,** Epidemiology of *Hymenolepsis nana* infection in a selected rural community, *J. Commun. Dis.*, 10, 170, 1978.
30. **Blanc, G.,** Leishmaniose viscérale généralisée observée chez un écureuil *(Xerus getulus)* au Maroc, *Bull. Soc. Pathol. Exot.*, 32, 756, 1939.
31. **Blumenthal, W., Kessler, R., and Ackerman, R.,** Uber die Durchseuchung der ländlichen Bevölkerung in der Bundesrepublik Deutschland mit dem virus der Lymphocytären choriomeningitis, *Zentrabl. Bakteriol. Parasitenkd. Infecktionskr. Hyg. Abt 1 Orig.*, 213, 1970.
32. **Boyer, K. M., Munford, R. S., Maupin, G. O., Pallison, C. A., Fox, M. D., Barnes, A. M., Jones, W. L., and Maynard, J. L.,** Tick-borne relapsing fever: an interstate outbreak originating at Grand Canyon National Park, *Am. J. Epidemiol.*, 105, 469, 1977.
33. **Bozeman, F. M., Shirai, A., Humphries, J. W., and Fuller, H. S.,** Ecology of Rocky Mountain spotted fever. II. Natural infection of wild mammals and birds in Virginia and Maryland, *Am. J. Trop. Med. Hyg.*, 16, 48, 1967.
34. **Bray, R. S.,** The zoonotic potential of reservoirs of leishmaniasis in the Old World, *Ecol. Dis.*, 1, 257, 1982.
35. **Brener, Z.,** Recent developments in the field of Chagas' disease, *Bull. W.H.O.*, 60, 463, 1982.
36. **Brooks, J. E., Naing, H., Walton, D. W., Myint, D. S., Tun, M. M., U Thaung, and Kyi, D. O.,** Plague in small mammals and humans in Rangoon, Burma, *Southeast Asian J. Trop. Med. Public Health,* 8, 335, 1977.
37. **Brown, A. W. A.,** Flea fauna of known and probable foci of plague, World Health Organization, Geneva, unpublished document, VBC/83.2, 1983.
38. **Brown, G. W., Robinson, D. M., Huxsoll, D. L., Ng, T. S., Lim, K. J., and Sannasey, G.,** Scrub typhus: a common cause of illness in indigenous populations, *Trans. R. Soc. Trop. Med. Hyg.*, 70, 444, 1976.
39. **Brown, R. J., Carney, W. P., Cross, J. H., and Sulianti, S.,** *Capillaria* hepatitis in the Indonesian house shrew *Suncus murinus, Southeast Asian J. Trop. Med. Public Health,* 6, 599, 1975.
40. **Brummer-Korvenkontio, M., Vaheri, A., Hovi, T., von Bonsdorff, C-H, Vuorimies, J., Manni, T., Penttinen, K., Oker-Blom, N., and Lähdevirta, J.,** Nephropathia epidemica: detection of antigen in bank-voles and serologic diagnosis of human infection, *J. Infect. Dis.*, 141, 131, 1980.
41. **Brygoo, E. R.,** Epidémiologie de la peste a Madagascar, *Arch. Inst. Pasteur Madagascar,* 35, 9, 1966.
42. **Burgdorfer, W.,** Ecology of tick vectors of American spotted fever, *Bull. W.H.O.*, 40, 375, 1969.
43. **Burgdorfer, W.,** Murine (flea-borne) typhus, in *Tropical Medicine,* 5th ed., W.B. Saunders, Philadelphia, Pa., 1976, 110.
44. **Burgdorfer, W.,** Tick-borne diseases in the United States: Rocky Mountain spotted fever and Colorado tick fever, *Acta Trop.*, 34, 103, 1977.
45. **Burgdorfer, W.,** Tick-borne diseases in the United States: Rocky Mountain spotted fever and Colorado tick fever, *Acta Trop.*, 34, 103, 1977.
46. **Burgdorfer, W.,** The spotted fever-group diseases, in *CRC Handbook Series in Zoonoses, A., Bacterial, Rickettsial and Mycotic Diseases, Vol. II,* Stoenner, H., Kaplan, W., and Torten, M., Eds., CRC Press, Boca Raton, Fla., 1980, 279.
47. **Burgdorfer, W. and Brinton, L. P.,** Mechanisms of transovarial infection of spotted fever rickettsiae in ticks, *Ann. N.Y. Acad. Sci.*, 266, 61, 1975.
48. **Burgdorfer, W., Newhouse, V. F., Pickens, E. G., and Lackman, D. B.,** Ecology of Rocky Mountain spotted fever in Western Montana. I. Isolation of *Rickettsiae rickettsiae* from wild mammals, *Am. J. Hyg.*, 76, 293, 1962.
49. **Burridge, M. J.,** Toxoplasmosis, *Comp. Contin. Educ. Pract. Vet.*, 2, 233, 1980.
50. **Burridge, M. J., Bigler, W. J., Forrester, D. J., and Hennemann, J. M.,** Serologic survey for *Toxoplasma gondii* in wild animals in Florida, *J. Am. Vet. Med. Assoc.*, 175, 964, 1979.
51. **Cabrera, B. D.,** *Schistosomiasis japonica* in field rats in Leyte, Philippines, *Southeast Asian J. Trop. Med. Public Health,* 7, 50, 1976.
52. **Caflisch, U., Tönz, O., Schaas, V. B., Aeschlimann, A., and Burgdorfer, W.,** Tick-borne meningo-radiculitis-a spirochetosis, *Schweiz Med. Wochenschr.*, 114, 630, 1984.
53. **Carney, W. P., Van Peenen, P. F. D., and Sudomo, M.,** A mammalian reservoir of *Schistosoma japonicum* in the Napu Valley, Central Sulawesi, Indonesia, *J. Parasitol.*, 64, 1138, 1978.
54. Centers for Disease Control, Rat-bite fever in a college student-California, *Morbid. Mort. Wkly. Rep.*, 33, 318, 1984.
55. Centers for Disease Control, Rocky Mountain Spotted Fever-United States, 1984 *Morbid. Mort. Wkly. Rep.*, 23, 195, 1985.
56. Centers for Disease Control, Update: Lyme disease and cases occurring during pregnancy-United States, *Morbid. Mort. Wkly. Rep.*, 34, 376, 1985.
57. **Chamberlain, R. W., Sudia, W. D., Coleman, P. H., and Work, T. H.,** Venezuelan equine encephalitis from south Florida, *Science,* 145, 272, 1964.

58. **Childs, J. E., Korch, G. W., Smith, G. A., Terry, A. D., and LeDuc, J. W.,** Geographical distribution and age related prevalence of antibody to Hantaan-like virus in rat populations of Baltimore, Maryland, USA, *Am. J. Trop. Med. Hyg.*, 34, 385, 1985.
59. **Combes, C. and Delattre, P.,** Principaux paramètres de l'infestation des rats *(Rattus rattus et Rattus norvegicus)* par *Schistosoma mansoni* dans un foyer de schistosomose intestinale de la région caraibe, *Acta Oecologia/Oecol. Applic.*, 2, 63, 1981.
60. **Coura, J. R., Ferreira, L. F., and Silva, J. R.,** Triatomineos no estado do Gunanabara e suas relacoes com o domicilio humano, *Rev. Inst. Med. Trop. Sao Paulo,* 8, 162, 1966.
61. **Daiter, A. B.,** Transovarial and transpermal transmission of *Coxiella burneti* by the tick *Hyaloma asiaticum* and its role in Q-rickettsiosis ecology, in Russian with English summary, *Parazitologiya,* 11, 403, 1977.
62. **D'Alessandro, A., Barreto, P., Saravia, N., and Barreto, M.,** Epidemiology of *Trypansoma cruzi* in the oriental plains of Colombia, *Am. J. Trop. Med. Hyg.*, 33, 1084, 1984.
63. **Danysz, J.,** Destruction des campagnols et de mulots par les cultures Artifielles des microbes pathogenes, *J. Agric. Prat.*, 57, 920, 1893.
64. **Davis, D. H. S.,** Plague in South Africa: a study of the epizootic cycle in gerbils *(Tatera brantisi),* in the Northern Orange Free State, *J. Hyg.*, 51, 427, 1953.
65. **Davis, D. H. S., Heisch, R. B., McNeil, D., and Meyer, K. F.,** Serological survey of plague in rodents and other small mammals in Kenya, *Trans. R. Soc. Trop. Med. Hyg.*, 62, 838, 1968.
66. **Dedet, J.-P., Derouin, F., Hubert, B., Schnur, L. F., and Chance, M. L.,** Isolation of *Leishmania major* from *Mastomys erythroleucus* and *Tatera gambiana* in Senegal (West Africa), *Ann. Trop. Med. Parasitol.*, 73, 433, 1979.
67. **Dubrovsky, Yu. A.,** Role of the great gerbil in Turan lowland desert parasitic systems, in Russian, *Fauna Ekol. Gruz.*, 52, 106, 1983.
68. **Dyer, R. E., Rumreich, A., and Badger, L. F.,** Typhus fever. A virus of the typhus type derived from fleas collected from wild rats, *Public Health Rep.*, 46, 334, 1931.
69. **Emmons, R. W.,** Colorado tick fever, in *CRC Handbook Series in Zoonoses, B, Viral Zoonoses, Vol. I,* Beran, G., Ed., CRC Press, Boca Raton, Fla., 1981, 113.
70. **Eyles, D. E.,** *Toxoplasma* in the Norway rat, *J. Parasitol.*, 38, 226, 1952.
71. **Faine, S., Ed.,** Guidelines for the control of leptospirosis, *WHO Offset Pub.*, No. 67, Geneva, 1982.
72. **Farhang-Azad, A.,** Ecology of *Capillaria hepatica* (Bancroft 1893) Nematoda. I. Dynamics of infection among Norway rat populations of the Baltimore Zoo, Baltimore, Maryland, *J. Parasitol.*, 63, 117, 1977.
73. **Farhang-Azad, A. and Schlitter, D. A.,** *Capillaria hepatica* in small mammals collected from Shoa Province, Ethiopia, *J. Wildl. Dis.*, 14, 358, 1978.
74. **Farhang-Azad, A. and Traub, R.,** Transmission of murine typhus rickettsiae by *Xenopsylla cheopis* with notes on experimental infection and effects of temperature, *Am. J. Trop. Med. Hyg.*, 34, 555, 1985.
75. **Farhang-Azad, A., Traub, R., Sofi, M., and Wisseman, C. L.,** Experimental murine typhus infection in the cat flea, *Ctenocephalides felis* (Siphonoptera: Pulicidae), *J. Med. Entomol.*, 21, 675, 1984.
76. **Farhang-Azad, A., Traub, R., and Wisseman, C. L.,** *Rickettsia mooseri* infection in the fleas *Leptopsylla segnis* and *Xenopsylla cheopis, Am. J. Trop. Med. Hyg.*, 32, 1392, 1983.
77. **Fendall, N. R. E.,** The spread of kala azar in Kenya, *East Afr. Med. J.*, 38, 417, 1961.
78. **Fisher-Hoch, S. P. and McCormick, J. B.,** Haemorrhagic fever with renal syndrome: a review, *Abstr. Hyg. Communic. Dis.*, 60, R-1, 1985.
79. **Frankel, J. K. and Ruiz, A.,** Human toxoplasmosis and cat contact in Costa Rica, *Am. J. Trop. Med. Hyg.*, 29, 1167, 1980.
80. **Fraser, D. W., Campbell, C. C., Monath, T. P., Goff, P. A., and Gregg, M. B.,** Lassa fever in the Eastern Province of Sierra Leone, 1970-1972. I. Epidemiological studies, *Am. J. Trop. Med. Hyg.*, 23, 1131, 1974.
81. **Fuentes, L. G.,** Primer caso de fiebre de los Montonas Rocosas en Costa Rica, América Central, *Rev. Lat. Am. Microbiol.*, 21, 167, 1979.
82. **Galvao, V. A.,** *Capillaria hepatica* estudo da incidencia em ratas de Salvador, Bahia e dados imunopatologicos preliminares, *Rev. Soc. Bras. Med. Trop.*, 10, 333, 1976.
83. **Gamboa, J.,** Dispersion de *Rhodnius prolixus* en Venezuela, *Bol. Inf. Dir. Malarid. Sanidad Ambiental.*, 3, 262, 1963.
84. **Gaon, J., Karlovac, M., Gresikova, M., Hlaca, D., Rukavina, J., Knezevic, V., Saratlic, D., and Vompotic, A.,** Epidemiological features of haemorrhagic fever, *Folio Med. Fac. Med. Univ. Saraviensis* (Sarajevo), 3, 23, 1968.
85. **Gear, J.,** The rickettsial diseases of Southern Africa, *S. Afr. J. Clin. Sci.*, 5, 158, 1954.
86. **Geigy, R. and Aeschlimann, A.,** Ratten als reservoir von *Borrlia duttoni, Tropenmed. Parasitol.*, 8, 96, 1957.
87. **Gerichter, Ch. B.,** Animal sources of *Salmonella* in Israel, *Isr. J. Med. Sci.*, 6, 413, 1970.
88. **Gilyard, R. T.,** A clinical study of Venezuelan virus equine encephalomyelitis in Trinidad, B.W.I., *J. Am. Vet. Med. Assoc.*, 151, 267, 1945.

89. **Gonzalez, N. R., Torrealba, J. W., Comach, G. A., and Alfaro, O.,** *Rattus rattus* hospedador natural de *Schistosoma mansoni* en Venezuela, *Bol. Inf. Dir. Malariol. Sanidad Ambiental,* 16, 126, 1976.
90. **Goubau, P. F.,** Relapsing fevers, a review, *Ann. Soc. Belge Med. Trop.,* 64, 335, 1984.
91. **Gratz, N. G.,** Mosquito-borne disease problems in the urbanization of tropical countries, *CRC Crit. Rev. Environ. Control,* 3, 455, 1973.
92. **Gratz, N. G.,** Urban rodent-borne disease and rodent distribution in Israel and neighbouring countries, *Israel J. Med. Sci.,* 9, 969, 1973.
93. **Gratz, N. G.,** Global public health importance of rodents, in *Proc. Conf. Organization and Practice of Vertebrate Pest Control,* ICI, Surrey, England, 413, 1984.
94. **Grayson, M. A. and Galindo, P.,** Ecology of Venezuelan equine encephalitis virus in Panama, *Am. Vet. Med. Assoc.,* 155, 2141, 1969.
95. **Greenberg, M., Pellitteri, O. J., and Jellison, W. L.,** Rickettsialpox-A newly recognized rickettsial disease. III. Epidemiology, *Am. J. Public Health,* 37, 860, 1947.
96. **Greenwood, B. M., Ajdukiewicz, A. B., Conteh, S., Hagan, P., Mabey, D. C. W., and Panton, L. J.,** Leishmaniasis in the Gambia. III. Is its incidence increasing?, *Trans. R. Soc. Trop. Med. Hyg.,* 78, 407, 1984.
97. **Gregg, M. B.,** Recent outbreaks of lymphocytic choriomeningitis in the United States of America, *Bull. W.H.O.,* 52, 549, 1975.
98. **Gresikova, M., Rajcani, J., Sekeyova, M., Brummer-Korvenkontio, M., Kozuch, O., Labuda, M., Turek, R., Weismann, P., and Lysy, J.,** Haemorrhagic fever virus with renal syndrome in small rodents in Czechoslovakia, *Acta Virol.,* 28, 416, 1984.
99. **Guineé, P. A. M., Kampelmacher, E. H., Van Keulen, A., and Ophof, A. J.,** Incidence of salmonella in brown rats *(Rattus norvegicus)* caught in and near slaughterhouses, farms and mink farms, *Z. Veterinarmed.,* 10, 181, 1963.
100. **Haas, V. S.,** A study of *Pseudotuberculosis rodentium* recovered from a rat, *Public Health Rep.,* 53, 1033, 1938.
101. **Haile, P. T. and Lemma, A.,** Isolation of *Leishmania* parasites from *Arvicanthis* in Ethiopia, *Trans. R. Soc. Trop. Med. Hyg.,* 71, 180, 1977.
102. **Hamidi, A. N., Saadatezadeh, H., Tarasevich, I. V., Arata, A. A., and Farhang Azad, A.,** A serological study of rickettsial infections in Iranian small mammals, *Bull. Soc. Pathol. Exot.,* 67, 607, 1974.
103. **Hammon, W. McD., Reeves, W. C., and Galindo, P.,** Epidemiologic studies of encephalitis in the San Joaquin Valley of California, 1943, with isolation of viruses from mosquitoes, *Am. J. Hyg.,* 42, 299, 1945.
104. **Heisch, R. D.,** On *Spirochaeta dipodilli* sp. nov., a parasite of pygmy Gerbils (*Dipodillus* sp.), *Ann. Trop. Med. Parasitol.,* 44, 260, 1950.
105. **Hoogstraal, H.,** Tickborne hemorrhagic fevers, encephalitis and typhus in U.S.S.R. and Southern Asia, *Exp. Parasitol.,* 21, 98, 1967.
106. **Hoogstraal, H.,** The epidemiology of tick-borne Crimean-Congo haemorrhagic fever in Asia, Europe and Africa, *J. Med. Entomol.,* 15, 307, 1979.
107. **Hopla, C. E.,** The ecology of tularemia, *Adv. Vet. Sci. Comp. Med.,* 18, 25, 1974.
108. **Hubert, W. T.,** Yersiniosis in mammals and birds in the United States — case reports and reviews, *Am. J. Trop. Med. Hyg.,* 21, 458, 1972.
109. **Huebner, R. J., Jellison, W. L., and Armstrong, C.,** Rickettsialpox-A newly recognized rickettsial disease. V. Recovery of *Rickettsia akari* from a house mouse *(Mus musculus), Public Health Rep.,* 62, 777, 1947.
110. **Jackson, E. B., Danauskas, J. X., Coale, M. C., and Smadel, J. E.,** Recovery of *Rickettsia akari* from the Korean vole, *Microtus fortis pelliceus, Am. J. Hyg.,* 66, 301, 1957.
111. **Jacobs, L. and Frenkel, J. K.,** Toxoplasmosis, in *CRC Handbook Series in Zoonoses, C, Parasitic Zoonoses, Vol. I,* Jacobs, L. and Arambulo, P., Eds., CRC Press, Boca Raton, Fla., 1982, 167.
112. **Johnson, D. E., Crum, J. W., Hanchalay, S., and Saengrachi, C.,** Seroepidemiological survey of *Rickettsia tsutsugamushi* infection in a rural Thai village, *Trans. R. Soc. Trop. Med. Hyg.,* 76, 1, 1982.
113. **Johnson, K. M., Kuns, M. L., Mackenzie, R. B., Webb, P. A., and Yunker, C. E.,** Isolation of Machupo Virus from wild rodent *Calomys callosus, Am. J. Trop. Med. Hyg.,* 15, 103, 1966.
114. **Johnson, K. M., Taylor, P., Elliot, L. H., and Tomori, O.,** Recovery of a Lassa related arbovirus in Zimbabwe, *Am. J. Trop. Med. Hyg.,* 30, 1291, 1981.
115. **Jonkers, A. H.,** Silent hosts of Venezuelan equine encephalitis (VEE) virus in endemic situations: mammals, in Venezuelan encephalitis, Pan American Health Organization, Sci. Pub. No. 243, Washington, D.C., 1972, 263.
116. **Jourdane, J. and Imbert-Establet, D.,** Etude expérimentale de la permissivité du rat sauvage *(Rattus rattus)* de Guadeloupe à l'egard de *Schistosoma mansoni.* Hypothèse sur le role de cet hote dans la dynamique des foyers naturels, *Acta Trop.,* 37, 41, 1980.
117. **Kaneko, K. I., Hamada, S., Kasai, Y., and Hashimoto, N.,** Smouldering epidemic of *Yersina pseudotuberculosis* in barn rats, *Appl. Environ. Microbiol.,* 37, 1, 1979.

118. **Kawashima, K., Katamine, D., Sakamoto, M., Shimada, M., Nojima, H., and Miyahara, M.,** Investigations on the role of wild rodents as reservoirs of human schistosomiasis in the Taveta area of Kenya, East Africa, *Jpn. J. Trop. Med. Hyg.*, 6, 195, 1978.
119. **Keenlyside, R. A., McCormick, J. B., Webb, P. A., Smith, E., Elliot, L., and Johnson, K. M.,** Case control study of *Mastomys natalensis* and humans in Lassa virus-infected households in Sierra Leone, *Am. J. Trop. Med. Hyg.*, 32, 829, 1983.
120. **Kritsky, D. C., Leiby, P. D., and Miller, G. E.,** The natural occurrence of *Echinococcus multilocularis* in the bushy-tailed woodrat, *Neotoma cinerea rupicola* in Wyoming, *Am. J. Trop. Med. Hyg.*, 26, 1046, 1977.
121. **Kulagin, S. M., Tarasevich, I. V., Kudryashova, N. I., and Gopachenko, I. M.,** On the natural focus of scrub typhus in the south of the Primorie area of the USSR, *Acta Med. Biol.*, 15(suppl.), 49, 1967.
122. **Kums, M. L.,** Epidemiology of Machupo virus infection. II. Ecological and cintrol studies of hemorrhagic fever, *Am. J. Trop. Hyg.*, 14, 813, 1965.
123. **Kunz, Ch.,** *Tick-Borne Encephalitis,* Facultas-Verlag, Vienna, 1981.
124. **Kziazek, T. C. and Yuill, T. M.,** Viremia and antibody response to LaCrosse virus in sentinel gray squirrels *(Sciurus carolinensis)* and chipmunks *(Tamias striatus), Am. J. Trop. Med. Hyg.*, 26, 815, 1977.
125. **Lainson, R.,** Leishmaniasis, in *CRC Handbook Series in Zoonoses, C., Parasitic Zoonoses,* Vol. I, Jacobs, L. and Arambulo, P., Eds., CRC Press, Boca Raton, Fla., 1982, 41.
126. **Lancastre, F., Nassi, H., and Poirot, J.-L.,** Sur la localisation pulmonire massive de *Schistosoma mansoni* chez des rats *(Rattus norvegicus)* en Guadeloupe, *Ann. Parasitol.*, 51, 349, 1976.
127. **Lane, R. S., Emmons, R. W., Dondero, D. V., and Nelson, B. C.,** Ecology of tick-borne agents in Calfornia. I. Spotted fever group rickettsiae, *Am. J. Trop. Med. Hyg.*, 30, 230, 1981.
128. **Larivière, M., Camerlynck, P., Ranque, P., and Villod, M. T.,** *Arvicanthis* sp. réservoir de virus naturel possible de *Leishmania tropica* au Sénégal, *C. R. Acad. Sci. Paris Sér. D,* 260, 4869, 1965.
129. **LeDuc, J. W.,** The ecology of California group viruses, *J. Med. Entomol.*, 16, 1, 1981.
130. **LeDuc, J. W., Smith, G. A., and Johnson, K. M.,** Hantaan-like viruses from domestic rats captured in the United States, *Am. J. Trop. Med. Hyg.*, 33, 992, 1984.
131. **LeDuc, J. W., Smith, G. A., Pinheiro, F. P., Vasconcelos, P. F. C., Rosa, E. S. T., and Maiztegui, J. I.,** Isolation of Hantaan-related virus from Brazilian rats serologic evidence of its widespread distribution in South America, *Am. J. Trop. Med. Hyg.*, 34, 810, 1985.
132. **Lee, H. W., Baek, L. J., and Johnson, K. M.,** Isolation of Hantaan virus, the etiologic agent of Korean hemorrhagic fever from wild urban rats, *J. Infect. Dis.*, 46, 638, 1982.
133. **Lee, H. W. and Lee, P. W.,** Korean haemorrhagic fever. I. Demonstrations of causative antigen and antibodies, *Korean J. Int. Med.*, 19, 371, 1976.
134. **Lee, H. W., Lee, P. W., and Johnson, K. M.,** Isolation of the etiologic agent of Korean hemorrhagic fever, *J. Inf. Dis.*, 137, 298, 1978.
135. **Lee, P. W., Amyx, H. L., Gajdusek, D. C., Yanagihara, R. T., Goldhaber, D., and Gibbs, C. J.,** New haemorrhagic fever with renal syndrome related virus in indigenous wild rodents in United States, *Lancet,* 2, 1405, 1982.
136. **Lee, P. W., Yanagihara, R., Franko, M. G., Amyx, H. L., Gibbs, C. J., Gajdusek, D. C., and Traub, R. T.,** Preliminary evidence that Hantaan or a closely related virus is enzootic in domestic rodents, *J. Med.*, 307, 624, 1982.
137. **Le Gac, P. and Arquié, E.,** Les facteurs endémiques du scrub-typhus Indochinois, *Bull. Soc. Pathol. Exot.*, 57, 277, 1964.
138. **Leiby, P. D., Carney, W. P., and Woods, C. E.,** Studies on sylvatic echinoccosis. III. Host occurrence and geographic distribution of *Echinococcus multilocularis* in the north-centrl United States, *J. Parasitol.*, 56, 1141, 1970.
139. **Leslie, P. H.,** The bacteriological classification of the principal cultures used in rat and mouse control in Great Britain, *J. Hyg.*, 42, 552, 1942.
140. **Levine, J. F., Wilson, M. L., and Spielman, A.,** Mice as reservoirs of the Lyme disease spirochete, *Am. J. Trop. Med. Hyg.*, 34, 335, 1985.
141. **Lhuillier, M., Pajot, F. X., Mouchet, J., and Robin, Y.,** Arboviruses en Amerique du sud et dans les Caraibes, *Med. Trop.*, 41, 73, 1981.
142. **Lim, B. L., Ow-Yang, C. k., and Lie, K. J.,** Natural infection of *Angiostrongylus cantonensis* in Malaysian rodents and intermediate hosts and preliminary observations on acquired resistance, *Am. J. Trop. Med. Hyg.*, 14, 610, 1965.
143. **Lindenbaum, I. and Eylan, E.,** Leptospirosis in *Rattus norvegicus* and *Rattus rattus* in Israel, *Isr. J. Med. Sci.*, 18, 271, 1982.
144. **Linnemann, C. C., Barber, L. C., Dine, M. S., and Body, A. E.,** Tick-borne relapsing fever in the eastern United States, *Am. J. Dis. Child.*, 132, 40, 1978.
145. **Loeffler, F.,** Die Feldmausplage in Teessalien und ihre erfolgreiche Bekaempfung mittels des *Bacillus typhi-murium, Z. Bakteriol.*, 12, 1, 1892.

146. **Lord, R. D., Calisher, C. H., Sudia, W. D., and Work, T. H.,** Ecological investigations of vertebrate hosts of Venezuelan equine encephalomyelitis virus in south Florida, *Am. J. Trop. Med. Hyg.*, 22, 116, 1973.
147. **Lumbreras, H.,** El problema de la enfermedad de Chagas en los diferentes departamentos del Peru, *Viernes Med.*, 23, 43, 1972.
148. **Macchiavello, A.,** Peste bubonica en el Peru, *Bol. Oficina Sanit. Panam.*, 24, 704, 1958.
149. **Macchiavello, A.,** Estudios sobre peste selvatica en América de sur. V. Peste selvatica en Bolivia. Consideraciones generales sobre la geografica e historia de la peste, *Bol. Oficina Sanit. Panam.*, 46, 509, 1959.
150. **Mackenzie, R. B., Beye, H. K., Valverde, L., and Garron, H.,** Epidemic hemorrhagic fever in Bolivia. I. A preliminary report of the epidemiologic and clinical findings in a new epidemic area in South America, *Am. J. Trop. Med. Hyg.*, 13, 620, 1964.
151. **Maiztegui, J. I.,** Clinical and epidemiological patterns of Argentinian haemorrhagic fever, *Bull. W.H.O.*, 52, 567, 1975.
152. **Malek, E. A.,** Presence of *Angiostrongylus costaricensis* Morea and Céspedes 1971 in Colombia, *Am. J. Trop. Med. Hyg.*, 30, 81, 1981.
153. **Marchette, N. J.,** Rickettsioses (tick typhus, Q fever, urban typhus) in Malaysia, *J. Med. Entomol.*, 2, 339, 1966.
154. **Margono, S. and Ilahude, H. D.,** *Angiostrongylus cantonensis* in rats and intermediate hosts in Jakarta and its vicinity, *Southeast Asian J. Trop. Med. Public Health,* 5, 236, 1974.
155. **Marinkelle, C. and Grose, E. S.,** Species of *Borrelia* from a Columbian bat *(Natalus tumidirostris), Nature,* 218, 487, 1968.
156. **McConnell, S. and Spertzel, R. O.,** Venezuelan equine encephalitis (VEE), in *CRC Handbook Series in Zoonoses, B, Viral Zoonoses, Vol. I,* Beran, G. W., Ed., CRC Press, Boca Raton, Fla., 1981, 59.
157. **McCoy, G. W. and Chapman, C. W.,** Studies of plague, a plague-like disease and tuberculosis among rodents in California. V. *Bacterium tularense* — the cause of a plague-like disease of rodents, U.S. Public Health Bull., 53, 17, 1912.
158. **McKiel, J. A., Rappay, D. E., Cousineau, J. G., Hall, R. R., and McKenna, H. E.,** Domestic rats as carriers of leptospirosis and salmonellae in Eastern Canada, *Can. J. Public Health,* 61, 336, 1970.
159. **Men Ruhe, J. K., Yuehuai, G., Xianlu, A., and Xinhua, Z.,** Studies on the isolation and identification of the etiologic agents of haemorrhagic fever with renal syndrome, in World Health Organization ad hoc meeting on Surveillance of Viral Haemorrhagic Fevers, abstr., Antwerp, Belgium, 1983.
160. **Mercado, R. R.,** Rodent control programmes in areas affected by Bolivian haemorrhagic fever, *Bull. W.H.O.*, 52, 691, 1975.
161. **Misonne, X.,** Les rongeurs des foyers de peste congolais, *Ann. Soc. Belge. Med. Trop.*, 39, 436, 1959.
162. **Misonne, X.,** Un foyer naturel de peste en Libye, *Ann. Soc. Belge. Med. Trop.*, 57, 163, 1977.
163. **Monath, T. P., Newhouse, V. F., Kemp, G. E., Setzer, H. W., and Cacciapuoti, A.,** Lassa virus isolation from *Mastomys natalensis* rodents during an epidemic in Sierra Leone, *Science,* 185, 263, 1974.
164. **Monson, M. H., Frame, J. D., Jahrling, P. B., and Alexander, K.,** Endemic Lassa fever in Liberia. I. Clinical and epidemiological aspects at Curran Lutheran Hospital, Zorzor, Liberia, *Trans. R. Soc. Trop. Med. Hyg.*, 78, 549, 1984.
165. **Mooser, H., Castaneda, M. R., and Zinsser, H.,** Rats as carriers of Mexican typhus fever, *J.A.M.A.*, 97, 231, 1931.
166. **Morita, C., Sugiyama, K., Matsurra, Y., Kitamura, T., Komatsu, T., Akao, Y., Jitsukawa, W., and Sakakibara, H.,** Detection of antibody against haemorrhagic fever with renal syndrome (HFRS) virus in sera of house rats captured in port areas of Japan, *Jpn. J. Med. Sci. Biol.*, 36, 55, 1983.
167. **Moulton, D. W. and Thompson, W. H.,** California group virus infections in small forest dwelling mammals of Wisconsin: some ecological considerations, *Am. J. Trop. Med. Hyg.*, 20, 474, 1971.
168. **Mull, I., Lim, B. L., and Walker, J. S.,** Scrub typhus infection in rats in four habitats in Peninsular Malaysia, *Trans. R. Soc. Trop. Med. Hyg.*, 71, 493, 1977.
169. **Mutinga, M. J.,** The animal reservoirs of cutaneous leishmaniasis on Mt. Elgon, Kenya, *E. Afr. Med. J.*, 52, 142, 1975.
170. **Nguyen-Van-Ai, Nguyen-Duc-Hanh, Le Tien-Van, Nguyen-Van-Le, Nguyen-Thi Lan-Huong, and Pham-Van-Dien,** Les salmonelles isolées des rats et des musaraignes dans le Région de Saigon-Giadinh, *Bull. Soc. Pathol. Exot.*, 67, 579, 1974.
171. **Niphadkar, S. M. and Rao, S. R.,** On the occurrence of *Raillietina celebensis* (Jericki, 1902) in rats of Bombay with special reference to its zoonotic importance, *Indian Vet. J.*, 46, 816, 1969.
172. **Otsuru, M.,** *Angiostrongylus cantonensis* and angiostrongyliasis in Japan, in *Studies on Angiostrongyliasis in Eastern Asia and Australia,* U.S. Navy Medical Research Unit 2, pub. NAMRU 2, Taipei, 1979, 74.
173. Pan American Health Organization, Plague in the Americas, Sci. Pub. No. 115, Pan American Health Organization, Washington, D.C., 1965.
174. Pan American Health Organization, Venezuelan encephalitis, Sci. Pub. No. 243, Pan American Health Organization, Washington, D.C., 1972.

175. Pan American Health Organization, Argentinian haemorrhagic fever, Epidem. Bull., Pan American Health Organization, 3, 1, Washington, D.C., 1982.

176. **Pesigan, T. B., Farooq, M., Hairston, N. B., Jauregue, J. J., Garcia, E. G., Santos, B. D., and Besa, A. B.,** Studies on *Schistosoma japonicum* infection in the Philippines, *Bull. W.H.O.*, 18, 345, 1958.

177. **Petrovic, Z., Bordyoski, A., and Savin, Z.,** Les résultats de recherches sur les réservoir de *Leishmania donovani* dans une région endémique de kala azar, 2nd Eur. Multi. Coll. Parasitol., Trogir, 97, 1975.

178. **Philip, R. N.,** Scrub Typhus, in *CRC Handbook Series in Zoonoses A, Bacterial, Rickettsial and Mycotic Diseases,* Vol. II, Stoenner, H., Kaplan, W., and Torten, M., Eds., CRC Press, Boca Raton, Fla., 1980, 303.

179. **Piesman, J., Donahue, J. G., Mather, T. N., and Spielman, A.,** Transovarially acquired Lyme disease spirochetes *(Borrelia burgdorferi)* in field-collected larval *Ixodes dammini,* (acari: Ixodidae), *J. Med. Entomol.*, 23, 219, 1986.

180. **Piesman, J., Mota, E., Sherlock, I. A., and Todd, C. W.,** *Typanosoma cruzi:* an association between seroactivity of children and infection rates in domestic, *Panstrongylus megistus* (Hemiptera: Reduviidae), *J. Med. Entomol.*, 22, 130, 1985.

181. **Poland, V. D. and Barnes, A. M.,** Plague, *CRC Handbook Series in Zoonoses, A, Bacterial Rickettsial and Mycotic Diseases, Vol. I,* Stoenner, H., Kaplan, W., and Torten, M., Eds., CRC Press, Boca Raton, Fla., 1979, 515.

182. **Pollitzer, R.,** Plague, World Health Organization Monograph Series, 22, Geneva, 1954.

183. **Pollitzer, R.,** A review of recent literature on plague, *Bull. W.H.O.*, 23, 313, 1960.

184. **Pospelova-Shtrom, M. V.,** Ornithdorini ticks and their epidemiological significance, U.S. Dept. of Commerce, Office of Technical Services, Joint Publication Research Service, Washington, D.C., 1962.

185. **Pozio, E., Gradoni, L., Bettini, S., and Gramiccia, M.,** Leishmaniasis in Tuscany (Italy). V. Further isolations of *Leishmania* from *Rattus rattus* in the Province of *Grosseto,* Ann. Trop. Med. Parasitol., 75, 393, 1981.

186. **Pratt, H. D.,** The changing picture of murine typhus in the United States, *Ann. N.Y. Acad. Sci.*, 70, 516, 1958.

187. **Ramos Diaz, A.,** Epidemiologia de la peste bubonica en la sierra del Departamento de Lambayeque, *Bol. Oficina Sanit. Panam.*, 17, 776, 1938.

188. **Reeves, W. C., Emmons, R. W., and Hardy, J. L.,** Historical perspectives on California encephalitis virus in California, in *California Serogroup Viruses,* Allen R. Liss, New York, 1983, 19.

189. **Rioux, J. A., Petter, F., Akalay, O., Lanotte, G., Quazzani, A., Sequignes, M., and Mochine, A.,** *Meriones shawi* (Duvernoy, 1842) (Rodentia, Gerbillidae) reservoir de *Leishmania major* (Yakomoff et Schokhov, 1914) (Kinetoplastida, Trypanosomatae) dans le sud marrocain, *C.R. Acad. Sci. Paris,* 294, 515, 1982.

190. **Rioux, J. A., Petter, F., Zahaf, A., Lanotte, G., Houin, R., Jarry, D., Perieres, J., Martini, A., and Sarhani, S.,** Isolement de *Leishmania major* Yakimoff et Schokhor, 1914 [Kinetoplastida-Trypanosomatidae] chez *Meriones shawi shawi* (Duvernoy, 1842) [Rodentia-Gerbillidae] en Tunisie, *Ann. Parasitol. Hum. Comp.*, 61, 139, 1986.

191. **Robinson, D. M., Gan, E., and Donaldson, J. R.,** The prevalence of scrub typhus antibodies in residents of West Malaysia, *Trop. Geograph. Med.*, 28, 303, 1976.

192. **Rodhain, F.,** *Borrelia* et fièvres récurrentes: aspects épidémiologiques actuels, *Bull. Inst. Pasteur,* 74, 173, 1976.

193. **Ruiz, Z. and Frankel, J. K.,** Intermediate and transport hosts of *Toxoplasma gondii* in Costa Rica, *Am. J. Trop. Med. Hyg.*, 29, 1161, 1980.

194. **Scherer, W. F., Dickerman, R. W., La Fiandra, R. P., Wong Chia, C., and Terrian, J.,** Ecologic studies of Venezuelan encephalitis virus in southeastern Mexico. IV. Infections of wild animals, *Am. J. Trop. Med. Hyg.*, 20, 980, 1971.

195. **Scherer, W. F., Dickerman, R. W., Ordonez, J. V., Seymour, C., Kramer, L. D., Jahrling, P. B., and Powers, C. D.,** Ecologic studies of Venezuelan encephlitis virus and isolations of Nepuyo and Patois viruses during 1968-1973 at a marsh habitat near the epicenter of the 1969 outbreak in Guatemala, *Am. J. Trop. Med. Hyg.*, 25, 151, 1976.

196. **Schluger, E. G. and Kudryashova, N. I.,** Red-bodied mites (Trombiculidae) and *Rickettsia tsutsugamushi,* in *Biological Relationships Between Bloodsucking Arthropods and Pathogens of Human Disease,* Meditsina, Moscow, 1967, 143.

197. **Schmid, G. P.,** The global distribution of Lyme Disease, *Yale J. Biol. Med.*, 57, 617, 1984.

198. **Schulze, T. L., Bowen, G. S., Bosler, E. M., Lakat, M. F., Parkin, W. E., Altman, R., Ormiston, B. G., and Shisler, J. K.,** *Amblyoma americanum* — a potential vector of Lyme disease in New Jersey, *Science,* 224, 601, 1984.

199. **Scrimgeour, E. M. and Welch, J. S.,** *Angiostrongylus cantonensis* in East New Britain, Papua New Guinea, *Trans. R. Soc. Trop. Med. Hyg.*, 78, 774, 1984.

200. **Sellers, R. F., Bergold, G. H., Suarez, O. M., and Morales, A.,** Investigations during the Venezuelan equine encephalitis outbreaks in Venezuela — 1962-1964, *Am. J. Trop. Med. Hyg.*, 14, 460, 1965.
201. **Sergent, A.,** Fièvre récurrente a *Spriochaeta hispanicum* en Algérie, *Arch Inst. Pasteur Algérie*, 16, 403, 1938.
202. **Sharma, K. D., Renapurkar, D. M., Bhopale, M. K., Nathan, J., Boroskar, A., and Chotani, S.,** Study of a focus of *Angiostrongylus cantonensis* in greater Bombay, *Bull. Haffkine Inst.*, 9, 38, 1981.
203. **Sidwell, R. W., Gebhardt, O. P., and Thorpe, B. D.,** Epidemiological aspects of Venezuelan equine encephalomyelitis virus infections, *Bacteriol. Rev.*, 31, 65, 1967.
204. **Silva, E. O. R., Souza, J. M. P., Andrade, J. C. R., Mello, C. S., and Ferreira, O. A.,** Preferencia alimentar (entre sangue humano e ave) dos *Triatoma sordida* encontrados em casas habitadas de regiao norte do estado de Sao Paulo, Brasil, *Rev. Saude. Publ. Sao Paulo*, 11, 258, 1977.
205. **Siongok, T. K. A., Njagi, A. M., and Masaba, S.,** Another focus of sylvatic plague in Kenya, *E. Afr. Med. J.*, 54, 694, 1977.
206. **Smith, W. W.,** Populations of the most abundant ectoparasites as related to the prevalence of typhus antibodies in farm rats in an endemic murine typhus region, *Am. J. Trop. Med. Hyg.*, 6, 581, 1957.
207. **Spielman, A., Levine, J. F., and Wilson, M. L.,** Vectorial capacity of North American *Ixodes* ticks, *Yale J. Biol. Med.*, 57, 507, 1984.
208. **Steele, J. H.,** Salmonellosis: a major zoonosis, *Arch. Environ. Health*, 19, 871, 1969.
209. **Steere, A. C., Broderick, T. F., and Malawista, S. E.,** Erythema chronicum migrans and Lyme arthritis: epidemiologic evidence for a tick vector, *Am. J. Epidemiol.*, 108, 312, 1978.
210. **Steere, A. C., Malawista, S. E., Snydman, D. R., Shope, R. E., Andiman, W. A., Ross, M. R., and Steele, F. M.,** Lyme arthritis: an epidemic of oligoarticular arthritis in children and adults in three Connecticut communities, *Arthritis Rheum.*, 20, 7, 1977.
211. **Stoenner, H. G.,** Q fever, in *CRC Handbook Series in Zoonoses, A, Bacterial, Rickettsial and Mycotic Diseases, Vol. II*, Stoenner, H., Kaplan, W., and Torten, M., Eds., CRC Press, Boca Raton, Fla., 1980, 337.
212. **Stovell, P. L.,** Pseudotubercular yersiniosis, in *CRC Handbook Series in Zoonoses, A, Bacterial, Rickettsial and Mycotic Diseases, Vol. II*, Stoenner, H., Kaplan, W., and Torten, M., Eds., CRC Press, Boca Raton, Fla., 1980, 209.
213. **Sudia, W. D., McLean, R. G., Newhouse, V. F., Johnston, J. G., Jr., Trevino, H., Bowen, G. S., and Sather, G.,** Epidemic Venezuelan equine encephalitis in North America in 1971: vertebrate field studies, *Am. J. Epidem.*, 101, 36, 1975.
214. **Tamura, M.,** Occurrence of epidemic hemorrhagic fever in Osaka City: first cases found in Japan with characteristic feature of marked proteinuria, *Bien J.*, 7, 49, 1964.
215. **Taylor, D. J., Lewis, A. L., Edman, J. D., and Jennings, W. L.,** California group arboviruses in Florida: host-vector relations, *Am. J. Trop. Med. Hyg.*, 20, 139, 1971.
216. **Taylor, J.,** Bacterial rodenticides and infection with *Salmonella enteritidis*, *Lancet*, 270, 630, 1956.
217. **Taylor, P., Gordon, D. H., and Isaacson, M.,** The status of plague in Zimbabwe, *Trop. Med. Parasit.*, 75, 165, 1981.
218. **Thiermann, A. B.,** Incidence of leptospirosis in the Detroit rat population, *Am. J. Trop. Med. Hyg.*, 26, 970, 1977.
219. **Tkachenko, E. A., Ivanov, A. P., Donets, M. A., Miasnikov, Y. A., Ryltseva, F. V., Gaponova, L. K., Bashkirtsev, V. N., Okulova, N. M., Drozdov, S. G., Slonova, R. A., Somov, G. P., and Astakhova, T. I.,** Potential reservoir and vectors of haemorrhagic fever with renal syndrome (HFRS) in the U.S.S.R., *Ann. Soc. Belge Méd. Trop.*, 63, 267, 1983.
220. **Torten, M.,** Leptospirosis, in *CRC Handbook Series in Zoonoses, A., Bacterial, Rickettsial and Mycotic Diseases, Vol. I*, Stoenner, H., Kaplan, W., and Torten, M., Eds., CRC Press, Boca Raton, Fla., 1979, 363.
221. **Traub, E.,** A filterable virus recovered from white mice, *Science*, 81, 298, 1935.
222. **Traub, R. and Wisseman, C. L.,** Current concepts of the ecology of chigger-borne rickettsiosis, (Scrub typhus), *Jpn. J. Med. Sci. Biol.*, 27, 1, 1973.
223. **Traub, R. and Wisseman, C. L.,** The ecology of chigger-borne rickettsiosis, (scrub typhus), *J. Med. Entomol.*, 11, 237, 1974.
224. **Traub, R., Wisseman, C. L., and Farhang-Azad, A.,** The ecology of murine typhus — A critical review, *Trop. Dis. Bull.*, 75, 237, 1978.
225. **Tsai, C. C., Kundin, W. D., and Fresh, J. W.,** The zoonotic importance of urban rats as a potential reservoir for human leptospirosis, *J. Formos. Med. Assoc.*, 70, 1, 1971.
226. **Turell, M. J., Reeves, W. C., and Hardy, J. L.,** Transovarial and trans-stadial transmission of California encephalitis virus in *Aedes dorsalis* and *Aedes melanimon*, *Am. J. Trop. Med. Hyg.*, 31, 1021, 1982.
227. **Twigg, G. I.,** The role of rodents in plague dissemination: a world-wide review, *Mammal Rev.*, 8, 77, 1978.

228. **Ubelaker, J. E. and Hall, N. M.,** First report of *Angistrongylus costaricensis* Morea and Céspedes 1971 in the United States, *J. Parasitol.*, 65, 307, 1979.
229. **Van der Groen, G., Trochenko, E. A., Ivarov, A. P., and Verhagen, R.,** Haemorrhagic fever with renal syndrome related virus in indigenous wild rodents in Belgium, *Lancet,* 2, 110, 1983.
230. **Van Peenen, P. F. D., Marshall, J. D., Cavanaugh, D. C., and Rust, J. H.,** Mammals of South Vietnam. II. Disease implications, *Mil. Med.*, 135, 391, 1970.
231. **Velimirovic, B.,** Plague in South-east Asia, *Trans. R. Soc. Trop. Med. Hyg.*, 66, 479, 1972.
232. **Vasileva, I. S., Sharipov, M. K., Ershova, A. S., Mansurov, A. A., Muckhitdinov, A. G., Ibragimov, Yu. I., and Usmanov, S. U.,** The contemporary state of foci and infection rate of tickborne relapsing fever in Uzbekistan, (in Russian), *Parazitologya,* 18, 10, 1984.
233. **Walder, R., Octavio, M. S., and Calisher, C. H.,** Arbovirus studies in the Guajira region of Venezuela: activities of Eastern equine encephalitis and Venezuelan equine encephalitis viruses during interepizootic period, *Am. J. Trop. Med. Hyg.*, 33, 699, 1984.
234. **Wallace, G. D. and Rosen, L.,** Studies on eosinophilic meningitis. I. Observations on the geographic distribution of *Angiostrongylus cantonensis* in the Pacific area and its prevalence in wild rats, *Am. J. Epidemiol.*, 81, 52, 1965.
235. **Walton, D. W., Brooks, J. E., Tun, M. M., and Naing, H.,** Observations on reproductive activity among female *Bandicota bengalensis* in Rangoon, *Acta Theriol.*, 23, 489, 1978.
236. **Ward, R. D.,** New World leishmaniasis: a review of the epidemiological changes in the last three decades, Proc. 15th Int. Cong. Entomology, Washington, 505, 1977.
237. World Health Organization, Joint FAO/WHO Expert Committee on Zoonoses, Tech. Rep. Series No. 378, World Health Organization, Geneva, 1967.
238. World Health Organization, Technical guide for a system of plague surveillance, *Wkly. Epidem. Rec.*, 48, 149, 1973.
239. World Health Organization, Parasitic Zoonoses, Tech. Rep. Ser. No. 637, World Health Organization, Geneva, 1979.
240. World Health Organization, Rickettsioses: a continuing disease problem, *Bull. W.H.O.*, 60, 157, 1982.
241. World Health Organization, Haemorrhagic fever with renal syndrome: memorandum from a WHO meeting, *Bull. W.H.O.*, 61, 269, 1983.
242. World Health Organization, The Leishmaniases, Tech. Rep. Ser. No. 701, World Health Organization, Geneva, 1984.
243. World Health Organization, The control of schistosomiasis, Tech. Rep. Ser. No. 728, Geneva, 1985.
244. World Health Organization, Arthropod-borne and rodent-borne viral diseases, World Health Organization, Tech. Rep. Ser. No. 719, Geneva, 1985.
245. World Health Organization, Viral haemorrhagic fevers, World Health Organization Tech. Rep. Ser. No. 721, Geneva, 1985.
246. World Health Organization, Chagas' disease in the region of the Americas, *Wkly. Epidem. Rec.*, 60, 37, 1985.
247. **Wodzicki, K.,** Prospects for biological control of rodent populations, *Bull. W.H.O.*, 48, 461, 1973.
248. **Wulff, H., Fabiyi, A., and Monath, T. P.,** Recent isolations of Lassa virus from Nigerian rodents, *Bull. W.H.O.*, 52, 609, 1975.
249. **Wulff, H., McIntosh, B. M., Hammer, D. B., and Johnson, K. M.,** Isolation of an arbovirus closely related to Lassa fever from *Mastomys natalensis* in Southeast Africa, *Bull. W.H.O.*, 55, 441, 1977.
250. **Yaeger, R. G.,** American trypanosomiasis, in *CRC Handbook Series in Zoonoses, C, Parasitic Zoonoses, Vol. I,* Jacobs, L. and Arambulo, P., Eds., CRC Press, Boca Raton, Fla., 1982, 105.
251. **Yen Yu-Chen, Kong Liang-Xiong, Lee Ling, Zhang Yu-Quin, Li Feng, Cai Bao-Jian, and Gao-Shou-Yi,** Characteristics of Crimean-Congo hemorrhagic fever virus (Xinjiong strain) in China, *Am. J. Trop. Med. Hyg.*, 34, 1179, 1985.
252. **Zeledon, R.,** Epidemiology, modes of transmission and reservoir hosts of Chagas' disease, *Ciba Found. Symp.*, 20, 51, 1974.
253. **Zeledon, R. and Rabinovich, J.,** Chagas' disease: an ecological appraisal with special emphasis on its insect vectors, *Ann. Rev. Entomol.*, 26, 101, 1981.
254. **Fedovov, Y. and Abgarian, G.,** Assignment Report, Plague Control Project in Libya, Unpublished WHO Report No. EM/PLA/5, World Health Organization, Alexandria, 1983.
255. Ministry of Health, unpublished report, Brazil, 1968.

Chapter 8

POPULATION PRINCIPLES

David E. Davis

TABLE OF CONTENTS

I. INTRODUCTION

The control of rats requires an understanding of population principles for permanent results. Without such knowledge results will be temporary or useless. This section describes principles that are useful to direct the kind of effort and to evaluate the results. The principles are derived from ecological experience with many species but here are described in reference to rodents. It is inevitable that the data comes from abundant and hence frequently-studied genera such as *Rattus* and *Microtus*. However, the principles apply to other genera and can be used to direct the management program of any rodent.[1-6]

II. POPULATION PRINCIPLES

A. Carrying Capacity

The number of rodents in an area reflects the conditions of the environment. The population reaches a level called carrying capacity which is defined as the number of rodents the habitat supports. Changing the factors of the habitat will increase or decrease the carrying capacity. Because the factors are continuously changing due to seasonal or other events a population may be stationary, increasing, or decreasing.

1. Stationary Populations

The number of rodents in an area may remain the same for short or long periods of time. Small changes in numbers cannot be detected by current census methods[2] and thus are generally ignored. While the number of rodents may be constant the individuals change as a result of births, deaths, and movements. The population is said to be at equilibrium of numbers but not of individuals.

The stationary state is more often found in theory than in fact for the simple reason that the environment changes frequently, causing increases or decreases of food, shelter, predation, and social behavior. Nevertheless some cases of equilibrium have been observed to last for several years.

Some populations may be stable in some characteristic, even though increasing or decreasing. For example, the sex ratio could be stable (not changing) even though the population was increasing. Similarly, the prevalence of a parasite might be stable even though the population was decreasing. Unfortunately the terms stationary and stable are frequently treated as synonyms.

2. Increasing Populations

Following a change in environmental conditions the number of rodents will increase if the change increases food, shelter, or some other factor that has been limiting the population. The number of rodents will adjust a new level dictated by the changed conditions. In practice, because conditions are changing most of the time, populations are changing most of the time.

3. Decreasing Populations

The comments above about increasing populations apply equally to decreasing populations. The changes are achieved by changes in rates which will be discussed below.

B. Regulation

A population adjusts to the environmental conditions by changes in rates of birth, death, or movements. These changes regulate the numbers up or down according to circumstances. Some of the conditions have a constant proportional effect irrespective of the number of individuals. For example, a flood may drown x percent of the rodents in an area whether the population is high or low. Such a relation is called density-independent. But other conditions may have a variable effect according to the number of rodents. For example, a parasite may cause the death of 10% at low numbers and 50% at high numbers. Such a relation is called density-dependent. A density-independent relation cannot regulate a population for the simple reason that a constant proportional effect merely alters the magnitude of the rate but cannot change the direction. In contrast, a density-dependent condition can alter the direction of a rate. The only truly regulatory factors are density-dependent. These are the important factors in rodent pest management.

C. Limitation

The number of rodents is limited by the quantity of some resource such as food or shelter, or some interaction such as parasites or social relations. A population, through the processes of regulation, adjusts to this limit and will remain at that level until conditions change. In nature, conditions are in continuous flux so that the actual factor that limits the population differs from time to time and place to place. For example, a microtine species may be limited in field A by food and in field B by cats, or it might be limited in the spring by social aggression and in the fall by a parasite. The frequency and rapidity of changes of factors is so great that from the viewpoint of managing a rodent pest it is futile to try to determine exactly which factor is limiting at a particular time and place. The only effective management is to alter as many factors as is feasible.

In rodents and many other animals the most pervasive density-dependent factor is social behavior. As numbers increase the number of interactions increases proportionally as the square of the number. Thus, the most significant regulatory factor is competition among numbers of the population. Competition, of course, concerns some object or condition such as food, shelter, or space. Thus the fundamental guide for the management of a rodent pest is to alter factors in a direction that will alter competition. To increase a rodent population, increase shelter which will reduce competition. To decrease a population, decrease shelter which will increase competition.[4] Observance of this rule will prevent mistakes of management.

D. Numbers

The numbers of rodents in an area has been estimated by many methods.[2] A count or census is difficult and expensive. An estimate may be simple and inexpensive if low precision is acceptable. But, in most cases neither a count nor an estimate is necessary; a determination of change is satisfactory. Various simple procedures can indicate that more rodents are here than there, or that less rodents are present now than previously. In management much expense can be avoided by the use of relative methods.

III. MORTALITY RATES

For the management of a pest population mortality rates would seem to be the most critical aspect for action. Increasing the rates would appear to result in decreased numbers. However, birth rates are equally important and increases in mortality rates may be compensated so that no change in numbers occurs.

A. Causes of Death

In nature the causes of death appear to be infinite. Accidents, predators, parasites, starvation, and competition take their toll. In most cases several factors act more or less synchronously so that it is better to refer to what the rodent died *with* rather than the cause of death. In practice, few dead specimens are found in good enough condition for examination, and adequate tests are difficult and expensive. Thus, efforts to determine the cause of death are unrewarding for management except in a few cases such as an epizootic of plague.

B. Death Rate

Mortality is quantitatively described by two definitions. The death rate is the number that die in a time period divided by the average number during that time. Thus if 60 rodents die during a year in a population which averages 100, then the death rate is 60/100 = 0.6. In contrast, the mortality rates or prevalence of dying is the number that die divided by the number in the cohort. Thus if 60 rats out of an initial group of 100 die then the mortality ratio is 0.6. While the numerical values of the two concepts are identical the population consequences differ.[1]

Death rates and prevalence of dying can be measured by various methods, often involving counts at two different times. Knowledge of these values may be useful in management to assess the virtues of a procedure or to determine causes of failures.

C. Sustained Yield

Often the removal of some individuals is compensated by increased proportional survival of the remainder.[5] Also, as will be described below, the birth rate may increase. The result of these two changes is a sustained yield; the number removed is balanced and therefore continues unabated. A sustained yield is the worst possible result of management. The removal of many rodents gives the appearance of success at a considerable cost. Unfortunately, a high proportion of the programs that kill rodents merely set up a sustained yield situation without reducing damage by rodents.

D. Turnover

Turnover, a popular concept of some value, is the time required for one generation to die and another to take its place. Because in nature generations overlap the concept becomes difficult to express quantitatively.

E. Extermination (Extirpation)

In some situations it is possible by intensive alteration of habitat and/or a persistent poisoning program to extirpate the pest from the area. If the habitat has been altered the pest will not increase or will increase very slowly until the habitat returns to its original condition. If the habitat is not altered, the pest, because it is so rare, will increase very slowly, if at all. Under such circumstances little, if any, pesticide is needed and therefore the use of a pesticide is not a threat to the environment. However, constant vigilance is essential to detect immigration and the start of a new population.

F. Use In Management

Knowledge of death rates surprisingly has little utility in management. The effectiveness of a reduction procedure is usually adequately known from census figures.

IV. BIRTH RATES

Since births replace deaths a management program must understand the birth rate. Some knowledge of anatomy and of physiology is necessary to evaluate the reproductive condition of individuals in a population. Any elementary text will provide more details.

A. Reproductive Characters

Male rodents have testes, seminal vesicles, and several glands. In young rodents the testes are small and contained within the body cavity. At maturity they descend into a sac which is readily visible without opening the body. Many rodents lack a seasonal change in the size and function of the testes.

Female rodents have ovaries, uterus, vagina, clitoris, and mammary glands. For management purposes external examination provides adequate information to determine sexual maturity and seasonal reproduction. If the vaginal orifice is perforated (open) the female is sexually mature. If the mammary glands are developed (visible) the female has had one or more litters. To determine the number of young the abdominal cavity must be opened. If pregnant, embryos can be easily counted. If not pregnant the scars (black spots) in the uterus can be counted. With some minor qualifications either figure can be considered the number born. The next number to determine is the number of pregnancies (I) in a season or year. First determine what proportion (P) of the females are pregnant and the duration (T) of collection of the sample (usually a year or 365 days). Then the number of pregnancies (I) equals the prevalence (P) times the duration (d) divided by 18 which is the number of days that pregnancy can be detected.

$$I = \frac{Pt}{d}, \text{ for example } I = \frac{(0.158)\,(365)}{18} = 3.2. \tag{1}$$

If the breeding season is less than 365 days, use the appropriate number.

B. Physiology

For management some understanding of function is helpful. The sexual maturity and seasonal changes in function depend upon a series of hormones. Stimulated by external conditions the lower part of the brain (hypothalamus) sends chemical messages to the pituitary which in turn sends messages to the testes or ovaries. These produce sperm or eggs. After fertilization the egg develops in the uterus under the guidance of several hormones until birth.

C. Birth Rate

The actual rate can be calculated from the data obtained above. The birth rate (b) is the number of young born during an interval of time divided by the average total population during that time. In contrast the reproductive rate (R) is the number of young born during an interval of time divided by the average number of mature females during that time. Many other definitions and nuances exist, but for management the number for birth rate is adequate. Fortunately this number can be obtained easily in contrast to some of the other numbers.

D. Management Use

In contrast to knowledge of the death rate, knowledge of the birth rate has obvious value to management. In the section on carrying capacity it was noted that most populations reach a certain level and remain there until some habitat change occurs. But at that level the birth rate is less than the physiologically possible maximum and the age (size) at sexual maturity is relatively high. When the population declines as a result of some removal technique (traps, poisons) then the birth rate rises and the age (size) at sexual maturity declines. Thus data on these two characteristics should be obtained before removal and at intervals thereafter. An increase in birth rate and a decline in age (size) at maturity indicate that the population is decreasing. At some time the technique will have removed all adults, leaving only sexually immature young. The birth rate then is zero. In practice, percentage pregnant rather than birth rate is adequate and much simpler. Unfortunately most control programs stop before these reproductive changes occur and thereby have little long-term effect.

In management by altering the habitat, these reproductive changes do not occur. As the conditions of the habitat become less suitable for the rodents the population declines but remains at carrying capacity which is declining due to management. Therefore the compensatory increase in birth rate does not occur. Thus a census is the only way to determine the extent of success of the management technique.

V. MOVEMENT RATES

Data on movements of rodents may be difficult to obtain for the simple reason that when a rodent moves it either comes from or goes to an area that is not being studied. In reality, with some exceptions, most of our knowledge of movements comes from rodents that did not move. Some definitions are desirable. Immigration is movement into an area; emigration is movement out of an area. Migration is a to-and-fro movement for the purpose of breeding. Dispersal is the departure permanently from an area, usually of birth. These definitions may be qualified in detail.

A. Relation to Population

As a population increases the types of movements change. At low populations some individuals (if available) may move into an area (immigration) but none move out of the area (emigration). However, as the population increases these movements reverse so that when the population has increased immigration may stop and some emigration occur. This phase of movement is called presaturation because the habitat is not yet saturated or at carrying capacity. When the population reaches capacity then emigration may be large (saturation).

The characteristics of the emigrants and immigrants has been intensively studied because of its importance to the theories of population regulation,[6] but for management purposes the question is not important. Several general ideas are significant. Both immigrants and emigrants suffer a high mortality from a variety of causes. An immigrant into a high population has a low chance of survival, but into a low or zero population has a chance to found a colony. Thus the level of population is the crucial aspect.

B. Home Range

When a rodent has found a suitable place with adequate food and shelter it remains there for life or until the habitat is altered. The rodent moves about a small area, called home range, usually less than a hectare in size. The size of the range has been determined for many species in many places but the details are not important in management. The critical aspect is that because rodents have a home range management can be based on small areas. Intensive application of the chosen techniques will reduce the damage by rodents in that area. Then the application can be moved to another, perhaps adjacent, area. The rate of movement into the area will vary with circumstances but will be low.

VI. HABITAT

The term habitat is used loosely to mean several things. A common meaning is the geographic or topographic location of a population. However, in this article habitat means the physical, chemical, and biological characteristics of the place where the population exists. These characteristics can be grouped into three broad categories: (1) Environment, (2) parasites and pathogens, and (3) competition, both intra- and interspecific.

A. Environment

The presence of a pest in significant numbers to cause damage is adequate evidence that the environment is suitable. Therefore, even a cursory inspection will reveal that food is abundant and that shelter is available. The management plan should include alteration of these supplies. Many examples are described in the pest control and game management literature. Here one essentially unpublished case will be described.

Ground squirrels *(Spermophilus beecheyi)* are severe pests in agriculture and are reservoirs (multipliers) of plague *(Yersinia pestis)*. Their optimum environment includes sloping banks of streams or fields with low grass or weeds. The squirrels can exist on edges of brush or woods and under stumps or tree roots. In late summer young squirrels may dig burrows in flat parts of grain fields. Populations thrive in heavily-grazed pastures, especially where rocks, tree roots, banks, or ditches exist. The squirrels will decrease, even to zero, in numbers if the plants are permitted to grow to about 0.5 m high. Elimination of shelter such as rocks, roots, or ditches will hasten the decline.

B. Parasites, Pathogens, and Predators

In some situations these agents reduce the population but generally they are ineffective for management. The reason is simple. The persistence of these agents depends on a certain level of the hosts or prey. That level generally is above the threshold of economic damage so no benefit is obtained. Exceptions, of course, exist. Plague can reduce or eliminate prairie dogs *(Cynomys)* in pasture or crop land. In fact, the correct view is that the prey or host population determines the parasite, pathogen, or predator population. In its simplest case, since the prey is the food of the predator, therefore the prey abundance determines the predator abundance.

C. Competition

Individuals compete with members of another species (interspecific) and with members of the same species (intraspecific). The results are similar since it matters little whether a cow or a vole eats the grass; the food is gone. However, the intraspecific competition is more severe for the simple reason that members of the same species have similar requirements and thus demand the same resources. For example, starlings *(Sturnus vulgaris)* compete with woodpeckers and other starlings for nest holes. When a starling drives a woodpecker away the woodpecker population is affected, but when a starling drives a starling away the starling population is restricted.

Competition is supremely density-dependent (see discussion of regulation). As numbers increase, competition increases and thus reduces the rate of increase.

VII. OBJECTIVES

The usual assumption of pest control management is that the population should be reduced. However, the real objective is to reduce damage from the pests, which may not require reduction of the pest. For example, barriers such as fences or a tilled strip may prevent damage. Planting or harvesting at somewhat different time may reduce damage. Consultation with a crop or disease specialist may reveal unexpected techniques of management to reduce damage.

REFERENCES*

1. **Davis, D. E. and Jackson, W. B.,** Rat Control, in *Advances in Applied Biology,* Cocker, T. H., Ed., Academic Press, New York, 1981, 332.

* The literature on rodent pest management is enormous and has been reviewed many times. Therefore this chapter on principles will cite only a few references, especially those that review principles or have historical significance. Numerous references appear in other chapters.

2. **Davis, D. E., Ed.,** *CRC Handbook of Census Methods for Terrestrial Vertebrates,* CRC Press, Boca Raton, Fla., 1982, 397.
3. **Davis, D. E.,** Advances in rodent control, *Z. Angew. Zool.,* 64, 193, 1977.
4. **Davis, D. E.,** The role of intraspecific competition in game management, *Trans. North Am. Wildl. Conf.,* 14, 225, 1949.
5. **Davis, D. E. and Christrian, J. J.,** Population consequences of a sustained yield program for Norway rats, *Ecology,* 39, 217, 1957.
6. **Tamarin, R. H., Ed.,** *Population Regulation,* Dowden, Hutchinson & Ross, Stroudsburg, Pa., 1978, 391.

Chapter 9

CHANGING PATTERNS OF RODENT POPULATIONS IN INDIA

Ishwar Prakash

TABLE OF CONTENTS

I. INTRODUCTION

The Indian subcontinent exhibits a spectacular diversity of habitats and animals occupying them. Rodents from 46 genera, 135 species, and over 300 subspecies have been reported.[7] Every bioclimatic zone of the region has some typical rodent taxon. Whereas marmots *(Marmota)* and hamster *(Cricetulus)* are commonly found in the Himalayas, the woolly flying squirrel *(Eupetaurus)* and a microtine genus, *Hyperacruis* (vole) are restricted to Kashmir, and the bamboo rats *(Cannomys, Rhizomys)* to northeastern Himalayas. In the plains *Gerbillus, Tatera,* and *Meriones* are the predominant field rodents in the desert, a number of species of *Rattus* are common in northeastern hilly tracks; whereas *Rattus meltada* and *Bandicota bengalensis* are relatively abundant species particularly in the irrigated crop fields and environs.[3] Moreover, throughout India, a mixed population of several rodent species is found in the grasslands, forests, and croplands.[37]

Usually all the rodents are incriminated as pests, but the fact is that, fortunately, only a few species,[5] which occur abundantly, inflict losses to standing crops, stored foodgrains, and other commodities. The success of a control operation depends not only on the efficiency in conducting it but also upon the species composition and density of a rodent population. Information on such aspects is, therefore, extremely necessary before initiating an actual control campaign. In a tropical and subtropical region, like the Indian subcontinent, not only the density of rodents in various habitats keep changing but the species composition also changes with time, land use pattern, and due to human actions. These changes are important to monitor since more destructive and aggressive rodent species have replaced relatively docile ones and have, moreover, enhanced the epidemiological hazards in certain regions. Fortunately, some quantitative data are available pertaining to Indian rodents of economic importance.

II. REPLACEMENT OF SPECIES

A. Due to Canal Irrigation

The Great Indian desert or the Thar in western India is separated into four administrative states: Gujarat (20.0%), Haryana (4.4%), Punjab (5.5%), and Rajasthan (70.0%). To cope with an increasing demand for food resources, an intensive canal system has been developed in parts of the desert for growing crops by bringing irrigation water from perennial rivers to the thirsty xeric environment. Irrigated agriculture is not a natural process in an arid ecosystem, and as a consequence a number of man-made problems have cropped up. One of them is the replacement of rodents adapted for desert survival by mesic species which are relatively more destructive to vegetation and which have enhanced epidemiological hazards.

In Punjab, the irrigation system is about 150 years old and a total transformation of desert grasslands into irrigated croplands has occurred; but in parts of Haryana and in the Sri-Ganganagar district, northern Rajasthan (Figure 1), where the irrigation is only 25 to 50 years old, the process of transformation is not complete. Prior to the development of the irrigation system, the flora and fauna of this region was more or less similar [2,3,7,25] to that in the Rajasthan desert where there has been no impact by any type of irrigation. The comparison of rodent communities in desert habitats with those found at present in the Sri Ganganagar district (irrigation 50 years old) and with those found at present in Punjab (irrigation 150 years old) indicates the successive changes in the rodent communities which have taken place due to canal irrigation, which has consequently transformed the desert grassland into irrigated croplands.

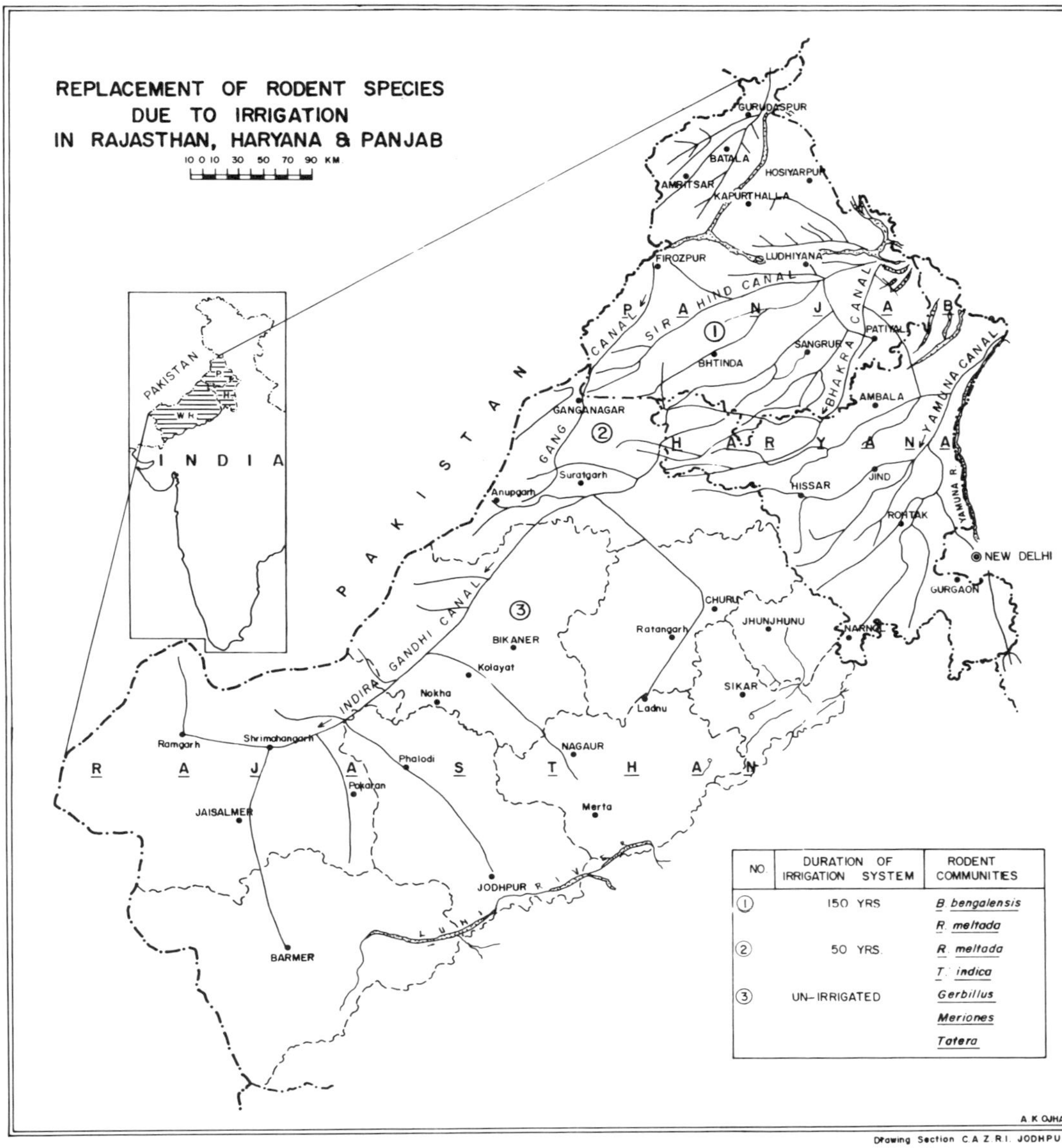

FIGURE 1. Replacement of rodent species due to irrigation in the Indian desert

1. Rodent Communities in the Continuous Desert Spread from Western Rajasthan to Punjab Prior to Irrigation

A systematic study on the relative abundance of rodents in the desert biome of Rajasthan[41] has revealed that among all the rodent species found, *Meriones hurrianae* and *Tatera indica* are most abundant, whereas *Rattus meltada pallidior* and *Gerbillus gleadowi* are the next common species (Table 1).

In the sandy habitat *M. hurrianae* is found to be the most predominant rodent (41.6%).[39] *T. indica indica* (19.0%), *Gerbillus gleadowi* (16.6%), and *R. meltada* (13.0%) are the next common rodents (Table 1). The rest of the species are found in low numbers.

Tatera i. indica and *Rattus m. pallidior* are the most common rodents in the ruderal habitat (village complex with cropland). *Meriones hurrianae* and *Gerbillus gleadowi* are

Table 1
RELATIVE ABUNDANCE OF RODENTS IN THE FIELD IN THE DESERT REGION PRIOR TO IRRIGATION

Rodent species	% Frequency[a]	% Collected in various habitats	
		Sandy	Ruderal
Funambulus pennanti	3.1	1.1	4.2
Gerbillus nanus indus	1.5	4.1	—
Gerbillus gleadowi	1.1	16.6	13.4
Tatera i. indica	24.7	19.0	39.0
Meriones hurrianae	26.0	41.6	16.4
Rattus c. cutchicus	12.0	—	—
Rattus meltada pallidior	13.1	13.0	20.1
Rattus gleadowi	0.6	1.1	—
Mus musculus bactrianus	0.2 (in field)	—	0.6 (in field)
Mus b. booduga	0.6	—	1.8
Mus cervicolor phillipsi	1.5	—	—
Mus platythrix sadhu	3.3	1.7	1.2
Golunda ellioti gujerati	1.7	1.1	3.0

[a] Percent of total number of rodents collected irrespective of their different habitats.

also found in this habitat in appreciable numbers, but usually when the villages are situated on sand dunes or in close proximity to sand dunes. Other rodent species occur in low numbers (Table 1).

It has been estimated that about 81% of the rodent fauna is typically of desertic type and bears Saharo-Rajasthani affinities.[31] It is composed of four species of gerbils, out of which three occur only in the xeric environment. The others are also micro-endemic to the region and are highly adapted to the xeric environment.[33]

2. *Rodent Communities as Influenced by 50 Years of Irrigation*

In the Sri Ganganagar district, the irrigation system is about 50 years old. The transformation of desert grasslands into irrigated croplands is not complete as scattered sand dunes and sandy patches still occur in the region. Our study[41] indicates an abundance of *Tatera indica* and *Rattus meltada pallidior* but *Meriones hurrianae,* a desert element, is still found in fair abundance (Table 2). However, the two species of gerbils, highly adapted to the xeric environment, *Gerbillus nanus indus* and *G. gleadowi,* have vanished from the region.[29,36]

3. *Rodent Communities as Influenced by 150 Years of Irrigation*

In Punjab, with the irrigation being almost 150 years old, the land use pattern has totally changed to irrigated agriculture. At present, the murid fauna in the croplands of Punjab is composed of eight species[20,51] (Table 3).

It is apparent that the predominant species in agricultural fields are the field rat, *Rattus meltada pallidior* and house mouse, *Mus musculus bactrianus.* Another field mouse, *Mus b. booduga,* is next in order of frequency of occurrence in the cereal crops.[20] *Bandicota bengalensis, Tatera indica,* and *Mus platythrix* are commonly found, but *Nesokia indica* and *Golunda ellioti gujerati* are not so common. However, Sood and Guraya[51] observed that in the "Punjab State during the early periods of the last decade, *Tatera indica* (desertic

Table 2
RODENT COMMUNITIES IN THE SANDY HABITAT PARTIALLY TRANSFORMED INTO IRRIGATED CROPLANDS IN THE SRIGANGANAGAR DISTRICT

Rodent species	Abundance index	% Occurrence
Funambulus pennanti	Rare	2.6
Tatera indica	Abundant	44.7
Meriones hurrianae	Abundant	15.8
Rattus meltada	Abundant	21.1
Mus musculus	Common	7.9
Nesokia indica	Common	7.9
Bandicota bengalensis	—	Present

Table 3
THE MURID FAUNA IN THE CROPLANDS OF THE PUNJAB

Species	Abundance index	Frequency of occurrence (%)
Tatera indica	Common	8.88
Rattus meltada	Abundant	39.07
Mus musculus	Abundant	26.14
Mus booduga	Abundant	15.33
Mus platythrix	Common	0.22
Nesokia indica	Not Common	—
Bandicota bengalensis	Common	8.15
Golunda ellioti	Not Common	2.0

form) was commoner but now the relative abundance of *Rattus meltada* and *Mus* spp. (Mesic forms) has been observed in all the seasons''. The authors further named the species in the following order of abundance: *Rattus meltada, Mus* spp., *Bandicota bengalensis, Tatera indica, G. ellioti,* and *Nesokia indica.*

4. Species Replacement Succession

Comparing the rodent communities in the above three periods and systems, a succession in the replacement of rodent communities due to change in land use pattern has been worked out (Table 4).[38] This succession pattern suggests that the desert adapted rodent species have been replaced by *Rattus meltada, Bandicota bengalensis, Mus musculus,* and *Mus booduga.* This indicates that no sooner had the desert environment changed to a wetter, irrigated system, the desert rodents *(Gerbillus, Meriones, Tatera),* having Saharo-Rajasthani affinities, were ousted by the mesic Oriental rodents. We do find a phase in between the two which is at present existing in the partially modified Sri Ganganagar district wherein both the desertic and mesic elements of rodents co-exist, though the frequency of occurrence of the former has been greatly reduced (Table 4).

5. Implications of Replacement of Rodent Species

a. Escalation in Damages to Crops

The mesic rodents are larger in size and have a higher rate of food consumption than the desert rodents. The bandicoot, *Bandicota bengalensis,* not only feeds upon the standing

Table 4
SUCCESSION OF RODENT COMMUNITIES IN THE DESERT LAND DURING THE LAST 150 YR BROUGHT ABOUT DUE TO THE CHANGE IN LANDUSE PATTERNS — FROM A DESERT GRAZING LAND TO AN IRRIGATED CROPLAND

Time sequence	Environment type	Rodent communities	Zoo-geographical affinities
At present	Mesic environment, irrigated cropland	*Bandicota bengalensis*	Oriental
		Rattus meltada	Oriental
		Mus spp.	—
50 yr before present intermediate stage	Irrigated croplands interspersed with native sand dunes	*Rattus meltada*	Oriental
		Tatera indica	Irano-Oriental
		Bandicota bengalensis	Oriental
		Meriones hurrianae	Irano-Rajasthani
150 before present (xeric)	Desert environment	*Gerbillus gleadowi*	Sindo-Rajasthani
		Gerbillus nanus	Saharo-Rajasthani
		Meriones hurrianae	Irano-Rajasthani
		Tatera indica	Irano-Oriental

crops but also hoards large quantities of ear-heads of paddy, wheat, and other crops, often collecting from 600 ear-heads (2.58 to 14 kg) per bandicoot burrow.[10,26,41,53,59]

b. Health Hazards

Several wild rodents occurring in the desert region are found to be positive with plague infection, namely, *Funambulus pennanti, Tatera indica, Rattus meltada,* and *Bandicota bengalensis*. Whereas the latter two species are highly susceptible to plague, *T. indica* is moderately susceptible. The nature of resistance in *T. indica* is temporary and the reduced susceptibility of this species suggests the possibility that it acts as a plague enzootic reservoir.[4,6] Prior to the advent of canal irrigation in the desert, plague bacillus had no suitable host for transfer from *T. indica,* but due to the change in land use pattern, the frequency of occurrence of *B. bengalensis* and *R. meltada* (both highly susceptible to plague infection) has considerably enhanced in the habitat of *T. indica* (epizootic host). The occupation of the same habitat by these rodent species is a dangerous situation. It is probably due to the low infestation of fleas (*Xenopsylla* spp.) that this dreadful disease is not spreading. But, a serious situation can develop any time.

Salmonellosis (of which the rodents are the carriers) deaths have become common in poultry birds, apparently the infection is being transmitted through rodent fecal pellets that are ingested by birds. This sort of contamination poses a health hazard to desert dwellers also. However, no data are available on the relationship of rodents and the so called ''viral'' fever prevalent in man in the arid regions.

There is no reason as to why such dreaded situation should be allowed when well proven and cost-worthy technologies for rodent pest management are available.[5,9,37]

B. Due to Urbanization

Urbanization has also influenced the composition of rodent populations. The field rodent *Bandicota bengalensis* has replaced the *Rattus* species in two metropolitan cities of India during the last 60 years (1906 to 1965). Based on the study of thousands of urban rodents collected per year, it was revealed[48] that in Calcutta *Bandicota bengalensis* out-numbered other sympatric rodent species. The collections during 1906 indicated that bandicoot rats constituted 60% of the rodent population at Calcutta, whereas *Rattus rattus* and *R. norvegicus* were 14 and 26%, respectively. Gradually, however, this proportion altered in 1965 to 91.6,

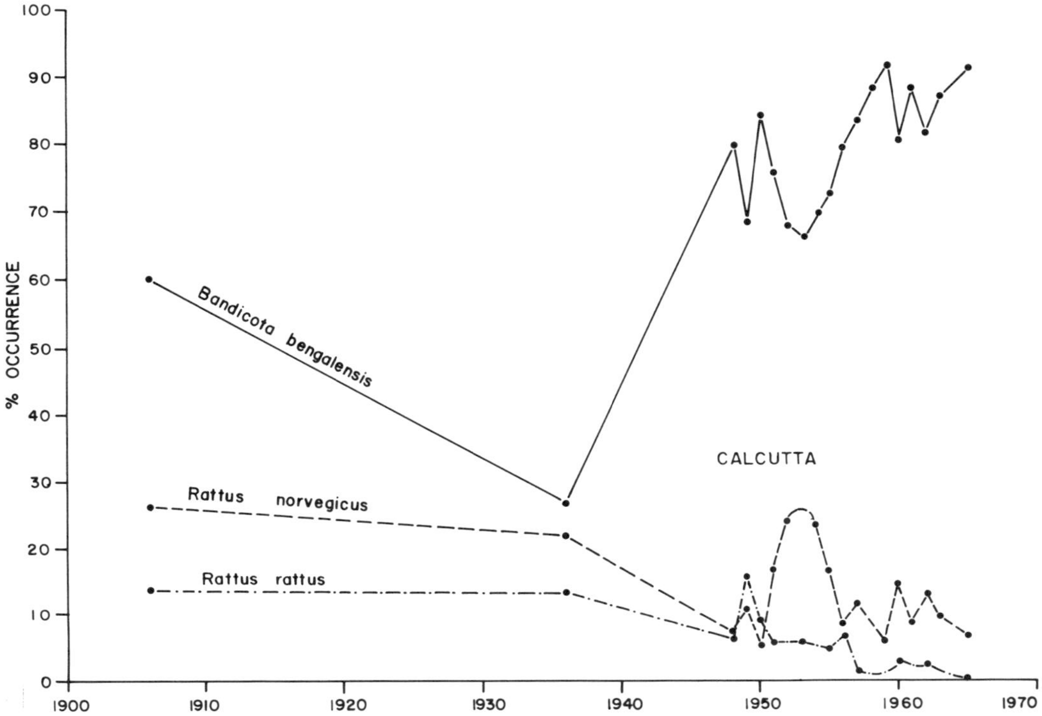

FIGURE 2. Rodent species replacement due to urbanization in Calcutta metropolitan town

0.8, and 7.1% (*B. bengalensis, R. rattus,* and *R. norvegicus,* respectively) indicating that the bandicoot more or less replaced *Rattus* (Figure 2). In Howrah (a suburban town of Calcutta situated on the other side of river Ganges), the changeover was more spectacular in even a shorter period. In 1952, *Rattus rattus, R. norvegicus, B. bengalensis,* and *T. indica* constituted 13.2, 13.2, 21.9, 61.0, and 3.1% of the rodent population, respectively.[48] The relative abundance of species changed to 0.0, 2.8, 96.9, and 0.0%, respectively, in 1965; *B. bengalensis* totally replaced *Rattus rattus* and the Indian gerbil, *Tatera indica,* and outnumbered *Rattus norvegicus.*

A similar species-replacement was reported by Deoras[10] in Bombay. In 1910, *R. rattus* was the predominant rodent in Bombay and the population of *Bandicota bengalensis* was about 1% (Figure 3). However, in 1956, the latter increased to 40.6%. The author conjectured that the preponderance of females in the population, a higher embryo index, the ability to breed all the year round, burrowing habits, availability of food, and absence of natural enemies were the factors that led to the increase of *B. bengalensis* in the city. It is, however, considered that the more aggressive nature of this rodent was responsible for its replacement of *R. rattus.* Another reason I assign to the phenomenon of this replacement of species is the expansion of the township over the natural grasslands and crop fields. When the city expanded the field rodent, *B. bengalensis,* inhabiting the open fields, quickly adapted to living in godowns, and colonized the entire cities of Bombay and Calcutta, thus replacing the commensal rodents.

An analogous example has been observed in the Rajasthan desert. Whenever a village is established on sand dune top, typical psammophile rodent species like the gerbils, *Gerbillus gleadowi* and *Meriones hurrianae,* which are the original inhabitants of this habitat, quickly adapt to stay very close to man in the village.[36]

Another scrub-grassland species, the Indian gerbil, *Tatera i. indica,* has totally colonized

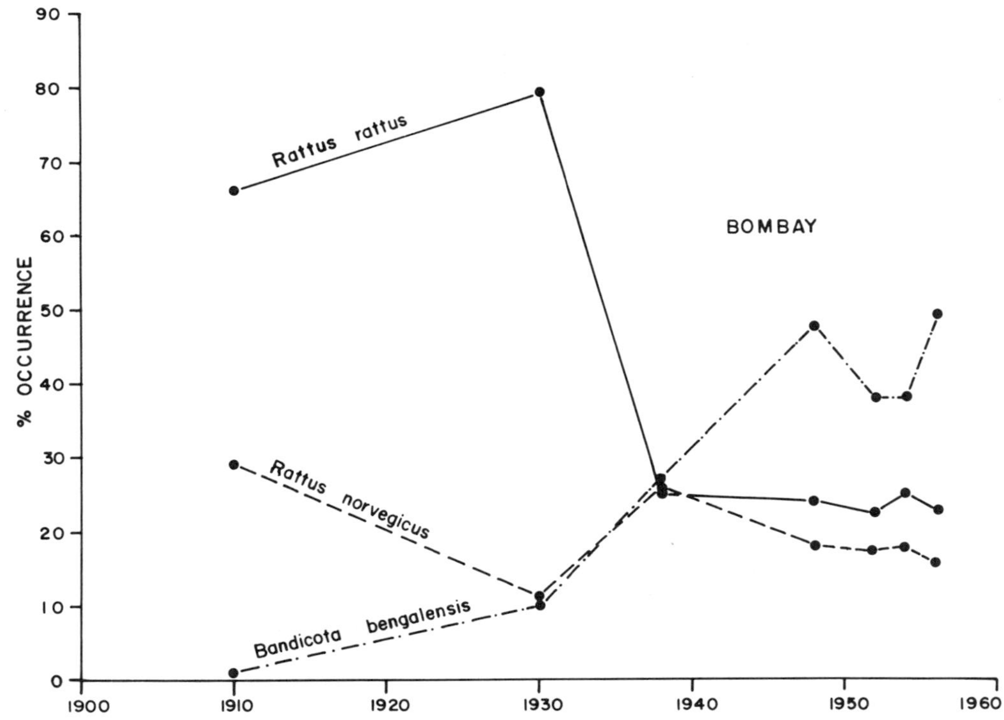

FIGURE 3. Rodent species replacement due to urbanization in Bombay metropolitan town

the Bikaner town in the Thar desert. Bikaner was established about 400 years ago over a flat sand dune top. Like all other sand dunes this one must have been occupied, prior to the settlement of human populations, by *Gerbillus nanus, G. gleadowi, Meriones hurrianae,* and *Tatera indica.*[34] Possibly due to human activities, all other species either perished or migrated and only *T. indica* survived, adapted to live within the town, and is now the most abundant rodent species. Scores of these gerbils can be witnessed at night. They venture out of their burrows situated in the streets, enter houses through open drains, and have virtually replaced *R. rattus,* the house rat. Only in recent years,[42] *B. bengalensis* has also been collected from the foodgrain bazaar at Bikaner. Their number is increasing. Both *T. indica* and *B. bengalensis* have adjusted to stay together. Hours of night observation did not yield any interaction between them, though both are fairly aggressive in their behavior. The bandicoot may have been transported by human agencies from the Sri Ganganagar district, where, during the last 100 years, it migrated to, from Punjab along the irrigation canals.[38] It is visualized that the bandicoot will further spread in vast areas of the desert along the Indira Gandhi Canal (Figure 1).

C. Impact of Drought

Climatic fluctuations have also been observed to influence rodent species composition in a habitat. In the Great Indian desert, the northern region of Rajasthan was under a continuous spell of drought from 1969 until 1974 when the summer rainfall was normal (290 mm). Rain water, as expected, plays an important role in the life of desert fauna by influencing (1) the soil moisture regime which in turn influences the microclimate inside the rodent burrows, (2) vegetation regeneration, thus affecting the availability of green food, and (3) indirectly, the breeding rate of the desert rodents.[30,32]

The results of our trapping studies[35] taken in a sand dune habitat, during 1969 to 1974,

Table 5
REPLACEMENT OF RODENT SPECIES DUE TO DROUGHTS IN THE SANDY HABITAT OF THE INDIAN DESERT

	Normal rainfall	Poor rainfall		Good rainfall
Species	**1969**	**1972**	**1973**	**1974**
Meriones hurrianae	20.0	6.6	—	1.1
Gerbillus nanus indus	40.0	—	—	1.9
Gerbillus gleadowi	40.0	93.4	100.0	95.3
Tatera indica	Present	—	—	1.1

about 10 km south of Bikaner, indicated the preponderance of two *Gerbillus* species (Table 5); *M. hurrianae* constituted a relatively low proportion and *Tatera indica* were not caught in our snap trap lines but were captured in live traps. During the next 2 years, facing acute drought, *Gerbillus gleadowi* was found to be the most abundant species as the number of other species declined with the increasing drought conditions. In 1974, a normal rainfall year, however, other species staged a comeback in very low numbers (Table 5).

Compared to years of normal rainfall in the desert region, *Meriones hurrianae* and *Tatera indica* are more abundant than *Gerbillus*.[35] The influence of droughts in replacing *Meriones* and *Tatera* by *Gerbillus* is explained due to the latter's ability to withstand the food, water, and temperature stresses.[12,14-16,44] The two gerbils, endemic to the Thar desert *(Gerbillus nanus indus* and *G. gleadowi)* are decidedly superior to *Meriones* and *Tatera* in their capability to survive under drought conditions, especially their tolerance to salt which is dissolved in the halophytes, their major food source.[43]

It has been reported that the Nile grass rat, *Arvicanthis niloticus* is invading barns and villages, apparently it has changed its habit from being a purely field rat to a commensal rat.[1]

III. OCCUPATION OF NEW HABITATS

Another surprising example of a rodent pest acquiring a new mode of life is exhibited by the house rat, *Rattus rattus*. It was probably introduced by man, through boats to the Lakshadweep islands situated off the western coast of south India. Since the soil on the islands is not suitable for burrowing, they climbed up the coconut palms and started living a totally arboreal life, in the course inflicting immense damage to coconuts (35 to 50% or about 6 million nuts worth $270,000 a year).[18,50] Acquiring such new habitats pose serious problems for rodent control operators.

In the Punjab, another commensal rodent *Mus musculus,* the house mouse, is now found in fair abundance in the crop fields, especially in sugar cane, probably because the altered environment is similar to its preresidential premises and much more conducive to survival. This is an interesting example of reversal of habit by a rodent species — a behavioral atavism.[40]

IV. UPSURGE IN RODENT POPULATIONS

Outbreaks in rodent populations are known in the Indian subcontinent but unfortunately no quantitative data are available. In southern India, usually a year after failure of monsoon,[19] outbreak in rodent populations were recorded (1826, 1878 to 1879, 1901 to 1902, 1913 to 1914, 1920 to 1921, and during 1974).[5] The major species involved were *Tatera indica, Bandicota bengalensis,* and *Rattus meltada.*[19,22] The magnitude of the outbreak could be assessed by the number of rodents killed in a control operation in 9 administrative districts, 14.97 million rodents. The damage cost due to this outbreak was enormous. *T. indica* cut off the ears and devoured the sorghum grains in only a few nights. Ravaging an entire crop land, *R. meltada* gnawed off the corn crops. The factors that induced the outbreak are not known. It is possible that the fields, left uncropped due to the failure of monsoon the previous year, provided a habitat conducive to rodent multiplication.

Two outbreaks of rats in North Gujarat, Rajasthan, South-East Sind (Pakistan) are on record in the years 1901 to 1902 and 1913 to 1914.[27] Both these were apparently caused by failure of monsoon during the respective years 1899 and 1911.[49]

Cyclic outbreaks of rodent population associated with bamboo flowering in the north-eastern hilly regions of India (Arunachal Pradesh, Mizoram, and Manipur) have been observed for over 100 years, 1863 to 1866, 1892 to 1893, 1911 to 1912, 1929 to 1933, 1960, and 1975 to 1977.[13] Several of these upsurges in rodent population assumed menacing proportions; down-the-hill migrating rodents devastated the paddy fields to the extent that famine conditions were caused in areas of bamboo flowering.[57] The magnitude of the population explosion can be judged by the number of rodents killed (650,000) in a rodent control operation in Ratu village in 1977.[28]

In northeastern India, bamboos flower in two cycles independent of each other and the gap between the two alternating cycles is 18 and 30 years. The first cycle is that of *Melocanna bambusoides* which flowers after 48 ± 2 years (period of flowering is known as *Mautam*). The second cycle is associated with the flowering of *Bambusa tulda* and *Dendrocalamus longispathus (Thingtam),* which also flower in a cycle of 48 ± 2 years. In each cycle the flowering continues for 4 to 5 years in one locality or the other. The associated upsurge in rodent population occurs at an interval of 25 and 50 years. Due to extreme damage to paddy fields in the *Jhum* cultivation, the farmers do not grow rice if they observe bamboo flowering in the region. Unfortunately, very little scientific work has been carried out on this phenomenon inasmuch as the major rodent species involved is not definitely known. However, Chauhan[8] suspects that the pest is *Rattus rattus brunneusculus.* This is one area of research in the field of rodent biology which needs the immediate attention of the rodentologists in India.

Outbreaks of rodent population synchronized with bamboo flowering are known from Burma,[58] Brazil,[11] Madagascar,[45] and Japan.[14,54,55] Population explosion of rodent numbers has also been reported in agricultural lands in Australia,[24] North Xinjiang, China,[46] Kenya,[56] Scandinavia,[23] and Thailand.[21]

From various studies it appears that during this century *Bandicota bengalensis* has taken over as the most predominant species in most parts of India. In certain newly acquired habitats it is abundant, with a maximum density of 0.9 rodent per m^2 of godown space.[52] It has replaced a number of commensal rodents in urban habitats and several field rodents from croplands during the last 80 years. Instances of replacement of other species and occupation of newer habitats by rodents have made it imperative that continued monitoring of population structure should be taken up for determining the relative abundance of various rodent pests, with a view to maximize control success. Moreover, such long-term studies should possibly forewarn of the sporadic outbreaks of rodent populations.

REFERENCES

1. **Ali, A. M.,** Impact of changing irrigation on agricultural pests and wildlife in Egypt, in *Arid Land Irrigation in Developing Countries, Environmental Problems and Effects,* Worthington, E. B., Ed., Pergamon Press, Oxford, 1977, 331.
2. **Adams, A.,** *Western Rajputana States,* Taylor and Francis, London, 1899.
3. **Anonymous,** *The Imperial Gazetteer of India,* Clarendan Press, Oxford, 1908.
4. **Baltazard, M. and Bahmanyar, M.,** Recherches sur la peste en Inde, *Bull. W.H.O.,* 23, 169, 1960.
5. **Barnett, S. A. and Prakash, Ishwar,** *Rodents of Economic Importance in India,* Arnold-Heinemann, New Delhi, 1975, 175.
6. **Bhatnagar, J. K.,** The role of rodents in the epidemiology of plague in Uttar Pradesh, *Proc. Indian Rodent Symp.,* U.S. Agency for International Development, Calcutta, 204, 1966.
7. **Blanford, W. T.,** *The Fauna of British India, including Ceylon and Burma, Mammalia,* Taylor and Francis, London, 1888-91, 122.
8. **Chauhan, N. S.,** Studies on population dynamics and biology of reproduction of the field rat, *R. rattus brunneusculus* (Hodgsen) associated with bamboo flowering in northeastern hilly region (Mizoram) of India, Ph.D. dissertation, University of Delhi, 1981, 1.
9. **Fitzwater, W. D. and Prakash, Ishwar,** *Handbook of Vertebrate Pest Management,* Indian Council of Agricultural Research, New Delhi, 1973, 96.
10. **Deoras, P. J.,** Significance of probable change of rat population in Bombay, *Proc. Indian Rodent Symp.,* U.S. Agency for International Development, Calcutta, 58, 1966.
11. **Derby, O. A.,** Rats in Brazil and their connection with the flowering of the bamboo, *Indian Forester,* 9, 529, 1879.
12. **Gaur, B. S. and Ghosh, P. K.,** Effect of water deprivation on faecal water loss in the Indian desert gerbil, *Meriones hurrianae* Jerdon, *J. Anim. Morphol. Physiol.,* 18, 121, 1971.
13. **Ghosh, A. K., Ed.,** *Studies on Rodents and Their Control,* The Meghalaya Science Society, Shillong, 1980, 136.
14. **Ghosh, P. K.,** Thermo-regulation and water economy in Indian desert rodents, in *Rodents in Desert Environments,* Prakash, I. and Ghosh, P. K., Eds., Dr. W. Junk, The Hague, 1975, 397.
15. **Ghosh, P. K. and Gaur, B. S.,** A comparative study of salt tolerance and water requirements in desert rodents, *Meriones hurrianae* and *Gerbillus gleadowi, Indian J. Exp. Biol.,* 4, 228, 1966.
16. **Ghosh, P. K., Purohit, K. G., and Prakash, I.,** Studies on the effects of prolonged water deprivation on the Indian desert gerbil, *Meriones hurrianae,* in *Environmental Physiology and Psychology in Arid Conditions,* UNESCO, Paris, 1962, 301.
17. **Ito, T.,** Outbreaks of the field vole, *Microtus montebelli* in Kansai and Chugoku districts, *Bull. Gov. Forest Exp. Stn. Meguro,* 271, 39, 1975.
18. **Kidavakoya, P. I. K.,** Rat hunt in Laccadives, *Bull. Indian Cent. Coconut Comm.,* 9, 104, 1955.
19. **Kinnear, N. B.,** Expected plague of field rats in 1920, *J. Bombay Nat. Hist. Soc.,* 26, 663, 1919.
20. **Mann, G. S.,** Crop preference of rodents at Ludhiana, *J. Bombay Nat. Hist. Soc.,* 72, 383, 1975.
21. **Marshall, J. T.,** *Rats and Mice of Thailand,* Manuscript, 1972.
22. **Millard, K. S.,** Expected plague of field rats in 1920, *J. Bombay Nat. Hist. Soc.,* 27, 165, 1920.
23. **Mullen, D. A.,** Reproduction in Brown lemmings and its relevance to their cycle of abundance, *Univ. Calif. Pub. Zool.,* 185, 24, 1969.
24. **Newsome, A. E. and Corbett, L. K.,** Outbreaks of rodents in semi-arid and arid Australia, in *Rodents in Desert Environments,* Prakash, I. and Ghosh, P. K., Eds., Dr. W. Junk, The Hague, 1975, 117.
25. **Parker, R. N.,** *A Forest Flora of the Punjab with Hazara and Delhi,* Government Printing Press, Lahore, Pakistan, 1924, 591.
26. **Parrack, D. W.,** A note on the loss of food to the lesser bandicoot, *Bandicota bengalensis, Curr. Sci,* 38:83, 1969.
27. **Pillai, R. S.** Outbreaks of rat populations in India with special reference to northeastern regions, in *Studies on Rodents and Their Control,* Ghosh, A. K., Ed., The Meghalaya Science Society, Shillong, 1980, 59.
28. **Prabhakaran, P. K. and Michael, R. G.,** Preliminary note on the rodent outbreak in Mizoram, in *Studies on Rodents and Their Control,* Ghosh, A. K., Ed., The Meghalaya Science Society, Shillong, 1980, 73.
29. **Prakash, Ishwar,** Extinct and vanishing mammals from the desert of Rajasthan and the problem of their preservation, *Indian For.,* 55, 154, 1958.
30. **Prakash, Ishwar,** Breeding of mammals in Rajasthan desert, India, *J. Mammal.,* 41, 386, 1960.
31. **Prakash, Ishwar,** Zoogeography and evolution of the mammalian fauna of Rajasthan desert, India, *Mammalia,* 27, 342, 1963.
32. **Prakash, Ishwar,** Breeding season and litter size of Indian desert rodents, *Z. Angew. Zool.,* 58, 441, 1971.
33. **Prakash, Ishwar,** The ecology of vertebrates of the Indian desert, in *Ecology and Biogeography in India,* Mani, M. S., Ed., Dr. W. Junk, The Hague, 1974, 369.

34. **Prakash, Ishwar,** The ecology and zoogeography of mammals, in *Environmental Analysis of the Thar Desert,* Gupta, R. K., and Prakash, I., Eds., English Book Depot, Dehradun, 1975, 448.
35. **Prakash, Ishwar,** Replacement of sympatric rodent species in the Indian desert, *Proc. All India Rodent Seminar,* Rodent Control Project, Sidhpur, 1975, 29.
36. **Prakash, Ishwar,** Population ecology of rodents in the Rajasthan desert, India, in *Rodents in Desert Environments,* Prakash, I. and Ghosh, P. K., Eds., Dr. W. Junk, The Hague, 1975, 751.
37. **Prakash, Ishwar,** *Rodent Pest Management, Principles and Practices.* Monograph 4, Central Arid Zone Research Institute, Jodhpur, 1976, 28.
38. **Prakash, Ishwar,** Impact of changing landuse patterns or the rodent communities in part of the Indus basin, *Proc. Land and Water Mgmt. in the Indus Basin, India,* National Symposium, Punjab Agricultural University, Ludhiana, 481, 1978.
39. **Prakash, Ishwar,** *Ecology of the Indian Desert Gerbil, Meriones hurrianae,* Monograph 10, Central Arid Zone Research Institute, Jodhpur, 1981, 87.
40. **Prakash, Ishwar,** Rodents acquire newer territories, in *Frontier of Research in Agriculture,* Roy, S. K., Ed., Indian Statistical Institute, Calcutta, 1983, 436.
41. **Prakash, Ishwar, Gupta, R. K., Jain, A. P., Rana, B. D., and Dutta, B. K.,** Ecological evaluation of rodent populations in the desert biome of Rajasthan, *Mammalia,* 35, 384, 1971.
42. **Prakash, Ishwar and Mathur, R. P.,** *Bandicota bengalensis* in Bikaner town, *Rodent Newsl.,* 3, 12, 1979.
43. **Prakash, Ishwar and Purohit, K. G.,** Some observations on the hairy-footed gerbil, *Cerbillus gleadowi* Murray in the Rajasthan desert, *J. Bombay Nat. Hist. Soc.,* 63, 431, 1967.
44. **Prakash, Ishwar and Rana, B. D.,** A study of field population of rodents in the Indian desert. III. Sand dunes in 100 mm rainfall zone, *Z. Angew. Zool.,* 60, 31, 1973.
45. **Rakotomanana, R.,** Peut-on prevoir les invasions de rats?, *Terre Malgache Tanarive,* 1, 239, 1966.
46. Research Group on Rodent pest in Xinjiang, Some data concerning mass outbreaks of house mice in the agricultural regions of north Xinjiang, Uygur Autonomous Region, *Acad. Sinica. Zool.,* 2, 86, 1985, in Chinese.
47. **Roy, S. K.,** Preharvest loss of rice due to field rodents, *Econ. Political Weekly,* 9, 66, 1974.
48. **Seal, S. C. and Banerji, R. N.,** Changing patterns of rodent population in Calcutta and Howarh, *Proc. Indian Rodent Symp.,* U.S. Agency for International Development, Calcutta, 69, 1966.
49. **Sedgwick, L. J.,** Expected plague of field rats in 1920, *J. Bombay Nat. Hist. Soc.,* 26, 661, 1919.
50. **Shah, D. R., and Subiah, K. S.,** Notes on rodent control in Lakshadweep Islands with *Rodafarin* (Warfarin) wax blocks, *Pestology,* 2, 36, 1978.
51. **Sood, M. L. and Guraya, S. S.,** *Rats and Their Control,* Punjab Agricultural University, Ludhiana, 1976, 31.
52. **Spillet, J. J.,** *The Ecology of the Lesser Bandicot Rat in Calcutta,* Bombay Natural History Society, Bombay, 1968, 223.
53. **Srivastava, A. S.,** Population of *Bandicota bengalensis* (Gray) under field conditions, *Labdev J. Sci. Technol.,* 5, 344, 1967.
54. **Tanaka, R.,** Fluctuation in vole population following the widespread synchronous flowering of bamboo-grasses on Mt. Turugi, *Bull. Kochi Womens' Univ. Ser. Nat. Sci.,* 4, 61, 1956.
55. **Tanaka, R.,** An ecological review of small mammal outbreaks with special reference to their association with the flowering of bamboo grasses, *Bull. Kochi Womens' Univ. Ser. Nat. Sci.,* 5, 20, 1957.
56. **Taylor, K. D.,** An outbreak of rats in agricultural areas of Kenya in 1962, *East African Agric. For. J.,* 34, 66, 1968.
57. **Trivedi, S. and Tripathi, R. S.,** Bamboo flowering and increase in rodent population in Mizoram: a preliminary report, in *Studies on Rodents and Their Control,* Ghosh, A. K., Ed., The Meghalaya Science Society, Shillong, 1980, 68.
58. **Unwin, R.,** *Notes on Forest Protection in Burma,* University of Rangoon, Rangoon, Burma, 1927, 70.
59. **Wagle, P. V.,** The rice rats of lower Sind and their control, *J. Bombay Nat. Hist. Soc.,* 32, 330, 1927.

Chapter 10

PREVENTION OF PLAGUES OF HOUSE MICE IN RURAL AUSTRALIA

T. D. Redhead

TABLE OF CONTENTS

I. INTRODUCTION

The rodent fauna of Australia comprises approximately 50 native species, all of them being within the family Muridae, and perhaps three cosmopolitan murid species: the brown rat *(Rattus norvegicus)*, the black rat *(Rattus rattus)*, and the house mouse (*Mus* sp).* [57] Of these, only four of the native species and the three cosmopolitan species can be regarded as economically important pests in either urban or rural environments.

A. Rodent Problems in Urban Australia

In urban Australia, problems caused by the three cosmopolitan species are completely undocumented. Referring to rodent problems in urban areas of the U.S., Jackson[23] states "The house mouse *(Mus musculus)* remains the most important rodent problem. It always was more difficult to control, being less susceptible to anticoagulants. As the Norway rat decreased in importance because of better attention to sanitation or construction and baiting campaigns, mice previously inapparent, became obvious and of concern. Increasingly we hear of populations impossible to control with anticoagulants. Shopping centers, with independent but interconnected stores and scatterred feeding foci, presented serious problems . . . "

Australia's problem with urban house mice may be similar to that of the U.S. and other countries which are socio-economically similar. Control of urban rodents in such societies is dealt with by Marsh in Chapter 1.

B. Rodent Problems in Rural Australia

Rodent problems in rural Australia, other than chronic commensal infestations by black rats, brown rats, and house mice, are listed in Table 1. The most notorious are the irregular and often widespread outbreaks of house mice. Within the past 85 years, there have been at least 14 occasions in parts of southern and south-eastern Australia when house mice have multiplied so prolifically that hordes of them were present. The term "mouse plagues" is applied in Australia to these widespread outbreaks.

Like some other occasional natural phenomena, most plagues in rural Australia are catastrophic. At the peak of a mouse plague, large numbers of mice are present in crops, in farm buildings, and in farm houses. The problem is not restricted to farms, however, and houses, shops, and other facilities, especially in small country towns, are also invaded. How the massive losses caused by mouse plagues in rural Australia might be prevented in the future is the subject addressed in this chapter. First, the economic, social, and environmental consequences of mouse plagues in Australia are described. Mouse plagues continue to recur, in spite of the extensive use of poisons to reduce mouse numbers during each plague. The suitability of these ad hoc control attempts are examined. Then, alternative strategies for mouse plague control in the future are developed, and the resources required for their implementation are identified.

II. MOUSE PLAGUES IN RURAL AUSTRALIA

A. Occurrence and Distribution

The first recorded mouse plague in southern and south-eastern Australia was in 1902. Since then there have been at least seven other severe and widespread plagues.[34,49] Recently, plagues of limited distribution occurred in 1974-75 and widespread plagues occurred in 1979-80 and 1984 (Figure 1). While almost all the losses caused by mouse plagues occur in the cereal belt of eastern and southern Australia, the formation of mouse plagues is not due entirely to agriculture as mouse plagues also occur in the nonagricultural arid inland of

* The taxonomy of *Mus* in Australia is discussed briefly below.

Table 1
RODENT PROBLEMS IN RURAL AUSTRALIA, OTHER THAN THOSE CAUSED BY CHRONIC COMMENSAL INFESTATIONS BY BROWN RATS, BLACK RATS, AND HOUSE MICE

Species	Industry	Approximate annual loss ($ millions)	Ref.
	Economically Important		
Mus sp.	See Table 2	6[a]	37
Rattus sordidus	Sugar cane	0.5—4	27, 35, 41
Melomys burtoni	Sugar cane		
	Possibly Economically Important		
Rattus sordidus	Coarse grain[b]	?	—
Rattus tunneyi	Forestry	?	13
Rattus villosissimus	Pastoral	?	12
	Nuisance Value Only		
Melomys cervinipes	Sugar cane		
Uromys caudimaculatus	Sugar cane	—	27
Hydromys chrysogaster	Irrigation and poultry	—	—

[a] Based on an approximate loss of $48 million for mouse plagues occurring every 8 yr on average.
[b] In 1984-85 only.

Australia.[20,32] For unknown reasons, mouse plagues are less frequent in Western Australia than in the eastern and southern States. Within the cereal-growing belt in those States, mouse plagues occur over many soil types, ranging from heavy clays to the light sandy Mallee soils.

B. The House Mouse

There has never been a major taxonomic survey of house mice in Australia, and their exact identity remains unknown. Although often referred to locally as *Mus musculus,* it is generally agreed that, in accordance with Schwarz and Schwarz,[51] they are *Mus domesticus* (after Sage[44]) which probably first arrived in Australia with British colonizers in the 18th Century. Certainly, examination of the H2 system[25] of a small number of house mice from the Victorian Mallee Region and the Murrumbidgee Irrigation Region of New South Wales indicates that these were *Mus domesticus.*[53] However, skulls of mice from various locations in Australia, and now housed in the Australian National Wildlife Collection at the Commonwealth Scientific and Industrial Research Organization (CSIRO), Division of Wildlife and Ecology, Canberra, exhibit a high degree of variation in morphology, particularly with respect to the vertical margin of the zygomatic plate (Figure 2). That character is important among those that have been used in distinguishing the different species of house mice in Europe and elsewhere.[28]

In the absence of better data, it is possible only to speculate on the major sources of genetical variation which may exist amongst Australian house mice. Presumably, house mice travelling with man and his goods have continued to arrive, and not only from throughout

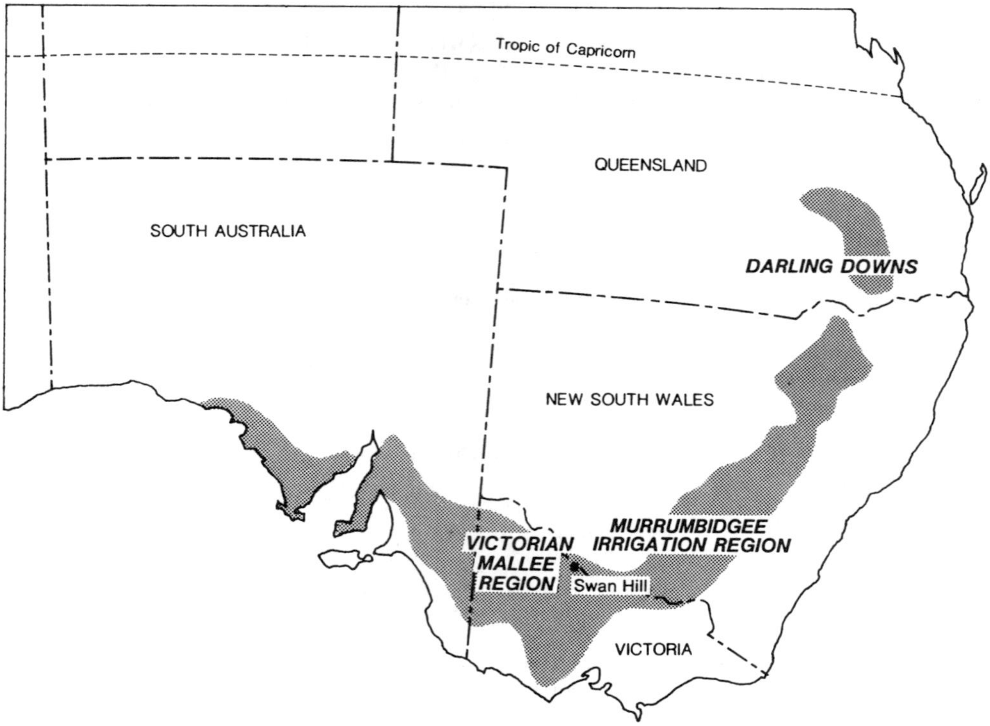

FIGURE 1. Map of south-eastern Australia showing the approximate extent of the grain belt throughout which plagues of house mice occasionally arise

Europe, where five species of house mice occur, but also from western and eastern Asia where another three species are present. Given the comparative ease and frequency of modern international transport, the possibility of diverse forms of house mice being present on the Australian continent appears to be high. Amongst the species from southern and south-east Asia is *M. castaneus,* which may be the most obligatory commensal species of all house mice.[28] If so, it might be the most likely candidate for passive dispersal. Further, nothing is known about the evolution of house mice since their first successful colonization of Australia. Certainly, they now inhabit an array of habitat types in which selection pressures may be markedly different. Moreover, the life strategy by which house mice may exist in temporary populations, sending out dispersers which may survive while their natal population becomes extinct,[6] results in the formation of local populations from one or a few immigrating progenitors, and is therefore conducive to both rapid evolution and genetic heterogeneity over the species range.

Nor is the taxonomy of Australian house mice just of academic importance, for the various species of house mice differ in their ecology. For example, in southern Europe, *Mus spretus* and *Mus domesticus* both breed in fields. *Mus domesticus* is commonly found there in barns and houses also, but *Mus spretus* is rarely found indoors.[58] Other examples of sympatric distributions are given by Marshal and Sage.[29] It is always assumed that the millions of mice that invade buildings during an Australian mouse plague are genetically similar to those found in fields. There has been but one study of genetic variation between mouse populations during plagues,[24] and it was not definitive because it was restricted to examination of variation at just four chemo-loci. If different genetic strains of mice were present, either at different times or at different places, then a need for different control activities could be necessary.

The remarkable ability of house mice, particularly *Mus domesticus,* to colonize most parts

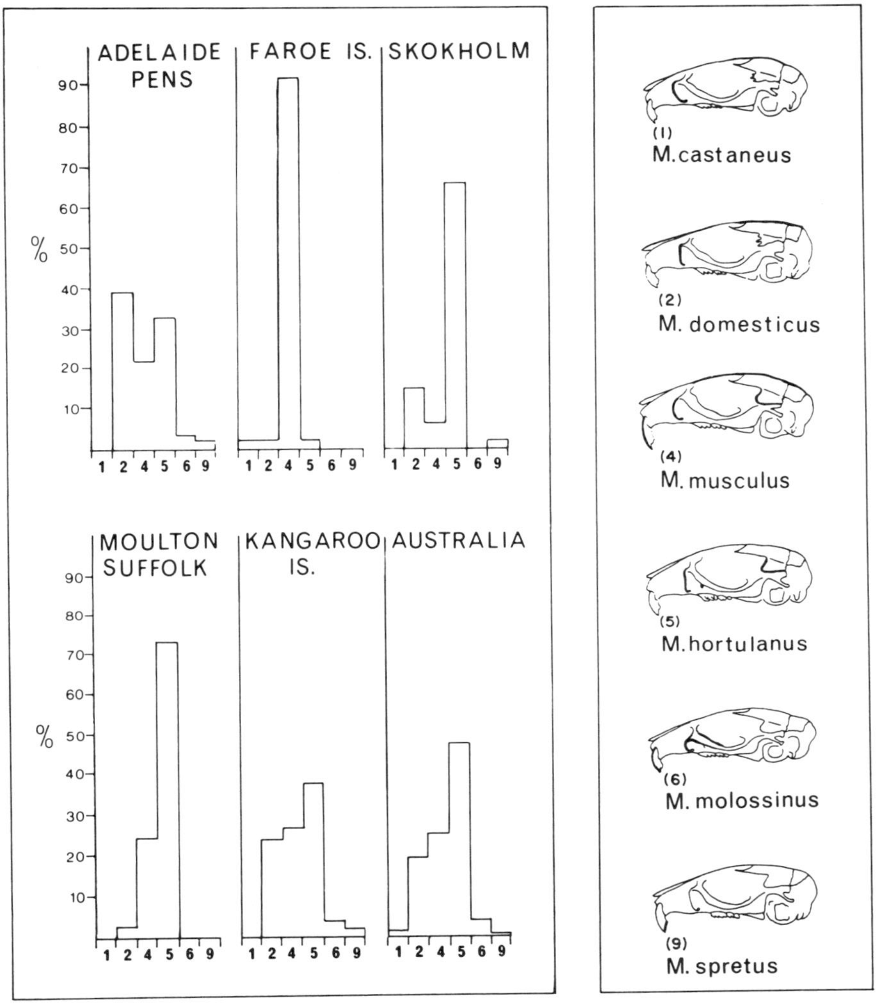

FIGURE 2. Some skull characteristics of house mice from parts of Australia and elsewhere. (Modified from Marshall, J. J., *The Mouse in Biomedical Research: History, Genetics and Wild Mice,* Vol. 1, Foster, H. L., Small, J. D., and Fox, J. G., Eds., Academic Press, N.Y., 1981.)

of the world and to utilize a vast array of habitats is summed up by the descriptive term "mammalian weed".[6] This success would appear to be due to the dispersive imperative (perhaps as a consequence of territorial aggression), social plasticity, and high fecundity of house mice.[6,11] Females as young as 6 weeks can give birth, and thereafter litters can be produced at approximately four-weekly intervals. The average number of young per litter can exceed nine.

Rarely, however, is this high reproductive potential achieved in the wild. Factors which operate to reduce reproductive productivity include autumnal cessation of breeding, submaximal litter size, sexual regression of adult females (sometimes independent of autumn cessation of breeding), and failure of young females to reach puberty.[6-11,14,16,17,33,36,38,55]

Various conceptual models have been proposed previously to explain the formation of mouse plagues in Australia.[20,30-32,36,38,49] While each of those models identifies a different mechanism, all of them identify realization of the reproductive potential of house mice, after one or more unusual weather events, as the major demographic change causing these massive increases in abundance.

Table 2
SUMMARY OF TYPES OF LOSSES CAUSED BY MOUSE PLAGUES IN RURAL AUSTRALIA

1. In-crop losses in cereals, coarse grains, and oil seeds
   A. Eating of seeds between sowing and germination
   B. Eating of plants between germinations and tillering
   C. Eating of grain from seed heads
   D. Felling of whole heads
2. On-farm post-harvest losses to grains
   A. Eating of seed grain stored in sheds on farm
   B. Eating and fouling of crop-product stored on farm
3. On-farm damage to vehicles, machinery, and equipment
4. Damage to farm buildings
5. Damage to personal and domestic goods and furniture
6. Damage in grain depots and silos
7. Losses in fruit and vegetable industries
8. Losses in intensive pig and poultry industries
9. Losses in shops in rural towns
10. Losses in the public sector
    A. Damage to electrical wiring
    B. Problems in hospitals, airports, etc.
    C. Loss of experimental agriculture trials

III. ECONOMIC CONSEQUENCES OF MOUSE PLAGUES

The wide array of situations in which mouse damage or contamination cause losses are listed in Table 2, from which it can be seen that there are a number of characteristics of Australian mouse plagues which distinguish them from most other agricultural pests:

1. Most agricultural pests do not cause significant losses outside of the agro-system. Mouse plagues inflict losses in nonagricultural operations.
2. Most agricultural pests are confined to one crop. Mouse plagues occur in, and cause significant damage in, a wide array of crops.
3. Most agricultural pests inflict losses at only one stage of the crop-product cycle. Mouse plagues can cause significant losses at any stage, from planting through to postharvest storage of the crop product.

By using questionnaires, the Victorian State Government estimated losses due to the 1979-80 plague to be in the order of $17 to 23 million in that State alone. Losses of similar magnitude may have occurred in South Australia, with somewhat smaller losses in southern New South Wales.

During the mouse plague of 1984, in-crop damage to wheat (and other winter cereals) in Victoria was assessed from field surveys. Losses due to damage at all stages of the crop, from planting to harvesting, were estimated to be $10.2 million. In New South Wales, in-crop losses to winter cereals were estimated at $1 million, while in South Australia, the plague appears to have subsided before economically-significant damage was inflicted.[37]

Two points about these estimates warrant discussion. First, drought conditions returned to eastern Australia in late-autumn of 1984. Some farmers, prepared to gamble that crop-saving rains would fall, planted at the normal time in May. Others delayed planting until the rain did arrive in August. The gamblers won! The August rains saved the crops, which produced average yields. In contrast, late-planted crops were stunted and yielded poorly. This bimodal distribution of the height and yield of the 1984 wheat crop had important effects on the losses caused by mice — most damage occurred in the crops that were planted at normal times, and little was recorded from the late-planted crops. Had there been even moderate rains in late autumn or early winter in 1984, mouse damage to the 1984 wheat crop would have been perhaps two or three times more severe.

Second, the 1984 plague was generally regarded as less severe, particularly in South Australia, than the 1979-80 plague. Even so, losses totalling $11 million occurred in wheat crops alone, indicating that the estimate of $15 to 20 million for all losses in Victoria in 1979-80 was a gross underestimation.

Unlike wheat crops, which are harvested in summer, coarse grains such as sorghum, maize, and rice, and oil seeds such as sunflower and soy bean may remain in the fields until autumn. During a plague, as mice continue to breed through the summer, peak mouse numbers are reached in autumn. These summer crops, which are grown mainly in New South Wales and Queensland, consequently suffer severe in-field damage during mouse plagues. Unfortunately, there is little documentation of the value of that damage, which could equal the damage to wheat crops. Ryan and Jones[43] used questionaires sent to 1100 rice farms in three districts in the Murrumbidgee Irrigation Region of New South Wales to obtain growers estimates of losses caused by the mouse plague in late 1969 and early 1970. For each of the three districts the average estimated loss per grower was $2398, $1988, and $1197, respectively. Based on inflation rates over the last 15 years, similar losses would be worth from $5000 to 10,000 per grower.

Known losses to vegetable growers in the Swan Hill district of Victoria (Figure 1) totalled $100,000 in 1984. If they occurred, losses in other States were not recorded. Intensive pig production and intensive poultry farming are affected by mouse plagues. In piggeries, losses occur because of decreased weight gain caused by diseases transmitted to the pigs by the mice (probably including salmonellosis, leptospirosis, erysipelas, and encephalo-myocarditis) and by mice disturbing the pigs. Such losses are difficult to quantify, and were estimated at between $0.9 and 3.4 million in Victoria in 1984.

Losses that occur on farms because of damage to equipment and vehicles gnawed and/or fouled by mice would appear impossible to quantify. It is the writer's opinion that such losses, together with damage to personal and domestic possessions in farm houses, probably exceed the losses occurring in crops.

Off-farm, severe losses occur in shops, especially in rural townships, where damage by gnawing and contamination by feces and urine destroys foodstuff and other goods. While most of these nonagricultural losses are not reported, some, because of their bizarre nature rather than their economic value, do receive the attention of the media. One newspaper article, under the heading of "The Byter Bitten", recounted the destruction, by mice, of computer print-outs accumulated in an unmanned mobile scientific laboratory in rural New South Wales.[1] Another article related how the death of an African elephant at a zoo may have been due to the mouse plague of 1984: "We lost the elephant because it died from the results of the mouse plague".[2] At post-mortem, gross autopsy and histopathology were indicative of the mouse-borne disease encephalomyocarditis.[21]

Stories such as those above have sufficient dramatic impact for them to be reported in newspapers. The real problems, losses in rural communities, a substandard domestic life for farming families during plagues, and environmental contamination with rodenticides, do not have such a popular appeal.

IV. SOCIAL CONSEQUENCES OF MOUSE PLAGUES

A. Domestic Consequences

During mouse plagues, mice invade houses, where they damage electrical wiring and insulation in refrigerators, stoves, etc., as well as furniture and fittings. Newspaper reports, such as the one which follows,[3] are not over-dramatizations of the stress that country folk have to endure during mouse plagues.

> "I've gone into the kids' room at night and seen mice eating at the corners of their mouths. It makes you sick."
> "They're so bad we'd pack up and leave tomorrow, but we've got nowhere to go," he explained.
> "Mice have eaten everything we had . . . everything. They've wrecked our home. They're slowly wrecking our lives."
> "The damage toll is almost unbelievable."
> "Mice have chewed through metres of electrical wiring, in the process destroying a new stove, a fridge, a freezer, and even a television set."
> "They've eaten the all the kids' bedclothes and a fair bit of our clothing".
> "When you lie in bed at night you feel them run across your face. Fair dinkum, it drives you crazy . . . the stench from droppings and dead mice was almost overpowering at times."
> "Sometimes you wonder how you put up with it. It's almost unbearable. You live on a knife's edge at night," she said.
> "But it's not much use worrying about it. We've got nowhere to go, so we'll just HAVE to put up with it."

B. Human Diseases Transmitted by Mice

Given the large range of zoonoses capable of transmission by house mice[59] it is somewhat surprising that mouse plagues appear not to play a significant role in the transmission of human diseases. It is certain that cases of salmonellosis occur during plagues, and dermatomycoses due to the mite *Trichophyton quinckeanum* and *T. mentagrophytes* have been recorded.[15,26] The possibility that mouse plagues could play a major role in the spread of an exotic disease appears not to have been considered at the practical level.

V. ENVIRONMENTAL CONSEQUENCES OF MOUSE PLAGUES

During mouse plagues, farmers desperately seek measures to reduce the damage occurring to their crops and possessions. No rodenticides are registered in any Australian State for in-crop use against house mice. This is because data required for registration for in-crop use, particularly those on residues in crop products, are not available. Desperate to stop the damage they can see being inflicted, people resort to whatever poison is available and massive amounts are used.[46] In 1981, the impact that such poisoning had on nontarget organisms was highlighted,[48] but the situation has not improved.

VI. STRATEGIES FOR REDUCTION OF LOSSES CAUSED BY MOUSE PLAGUES

A. Critical Features of Mouse Plagues

Pests of agriculture can be classified as either "occasional pests" or "key pests".[54] Occassional pests are those that cause significant damage only in certain places or at certain times. In contrast, key pests, in the absence of deliberate control by man, usually cause significant damage each year. There is no doubt about the status of Australian mouse plagues. During years between plagues mice are rare or extinct throughout much of the country over which an earlier plague had erupted. Mouse plagues in Australia are clearly an occasional pest problem.

As occasional pests, mouse plagues can be contrasted to outbreaks of rats *(Rattus rattus mindanensis, Rattus argiventer,* and *Rattus exulans)* which occur in some years in parts of south-east Asia. So severe are these outbreaks and so great is the amount of food eaten by the rats that people have been threatened with starvation because of the resulting food shortages.[45] These spectacular outbreaks, which attract much public attention, may mask the fact that the rats are, in fact, key pests rather than occasional ones. This view is expressed by Fall[18] who writes "Relatively low-level, chronic damage which passes unnoticed year after year, or which is simply tolerated by farmers, probably plays a more important role in reducing available food than the spectacular losses associated with occasional rodent population irruptions". In rural southern Australia, mouse plagues do not conform to Fall's description. Except during plagues years, losses caused by house mice in Australia are neglible. Neither do mouse plagues accord with the statement by Armstrong that "Rodent problems in Australia are not unlike those experienced in other countries of the world".[4] Mouse plagues apparently similar to those in Australia do occur in the North Xinjiang Province of the People's Republic of China[61] (but it is interesting to note that house mice in Xinjiang are *Mus musculus,* according to the map presented by Marshall[28] who considers Australian house mice to be *Mus domesticus*).

B. Ad Hoc Control: The Current Practice

In spite of the massive use of poisons during mouse plagues, the problem recurs at irregular intervals and serious losses occur. The main reason for the lack of any improvement in the situation since the turn of the century may be that the approaches to the problem have not been suited to an occasional pest. Attempts to alleviate the problem have consisted almost entirely of poisoning to reduce mouse numbers after they have increased to such a level that mice are apparent. Fewer than 40 mice per hectare of wheat field are inapparent to farmers.[30] By the time mouse numbers exceed that level, and farmers become aware of the presence of mice, the problem is probably out of hand. Mice that are killed by poisons will be quickly replaced by mice from nearby.* If the losses caused by mouse plagues are to be prevented in the future, a control strategy which takes into account the occasional and acute nature of the plagues, as well as the features noted in Section III, above, will have to be employed. Two such strategies which have been proposed previously,[39] are discussed in the remainder of this chapter.

C. The PICA Strategy

One way of taking into account the sporadic and acute nature of mouse plagues is to predict them. The initial letters of the four essential steps in a control strategy based on prediction — Predict, Inform, Control, and Assess — are used here to coin the term PICA control strategy. Each of the four steps is now discussed.

1. PICA Step 1: Prediction

a. Various Approaches to Predicting Mouse Plagues

There are three ways by which mouse plagues could be predicted. By monitoring mouse populations at various sites throughout the regions which have previously experienced plagues, equations relating the future abundance of house mice to various population parameters such as sex ratio and number of female mice could be developed, as was done recently in the Peoples Republic of China.[62,63] Under Australian conditions, however, this method of prediction would seem to be impractical for it would be difficult to maintain the monitoring operation during the sequence of years between plagues, when mouse numbers are extremely low. A more optimistic view is held in southern Queensland, where there is currently a pilot attempt to use farmers to monitor mouse populations.[60]

A second method of prediction is based on empirical relationships between past plagues

* There is one Australian example of apparently successful control of mouse damage after mice became obvious.[46,50] That outbreak, however, was most unusual in so far as it peaked in spring, compared to the more usual autumn peak. Also, that outbreak was confined to a relatively small area around which there was no outbreak. Success of the control operation may have been enabled by the lack of immigration of mice from surrounding fields into the poisoned area.

and past weather conditions. For example, it has been demonstrated that all plagues since 1901, with the possible exception of the 1969-70 plague, were preceded by at least two consecutive winter droughts.[49] While previously suggested causal relationships linking mouse plagues to preceding droughts (such as reduced efficiency of predators or parasites which would otherwise regulate mouse densities below plague levels) appear improbable, the relationship may prove to be a good predictor of mouse plagues.

Mechanistic prediction models constitute a third method for predicting mouse plagues. This approach requires a thorough understanding of the critical environmental trigger and subsequent population mechanisms which lead to the formation of mouse plagues. This type of model, once the critical features are known and understood, is the type most likely to give accurate, site-specific predictions of mouse plagues. In the next section, some mechanistic models of mouse plagues in Australia are discussed.

b. The Causes and Mechanisms of Mouse Plagues Formation

During the years between plagues, house mice are rare in rural Australia and, over much of the habitat which they occupy during plagues, they are absent. The environmental trigger and the population mechanisms which transform mouse populations from this state of absence or rarity to plague proportion are not fully understood. Most experts agree that rainfall is the key environmental factor, but there are a number of views on how rainfall patterns result in mouse plagues. Mouse plagues are not confined to agricultural lands, so it can be concluded that their ultimate cause is not a factor related directly to cropping.

Finlayson[20] proposed that plagues of house mice and native rodents *(Rattus villosissimus* and *Notomys alexis)* in nonagricultural, arid South Australia were due to the availability of high-quality food after unusually good rains. A different common cause for plagues of these three rodents is proposed by the Newsome-Corbett model which is based on the premise that, in most years, breeding occurs only during the summer, and then only in small refuge populations.[32] Although food is abundant over vast areas, such as the wheat crops, the soils are too hard for mice to burrow into, so they are unable to build nests there. In winter, although the soils are softened by rainfall, food is not sufficiently abundant to allow mice to breed. According to this model, the environmental trigger of a mouse plague is good rainfall in early summer. Such rain, it is postulated, softens the soils of the majority habitat (the wheat fields) where food, in the form of the ripening wheat crop, is abundant. The soft soils allow an abundance of nest sites also. If there is a concurrent abundance of these two resources, nest sites and food, for a period of 3 to 5 months, a plague will erupt.

This model would account for the timing of most plague peaks, in autumn. However, it fails to account for (1) the absence of mouse plagues in irrigated summer crops except during a generalized plague. Such summer irrigation, of crops such as rice, sunflower, maize, sorghum, etc. provides a concurrent abundance of food and soft soil every summer; and (2) the absence of mouse plagues most years in the Mallee soil regions of southern Australia. This region is one of the most severely affected during the mouse plagues of 1969-70, 1979-80, and 1984. There, however, the soils are sandy, and do not harden during summer. Some factor other than soil hardness must prevent extensive breeding by mice during summers in the Mallee soil region.

Within its general extent, the mouse plague of 1979-80 occurred in the Murrumbidgee Irrigation Region (MIR). From a study of house mice on an irrigated cereal farm in the MIR, a "food-quality, spacing-behavior" model has been proposed to explain the formation of mouse plagues.[36,38] The critical features of that model, shown in Figure 3, are as follows:

1. The distal cause, or environmental trigger, of the outbreak was identified as above-average rains in the autumn of early 1978, nearly 2 years before the outbreak. During the winter of 1978 (Phase 1 increase, Figure 3), some mice continued to breed,

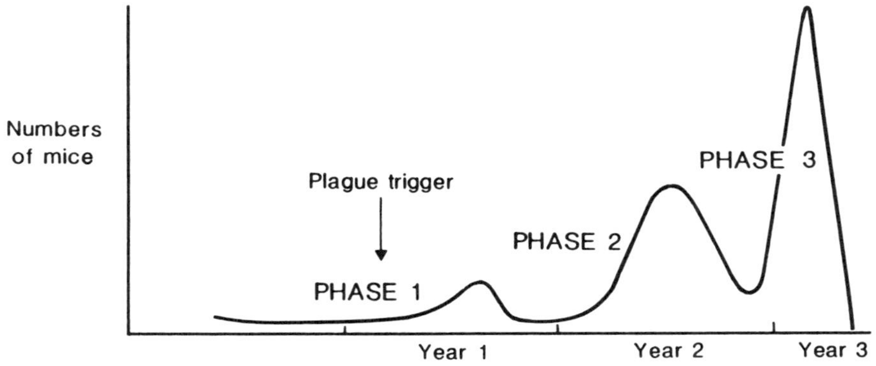

FIGURE 3. A schematic representation of changes in abundance of house mice during a mouse plague. The mechanism is triphasic, with the first peak in late winter or spring, the second peak in winter, and the third peak in late summer or autumn. The relative heights of the peak may vary between plagues, and drought in Year 2 may prevent Phase 3 from occurring.

compared to normal autumn cessation of breeding. This extended breeding season was attributed, in turn, to an extension of the period for which high quality food was available to mice. By examination of the effects of a range of qualitatively different foods on the breeding performance of wild house mice in the laboratory, protein decline was subsequently identified as the factor likely to be responsible for cessation of breeding in most autumns on irrigated farms.[7,8] This was then tested in a field experiment in which a protein-enriched diet (sunflower seeds) was added to the existing available food (mature rice grain) on 8 of 16 trapping grids in autumn 1983.[7,9] The proportion of females continuing to breed into winter on grids with supplementary protein-enriched food was nearly twice that on grids with supplementary low protein food.

2. The proximal cause of the outbreak was the commencement of breeding, in early summer at the end of 1979 (Phase 3 increase), of an unusually large number (between 3 and 4 times as many as at the corresponding time in 1978) of nulliparous adult females which had survived the winter of mid-1979.
3. The unusual abundance of adult females at the end of the 1979 winter was due to the high reproductive productivity of females during the 1978-79 summer breeding season (Phase 2 increase). This high productivity was attributed to a qualitative improvement in the mice which accompanied an extension of the 1978 breeding into the winter of that year and was manifested by an average litter size at the beginning of the breeding season in late 1979 of 9.6, compared to 6.4 and 7.2 at the beginning of the 1977-78 and 1979-80 breeding seasons.

c. A Recent Attempt at Predicting a Mouse Plague

The food-quality spacing-behavior model was used in 1983 to predict a mouse plague with its first peak in autumn-winter of 1984, and a second peak a year later. Above average rains fell in the autumn of 1983. According to the model, mice should have bred during the winter of 1983, which they did. As we have already seen, mice reached plague proportions in autumn and winter of 1984. With the rare exception of isolated areas with summer irrigated grain farming, the second predicted peak (Phase 3) failed to occur, probably because of the dry conditions of the 1984 winter, as discussed above.

Consequently, it would seem necessary that the predictive model should incorporate rainfall during the second winter after the autumn rains trigger a plague, thereby giving a better prediction with respect to Phase 3.

2. PICA Step 2: Information

Once a plague is predicted, those who are to undertake the control operations have to be informed. Information on the timing and distribution of the plague, and of activities which individuals can undertake to reduce or prevent losses, will be required, as will information on the hazards associated with control by poisoning.

3. PICA Step 3: Control

In contrast to current ad hoc operations after high mouse numbers become apparent, control in the context of the PICA Strategy aims at preventing a predicted mouse plague from forming. This prevention will require tactical intervention, using existing or new chemical or biological agents, during Phase 1. It will be necessary to carry out control activities in various habitats, including crop and noncrop areas, into which mice disperse during the Phase 1 increase. Such tactical use of a biological control agent has never been attempted against mouse plagues in Australia. Cultures of the bacterial pathogen Salmonella have been used in other countries, but the practice is now condemned.[5] Recently, tactical use of the nematode *Capillaria hepatica* to control populations during the build-up of a mouse plague has been proposed.[52,56] Possible use of Ectromelia virus (mousepox virus) for the same purpose has been examined. It was concluded that mousepox has only limited potential because of innate resistance in some of the wild mice tested.[42] Further examination, using different strains of Ectromelia virus, could, however, yield different results (See also EGEA Strategy, below.)

4. PICA Step 4: Assessment of the Operation

Because the aim of the control operations within the PICA strategy is to prevent an impending plague, it will be important for the efficacy of the control operations to be assessed accurately. Farmers and others could then be informed of the revised possibility of a plague eventuating.

D. The EGEA Strategy

The PICA Strategy discussed above adequately addresses the problems caused by the occasional and acute nature of mouse plagues. For its implementation, however, it would require not only a proven ability to predict mouse plagues and efficacious control agents, but also an infrastructure of personnel to do the predicting, to advise farmers of impending plagues and on when, how, and where to bait. Any loss-prevention strategy which would operate without the requirement for such an infrastructure would be desirable, particularly as the preventative nature of the PICA Strategy imposes a necessity for farmers to commence control activities before high mouse numbers become obvious. Most Australians are aware of the dramatic reduction in the abundance of the European rabbit *(Oryctolagus cuniculus)* after the introduction of the myxoma virus in 1950. Consequently, there is, from time to time, a popular demand for a myxoma-like control mechanism to prevent mouse plagues.

The control of rabbit numbers by myxomatosis remains, however, the only known example of control of a mammalian pest by an epizootic viral disease. That it is an "accident of nature"[19] is probably due to a unique combination of three properties of the myxoma virus:

1. The myxoma virus is highly mutable and, hence, able to adapt to changing environmental conditions.
2. Transmission of the myxoma virus is by insect vectors, which appear to selectively transmit isolates which allow a significant proportion of the infected rabbits to survive, and so act as a source of infection of other rabbits.
3. The virus has a narrow range of susceptible hosts.

While there has never been an exhaustive attempt to find a house-mouse analog of myxomatosis, it seems unlikely, given the vast knowledge about viruses of mice because of their involvement in medical research, that such an organism will be found in nature. With the recognition of this poor prognosis for prevention of mouse plagues by naturally-ocurring organisms, it has been proposed that genetic engineering could be used to produce such an organism.[22] The term EGEA Strategy, standing for Epizootic Genetically Engineered Agent, has been coined for this strategy.[39] Clearly, an intensive program of research is required if the EGEA Strategy is to be implemented to prevent mouse plagues in the future. As well as leading to the production of the new organism, that research program will also have to lead to a thorough understanding of the population biology of house mice in rural Australia.

VII. CONCLUSION

Plagues of house mice recur at irregular intervals in rural Australia, despite the use of massive amounts of poisons during each plague. Failure to prevent severe losses, as well as social and environmental damage, appears to be due to the fact that ad hoc control is implemented too late, after mice have become abundant. Two new strategies, to prevent the irruption of mouse plagues, are proposed. One (The PICA Strategy) is based on prediction of plagues followed by tactical intervention. The other (the EGEA Strategy) is based on the genetic engineering of an epizootic organism, somewhat analogous to the myxoma virus which has significantly reduced rabbit abundance. Both strategies require concerted programs of research if they are to be implemented in the future.

REFERENCES

1. **Anon,** The Byter bitten, *CoResearch,* 230, 4, 1980.
2. **Glascott, J.,** Questions about dead foal at Dubbo zoo, *The Sydney Morning Herald,* December, 11, 1984.
3. **Davies, O.,** The mouse house of horrors, *The Herald,* Melbourne, April 14, 1980, 1.
4. **Armstrong, D. J.,** Rodents in Australia, in *Proc. Conf. Organization and Practice of Vertebrate Pest Control,* Dubock, A. C., Ed., Imperial Chemical Industries PLC, Haslemere, U.K., 1984, 49.
5. **Barnett, S. A. and Prakash, I.,** *Rodents of Economic Importance in India,* Arnold-Heineman, New Delhi, 1975, 33.
6. **Berry, R. J.,** Population dynamics of the house mouse, *Symp. Zool. Soc. London,* 47, 395, 1981.
7. **Bomford, M.,** Food quality, diet and reproduction of house mice on irrigated cereal farms, Ph.D. thesis, Australian National University, Canberra, 1985.
8. **Bomford, M.,** Food and reproduction of wild house mice. III. Experiments on the breeding performance of caged house mice fed rice-based diets, *Aust. Wildl. Res.,* 14, 207, 1987.
9. **Bomford, M. and Redhead, T. D.,** Food and reproduction of wild house mice. IV. A field experiment to examine the effects of food quality and population density on breeding, *Oikos,* 48, 304, 1987.
10. **Breakỳ, D. R.,** The breeding season and age structure of feral house mouse populations near San Fransisco Bay, California, *J. Mammal.,* 44, 153, 1963.
11. **Bronson, F. H.,** The reproductive ecology of the house mouse, *Q. Rev. Biol.,* 54, 265, 1979.
12. **Carstairs, J. L.,** Population dynamics and movements of *Rattus villosissimus* (Waite) during the 1966-69 plague of Brunett Downs, Northern Territory, *Aust. Wildl. Res.,* 3, 1, 1976.
13. **Covacovich, J. and Easton, A.,** *Rats and Mice in Queensland,* Queensland Museum Booklet Number 9, Brisbane, 1974, 46.
14. **DeLong, K. T.,** Population ecology of feral house mice, *Ecology,* 48, 611, 1967.
15. **Donald, G. F. and Brown, G.,** *Aust. J. Dermatol.,* 3, 133, 1964.
16. **Drickamer, L. C.,** Acceleration and delay of first estrus in wild *Mus musculus, J. Mammal.,* 60, 215, 1979.
17. **Evans, F. C.,** A population study of house mice *(Mus musculus)* following a period of local abundance, *J. Mammal.,* 30, 351, 1949.

18. **Fall, M.,** Management strategies for rodent damage problems in agriculture, in *Proc. Symp. Small Mammals: Problems and Control,* Sanchos, F. F., Ed., Biotrop Special Publications No 12, Bogor, Indonesia, 1980, 177.
19. **Fenner, F. and Myers, K.,** Myxoma virus and myxomatosis in retrospect: the first quarter century of a new disease, in *Viruses and the Environment,* Kurstak, E. and Maranovosch, K., Eds., Academic Press, New York, 1978, 539.
20. **Finlayson, H. H.,** On mammals of the Lake Eyre Basin. IV. General remarks on the increase of murids and their population movements in the Lake Eyre Basin during the years 1930-1936, *Trans. R. Soc. South Aust.,* 63, 348, 1939.
21. **Finney, E. P.,** personal communication, March 11, 1985.
22. **Gibbs, A. J.,** Genetic manipulation, viruses and biological control, *Rev. Rur. Sci.,* 1986.
23. **Jackson, W. B.,** Urban rodent control in the United States, in *Proc. Conf. Organization and Practice of Vertebrate Pest Control,* Dubock, A. C., Ed., Imperial Chemical Industries PLC, Haslemere, U.K., 1984, 66.
24. **Kirby, G. C.,** The Genetical Structure of Subdivided Populations, Ph.D. thesis, University of Adelaide, Adelaide, 1974.
25. **Klein, J., Gotze, D., Nadeau, J. H., and Wakeland, E. K.,** Population immunogenetics of Murine H-2 and T systems, *Symp. Zool. Soc. London,* 47, 439, 1981.
26. **Lawrence, H.,** Dermatomycosis in mice and men, *Med. J. Aust.,* 1, 146, 1918.
27. **McDougall, W. A.,** An investigation of the rat pest problem in Queensland canefields. II. Species and general habits, *Queensl. J. Agric. Sci.,* 1, 1, 1944.
28. **Marshall, J. J.,** Taxonomy, in *The Mouse in Biomedical Research: History, Genetics and Wild Mice,* Vol. 1, Foster, H. L., Small, J. D., and Fox, J. G., Eds., Academic Press, New York, 1981, chap. 2.
29. **Marshal, J. J. and Sage, R. D.,** Taxonomy of the house mouse, *Symp. Zool. Soc. London,* 47, 15, 1981.
30. **Newsome, A. E.,** The Ecology of the House Mouse in South Australia, Ph.D. thesis, University of Adelaide, Adelaide, Australia, 1966.
31. **Newsome, A. E.,** A population study of house-mice temporarily inhabiting a South Australian wheatfield, *J. Anim. Ecol.,* 38, 341, 1969.
32. **Newsome, A. E. and Corbett, L. K.,** Outbreaks of rodents in semi-arid and arid Australia: causes, preventions and evolutionary considerations, in *Rodents in Desert Environments,* Prakash, I. and Ghosh, P. K., Eds., Dr. W. Junk, The Hague, 1975, chap. 6.
33. **Pelikan, J.,** Patterns of reproduction in the house mouse, *Symp. Zool. Soc. London,* 47, 205, 1981.
34. **Plomley, N. J. B.,** Some notes on plagues of small mammals in Australia, *J. Nat. Hist.,* 6, 363, 1972.
35. **Redhead, T. D.,** A relationship between rainfall and rat damage to sugar cane in North Queensland, *Queensl. J. Agric. Anim. Sci.,* 37, 181, 1980.
36. **Redhead, T. D.,** Reproduction, growth and population dynamics of house mice on irrigated and non-irrigated cereal farms in New South Wales, Ph.D. thesis, Australian National University, Canberra, 1982.
37. **Redhead, T. D.,** Unpublished report to the Standing Committee on Agriculture, Canberra, 1985.
38. **Redhead, T. D.,** Demography of an outbreak of house mice on a summer-irrigated cereal farm in New South Wales, in manuscript, 1986.
39. **Redhead, T. D.,** Rodent control research in rural Australia, Unpublished report, Commonwealth Scientific and Industrial Research Organization, Division of Wildlife and Rangelands Research, Canberra, 1986.
40. **Redhead, T. D., Enright, N., and Newsome, A. E.,** Causes and predictions of outbreaks of *Mus musculus* in irrigated and non-irrigated cereal farms, *Acta Zool. Fenn.,* 173, 123, 1985.
41. **Redhead, T. D. and Saunders, I. W.,** Evaluation of thallium sulphate against rats in Queensland (Australia) sugar-cane fields adjacent to different types of vegetation, *Prot. Ecol.,* 2, 1, 1980.
42. **Roberts, J. A.,** Susceptibility of Australian feral house mice to Ectromelia virus, *Aust. Wildl. Res.,* 13, 49, 1986.
43. **Ryan, G. E. and Jones, E. L.,** A report on the mouse plague in the Murrumbidgee and Coleambally Irrigation Areas, 1970, unpublished report, New South Wales Department of Agriculture, Sydney, 1972.
44. **Sage, R. D.,** Wild mice, in *The Mouse in Biomedical Research History, Genetics and Wild Mice,* Vol. I, Foster, H. L., Small, J. D., and Fox, J. G., Eds., Academic Press, New York, chap. 2.
45. **Sanchez, F. F. and Reidinger, R. F.,** Control of agriculture rodents through sustained baiting method, in *Proc. Symp. on Small Mammals: Problems and Control,* Sanchez, F. F., Ed., Biotrop Special Publication No 12, Bogor, Indonesia, 1980, 183.
46. **Saunders, G. R.,** Evaluation of mouse plague control techniques in irrigated sunflower crops, *Crop Prot.,* 2, 437, 1983.
47. **Saunders, G. R.,** Mice and their control, *Agfacts,* No A.O.1., Department of Agriculture, Sydney, Australia, 1982.
48. **Saunders, G. R. and Cooper, K.,** Pesticide contamination of birds in association with mouse plague control, *Emu,* 82, 227, 1981.

49. **Saunders, G. R. and Giles, J. R.,** A relationship between plagues of house mice, *Mus musculus,* (Rodentaia, Muridae) and prolonged periods of dry weather in south-eastern Australia, *Aust. Wildl. Res.*, 4, 241, 1977.
50. **Saunders, G. R. and Robards, G. E.,** Economic considerations of mouse plague control in irrigated sunflower crops, *Crop Prot.*, 2, 153, 1983.
51. **Schwarz, E. and Schwarz, H. K.,** The wild and commensal stocks of the house mouse, *Mus musculus, J. Mammal.*, 24, 59, 1943.
52. **Singleton, G. R. and Spratt, D. C.,** The effects of *Capillaria hepatica* (Nematoda) on natality and survival to weaning in BALB/c mice, *Aust. J. Zool.*, 677, 1986.
53. **Singleton, G. R.,** personal communication, 1986.
54. **Smith, R. F. and van den Bosch, R.,** Integrated control, in *Pest Control: Biological, Physical and Selected Chemical Methods,* Kilgore, W. W. and Doutt, R. L., Eds., Academic Press, New York, 1967, chap. 9.
55. **Smith, W. W.,** Reproduction in the house mouse *Mus musculus* L. in Mississippi, *J. Mammal.*, 35, 509, 1954.
56. **Spratt, D. C. and Singleton, G. R.,** Studies of the life cycle, infectivity and clinical effects of *Capillaria hepatica* (Bancroft, 1893) (Nematoda) in mice *(Mus musculus), Aust. J. Zool.*, in press.
57. **Strahan, R., Ed.,** *The Australian Museum Complete Book of Australian Mammals,* Angus and Robertson, Sydney, 1983.
58. **Thaler, L., Bonhomme, F., and Britton-Davidian, J.,** Processes of speciation in the house mouse, *Symp. Zool. Soc. London,* 47, 27, 1981.
59. **Weber, W. J.,** *Diseases Transmitted by Rats and Mice,* Thomas Publications, Fresno, Calif., 1982.
60. **Wilson, J.,** personal communication.
61. **Yan Zhitang,** Some data concerning mass outbreaks of house mice in the agricultural regions of North Xinjiang Uygur autonomous region, source unclear, in Chinese with English summary, 1985.
62. **Yan Zhitang, Li Chinqui, and Zhu Shengkan,** Investigations of age structure in population of house house (*Mus musculus* L.) and the prediction to the mouse number, *Acta Theriol. Sinica,* 3, 53, 1983, in Chinese, with English summary.
63. **Yan Zhitang and Zhong Mingming,** The prediction of fluctuations in home mouse *(Mus musculus)* populations and studies of mechanisms, *Acta Theciol. Sinica,* 4, 140, 1984, in Chinese, with English summary.

Chapter 11

REPRODUCTION

G. I. Twigg

TABLE OF CONTENTS

I. INTRODUCTION

One of the chief features of rodents, which results in their pest status, is the ability to produce large populations in relatively short periods of time. This basic capacity is probably a strategy of some value in the members of an Order of mammals which is heavily preyed upon and it has been exploited even further in commensal species to take advantage of constant food and favorable microclimates.

Rodents comprise approximately one-third of named Eutherian species, inhabit most terrestrial niches, and range in size from the tiny *Micromys minutus* at 6.1 g[56,63] to the Capybara, *Hydrochaeris hydrochaeris,* which may weigh up to 79 kg.[45] They exhibit a wide range of reproductive patterns which are capable of modification as habitat needs dictate. Many, if not most, species are polyoestrus, a factor which is of importance if adjustments to breeding rate are to be made. Along with this ability to produce young at short intervals is the additional factor that litters may be large, especially in the genera *Rattus* and *Mus.* Indeed, it is amongst the Myomorpha that the greatest fecundity is observed and, with the exception of the squirrels, the most serious economic pests are found.

The term "reproduction" is one which needs to be defined since different workers have tended to refer to different components of the total process. Strictly speaking, it means the process of regeneration and so it is probably more useful to attempt comparisons of the number of animals being weaned into the population since, of the number of ova ovulated, there will be losses in the uterus, at birth and while suckling. It is therefore more precise, and of more value to those involved in rodent control, to speak of recruitment to the population, i.e., the weaning figure. This is not always easy to define because it is a difficult figure to acquire for rodents where the number of young which leave the nest to fend for themselves is rarely known, unlike those mammals in which the young are at heel while suckling. Thus, as most studies have produced data from dead females there is a tendency for intra-uterine evidence to predominate and many studies have produced useful data on the percentage of females pregnant and the number of embryos per pregnant female. If, then, the number of litters per year or per breeding season is known, some preliminary assessment of possible recruitment can be attempted. Table 1 gives some examples of the reproductive powers of some myomorph rodents. Many more examples could be included but this short list serves to make the main points of reproductive potential. On a cosmopolitan scale the commensal rodents *Rattus* and *Mus* are the most widespread and at the same time, given favorable circumstances, among the most productive. It is scarcely surprising, therefore, that a large proportion of rodent control effort is directed towards members of these genera.

II. COMMENSAL SPECIES

A. Reproductive Potential

1. Rattus norvegicus — The Brown or Norway Rat

After spreading from Asia to western Europe in the first half of the eighteenth century this rat has since spread to all parts of the world. In temperate climates it is the common rat of both urban and rural habitats while in the tropics it is frequently restricted to ports, *Rattus rattus* being the common rat of countryside and town. In ports such as Bombay it is usual to find dwellings inhabited by *R. rattus,* while *R. norvegicus* roams between buildings and is more frequently found on the streets than *R. rattus.*

The number of young recruited into the population is the result of various factors, one of which is the amount of time in the year that the rats can breed. In general, the brown rat can breed all the year, although varying periods of productivity can be seen. Male rats in Baltimore were producing sperm and had secondary sex-organs in a mature state throughout

Table 1
REPRODUCTION IN SOME MYOMORPH RODENTS

Species	Breeding season	No. of young (range)	No. of litters/ breeding season/ yr, where known
Mastomys coucha A. Smith (Multimammate rat)	Polyoestrus — can breed all year but most pregnant ♀♀ found Oct. and Nov. in Sierra Leone; Oct.—Dec. in N. Cameroon	11.5 (7—17)	Variable
Neotoma floridana (Florida woodrat)	Polyoestrus all year	3—4	—
Cricetus cricetus (Common hamster)	March—September	4—18	—
Arvicola terrestris (Water vole)	March—October	6.0—6.7 (N. Scotland) 4.1—5.6 (Netherlands)	1.5—2.3 (N. Scotland)
Microtus agrestis (Field vole)	Polyoestrus April—September	3—7	Succession during breeding season
Mus musculus (House mouse)	Continuous in most urban habitats	5—6 (4—8)	5—10 (urban habitats)
Rattus rattus (Black rat, roof rat)	London — mid March to mid November Tropics — perhaps all year round	London c. 7.0	London 3—5
Rattus norvegicus (Brown rat, Norway rat)	May be continuous in good environment. Mainly summer and autumn in less favorable habitats.	7—8 (6—11)	5.0

the year,[15] unlike many rodent species which undergo marked regression in both gonads and accessory structures in males and a pronounced and extended period of anoestrus in females.

Data used to assess the seasonal and comparative productivity of female rats are various, but perhaps the most widely used yardstick is visible pregnancies. Embryos may be seen by eye in the uterus for around 18 of the 25 days of pregnancy and although this means that 28% of pregnancies will be missed through inability to see them, the method is widely used because field workers rarely have either the time or the facilities to carry out detailed histological studies.

Breeding in *R. norvegicus* may be continuous in an unchanging environment where there is adequate food and in such conditions in England as many as 30% of females may be pregnant throughout the year.[35] This study on the fertility of rats in corn ricks and some other habitats showed a bimodal curve for rick rats as well as for "non-rick" rats. Bimodal peaks of pregnancy have been seen in San Francisco,[59] Georgia,[39] Norfolk,[25] and in Baltimore warehouses.[16] In this latter habitat the two peaks, in June and October, were due to pregnancies in small rats only while large females bred at a uniform rate throughout the year. In contrast, Norway rats caught indoors in San Antonia had a single pregnancy peak in the spring.[13]

In Bombay, despite a fairly constant pregnancy rate, there was a low point in January, the coolest part of the year, and a peak in March when temperatures were on the increase.[26] It seems hardly likely that the effect of temperature can be marked since the lowest temperatures of the year are approximately 26 to 28°C. (79 to 82°F). It is more probable that there is a need for a physiological respite from the demands of continuous breeding.

From the evidence available it appears that the double peak of pregnancy is more common than a single peak in this species.

An examination of many rats in England indicated that the individual rat did not rear a succession of litters throughout the year but that a period of intensive breeding was followed by a resting period.[47] The seasonal variation in the pregnancy rate suggested that either the bursts of breeding activity in individuals lasted longer during the period March to June or that a greater proportion of them coincided then than at other times of the year.

The Brown rat ovulates around ten ova, which is very high for a placental mammal. Over half of the litters conceived suffer some loss before reaching full term so that the average number in a litter at birth is 7 to 8 in the British Isles. In the U.S. data from a variety of studies[14] showed that visible pregnancies varied between 8.0 and 11.2 per female.

More recent work in Korea[8] has shown a seasonal difference in the number of embryos, with higher numbers per female during the peak of the breeding season than at the beginning and the end, although this may be related to female weight. In March, when 15% of females were pregnant, the number of embryos per female was 7.8 (range 6 to 9). At the peak of breeding in June and July, with 58 and 55% of females pregnant, respectively, the number of embryos per female was 9.1 (range 5 to 14) and 9.0 (range 5 to 14). No data are available as to the age of embryos in the Korean samples so that it is difficult to compare the birth figure of 7 to 8 given in the English study.[47]

The relationship of fertility to body weight has an interesting bearing on rat control. In heavier females pregnancies are more frequent and litters are larger. The fertility is therefore considerably higher in a population with a large proportion of heavy animals than in a population with a large proportion of light animals, and the structure of populations may be dependent on the type of control being practiced and its efficiency.

2. Rattus rattus — The Black, House, or Roof Rat

This species is found inhabiting dwellings where it lives in close proximity to people in the tropics. It has been transported to temperate latitudes where populations exist in ports and inland towns and probably depend on fresh importations from time to time to increase depleted populations. There has been little research on the reproduction of this species in cold climates, the only detailed study being in the Port of London.[65] Data were collected with respect to populations on ships and in warehouses. Most breeding in London occurred in the period mid-March to mid-November and there was a litter rate of 3 to 5/year. The average litter size was 6.9 on shore and 7.5 on ships in the docks. As in *Rattus norvegicus* the number of embryos was positively correlated with the body weight of the female. Intra-uterine mortality depleted the number of ova ovulated by about 25%.

In Cyprus, two pregnancy peaks have been demonstrated[66] and various workers elsewhere have shown both bimodal and unimodal curves for pregnancy in *Rattus rattus*. In India, this species tends towards a low level of breeding in the colder, drier parts of the year and with a peak of productivity in the hotter months from February to June. In the Punjab an additional breeding peak was observed during the monsoon. For details of these Indian studies the Indian Plague Commission reports[26-29] should be consulted.

3. Rattus exulans — The Polynesian Rat

This rat is widely distributed in the Pacific. On the Hawaiian islands pregnant females are found from spring to autumn and produce on average 4.3 litters per year with an average number of 4.0 live embryos per litter.[60] In the same habitat *Rattus rattus* has only 2.3 litters per year with 5.1 live embryos per litter. It was observed that the productivity of both species was generally low in comparison with those in other geographical areas.

4. Mus musculus — The House Mouse

Like the brown rat this species spread from Central Asia along the early trading routes and is established in most parts of the world. It is, as the name implies, particularly prevalent in urban habitats and in such conditions breeding is continuous. Females produce between 5 and 10 litters of between 4 and 8 per litter each year. The mouse can breed in complete darkness and cold environments, fertility being little affected by the low temperatures, providing there is adequate food and a supply of nesting material.

Gestation is 19 to 20 days and fertilization can occur at the post-partum oestrus. When this takes place implantation is delayed for between 2 and 16 days. After the embryos have implanted loss is low, about 5%. The young are weaned at between 18 and 20 days and the overall average litter size quoted is 5 to 6,[9] although there is considerable variation depending upon the environment.

B. Reproduction and Environment

The ability of some pest species to produce larger litters more often when conditions are good is one of the main reasons why they are pests. Among the commensal rodents, *Mus musculus* provides a striking example of the adaptability of a species to capitalize on favorable conditions, as was seen in a study of the reproduction of the mouse in four different English habitats.[33] These habitats included:

1. Corn ricks where cereal crops such as wheat, oats, and barley were stacked on the stem at harvest time and not dismantled until the following spring. This environment provided an abundance of food and cover, was generally free from predators, and had a comfortable microclimate.
2. Urban habitats which included houses, gardens, sheds, shops, and factories.
3. Buffer depots where food was stored in bulk as a reserve. These housed grain in quantity and were prone to mouse infestation.
4. Cold stores with deep frozen carcasses of meat, including poultry.

In these habitats *Mus musculus* bred throughout the year and there appeared to be little evidence for seasonal differences in either the percentage of adult females pregnant or in the number of embryos per litter.

The maximum theoretical number of litters per breeding female per year was 13 or 14 and the observed annual productivity in the four habitats was as follows:

Habitat	No. litters/year
Corn ricks	10.22
Buffer depots	7.97
Cold stores	6.68
Urban areas	5.52

Clearly, the litter productivity of the rick mice was nearer the theoretical maximum than that of mice in the other three habitats.

The average number of embryos per pregnant female was not significantly different between the urban, buffer depot, and rick samples, with an average of 5.60, although the cold store females produced on average 6.37. Embryo productivity, obtained by multiplying the number of litters by the average embryo number gave figures of:

Corn ricks	57.2
Buffer depots	44.6
Urban areas	30.9
Cold stores	42.5

Perhaps the most surprising feature of this survey was the low productivity of the urban mice since this habitat was regarded as the optimum for the species. The reason for the reduced productivity was considered to be the fact that the urban situation tended to be unstable with plenty of disturbance and a scanty and unreliable diet.

The cold stores provided an environment with a good microclimate in the nests. The animals gathered hessian from the wrappings of the carcasses and feathers from the poultry, and this more than offset the low temperatures. There was obviously no sterility in male mice in this habitat, although it has been reported in male rats fed solely on protein.

The pregnancy rate in buffer depots was higher than that found in cold stores or the urban environment and this was attributed to the fact that there was little human disturbance and no predators, while the rick environment was the most stable with an even temperature, unlimited food, and no predators. The survey revealed that a 9-month-old rick had a population of 2368 mice (15/m^3). If we consider the average food consumption of a mouse to be around 3.0 g/day then 2368 would eat 7104g (7.1 kg)/day. The results of such an infestation naturally resulted in greatly lowered yields of grain when threshing took place.

The use of ricks as a means of storing unthreshed cereal crops during the winter months has almost entirely ceased in Britain and most other cereal growing countries. When ricks were threshed mice escaped into nearby fields and hedgerows from where they could reinfest the next generation of ricks. Today, while mice in hedgerows are few, they are found in farm buildings, especially granaries, dairy units, mixed-food stores, and ancillary units, the smallest numbers of mice being found in the latter. In those habitats in southern England breeding continued without interruption during the winter months, due no doubt to the bulk storage of harvested crops and the indoor housing and feeding of livestock.[53] Of fecund females 26.7% were pregnant. This compares favorably with a rate of 21.9% in buildings in urban areas[33] but was much less than in corn ricks where 50.5% of females were pregnant.[57] The average litter size on the farms (5.37 ± 0.10) was also greater than that in female mice from urban habitats (5.18 ± 0.18) but less than that for those living in ricks (5.66 ± 0.07). In the farm buildings, females from the dairy units produced the largest litters (5.86 ± 0.16) and those from the granary the smallest (4.96 ± 0.18).

None of the populations considered above, not even the ricks, had reached a degree of crowding sufficient to bring about a measure of population limitation. *Mus musculus,* under extreme overcrowding, can achieve this by suppressing reproduction through a lowering of female fecundity. In experimental colonies,[10-12] as the number of females increased so the average number of young born per adult female decreased, viz;

Av. no. per litter	No. females per colony
8.5	2
7.2	5
3.8	19
2.9	40

In another colony comparable data were:

12.5	3
4.2	27
0	67
0	69

In confined mice the percentage litter resorption varied between 4.8 and 13.8 compared with only 2.7 in free living mice. No drop in male fertility occurs in these situations.

Table 2
REPRODUCTION AND ENVIRONMENT IN *RATTUS* SP.

Species and locality	Pregnancy rate (%) Crude	Pregnancy rate (%) Adjusted	Reproduction figure
R.r. diardii			
Town	13.8	26.3	47.5
R.r. jarak			
Jarak Island	11	20	12.1
R. tiomanicus[a]			
Oil palm	13.6	15.4	21.8
Forest	21	23	55.7
Wasteland	39	43	—
R. argentiventer[b]			
Wasteland	5.9	6.5	13.7
R. exulans			
Forest	20	21	—
Town and waste	26	27	—

[a] *R. tiomanicus* was formerly *R.r. jalorensis*.
[b] *R. argentiventer* was formerly *R.r. argentiventer*.

The productivity of *Rattus rattus diardii, R. r. jarak, R. tiomanicus, R. argentiventer,* and *R. exulans* from a variety of habitats in Malaya was studied.[24] Crude pregnancy rates were calculated as a percentage of all females trapped and adjusted rates on the number of females of the average weight of first pregnancy and above. In addition, a ''reproduction figure'' was calculated which was the number of embryos per 100 head of population. These data are presented in Table 2.

The differences in productivity between the various habitats were, it was believed, due mainly to the level of predation. The highest figure was that of the forest dwelling *R. tiomanicus,* almost equalled by *R. r. diardii* in towns. It was thought that the variety of carnivores in the forest and the predation of man and domestic animals in town were responsible for the high reproduction figures. At the other end of the scale, *R. r. jarak* was a rat which dominated an island almost devoid of predators.[23] *R. argentiventer,* too, lived in areas of lalang *(Imperata cylindrica)* which appeared to be almost free of predators and also had a low reproduction figure.

More recently, the reproduction of *R. argentiventer* has been investigated[31,32] in rice fields in Malaysia, where this rodent is an important pest. Where the rice was double cropped as in Bumbong Lima the breeding activity was sharply bimodal, but in areas of single cropping such as Kota Kuala Muda it was unimodal. Rat breeding began when the rice plant began to flower, and litters were found during the early grain filling stage. In the double-cropped areas the reproductive potential for females was estimated to be five times that of females from single-cropped areas. Mean litter sizes were roughly equal in both crop regimes. The author was of the opinion that nutritional factors associated with the rice plants were important factors in this pattern of seasonal breeding. Such factors are known for some rodents and this will be discussed later (Section V).

III. NONCOMMENSAL RODENTS

A. Cycles and High Numbers

Many rodents may, under certain conditions, assume pest status. Within the confines of a short review it is clearly impossible to cover all situations and I have therefore selected a sample of some of the more important field pests to illustrate features of rodent reproduction as it affects crops.

In north temperate latitudes voles of the genus *Microtus* undergo years of cyclic abundance on a 3 to 4 year basis and may damage severely fodder crops such as grass, to the detriment of the livestock needing it. Voles are also serious pests of forestry.[62] The high numbers at the peak of the cycle are brought about by changes in the length of the breeding season and these are a major driving force in causing populations to cycle. In normally summer breeding species there may be winter breeding during a period of population increase but not during the decline phase, and in the Norwegian lemmings *(Lemmus lemmus)* it is believed that this plays a major role in the rapid population growth.

In addition to this, summer breeding often starts early during the phase of population increase and ends late. As populations reach a peak there is a shortening of the summer breeding season and an absence of winter breeding, while during the decline phase of the cycle there is a delay in the onset of summer breeding.

Another factor which may enhance productivity is the age at sexual maturity. In microtine rodents this is variable, and changes in the rate of sexual maturation of young voles and lemmings are a major driving force behind population cycles. Young Norwegian lemmings about 20 days old (25 g weight) become pregnant in the summer of an increase phase but none of the summer-born young become mature in the year of peak density, and in *Clethrionomys rufocanus* the maturation rate of early summer young is affected markedly by population density. During the increase phase almost all the summer young of both sexes become mature, unlike the peak phase where hardly any males mature.

Thus, in these cyclic species of northern latitudes the length of the breeding season and the age at sexual maturity are the two components of reproduction to change during a population cycle. It is of some considerable practical value that predictive models for vole cycles, based mainly on climatic parameters, are now available.[41]

Even without these cyclic changes causing a rapid increase in numbers, the reproductive potential of microtine rodents is considerable. *Microtus agrestis,* the field vole, has a succession of litters of around 4 to 6 from April to early September and animals born at the beginning of the season usually reproduce in the same year. The number of embryos per female is 3 to 7, a maximum being reached in June. Females are polyoestrus, with a 3 to 4 day cycle which continues during lactation.[7] The young are weaned at between 14 and 28 days and the females become sexually mature at 3 weeks (12 g weight) and may mate at 6 weeks. This factor alone ensures maximum recruitment without any extension of the breeding season.

Larger voles may also achieve pest status but in a localized way,[30] and while they may cause damage to root crops and by burrowing in canal banks their lower rate of population increase and noncyclic behavior serve to ensure that the recruitment rate is slower. For example, *Arvicola terrestris,* the water vole, has between 1.5 and 2.3 pregnancies per year per adult female in northern Europe,[9] between 6.0 and 6.7 in a litter in the north of Scotland,[58] and 4.1 to 5.6 embryos per pregnancy in the Netherlands.[67] In this species losses from immediately preparturition until about 3 weeks old are high, 64.3% and 86.1% in two successive years in Aberdeenshire, Scotland.[58] In European Russia the water vole has as many as four pregnancies per year per adult female.[46]

B. Noncyclic Abundance

It is important to distinguish between the high numbers arising from the regular cyclic nature of some rodent species and the years of abundance which have their origin in local factors such as agricultural practices or climatic factors.

In tropical regions years of high numbers occur on an irregular basis for the most part and while damage to crops may be severe locally the periods of abundance tend to be far apart. Members of the genus *Holochilus* (marsh rats) multiply rapidly and attain plague proportions in parts of South America. Such plagues are called "ratadas". In Uruguay *Holochilus* is the most abundant rat-sized rodent.[17] A "ratada" of several cricetine species, of which *Holochilus* was one, was observed in Southeast Brazil in 1946.[19] The rodents attacked cultivated plants and stored food when they had exhausted the supply of seeds of the "taquara lisa" (smooth bamboo; *Merostichis* sp.). A previous "ratada" had occurred 6 or 7 yr previously at the time of the fruiting season of the rough bamboo. In the sugar cane plantations of the Rio San Francisco Valley, northern Argentina, an irruption of marsh rats was recorded; a great increase in the rodent population began in February, 1944 and reached a peak in April to be followed by a sharp decline in numbers.[36]

A high rate of reproduction was responsible for the overpopulation of a planted lake bed in the Buena Vista Lake Basin, California, in 1926. Both *Mus musculus* and *Microtus* sp. colonized the lake and all crop food and cover was eaten. The two species then moved in three waves to nearby farms. Some indication of the population density may be gauged by the fact that in the center of the damaged area 80,000 rodents per acre were recorded.

Rats *(Rattus)* and *Mus musculus* have been introduced into Australia and the latter has been responsible for mouse irruptions in South Australia, Victoria, and New South Wales, notable wheat growing areas. Between 1910 and 1932 there were three serious outbreaks when numbers rose to very high levels. In 1911 in northern Victoria there was an outbreak and in the years 1916 to 1917 in South Australia, Victoria, and New South Wales damage was very severe. Between February and December, 1917 the recorded total weight of mice caught was 544 tons (approximately 32 million mice). In Victoria in 1932 the wheat areas were again heavily damaged and 2 tons of mice were caught in one night.

In addition to the introduced *Mus musculus* five species of native Australian rodents irrupt at irregular intervals over large parts of Australia.[44,51] Various factors serve to produce conditions favoring high populations of *Mus musculus,* one of which, the improved quality of food due to winter rains which extend the period during which phyto-oestrogens or vitamins are available, has the effect of increasing fertility.[52]

The alteration of a stable habitat by man in the development of commercial crops, as in the case of sugar cane in Guyana, may seriously unbalance the ecosystem in favor of a rodent species which then assumes pest status. Marsh rats of the genus *Holochilus* extend throughout the Amazon drainage area and into the savannahs of Guyana. The coastal plain is very flat, crossed by many slow moving rivers and creeks, and is grass covered. At times of heavy rain (in May-July and December-January between 90 and 130 in. of rain fall per year) the savannah is flooded to a depth of 6 to 10 in. and dry eminences are few. Sugar cane was first grown in the mid-nineteenth century and in recent times there have been up to 18 estates with an average size of 6000 acres (2428 ha). Large estates may have over 300 miles (483 km) of transport, irrigation, and drainage canals and are surrounded by man-made earth banks to prevent flooding from the savannah. The estates, therefore, provide oases, free from the dangers of flooding and with unlimited cover not only on the grass covered roadways but, principally, in the dense stands of sugar cane.

In a study of *Holochilus sciureus berbicensis* on an estate alongside the Berbice River in 1959 to 1960[64] the reproductive potential was examined in the hope that control measures might be made more effective. The species is highly productive, at all times of the year there is never less than 70% of the adult males fertile and in some months as many as 98% have sperms present.

At no time during the 19 month study was there less than 27% of adult females (those above the average weight at first pregnancy) pregnant. There were four well-marked peaks of pregnancy during one of which 72% were pregnant and in two others over 60% (Figure 1). As with male fertility there were high pregnancy rates in June-July and, in addition, a high rate in October, 1959 and February-March, 1960. The peaks occurred approximately

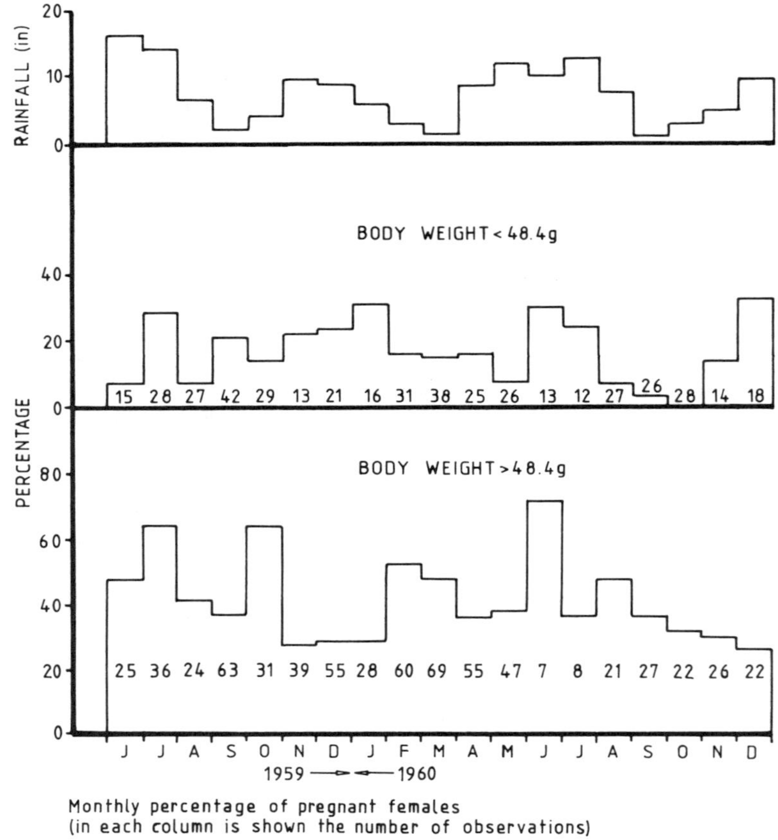

FIGURE 1. Seasonal pregnancy in *Holochilus*.

every 3 months. Young females below the average weight at first pregnancy showed higher pregnancy rates at the same times as there was increased fertility in young males below the mean weight at fertility and these periods coincided with high rainfall.

The mean number of embryos per litter was 3.12 (range 1 to 8) in a sample of 354 pregnancies and there were 49.8 embryos per 100 head of the population. This rate of productivity is indicative of high predation and is comparable with the reproduction figure of 55.7 for *R. tiomanicus* and 47.5 for *R. r. diardii* in Malaya where predation was also high. In Guyana *Holochilus* is taken in considerable numbers by various reptiles and raptors.

In addition to this capacity for high production all the year round there is immigration on to higher ground around the perimeter of estates at times of high rainfall when the savannah is flooded and these rodents soon find their way into the cane fields to augment the numbers already in occupation. It is interesting that estate records show heavy cane damage in those fields bordering the savannah in years of exceptionally heavy rainfall.

In Africa the multimammate rat *(Mastomys natalensis)* has been studied extensively because of its importance, as a semi-commensal of man, in the transmission of bubonic plague. Studies in Elisabethville, Belgian Congo,[50] Tanganyika[6] and on *Mastomys erythroleucus* in Sierra Leone[5] have shown that in general, substantial breeding occurs at the end of the rains and into the dry season. In the Congo breeding continues through much of the dry season. This extension of breeding in a semicommensal is in contrast with those African species which have no attachment to man and in whom the peak of breeding comes to an end when the dry season has begun.

In the Kenya Highlands a study on populations of *Mastomys natalensis, Rhabdomys pumilio, Arvicanthis niloticus,* and *Otomys angoniesis*[61] has demonstrated strong circumstantial evidence for a link between food supply and reproduction. An extension of the rainy season prolonged the period when seeds and cereals were available to rodents and this led to an extension of the breeding season and resulted in high rodent numbers and increased crop damage. The "opportunist" breeders *Mastomys, Rhabdomys,* and *Arvicanthis* had large litters and rapid population changes would be expected from these three species compared with *Otomys* with a more stable food source and smaller litters.

Severe infestation problems may result from the accidental or deliberate introduction of rodents into a new zoogeographical area, especially when species with a high reproductive potential are involved. The British Isles has suffered from two such incidents and the effects of one are still being experienced in continental Europe. The muskrat *(Ondatra zibethicus),* a large vole native to North America inhabits large natural marshes and is not regarded as a pest species but a natural resource for its fur. This rodent was introduced into Europe as a fur animal. Some were released and some farmed in captivity but there were the inevitable escapes from fur farms in France, Belgium, England, and Ireland.

The species has a high reproductive potential, females having an average litter size of 8.0 and producing 6 to 7 litters per year. The young disperse rapidly to colonize new areas. The escaped animals established large populations, which in continental areas are still spreading despite intensive control campaigns. As an example of the ability to breed and colonize, one pair escaped in County Tipperary, Ireland in 1927 and resulted in an infestation covering 150 mi^2 with 487 animals being trapped in the winter of 1933-34 alone. In the same year 5 males and 4 females escaped from a farm in Perthshire, Scotland and 890 were killed by trapping in the next 6 years.

Despite an intensive official extermination campaign based on trapping there was little reduction in the populations, trapping being shown to be an ineffective method of control. The species was exterminated in the British Isles shortly before the outbreak of World War II with the aid of an epizootic from an unidentified disease organism. This was not a "crash" disease such as is seen in some cyclic species, although *Ondatra* appears to cycle with peaks and crashes like other boreal voles but with a periodicity of approximately 10 years.

The coypu *(Myocastor coypus)* was introduced into the Norfolk Broads in eastern England during the 1930s and rapidly became a serious pest by inflicting damage to land drainage systems, crops, and native plants. This animal is susceptible to cold weather which causes depleted fat reserves and reduces the food supply to a critical level which leads to abortion and a reduced frequency of litter production.[20,21] Gosling[22] suggested that the coypu has the capacity for selective abortion of entire litters and the capability to determine the sex ratio of their offspring by preferential investment in offspring of the sex most likely to benefit.

IV. REPRODUCTION AND CONTROL

The rate at which a species can increase its numbers following control measures of varying degrees of effectiveness is basic information for control operatives and, among other things, may enable them to time poison treatments for maximum effectiveness. Most of the work on this aspect has been on the Norway rat and so most of the examples are biased towards this species. Some of the earliest work of a detailed nature was carried out on the rate of recovery of common rat in residential blocks in Baltimore.[18]

It was found that where populations were moderately reduced (between 50 and 90%) the rats showed early signs of recovery and increased at rates of about 4% (2 to 6%) of the capacity level each month. As they approached capacity they slowed down to a rate of about 2% per month or less.

Populations reduced more than 90% recovered at a slower rate of between 1 and 3% per

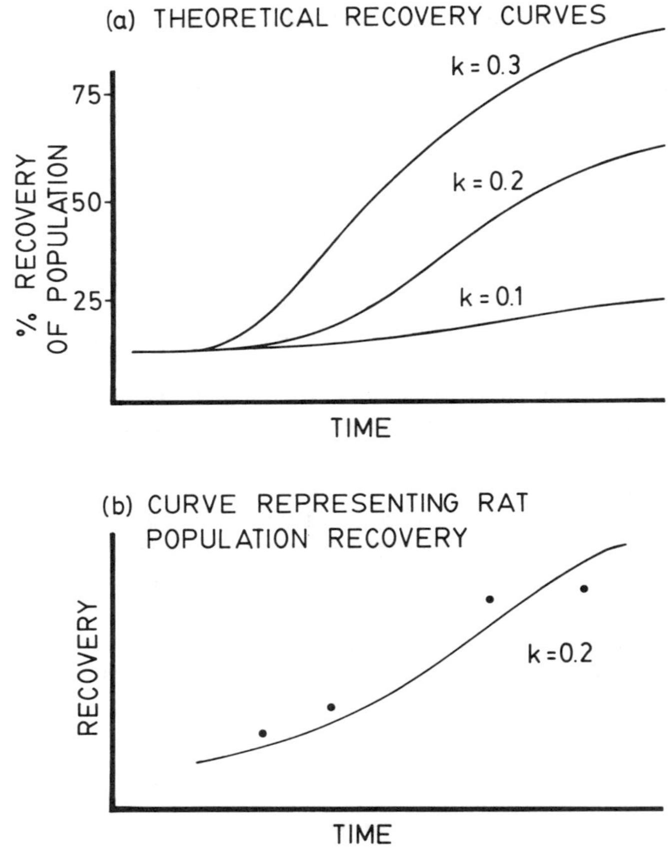

FIGURE 2. Population recovery curves in *Rattus norvegicus*.

month until they had advanced above the 10% level. In one case where complete clearance was obtained, no recovery occurred for over 2 years; progress then followed the usual course, slow at the start and accelerating gradually as the population gained a foothold.

The data from these operations suggest close conformity to the characteristics of the Verhulst-Pearl-Reed logistic which is widely applicable to growth curves of human and other populations:

$$y = \frac{100}{1 + e^{-kt}}$$

Where y = population; k = rate constant; and t = time measured from the point at which y reaches the midpoint of its growth.

Various theoretical recovery curves are possible (Figure 2a). With the data from the Baltimore blocks the rate of recovery fitted the recovery curve k = 0.2 (Figure 2b). The progress of a population over the central range of this curve may, for practical reasons, be considered to be roughly linear. Slopes obtained by the process of curve fitting indicate that a population reduced to 10% of its original level should climb to 90% of complete recovery in about 15 months when the k value of the selected curve is 0.3, 22 months when it is 0.2, and 44 months when it is 0.1. The typical curve of recovery of these rat populations suggests that of a logistic.

Other workers[1] studied the rate of increase of rats in London sewers following poison treatments. They found that whenever the population was reduced to less than 10% of the maximum the rate of increase was initially high but tended to flatten 16 to 20 weeks after poisoning. These data also suggested that the growth curve was of the S-shaped or logistic type. In the sewers 27.3% of females collected were pregnant. This was in January, one of the months when the breeding of surface living rats is at its lowest. The rate of natural increase in the sewers was calculated to be 0.1158 per week per head of the population, a rate so much higher than the 0.1040 calculated as the inherent rate of natural increase for *Rattus norvegicus*[34] as to suggest that some immigration from surface living rat populations was taking place.

In the Baltimore city blocks the isolation of populations was such that the observed post–treatment increase was due entirely to breeding. Where immigration, in addition to breeding, increases numbers it is clear that recovery may be much quicker. In these circumstances it is especially important to ensure as high a kill as possible. In English villages, after a single poison strike in the autumn the rat population had almost completely recovered by the following spring.[2] A further full scale treatment with three strikes was carried out and 6 months later, in the autumn, the population was only 17% of the prepoison level, despite the intervening breeding season. The point was made that in terms of cost per rat killed it was much more expensive to reduce a population numbering 10% of the maximum to 1% than to reduce it from 100% to 10%.

More recently the effect of poisoning on the breeding recovery of the Norway rat in Korean villages has been investigated.[8] Few commensal rats and mice can survive the winter in the fields and most of them use the villages as winter refuges. Breeding took place mainly between March and October so that for many reasons the villages represented an ideal target on which to focus control measures. Three farm villages having 20 to 30 houses each were treated with 30 g of pyriminyl cereal bait per house applied in December, April, and August.

The rate of population recovery was different depending on the time at which poisoning was carried out. When applied in December the population level was kept low for about 6 months. Because of the severe winter weather where the average temperature was −4°C the rodents were limited to the farm houses and this localization may have enhanced the efficacy of control so that the surviving population level and consequent breeding activity remained low for several months (Figure 3a). When poison was applied in April and August the population recovered within 3 months of poisoning and surviving populations were relatively higher than those following the December treatment (Figures 3b,c). It is likely that the population levels following the April and August treatments were inflated by the effect of mobile populations since there was considerable movement between farms and fields and vice versa. It is clear that mobile animals should be taken into consideration when planning the timing of control measures in habitats where normal reproductive processes are not the sole force in population increases.

Rattus tiomanicus (Miller) is almost the sole rat species damaging oil palm in the Malay Peninsula. A long-term study of this species[68] has yielded information on the recovery rate of poisoned populations and shown that, as in *Rattus norvegicus*, the shape of the increase curve indicates the need for complete elimination. Poisoned populations remained at low levels for about 6 months and reached prepoison levels in 12 to 18 months. Like *R. norvegicus*, *R. tiomanicus* also shows a logistic pattern of increase after control with a k-value of 0.4 per month against one of 0.2 to 0.3 per month for *R. norvegicus*.

V. REPRODUCTION AND PLANT DERIVATIVES

As early as 1926 plant components with apparent oestrogenic properties were known. Some of these substances have been found to cause reproductive dysfunction in domestic animals grazing on varieties of clover[40,54] but increased levels of fertility in microtine rodents

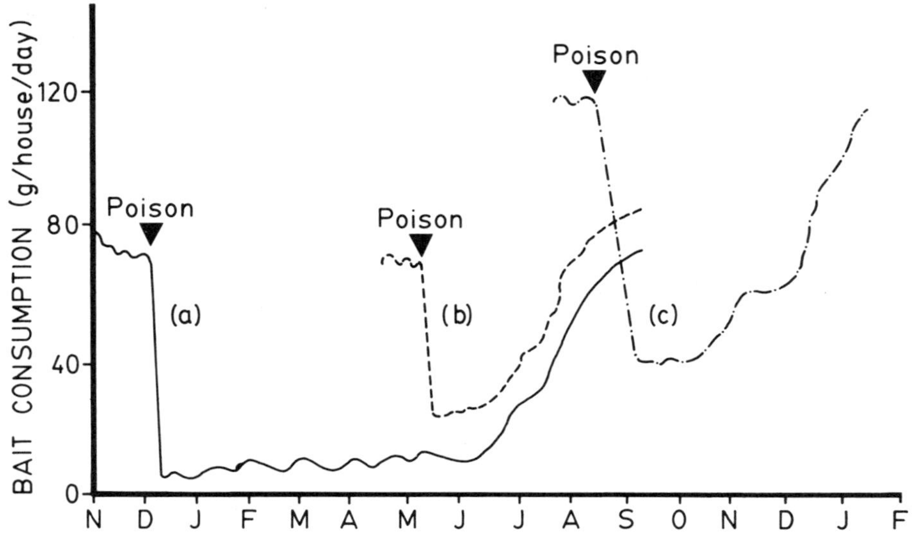

FIGURE 3. Rat population recovery after poison treatments in Korean villages.[8]

which ate oestrogen-containing plants have been shown.[43,48,49] In 1977 field experimental evidence was presented which showed that *Microtus montanus* was cued to its reproductive effort by chemicals in the plant food resources[3] and subsequently a cyclic carbamate, 6 methoxybenzoxazolinone, which stimulated reproductive activity in *Microtus montanus* was isolated from winter wheat.[55]

This nonestrogenic substance, known as 6-MBOA, when administered to experimental animals during the winter nonbreeding period resulted in a large proportion of the females becoming reproductively active with resultant pregnancies and in males a marked increase in testis weight. It is suggested that whereas photoperiod is of little predictive value for animals which inhabit highly unpredictable environments, the presence of 6-MBOA in plants enables close monitoring of the food resources and an optimal timing of reproductive effort. Field studies have confirmed that *Microtus montanus* populations synchronize reproductive effort when vegetative growth begins, cease breeding during drought, and begin again immediately after rainfall has induced vegetative growth. Observational data on rabbits, microtine rodents, and desert rodents suggest that some species of these may respond to cues from their vegetative food sources.

There is some evidence that plant oestrogens may enhance fertility and/or fecundity and an increase in litter sizes during vole *(Microtus guentheri)* irruptions has been found.[4]

An awareness of the potential role of plant substances in the reproduction of rodents and a knowledge of those plants implicated in these procedures may be of value to rodent control personnel in certain areas. The fact that 6-MBOA from winter wheat can induce both male and female voles to enter breeding condition despite both short day length and low temperatures might well be borne in mind by those attempting control of Norway rat in wheat growing regions. So far there have been no investigations to find out whether the substance has any effect on rats but it is known that extensive rat control is necessary in those English counties with extensive cereal cultivation although the extra quantities of grain may equally well be responsible for the high rat numbers. It is also interesting that the celebrated mouse irruptions of Australia have often been in wheat growing areas.

VI. CHEMOSTERILANTS

The success of chemosterilization in the control of insect pests has stimulated interest in the possibility of reproductive manipulation in rodent control. The method involves the administration of a substance which renders animals sterile so that their reproductive success becomes either partially or totally reduced, such sterilized individuals being reproductively dead so that sterilization is equivalent to conventional control in reducing the number of reproducing animals. In addition, if sterilized individuals also compete for mates with fertile individuals their reproduction is reduced further since fertile individuals will mate with sterile animals. If sterilization is irreversible and the lifespan and mating behavior are not adversely affected then sterile individuals will continue to compete for mates for several generations and suppress reproduction over a period of time.

Despite an extensive literature and much research the technique has found little support as a practical means of eradicating rodent populations. It is likely that chemosterilization and genetic controls could be of greater value in limiting populations than eradicating them[37,38] in environments where this could be tolerated, but for most situations the existence of rodents, even though they may not be adding greatly to their numbers, still results in damage and disease transmission and these factors, particularly the latter, may be unacceptable at any level.

ACKNOWLEDGMENTS

I am especially grateful to Dr. Alan P. Buckle who read the manuscript and made many helpful suggestions for its improvement and to Mrs. Sarah Wroot for the preparation of the figures.

REFERENCES

1. **Barnett, S. A. and Bathard, A. H.,** Population dynamics of sewer rats, *J. Hyg.*, 51, 483, 1953.
2. **Barnett, S. A., Bathard, A. H., and Spencer, M. M.,** Rat populations and control in two English villages, *Ann. Appl. Biol.*, 38,444, 1951.
3. **Berger, P. J., Negus, N. C., Sanders, E. H., and Gardner, P. D.,** Chemical triggering of reproduction in *Microtus montanus*, *Science*, 214, 69, 1981.
4. **Bodenheimer, F. S. and Sulman, F.,** The estrous cycle of *Microtus guentheri* A. and S. and its ecological implications, *Ecology*, 27, 255, 1946.
5. **Brambell, F. W. Rogers, and Davis, D. H. S.,** Reproduction in the multimammate mouse *(Mastomys erythroleucus)* of Sierra Leone, *Proc. Zool. Soc. London*, 111B, 1, 1941.
6. **Chapman, B. M., Chapman, R. F., and Robertson, I. A. D.,** The growth and breeding of the multimammate rat, *Rattus (Mastomys) natalensis* (Smith) in Tanganyika Territory, *Proc. Zool. Soc. London*, 133, 1, 1959.
7. **Chitty, H.,** The oestrus cycle and gestation period in the lactating field vole *Microtus agrestis*, *J. Endocrinol.*, 15, 279, 1957.
8. **Cho, W. S.,** Effect on rodent population following acute poisoning with emphasis on the breeding pattern of the Norway rat in Korea, in *Proc. Conf. The Organisation and Practice of Vertebrate Pest Control*, Dubock, A. C., Ed., I.C.I., Plant Protection Division, Haslemere, England, 1984, 353.
9. **Corbet, G. B. and Southern, H. N., Eds.,** *The Handbook of British Mammals*, Blackwell, Oxford, 1977.
10. **Crowcroft, P. and Rowe, F. P.,** The growth of confined colonies of the wild house-mouse (*Mus musculus* L.), *Proc. Zool. Soc. London*, 129, 359, 1957.
11. **Crowcroft, P. and Rowe, F. P.,** The growth of confined colonies of the wild house-mouse (*Mus musculus* L.): the effect of dispersal on female fecundity, *Proc. Zool. Soc. London*, 131, 357, 1958.
12. **Crowcroft, P. and Rowe, F. P.,** The weights of wild house mice (*Mus musculus* L.) living in confined colonies, *Proc. Zool. Soc. London*, 136, 177, 1961.

13. **Davis, D. E.,** Observations on rats and typhus in San Antonia, Texas, *Public Health Rep.*, 63, 783, 1948.
14. **Davis, D. E.,** The characteristics of rat populations, *Q. Rev. Biol.*, 28, 373, 1953.
15. **Davis, D. E. and Hall, O.,** The seasonal reproductive condition of male Brown rats in Baltimore, Maryland, *Physiol. Zool.*, 21, 272, 1948.
16. **Davis, D. E. and Hall, O.,** The seasonal reproductive condition of female Norway (Brown) rats in Baltimore, Maryland, *Physiol. Zool.*, 24, 9, 1951.
17. **Devincenzi, G. J.,** Mamiferos del Uruguay, *Ann. Mus. Hist. Nat. Montevideo*, 4 (2), 10, 1935.
18. **Emlen, J. T., Stokes, A. W., and Windsor, C. P.,** The rate of recovery of decimated populations of Brown rats in nature, *Ecology*, 29, 133, 1948.
19. **Giovannoni, M., Velloza, L. G. C., and Kubiak, G. V. L.,** Sôbre as "ratadas" do primeiro planalto paranaense, *Arch. Biol. Curitiba*, 1, 185, 1946.
20. **Gosling, L. M.,** Climatic determinants of spring littering by feral coypus, *Myocastor coypus*, *J. Zool. London*, 195, 281, 1981.
21. **Gosling, L. M.,** Coypus in East Anglia (1970 to 1984), *Trans. Norfolk Norwich Nat. Soc.*, 27, 151, 1985.
22. **Gosling, L. M.,** Selective abortion of entire litters in the coypu: adaptive control of offspring production in relation to quality and sex, *Am. Nat.*, 127, 772, 1986.
23. **Harrison, J. L.,** "The animals", in Audy, J. R., Harrison, J. L., and Wyatt-Smith, J., A survey of Jarak Island, Straits of Malacca, *Bull. Raffles Mus.*, 23, 238, 1950.
24. **Harrison, J. L.,** Reproduction in rats of the subgenus *Rattus*, *Proc. Zool. Soc. London*, 121, 673, 1951.
25. **Hasseltine, H. E.,** Rat-flea survey of the port of Norfolk, Va., *Public Health Rep.*, 44, 479, 1929.
26. India Plague Commission, The epidemiological observations made by the commission in Bombay City, *J. Hyg.*, 7, 724, 1907.
27. India Plague Commission, Observations on the Punjab villages of Dhand and Kasel, *J. Hyg.*, 8, 895, 1908.
28. India Plague Commission, Observations on plague in Belgaum 1908-1909, *J. Hyg.*, 10, 446, 1910.
29. Indian Plague Commission, Observations on plague in Poona, *J. Hyg.*, 10, 483, 1910.
30. **Kildemoes, A.,** The impact of introduced stoats *(Mustela erminea)* on an island population of the water vole, *Arvicola terrestris*, *Acta Zool. Fenn.*, 173, 193, 1985.
31. **Lam, Y. M.,** Reproductive behaviour of the rice field rat, *Rattus argentiventer* and implications for its control, in *Proc. Nat. Rice Conf.*, Malaysian Agricultural Research and Development Institute, Kuala Lumpur, Malaysia, 1980, 243.
32. **Lam, Y. M.,** Reproduction in the rice field rat, *Rattus argentiventer*, *Malay. Nat. J.*, 36, 249, 1983.
33. **Laurie, E. M. O.,** The reproduction of the house mouse *Mus musculus* living in different environments, *Proc. R. Soc. London Ser. B*, 133, 248, 1946.
34. **Leslie, P. H.,** On the use of matrices in certain population mathematics, *Biometrika*, 33, 183, 1945.
35. **Leslie, P. H., Venables, U. M., and Venables, L. S. V.,** The fertility and population structure of the brown rat *(Rattus norvegicus)* in cornricks and some other habitats, *Proc. Zool. Soc. London*, 122, 187, 1952.
36. **Llanos, A. C.,** Appreciaciones de campo con motivo de una concentracion de roedores en las provincial de Salta y Jujuy, *Rev. Argent. Zoogeogr.*, 4, 51, 1944.
37. **Marsh, R. E. and Howard, W. E.,** Chemosterilants as an approach to rodent control, *Proc. 4th Vert. Pest Conf.*, University of California at Davis, Davis, Calif., 55, 1970.
38. **Marsh, R. E. and Howard, W. E.,** Prospects of chemosterilant and genetic control of rodents, *Bull. W.H.O.*, 48, 309, 1973.
39. **Morland, H. B., Utterback, B. C., and Dent, J. E.,** Domestic rats in relation to murine typhus control, *Public Health Monogr.*, 5, 1, 1952.
40. **Moule, G. R., Braden, A. W. H., and Lamond, D. R.,** The significance of oestrogens in pasture plants in relation to animal production, *Anim. Breed. Abstr.*, 31, 139, 1963.
41. **Myllmäki, A., Hansson, L., and Christiansen, E.,** Models for forecasting population trends in two species of microtine rodent, *Microtus agrestis* and *Clethrionomys glareolus*, *Acta Zool. Fenn.*, 173, 93, 1985.
42. **Negus, N. C. and Berger, P. J.,** Experimental triggering of reproduction in a natural population of *Microtus montanus*, *Science*, 196, 1230, 1977.
43. **Negus, N. C. and Pinter, A. J.,** Reproductive responses of *Microtus montanus* to plants and plant extracts in the diet, *J. Mammal.*, 47, 596, 1966.
44. **Newsome, A. E. and Corbett, L. K.,** Outbreaks of rodents in semi-arid and arid Australia: causes, preventions and evolutionary considerations, in *Rodents in Arid Environments*, Prakash, I. and Ghosh, P. K., Eds., W. Junk, The Hague, 1975, 117.
45. **Nowak, R. M. and Paradiso, J. L., Eds.,** *Walker's Mammals of the World*, Vol. II, John Hopkins University Press, Baltimore, Md., 1983.
46. **Panteleyeev, P. A.,** *Population Ecology of the Water Vole*, Izdatel'stro Akademi Nauk 'SSSR, Moskva, 1968.
47. **Perry, J. S.,** The reproduction of the wild Brown rat (*Rattus norvegicus* Erxleben), Proc. Zool. Soc. *London*, 115, 19, 1945.

48. **Pinter, A. J.,** Effects of diet and light on growth, maturation, and adrenal size of *Microtus montanus*, *Am. J. Physiol.*, 215, 461, 1968.
49. **Pinter, A. J. and Negus, N. C.,** Effects of nutrition and photoperiod on reproductive physiology of *Microtus montanus*, *Am. J. Physiol.*, 208, 633, 1965.
50. **Pirlot, P. L.,** Pourcentages de jeunes et périodes de reproduction chez quelques rongeurs, *Ann. Mus. Congo Belge Tervueren Sci. Zool.*, 1, 41, 1954.
51. **Plomley, N. J. D.,** Some notes on plagues of small mammals in Australia, *J. Nat. Hist.*, 6, 363, 1972.
52. **Redhead, T. D., Enright, N., and Newsome, A. E.,** Causes and predictions of outbreaks of *Mus musculus* in irrigated and non-irrigated cereal farms, *Acta Zool. Fenn.*, 173, 123, 1985.
53. **Rowe, F. P., Swinney, T., and Quy, R. J.,** Reproduction of the House mouse (*Mus musculus*) in farm buildings, *J. Zool. London*, 199, 259, 1983.
54. **Samuel, D. E.,** A review of the effects of plant estrogenic substances on animal reproduction, *Ohio J. Sci.*, 67, 308, 1967.
55. **Sanders, E. H., Gardner, P. D., Berger, P. J., and Negus, N. C.,** 6-Methoxybenzoxazolinone: a plant derivative that stimulates reproduction in *Microtus montanus*, *Science*, 214, 67, 1981.
56. **Southwick, C. H.,** The abundance and distribution of harvest mice (*Micromys minutus*) in corn ricks near Oxford, *Proc. Zool. Soc. London*, 126, 449, 1956.
57. **Southwick, C. H.,** Population characteristics of house mice living in English corn ricks: density relationships, *Proc. Zool. Soc. London*, 131, 163, 1958.
58. **Stoddart, D. M.,** Breeding and survival in a population of water voles, *J. Anim. Ecol.*, 40, 487, 1971.
59. **Storer, T. I. and Davis, D. E.,** Studies on rat reproduction in San Francisco, *J. Mammal.*, 34, 365, 1953.
60. **Tamarin, R. H. and Malecha, R.,** Reproductive parameters in *Rattus rattus* and *R. exulans* of Hawaii, 1968 to 1970, *J. Mammal.*, 53, 513, 1972.
61. **Taylor, K. D. and Green, M. G.,** The influence of rainfall on diet and reproduction in four African rodent species, *J. Zool. London*, 180, 367, 1976.
62. **Teivainen, T.,** The damage risk and the extent of vole damage in forest plantations in Finland, in *Proc. Conf. Organisation and Practice of Vertebrate Pest Control*, I.C.I. Plant Protection Division, Haslemere, England, 1984, 573.
63. **Trout, R. C.,** Harvest mouse, in *The Handbook of British Mammals*, Corbet, G. B. and Southern, H. N., Eds., Blackwell, Oxford, 1977, 220.
64. **Twigg, G. I.,** Studies on *Holochilus sciureus berbicensis*, a Cricetine rodent from the coastal region of British Guiana, *Proc. Zool. Soc. London*, 145, 263, 1965.
65. **Watson, J. S.,** Some observations on the reproduction of *Rattus rattus* L., *Proc. Zool. Soc. London*, 120, 1, 1950.
66. **Watson, J. S.,** The rat problem in Cyprus. A report of investigations made in carob-growing areas, *Col. Res. Pub.*, 9, 1, 1951.
67. **Wijngaarden, A. van,** Biologie en Bestrijding van de Woelrat, Arvicola terrestris terrestris (L.) in Nederland, *Meded. plziektenk. Dienst Wageningen.*, 123, 1, 1954.
68. **Wood, B. J. and Liau, S. S.,** A long-term study of *Rattus tiomanicus* populations in an oil palm plantation in Johore, Malaysia. II. Recovery from control and economic aspects, *J. Appl. Ecol.*, 21, 465, 1984.

Chapter 12

BEHAVIOR

Robert M. Timm and Terrell P. Salmon

TABLE OF CONTENTS

I. INTRODUCTION

Preventing or minimizing rodent damage requires thorough knowledge of the animal and its environment. An important aspect of this is a solid understanding of animal behavior. This information allows us to develop rodent control programs in concert with appropriate ecological considerations.

Conflicts between rodents and humans are largely a result of the behavior of rodents as they manipulate resources within their environment. What we term "damage" is often the consequence of rodent feeding, gnawing, burrowing, nesting, or excreting. Because species have typical behavioral patterns, the damage a species causes in a given situation is generally predictable. Knowledge of a rodent's capabilities and behavioral patterns is often useful in preventing damage from occurring.

The success of a rodent control program depends upon the methods chosen and their application to the specific situation. A rodent's behavior toward control tools and materials may determine whether the rodent survives and continues to cause damage, or succumbs. Certain behavioral traits allow rodents to adapt to changing environments. We can use this information to define the environment where particular species will likely occur, and to predict rodent responses to habitat manipulation as a control strategy. Knowledge of the behavior of the target species is therefore a prerequisite to designing an efficacious control strategy. Species–specific behaviors can also be used to predict and/or reduce nontarget effects resulting from certain control methods and materials.

A complete discussion of animal behavior would require volumes. Our intent is to present basic information on rodent behavior and, through examples, demonstrate the importance of this subject to rodent control. Various aspects of animal behavior are dealt with in greater detail elsewhere in this volume. This chapter discusses the various behavioral components that are important in predicting, understanding, preventing, and controlling rodent damage to resources. These include physical capabilities, perception of and response to stimuli, activity and movement, feeding behavior, and social behavior.

II. PHYSICAL CAPABILITIES

A rodent's physical capabilities are some of the most basic behavioral traits that need to be understood to effectively solve rodent problems. This information is valuable for identifying the species in question, its potential for damage, and for defining potential control strategies.

Common to all rodents is their ability to gnaw. Their incisors, which grow continuously throughout their lifetime, are an adaptation that allows them to obtain and feed on hard materials such as nuts and seeds. Further, their gnawing ability allows the commensal rodents to gain access to structures by gnawing through building materials such as wood, soft metals, and even cinder block. Although a commonly held belief, this gnawing behavior is not necessary to maintain their teeth in good condition, because the upper and lower incisors can be worked against each other to keep the teeth sharp and prevent them from becoming excessively long.[1]

A considerable variety of damage can be attributed to rodents' gnawing ability. Commensal rats (*Rattus* spp.) and mice *(Mus musculus)* typically gnaw through packaging materials in order to gain access to foodstuffs. Further, they utilize fibrous materials such as burlap feed sacks, paper, and cardboard for construction of their nests. Tree squirrels (*Sciurus* spp.) strip bark from trees, both in feeding and nest-building. Pocket gophers (*Thomomys* spp. and *Geomys* spp.) consume the roots of plants, even severing tree roots several centimeters in diamater. Often, only the outer portion of the root is consumed but this is sometimes lethal to the plant. Pocket gophers and other rodents gnaw into and severely damage plastic irrigation pipes as well as buried electrical cables, causing disruption of telephone and electrical service.[16,29] Beavers *(Castor canadensis)* gnaw through the base of trees, which, when felled, they then use as food and for constructing their lodges and dams.

The excellent climbing ability of many rodents enables them to utilize food resources in relatively inaccessible locations. While some rodents climb naturally and live arboreally, others are atypical climbers that capitalize on a particular food resource out of their normal foraging strategy. Tree squirrels (*Sciurus* spp.), porcupines *(Erethizon dorsatum)*, and roof rats *(Rattus rattus)* are in the first category. Other rodents such as California ground squirrels *(Spermophilus beecheyi)* and Norway rats *(Rattus norvegicus)* are not considered climbing species but can certainly do so when the opportunity to obtain quality food presents itself. For example, Norway rats *(Rattus norvegicus)* have been found to gain access to buildings through windows or roof-level ventilation openings by climbing adjacent trees. The ability of commensal rodents *(Rattus* spp. and *Mus musculus)* to jump as well as climb further enables them to gain access to structures which supply food or shelter.

Some species of rodents are well-adapted to burrowing. Pocket gophers (*Thomomys* spp. and *Geomys* spp.) build extensive burrow systems, which under certain conditions may serve as channels for water, disrupting irrigation and contributing to significant increases in erosion of soil. Even less extensive rodent burrows, such as those constructed by ground squirrels (*Spermophilus* spp.) and prairie dogs (*Cynomys* spp.), may channel water and thereby contribute to erosion or cause other structural problems at certain sites.

III. PERCEPTION OF AND RESPONSES TO STIMULI

The rodent's behavior toward various stimuli will often be important in designing appropriate control strategies. Information on their ability to perceive certain signals is used in developing control methods that are either undetectable (i.e., blind set traps) or highly detectable (i.e., attractive odors for baits). The rodent's response to stimuli is also used to develop and improve control methods and strategies. For example, if the rodent responds negatively to new objects, getting the animal accustomed to that object might be incorporated into the technique in order to improve efficacy. Clearly, we must understand the rodent's sensory abilities in order to develop control methods and materials.

A. Vision

Many rodents rely upon their senses of smell, taste, hearing, and touch more than they do their sight. Some rodents are not believed to have high visual acuity, especially fossorial species such as pocket gophers (*Thomomys* spp. and *Geomys* spp.) that live in dark, underground tunnels. Gophers are, however, sensitive to light as well as to air movement, and will quickly detect and work to repair openings in their burrow systems. Other species, such as tree squirrels (*Sciurus* spp.) and ground squirrels (*Spermophilus* spp.), have relatively good vision and utilize this sense in detecting predators at a distance. Persons seeking to control such species by use of firearms often discover that the rodents quickly learn to associate the sight of humans, particularly humans with rifles, with danger and will quickly retreat to cover when humans approach. The sound of gunfire also appears to be used by rodents as a cue, as it has been observed that use of low-powered (and less noisy) firearms seems to frighten rodents less quickly than high-powered ones. The effectiveness of shooting rodents in a given population rapidly decreases over time. Although most rodents are assumed not to possess color vision, there is evidence that laboratory rats *(Rattus norvegicus)* can discriminate among some colors[60] and black-tailed prairie dogs *(Cynomys ludovicianus)* have been shown to have good color vision,[10] although the adaptive significance of this trait has not been explored. Many rodent baits are dyed for identification and to repel birds.[15] Dyes are assumed not to affect bait acceptance unless they have an objectionable taste.

B. Sound

Sounds may serve to warn rodents of danger, and thus many species are attuned to quickly reacting to loud or unusual noises. Rats *(Rattus norvegicus)* exhibit "freezing" behavior when they hear certain sounds, which is thought to be an adaptation for preventing detection by predators who locate prey particularly on the basis of motion.[27] The auditory sense of some species is specialized to detect sounds made by their predators; certain heteromyid rodents, such as kangaroo rats (*Dipodomys* spp.) can detect the sounds made by owls in flight.

Rodents communicate using a variety of vocalizations[61] both in the sonic and ultrasonic ranges. Rats (*Rattus* spp.) also are known to use ultrasonic sounds for echolocation.[33,50] Alarm calls are a part of some species' vocal repertoire and are instrumental in warning of approaching predators.

Attempts to use sound in rodent control programs have not generally resulted in useful techniques. Although rodents will avoid certain sounds, particularly loud or novel noises, they will also quickly habituate to repeated sounds. It is common to find commensal rodents living in very noisy locations, such as factories or feed mills. While ultrasonic sound can be demonstrated to be effective in repelling commensal rodents, the effect is short-lived, and ultrasonic sound-producing devices have not been shown to provide adequate levels of rodent control to date.[54]

C. Odor

Most rodents have a keen sense of smell, and gain a great deal of information about their environment from this sense. Rodents use odor to gain information about objects, and odor plays a key role in locating food; for example, deer mice *(Peromyscus maniculatus)* locate buried seed by olfactory means.[28]

Persons involved in trapping rodents have noted that previously-used traps often are more effective than new ones.[58] Recent research has shown that such factors as the reproductive condition of the target animal and the social dominance of the odor donor may be important in determining whether such scents are attractive, neutral, or even repellent.[20,64]

Odors produced by rodents' scent glands are important in communication. Such chemical signals may convey information about individual or group identity, facilitate maternal-offspring bonds, or be used in marking territory. Additionally, species-specific odors (often termed pheromones) may have a role in eliciting or inhibiting aggression, evoking fear or alarm, controlling sexual development in juveniles, regulating estrus cycles, or in disruption of pregnancy.[13]

Research on rodents' response to odors, including pheromones, has the potential for being increasingly important in the development of control materials and methods.[41] To date, researchers have developed techniques for quantifying rodents' responses to odors[14,52] and a number of candidate attractant and repellent odors have been investigated with selected species. There are few compounds which have been shown to be effective or practical repellents for any rodent species, but recent work indicates that predator odors may possess some characteristics which might lead at least to limited practical use.[56] Future work may lead to practical applications of odor for enhancing bait attractancy, increasing specificity of traps or toxicants, or disrupting reproductive processes within populations.

D. Taste

Most rodent species possess an excellent sense of taste. While odor often is important in an animal locating a potential food source, taste may play the major role in determining whether the item will be consumed. The taste of a substance appears to be a key element in the formation of learned food aversions in rodents.[47] Foods which have objectionable taste may be rejected after being sampled. Norway rats have been shown to detect certain compounds, which are contaminants in warfarin, at a concentration as low as 250 ppb;[8] baits containing such contaminants may be rejected or fed on at reduced levels. For this reason, rodent baits should always be fresh and in good condition. Use of spoiled or old ingredients, or storage of baits in locations where they may absorb foreign odors or contaminants, can lead to reduced bait acceptance or bait rejection, resulting in poor control success.

E. Reaction to Objects

Rodents typically explore their home ranges intensively, using their senses of touch, smell, and taste, in particular, to learn about objects within their main areas of activity. The commensal rodents *(Rattus norvegicus, Rattus rattus, Mus musculus)* have been studied most intensively in this regard. Their travels throughout their home range become repetitive and memorized, and are executed as repeated sequences of muscular movements which have been termed the "kinesthetic sense" as described by Chitty and Southern.[12] The adaptive advantage of such behavior is that it allows the rodent to move to shelter quickly, even in total darkness, in the event of danger.[4,19]

Despite their curiosity and tendency to examine things within their environment, some rodents are wary of new objects. This "new object reaction" or neophobia[3,12] is more pronounced in some species than in others. It is particularly notable in Norway rats *(Rattus norvegicus),* which may in some circumstances avoid new objects placed within their home range for days or even weeks. Within populations, certain individuals may be more neophobic than others; male house mice typically enter traps more readily than do females.[18] While some individuals are trap-shy, others are apparently trap-prone.[58] Because of neophobia in certain species or individuals, bait stations or traps may achieve little control success until the rodent population becomes familiarized with them. For this reason, prebaiting bait boxes with nontoxic food is often practiced prior to use of single-dose toxicants, and leaving unset traps baited for several days is often done to overcome neophobia and thereby enhance later trapping success. Prebaiting has a greater effect on some rodent species than on others.[26]

Most rodents have developed specific antipredator strategies. As we search for new and improved control methods, an understanding of these becomes important. Hiding is the most common antipredator strategy. Information on specific hiding strategies is important for many aspects of a control program including detection and monitoring, habitat modification, and trap and bait placement. We may also be able to circumvent certain hiding strategies, forcing the animals to be more susceptible to predation (human or nature induced).

Spotting predators is another important antipredator strategy. Certain rodents have a keen ability to spot predators. Some even communicate the presence of predators to others in their social group. Work on visual screens to minimize the rodent's ability to spot predators has been conducted.

IV. ACTIVITY AND MOVEMENT

Rodent control specialists constantly deal with questions about rodent activity and movement. On a daily basis, rodents often have periods of high or low activity. Some species are primarily active at night, others during daylight hours, while yet others may exhibit peak activity at dawn and dusk. Knowledge of the expected activity peaks of a species can be helpful in successfully applying control measures. For example, application of a single-dose bait is usually best done immediately prior to the time of most intensive feeding activity.[42]

On a seasonal basis as well there may be changes in rodent activity. Species such as ground squirrels (*Spermophilus* spp.) may hibernate during winter or may aestivate during the hottest part of summer. Control operations using baits or even burrow fumigants at these times may not be efficacious. Timing of emergence from hibernation may vary among sexes or age classes, as has been shown for Richardson ground squirrels *(Spermophilus richardsonii)*.[42,43] Toxic bait applied early in spring may affect only that portion of the population that is active.

An individual rodent's home range may be quite limited in size. Norway rats *(Rattus norvegicus)*, for example, have been found to use an area with a diameter of 30 to 50 m[21] while house mice may restrict their usual activities to a space less than 3 to 10 m across.[23] This means that control tools such as bait stations or traps, in order to be useful in affecting all members of the population, must be numerous and placed at many locations through the area occupied by the population.

Dominant, reproductive individuals of a rodent population may be less susceptible to control efforts than other individuals.[35] For example, house mice living in prime areas of habitat (such as in the middle of a stack of sacked grain) may seldom venture far enough to encounter traps or baits; subordinate individuals which have been socially restricted to less favorable parts of the habitat may be far more susceptible to control. In some species, the tendency to explore and gain new territory may make animals vulnerable to certain control measures. The burrow builder machine, used for pocket gopher (*Thomomys* spp. and *Geomys* spp.) control in North America, creates an artificial burrow, which resident gophers are quick to investigate, encountering and feeding on toxic grain in the process.[40] This behavior may have importance in the future development of larger toxic baits to control invading pocket gophers following control of the original burrow resident.[59]

Movement of rodents is sometimes predictable. Seasonal movement of house mice *(Mus musculus)* to and from agricultural fields has been noted,[51] as has similar movement of Norway rats *(R. norvegicus)* in Wisconsin[24] and in Britain[57] where individual animals were found to travel as far as 0.5 km from farmlands to reach farm buildings. Various species of Indian rodents were found to make similar movements away from agricultural fields after harvest, and return to fields the next year when food became available.[2] In many rodent populations, dispersal of juvenile or young adult individuals occurs on a seasonal basis. Where such movements can be predicted, control can be initiated before movement occurs. For example, *Microtus* spp. overwintering in weedy ditchbanks or similar areas can be controlled using toxic grain in late winter or early spring, prior to the time the adjacent agricultural fields yield a source of food.[25] When control can be conducted prior to such movement or dispersal, it is usually more cost-effective than control attempted at a later time.[42]

V. FEEDING BEHAVIOR

Because of their well-developed sense of taste, rodents often exhibit marked food preferences. They may select certain items within their environment to utilize heavily, feed on others to a lesser extent, and completely reject other items. A rodent's feeding behavior influences many important aspects of rodent control. These include: (1) identifying potential food items (crops or stored feed that may be subject to damage); (2) developing alternate, more attractive food sources; (3) altering foods to make them less acceptable to the rodent, either by developing taste repellents, aversive agents, or by selecting and breeding less palatable varieties; and (4) development and use of acceptable food items as baits for toxicants and traps.

Such feeding preferences are based on the species innate traits as well as on learned behavior. Food preferences can be transmitted from parents to offspring[37] and within social groups from one member to another.[45]

When rodents exhibit rejection of food items, it may be simply due to taste. Or it may be because of a previous experience with similar foods which have produced an unpleasant sensation and thereby led to development of a learned aversion. Natural taste aversions which lead to avoidance of toxic, but natural, food items occur. For example, Jacobs and Labows[32] showed that wild guinea pigs *(Cavies aperea)* form conditioned aversion to certain toxic plants. Innate aversions or rejections of novel foods may offer possibilities for protection of resources from rodent feeding; such food items may have properties which can be developed to reduce damage. For example, pocket gophers (*Thomomys* spp.) demonstrate preference for some varieties of ponderosa pine over others, and this preference has been correlated with the presence of some components of oils in stems and roots. Further research may reveal that certain varieties of ponderosa pine are resistant to gopher feeding, or at least may be less likely to sustain damage.[46]

Jackson[31] reports that commensal rodents have taste patterns similar to humans, and when given a choice of foods they will select a nutritionally balanced diet. Food preferences may change seasonally, and such shifts may be based in part upon the availability of certain food items in the environment.[42] In general, rodents prefer foods with which they are familiar,[44] and therefore toxic baits often are formulated using ingredients that rodents readily recognize and accept. Bait acceptance trials are an important step in selecting ingredients for bait formulation.[38]

Feeding behavior varies not only among species, but also among individuals within a population. Some generalities can be stated, however. House mice *(Mus musculus)* are considered to be light and intermittent feeders as compared to rats (*Rattus* spp.). They may visit many feeding locations during a single foray, and they exert significant amounts of selection in choosing among available foot items.[17] The types and amounts of food an individual rodent is likely to consume are important factors to consider in formulating baits, as they affect the bait ingredients as well as the concentration of active ingredient selected.

Rodents are apt to feed in locations where they feel comfortable and secure, rather than in places where they sense danger. House mice *(Mus musculus)* are more likely to consume feed inside small bait boxes than from open bait trays; thus, mouse-size bait stations enhance bait acceptance and therefore in some situations contribute to better levels of control.[7] Further, mice prefer bait stations of certain shapes and materials. For example, they feed more readily from cardboard bait boxes than from plastic ones of the same size and shape.[7] Commensal rodents have been known to carry food items from one site to another. It has been suggested that this is not food hoarding or storage behavior, but simply an attempt by the rodent to feed on the items in a more secluded or comfortable location.[9] Transporting of rodent baits can generally be prevented by using finely ground baits, as opposed to bait in pellets or large particles.[9] Food hoarding or caching is a more pronounced behavior in certain other rodents, for example deer mice *(Peromyscus maniculatus),* than it is in the commensal rodents of North America.

The hesitancy of rodents to feed on novel materials can be overcome to an extent by prebaiting, or conditioning rodents to utilize a particular food material.[12,22] This practice has often been followed to increase bait acceptance prior to use of fast-acting, single-dose toxicants. Wing and Birch[63] demonstrated that contrasting tastes between nontoxic and toxic food items enhances bait shyness. Obviously this information is important in developing better accepted baits. Roof rats *(Rattus rattus)* have the ability to detect bait texture such as whole versus hulled grain. This has been shown to influence bait preferences.[6] The more closely the prebait resembles the toxic bait to be used, the more likely the rodent will consume a toxic dose of bait when it is offered in place of the nontoxic prebait.[49] Rodents can distinguish between the taste of the food item and the toxicant, and shyness to either or both has been demonstrated.[5]

Learned food aversions, which are discussed in more detail elsewhere in this volume, are important behavioral adaptations which rodents have developed. In brief, these aversions protect rodents from consuming foods which may cause adverse physical reactions (illness or death), and as such are a basic survival adaptation. Some species form learned aversions more readily than others.[48] Learned aversions can be transmitted from parent to young probably via mother's milk.[37] The characteristics of food aversions, as well as those of neophobia, offer potential for the development of new strategies in controlling rodents and rodent damage.[47] One strategy for using toxicants in rodent control, which avoids problems of toxicant aversion or bait aversion, is to administer the toxicant as a tracking powder or dust. A rodent travels across the dust patch, placed along its usual route of travel, and some of the toxicant adheres to its fur. In its normal process of self-grooming, the rodent ingests a lethal dose of the toxicant. This grooming behavior, which is more pronounced in some species (e.g., house mice, *Mus musculus*) than in others, provides a means of controlling rodents with toxicants that might be rejected, at least by a portion of the population, were they formulated into food baits.[39]

Repellents for rodent control rely on the animal's behavior toward certain tastes and odors. Innate aversions toward some food items can be used as taste repellents. Many plants develop chemicals toxic to rodents. In some cases, plant breeders have capitalized on these chemicals to develop plant varieties resistant to rodent attack. The entire concept of taste aversion learning can be used to develop potential control strategies by conditioning the rodent to reject a particular food item (e.g., the crop). While theoretically possible, this method for reducing rodent damage is relatively unexplored.

VI. SOCIAL BEHAVIOR

Rodent social behavior ranges from solitary existence to highly organized communal groups. Conspecific interactions effect rodent control in many ways. Their impact on population dynamics, individual and group spacing, and reactions toward control measures have all been demonstrated. Most rodent control programs use this information to better understand the animal and improve efficacy of the control technique.

Social behaviors exhibited within groups of conspecifics are important in affecting rodent numbers, and consequently may affect the severity of damage the rodents may cause. Such behaviors as aggression, territoriality, mating, and rearing of young influence the population density a given species may attain within a particular habitat. Social interactions (e.g., aggression) may limit the density of some species, particularly where populations are confined by conditions of the habitat.[11,19] Where suitable habitat is not a limiting factor, high population densities may influence the tendency for individuals to emigrate or disperse to surrounding areas. Some species have adaptations for existing at high densities. Female house mice *(Mus musculus),* for example, share communal nests when nesting sites become scarce,[51] thereby continuing to reproduce successfully. Uncontrolled house mouse populations can reach denser and therefore more damaging levels than they could if this adaptation were not present.

Interactions among conspecifics can affect bait consumption and thereby influence the control program. West et al.[62] compared bait consumption by rats *(R. rattus)* from bait stations where one large or several small containers were available. They found significant increases in consumption at the "several small station" sites and hypothesized that reduced interaction among rats was responsible for their observed improved bait acceptance.

Dominance hierarchies which occur within rodent populations often limit individual access to resources such as food and water. Dominant animals typically utilize these resources first, while subordinate animals are forced to utilize less desirable resources or resources at less desirable sites. This has a number of implications for rodent control. For instance, when desiring quick control of a population of Norwy rats *(R. norvegicus),* it is important to place bait at enough locations so that the majority of individuals have the opportunity to feed. Even so, dominant individuals may exclude subordinates from baits initially, until they succumb and subordinates move in to occupy the more desirable territories. In instances when quick control is not necessary, placing bait stations containing anticoagulant rodenticides at locations of prime rodent habitat will give eventual control of the population, as subordinate individuals move in to occupy territories vacated by more dominant animals which have succumbed.[30]

In species with highly developed social systems, such as black-tailed prairie dogs *(Cynomys ludovicianus)*, maintenence of the social structure is important for the continued survival and success of the colony or "town". When the social structure is disrupted during breeding season, as it can be by intense shooting, decreased reproductive success will result.[55] Disruption of the social system may also cause a break-down in communications among animals which are important in warning the colony about approaching predators; this may lead to increased predation upon individuals in the town, at least temporarily.

In a species such as the beaver *(Castor canadensis)*, there may exist some situations in which damage is a concern principally because offspring of one pair of animals disperse to adjacent sites where they are not desired. Because this species is territorial, monogamous, and long-lived, sterilization of the parent individuals is at least theoreticaly an effective way to deal with the problem. The social behavior of the individuals is not impaired by sterilization, so they maintain their territory while excluding other conspecifics and producing no young. Were there an efficient means of sterilizing the adults, this strategy would have practical use.[34]

VII. CONCLUSION

The most successful rodent pests are those species that have adapted to a wide variety of situations. Behavioral adaptations are important in a species' ability to reach damaging population levels while avoiding certain control methods. The house mouse *(Mus musculus)*, for example, is a highly successful species world-wide because, among other reasons, it possesses a network of behavioral adaptations which allow it to exploit a variety of living situations through a system which allows social flexibility.[36]

Rodents, as a group, are highly successful because of their biological and behavioral adaptations which allow them to exploit resources. In considering how rodent behavior affects rodent damage as well as rodent control, it is important to remember that rodent populations are not homogeneous. No matter how attractive and palatable a toxic bait may be to the majority of the population, there will likely be some individuals which will not feed sufficiently to consume a lethal quantity. No matter how well traps are placed, some individuals may not be captured. Despite considerable effort which may be taken to make an area rodent-proof by means of mechanical exclusion, some rodents will likely find their way in.

Behaviors present within a rodent population may depend on a variety of factors including species, habitat, age of individuals,[53] and prior attempts at control. A rodent control strategy which uses a variety of techniques and is "tailored" to the particular damage situation is most likely to be efficacious.

REFERENCES

1. **Addison, W. H. F. and Appleton, J. L.,** The structure and growth of the incisor teeth of the albino rat, *J. Morphol.*, 26, 43, 1915.
2. **Advani, R. and Mathur, R. P.,** Experimental reduction of rodent damage to vegetable crops in Indian villages, *Agro-Ecosystems*, 8, 39, 1982.
3. **Barnett, S. A.,** Experiments on "neophobia" in wild and labortory rats, *Br. J. Psychol.*, 49, 195, 1958.
4. **Barnett, S. A.,** *The Rat: A Study in Behaviour, Revised Edition,* University of Chicago Press, Chicago, Ill., 1975.

5. **Bhardwaj, D. and Khan, J. A.,** Mitigating poison and bait-shyness developed by wild rats (*Rattus rattus* L.): effect of poisoning at short intervals, *Indian J. Exp. Biol.,* 15, 624, 1977.
6. **Bhardwaj, D. and Khan, J. A.,** Effect of texture on the food preferences of bait-shy wild rats (*Rattus rattus* L.), *Proc. Indian Acad. Sci.,* 87B, 77, 1978.
7. **Bohills, S. T., Meehan, A. P., and Leonard, S. P.,** Advantages of bait boxes in house mouse control, *Int. Pest Control,* March-April, 34, 1982.
8. **Bowerman, A. M. and Brooks, J. E.,** What makes a palatable warfarin?, *Pest Control,* 40, 22, 1972.
9. **Brooks, J. E. and Rowe, F. P.,** Commensal rodent control, Worth Health Organization Publication VBC 79.726, 1979.
10. **Cain, R. E. and Carlson, R. H.,** Evidence for color vision in the prairie dog *(Cynomys ludovicianus), Psychon. Sci.,* 13, 185, 1968.
11. **Calhoun, J. D.,** The ecology and sociology of the Norway rat, U.S. Dept. of Health, Education and Welfare, Public Health Ser. Publ. No. 1088, Washington, D.C., 1962.
12. **Chitty, D. and Southern, H. N.,** *The Control of Rats and Mice,* Vols. 1, 2, and 3, Clarendon Press, Oxford, 1954.
13. **Christiansen, E.,** Pheromones in small rodents and their potential use in pest control, in *Proc. 7th Vert. Pest Conf.,* Siebe, C. C., Ed., University of California-Davis, 1976, 185.
14. **Christiansen, E., Doving, K. B., and Mork, E.,** A device for bioassay of response to olfactory stimuli in wild rodents, in *Test Methods for Vertebrate Pest Control and Management Materials, ASTM STP 625,* Jackson, W. B. and Marsh, R. E., Eds., American Society for Testing and Materials, Philadelphia, Pa., 1977, 27.
15. **Clark, D. O., Ed.,** *Vertebrate Pest Control Handbook,* California Dept. of Food and Agriculture, Division of Plant Industry, Sacramento, Calif., 1975.
16. **Connolly, R. A. and Landstrom, R. E.,** Gopher damage to buried cable materials, *Mater. Res. Stand.,* 9, 13, 1969.
17. **Crowcroft, P.,** *Mice All Over,* Foulis, London, 1966.
18. **Crowcroft, P. and Jeffers, J. N. R.,** Variability in the behavior of wild house mice (*Mus musculus* L.) toward live traps, *Proc. Zool. Soc. London,* 137, 573, 1961.
19. **Crowcroft, P. and Rowe, F. P.,** Social organisation and territorial behaviour in the house mouse (*Mus musculus* L.), *Proc. Zool. Soc. London,* 140, 44, 1963.
20. **Daly, M., Wilson, M. I., and Behrends, P.,** Factors affecting rodents' responses to odours of strangers encountered in the field: experiments with odour-baited traps, *Behav. Ecol. Sociobiol.,* 6, 323, 1980.
21. **Davis, D. E., Emlen, J. T., and Stokes, A. W.,** Studies on home range in the brown rat, *J. Mammal.,* 29, 207, 1948.
22. **Doty, R. E.,** The prebaited feeding station method of rat control, *Hawaiian Plant Rec.,* 42, 39, 1938.
23. **Emlen, J. T., Young, H., and Strecker, R. L.,** Demographic responses of two house mouse populations to moderate suppression measures with 1080 rodenticide, *Ecology,* 39, 200, 1958.
24. **Errington, P. L.,** Wintering of field living Norway rats in south central Wisconsin, *Ecology,* 16, 122, 1935.
25. **Flint, M. L.,** Integrated pest management for tomatoes, University of California, Statewide Integrated Pest Management Project, Division of Agricultural Sciences, Publ. 3274, Berkeley, Calif., 1982, 104.
26. **Gurnell, J.,** Observations on trap response in confined populations of wood mice, *Apodemus sylvaticus, J. Zool. London,* 185, 279, 1978.
27. **Hansen, S. and Hard, E.,** Ontogeny of the audiogenic immobility reaction in the rat, *Biol. Behav.,* 5, 163, 1980.
28. **Howard, W. E., Marsh, R. E., and Cole, R. E.,** Food detection by deer mice using olfactory rather than visual cues, *Anim. Behav.,* 16, 13, 1968.
29. **Howard, W. E.,** Tests of pocket gophers gnawing electric cables, *J. Wildl. Manage.,* 17, 296, 1953.
30. **Howard, W. E.,** Where to place rodent bait boxes, in *Proc. 2nd Symp. Recent Advances in Rodent Control,* Ministry of Public Health, Kuwait, 1986.
31. **Jackson, W. B.,** Feeding patterns in domestic rodents, *Pest Control,* 33, 12, 1965.
32. **Jacobs, W. W. and Labows, J. N.,** Conditioned aversion, bitter taste and the avoidance of natural toxicants in wild guinea pigs, *Physiol. Behav.,* 22, 173, 1979.
33. **Kaltwasser, M. and Schnitzler, H.,** Echolocation signals confirmed in rats, *Z. Saugetierkunde,* 46, 394, 1981.
34. **Kennelly, J. J.,** Evaluation of induced sterility for beaver *(Castor canadensis)* management problems, *Proc. 1st Eastern Wildl. Damage Conf.,* Decker, D. J., Ed., Cornell University Cooperative Extension, Ithaca, N.Y., 169, 1983.
35. **Klimstra, W. D.,** House mouse behavior and its significance to control, in *Proc. 5th Vert. Pest Conf.,* Marsh, R. E., Ed., University of California-Davis, 149, 1972.
36. **Mackintosh, J. H.,** Behaviour of the house mouse, *Symp. Zool. Soc. London,* 47, 337, 1981.

37. **Mainardi, M., Pasquali, A., and Pasquale, E.,** Food choice in *Mus musculus:* effects of experience on aversion to an unpleasant flavour, *Biol. Behav.,* 10, 67, 1985.
38. **Marsh, R. E.,** Principles and techniques of formulating effective rodent baits — present and future, in *Proc. 2nd Symp. Recent Advances in Rodent Control,* Ministry of Public Health, Kuwait, 1986, 211.
39. **Marsh, R. E.,** Tracking powders: providing effective rodent control, *Pest Control,* 51, 42, 1983.
40. **Marsh, R. E. and Cummings, M. W.,** Pocket gopher control with mechanical bait applicator, Leaflet 2699, Division of Agricultural Sciences, University of California, Berkeley, Calif., 1976.
41. **Marsh, R. E. and Howard, W. E.,** Pheromones (odors) for rodent control?, *Pest Control Technol.,* 7, 22, 1979.
42. **Matschke, G. H.,** Problems associated with designing field studies when testing rodenticides for efficacy, with special reference to ground squirrels, in *Proc. Conf. Organisation and Practice of Vertebrate Pest Control,* Imperial Chemical Industries, Surrey, England, 1984, 95.
43. **Michener, G. R.,** Spring emergence schedules and vernal behavior of Richardson's ground squirrels: why do males emerge from hibernation before females?, *Behav. Ecol. Sociobiol.,* 14, 29, 1983.
44. **Partridge, L.,** Increased preferences for familiar foods in small mammals, *Anim. Behav.,* 29, 211, 1981.
45. **Posadas-Andrews, A. and Roper, T. J.,** Social transmission of food-preferences in adult rats, *Anim. Behav.,* 31, 265, 1983.
46. **Radwan, M. A., Crouch, G. L., Harrington, C. A., and Ellis, W. D.,** Terpenes of ponderosa pine and feeding preferences by pocket gophers, *J. Chem. Ecol.,* 8, 241, 1982.
47. **Reidinger, R. F., Jr. and Mason, J. R.,** Exploitable characteristics of neophobia and food aversions for improvements in rodent and bird control, in *Vertebrate Pest Control and Management Materials, ASTM STP 817,* Kaukeinen, D. E., Ed., American Society for Testing and Materials, Philadelphia, Pa., 1983, 20.
48. **Robbins, R. J.,** Poison-based taste aversion learning in deer mice *(Peromyscus maniculatus bairdi), J. Comp. Physiol. Psych.,* 92, 642, 1978.
49. **Robbins, R. J.,** Taste-aversion learning and its implications for rodent control, *Proc. 9th Vert. Pest Conf.,* Clark, J. P., Ed., University of California-Davis, 1980, 114.
50. **Rosenzweig, M. R., Riley, D. A., and Drech, D.,** Evidence for echolocation in rats, *Science,* 121, 600, 1955.
51. **Rowe, F. P.,** Wild house mouse biology and control, *Symp. Zool. Soc. London,* 47, 575, 1981.
52. **Salmon, T. P. and Marsh, R. E.,** Equipment and methods available for measuring the behavioral response of rodents to odors, in *Test Methods for Vertebrate Pest Control and Management Materials, ASTM STP 625,* Jackson, W. B. and Marsh, R. E., Eds., American Society for Testing and Materials, Philadelphia, Pa., 1977, 86.
53. **Salmon, T. P. and Marsh, R. E.,** Age as a factor in rodent susceptibility to rodenticides — a review, in *Vertebrate Pest Control and Management Materials, ASTM STP 680,* Beck, J. R., Ed., American Society for Testing and Materials, Philadelphia, Pa., 1979, 84.
54. **Shumake, S. A.,** Efficacy evaluation of ultrasonic rodent repellent devices, in *Proc. 6th Gt. Plains Wildl. Damage Contr. Wkshp.,* (Abstr.), Andelt, W. F., Ed., Kansas State University, Manhattan, Kan., 1983, 110.
55. **Stockrahm, D.,** Comparison of population structures of black-tailed prairie dog, *Cynomys l. ludovicianus* (Ord), towns in southwestern North Dakota, M.S. thesis, University of North Dakota, Grand Fords, N.D., 1979.
56. **Sullivan, T. P. and Crump, D. R.,** Influence of predator odours on the feeding behaviour of herbivores, presented at 4th Int. Theriological Congr., Edmonton, Alberta, August 13 to 20, 1985.
57. **Taylor, K. D.,** Range of movement and activity of common rats *(Rattus norvegicus)* on agricultural land, *J. Appl. Ecol.,* 15, 663, 1978.
58. **Temme, M.,** House mouse behavior in multiple-capture traps, *Pest Control,* 48, 16, 1980.
59. **Tunberg, A. D., Howard, W. E., and Marsh, R. E.,** A new concept in pocket gopher control, in *Proc. 11th Vert. Pest Conf.,* Clark, D. O., Ed., University of California-Davis, 1984, 241.
60. **Walton, W. E.,** Color vision and color preference in the albino rat. I. Historical summary and criticism, *J. Comp. Psychol.,* 15, 359, 1933.
61. **Watts, C. H. S.,** Vocalizations of nine species of rat (Rattus; Muridae), *J. Zool. London,* 191, 531, 1980.
62. **West, R. R., Fall, M. W., and Libay, J. L.,** Reducing interactions among rats to improve bait acceptance, *Philipp. Agric.,* 59, 31, 1975.
63. **Wing, J. F. and Birch, L. A.,** Relative cue properties of novel-tasting substances in avoidance conditioning, *Anim. Learn. Behav.,* 2, 63, 1974.
64. **Wuensch, K. L.,** Effect of scented traps on captures of *Mus musculus* and *Peromyscus maniculatus, J. Mammal.,* 63, 312, 1982.

Chapter 13

METHODS FOR ESTIMATING THE LOSSES CAUSED IN RICE AND OTHER CROPS BY RODENTS

B. D. Rennison and A. P. Buckle

TABLE OF CONTENTS

I. INTRODUCTION

Too little is known worldwide about the amounts and value of damage that rodents inflict annually on crops. Consequently rodent control is too frequently carried out in response to public or political pressure without consideration of the contribution that is being made to crop protection or to the economic benefits, if any, that are achieved. This unplanned approach has been referred to as rat farming and both the Food and Agriculture Organization (FAO)[6] and the World Health Organization (WHO)[16] have advocated the need to consider control in broader terms by directing attention to measuring the financial benefits.

The benefits cannot, however, be assessed unless the crop losses caused by rodents can be measured and given monetary value. Measuring crop damage on anything but a very small scale can only be practically achieved by taking samples from the crop. Various methods of sampling are therefore reviewed and discussed below. Rodent damage does not necessarily result in crop loss however. The amounts by which yields are reduced by damage depend significantly on the ability of crops to compensate for the damage. Generally, the earlier in the crop's development the damage occurs, the greater is the time available for compensatory growth and the less is the actual loss incurred. Thus yield losses tend to have direct relationships with the amounts of damage only when that damage is inflicted on the ripening or ripe crops. Rat damage of cocoa is an example. Because the rats only attack cocoa pods that are beginning to ripen, the percentage loss of cocoa is equal to the percentage of damaged pods in the harvested crops.[17] By contrast rats only attack coconuts that are between 3 and 8 months old and consequently compensatory development of undamaged nuts reduces the copra loss by about 50%.[19]

Rice is a crop that can be subjected to rat attack at all stages of growth. Early damage to tillers results in negligible yield loss because compensatory growth gives complete yield recovery. There is some compensation for midseason damage and this, therefore, results in only partial yield loss in relation to the amount of damage inflicted. Damage close to harvest results in yield losses of equivalent proportions. Assessing or predicting yield loss from damage in an immature crop is thus not always simple or straightforward. Yet the relationships that exist between damage occurring at different stages of development and final yield loss need to be known if control is to be carried out to the maximum economic effect.

Methods of deriving these important relationships are considered and methods of putting the results obtained from surveys and research to practical use for crop protection are discussed.

II. METHODS OF SAMPLING TO MEASURE DAMAGE

Planning a crop damage sampling scheme is usually a three stage process. The first stage is to choose a suitable sampling plan. The next step is to decide how large a sample will be needed. The third step is to plan how to take the sample so that it conforms with the requirements of the sampling plan. The objective of the planning should always be to combine the three components of the scheme to produce results that will be neither more nor less accurate than required. To achieve this objective it is necessary to consider the statistical properties of sampling plans as well as the methods that can be used to determine sample size and draw valid samples.

Snedecor and Cochran's *Statistical Methods*[14] is the text that has been most frequently consulted with regard to the sampling methods that are discussed. Statistical formulae have therefore been presented as they are in that excellent book.

A. Sampling Plans

1. Simple Random Sampling

In simple random sampling, each individual in the sample has to be drawn independently and with equal probability from the population that is under investigation. In other words the selection of each individual in the sample has to be left entirely to chance.

If this rule is obeyed when a sample of n variables is drawn from a population of N variables of Y_i (i = 1,2 . . . N), the sample mean $\bar{y}$ will be a valid estimate of the population mean $\bar{y}$. The population standard deviation

$$\sigma = \sqrt{\frac{\Sigma (Y_i - \bar{Y})^2}{N - 1}} \quad (1)$$

will be estimated by the sample standard deviation

$$S = \sqrt{\frac{\Sigma (Y_i - \bar{y})^2}{n - 1}} \quad (2)$$

and the standard error $\sigma_{\bar{y}}$, of the sample mean $\bar{y}$* will be estimated by

$$S_{\bar{y}} = \frac{S}{\sqrt{n}} \sqrt{(1 - \phi)} \quad (3)$$

The term $\sqrt{(1-\phi)}$, in which ϕ = n/N' the fraction of the population in the sample, is the finite population correction. This correction can be ignored when $n/N < 1/10$ which in practice is nearly always.

When the items in a simple random sample of size n are not variable values but units that fall into one of two classes (e.g., damaged or not damaged), the value that is measured is the proportion p in one of the classes (e.g., damaged). The sample standard deviation is then

$$S_p = \sqrt{\frac{p(1 - p)}{n}} \quad (4)$$

when p is the proportion of items in that class. Again, the finite population correction can be ignored if n is less than 10% of the population.

Simple random sampling is not a practical method for sampling cereals, sugar cane, or other densely planted crops from which it is impractical to draw samples of individuals. It is much more practical when sampling such crops to take clusters of individuals and to estimate the proportions damaged in each cluster.

2. Cluster Sampling

The clusters that form sampling units may be short row lengths of crops,[10] quadrats,[11] individual hills,[15] or groups of rice hills.[2] As in simple random sampling, the sample clusters have to be randomly chosen so that all the clusters in the population have the same chance of being included in the sample. If p_i is the proportion of damaged individuals in the ith cluster of n clusters in a sample, the mean proportion of damaged individuals in the population is estimated as

$$p = \frac{\Sigma p_i}{n} \quad (5)$$

* Lower case signifies the sample mean as opposed to population means.[14]

The sample standard deviation is

$$S = \sqrt{\frac{\Sigma (p_i - p)^2}{n - 1}} \tag{6}$$

and the standard error of p is

$$S_p = \frac{S}{\sqrt{n}} \tag{7}$$

If more than 10% of the clusters in the population of N clusters have been included in the sample, s_p must be corrected by being multiplied by the finite population correction.

Benigno[2] found that, because of the often clumped distribution of damaged plants, sampling precision improved as cluster size was increased. He found the optimum cluster size for sampling rat-damaged rice was 25 hills (5 × 5 hills). The same size of cluster was found to be optimal for damage in corn, but rows of ten plants were best for sampling damage in pineapples.

3. Stratified Sampling

A stratified sampling plan should be used when damage within the population under investigation is clearly not randomly distributed but heavier in some zones than in others. This is not uncommon when areas of cultivation are bounded on one or more sides by uncultivated land which is a refuge and habitat for the pest species. In such situations damage is often more intense on the edges of cultivated areas than at the centers. Benigno[2] records this particularly frequently in the case of rat damage in maize and also, but less frequently, in rice. Cases have also been recorded of rice rat damage being much more intense at the centers of fields than at the edges.[1]

To sample with the greatest precision when damage is clearly thus stratified, one should take a separate random sample from each stratum. The sample estimate of the population mean will then have the value

$$\bar{y}_{st} = \frac{\Sigma N_h \bar{y}_h}{N} \tag{8}$$

when N_h is the total number of sampling units in the hth stratum, Y_h is the sample mean in the hth stratum, and $N = \Sigma N_h$ is the total population size in all strata combined.

If the proportion of damaged units is being measured, the mean is estimated by

$$p_{st} = \frac{\Sigma N_h p_h}{N} \tag{9}$$

In practice it is easier to calculate the relative weight for each stratum

$$W_h = \frac{N_h}{N} \tag{10}$$

and to compute the weighted means as

$$\bar{y}_{st} = \Sigma\, W_h \bar{y}_h \tag{11}$$

or (12)

$$p_{st} = \Sigma\, W_h p_h$$

The standard error of $\bar{y}_{st}$ is

$$S(\bar{y}_{st}) = \sqrt{\Sigma\, W_h^2 \frac{S_h^2}{n_h}} \tag{13}$$

in which

$$S_h^2 = \frac{\Sigma\,(Y_{hi} - \bar{y}_h)^2}{n_h - 1} \tag{14}$$

when Y_{hi} is the ith member of the sample in the hth stratum and n_h is the sample size taken in the hth stratum.

The standard error of p_{st}, the weighted proportion of damaged items, is

$$S(p_{st}) = \Sigma\, W_h^2 p_h (1 - p_h)/n_h \tag{15}$$

The weighted means are the correct estimates of the population means. The arithmetic means of the samples, $\bar{y}$, and p are not, except with proportional allocation when the same fraction of the population in each stratum is sampled.

When that is the case, the sample is self-weighting and $\bar{y} = \bar{y}_{st}$. This is useful to know because it means that one can use proportional allocation as a contingency measure in sample surveys. If there is any possibility that some plots in an area are likely to suffer heavier damage by virtue of being closer to rat habitat than other plots, equal proportions of both types of plot should be sampled. It will not then matter if damage is unevenly distributed in this way, the arithmetic mean will be the correct estimator of the population mean.

4. Diagonal Sampling

Diagonal sampling provides a contingency against damage within plots being heavier at plot edges than at plot centers or vice versa. If this pattern of rodent damage was predictable one could derive an appropriate stratified sampling plan to measure it accurately. Unfortunately, the damage is not consistently stratified within plots[2] and it is necessary to compromise. Diagonal samples will, in the long run, measure damage that is stratified parallel to plot edges more accurately than simple random samples will. On the other hand they will be less effective than properly stratified samples.

In diagonal sampling, transects are taken from corner to corner along plot diagonals and plants or clusters of plants sampled en route. Each sample unit should be independently randomly selected, as indeed should the plot diagonal itself when only one diagonal is traversed. Provided these precautions are observed, the samples can be treated in the same way as simple random or cluster samples taken within plots. The method is thus suitable for sampling within plots in surveys or replicated experiments, but is not recommended where precise estimates of damage within individual plots are wanted. For the latter purpose investigators should draw simple random or, if more appropriate, stratified samples.

5. Systematic Sampling

Often, to save time, when drawing a sample from a listed population, a start is made at a randomly selected sampling unit near the beginning of the list after which every nth unit is included in the sample. The value given to n in such cases is often calculated to spread the units evenly over the population and leave the smallest remainder unsampled. In crop sampling a transect may be started at a random plant out of the first five or ten on a transect line and every nth plant after it sampled. Another variation is to choose random rows along both the length and breadth of a field and to sample the plants at the intersections of the grid so created.[15] All these methods of systematic selection tend to spread the sampling more evenly through the population than simple random sampling does, and the results are consequently more accurate than simple random sampling results.[14] However, there are two potential disadvantages.[14] First, if the population contains a periodic type of variation and the sampling interval (n) happens to coincide with it, the sample will be badly biassed. The second disadvantage is that there is often no reliable method of estimating the standard error of the sample mean. Systematic sampling plans should therefore only be used with care and forethought. They can often be successfully incorporated into more complex, multiple stage, plans.

6. Multiple-Stage Sampling

It is not often practical when faced with a large area of crop to draw a random sample in one step. Usually, it is necessary as a first step to draw a sample of primary units and to complete the process by drawing random samples within the chosen primary units. The second step is often referred to as subsampling. More complex plans, involving several stages, may have to be employed to sample crops on a national scale. Different sampling plans, including systematic ones, can be used for each stage.

When a simple random sample of n_1, primary units is drawn and n_2 subunits are sampled in each primary unit, the standard error of the sample mean $\bar{y}$ per sub-unit is estimated by

$$S_{\bar{y}} = \frac{1}{\sqrt{n_1}} \sqrt{\frac{\Sigma\,(\bar{y}_i - \bar{y})^2}{n_1 - 1}} \tag{16}$$

in which $\bar{y}_i$ is the mean subunit in the ith primary unit.

The data in Table 1 were collected by a three-stage process. As a first stage, the rice growing land was divided systematically into sampling areas of approximately equal size.[4] A random sample of three rice plots was then chosen in each area and a subsample of 25 hills chosen randomly on a diagonal of each plot. The primary sample was not random but systematic and so the standard error of the mean $\bar{y}$ per subunit is found from the analysis of variance (anovar) given under Table 1. The between areas component of variance, σ_1^2, is estimated by

$$S_1^2 = \frac{18.77 - 8.09}{3} = 3.56 \tag{17}$$

(the divisor $= n_2 = 3$ plots per area) and the between plots within areas component of variance, σ_2^2, is estimated by

$$S_2^2 = 8.09 \tag{18}$$

Table 1
THE MEAN NUMBERS OF RIPE TILLERS PER RICE HILL RECORDED IN A PILOT SURVEY IN 19 SAMPLING AREAS[4]

Sampling areas	Plots within areas		
1	14.0	6.7	16.0
2	16.7	14.2	19.6
3	18.2	17.6	18.1
4	17.4	13.9	17.0
5	16.0	10.6	12.0
6	17.2	16.1	15.8
8	17.8	14.1	18.6
9	11.9	11.6	11.4
10	16.8	22.1	17.5
11	14.2	13.0	17.4
12	22.7	21.1	17.9
13	15.1	8.8	16.8
14	12.1	13.8	16.1
15	15.2	16.4	17.4
16	9.8	19.1	20.8
17	12.7	13.0	12.8
18	13.8	15.9	14.7
19	19.4	17.8	19.6
20	8.9	12.3	18.7

Note: Analysis of variance.

Source of variation	Degrees of freedom	Mean sq	Parameters estimated
Between areas	18	18.77	$\sigma_2^2 + 3\sigma_1^2$
Between plots within areas	38	8.09	σ_2^2

The variance of the sample mean $\bar{y}$ which has the value

$$\sigma_{\bar{y}}^2 = \frac{\sigma_1^2}{n_1} + \frac{\sigma_2^2}{n_1 n_2} \tag{19}$$

is therefore estimated by

$$S_{\bar{y}}^2 = \frac{s_1^2}{n_1} + \frac{s_2^2}{n_1 n_2} = \frac{3.56}{19} + \frac{8.09}{(19)(3)} = 0.33 \tag{20}$$

so that the standard error is

$$S_{\bar{y}} = \sqrt{0.33} = 0.57$$

The components of variance σ_1^2 and σ_2^2 will be referred to again in the next section when considering the sizes of n_1 and n_2.

B. Choosing the Sample Size

1. Simple, Cluster, and Diagonal Random Samples

When we draw a random sample of n units from a population of units whose mean value is $\overline{Y}$, we know with 95% probability of being right, that the sample mean $\overline{y}$ will lie within the range $\overline{Y} \pm 2\sigma/\sqrt{n}$. Knowing this means that we can determine in advance how large a sample to take in order to estimate $\overline{y}$ with specified accuracy. For example, if we want to be 95% sure of estimating $\overline{Y}$ with an allowable error of $\pm$ L, we say that L must therefore have the value,

$$L = 2\sigma/\sqrt{n} \tag{21}$$

which gives:

$$n = 4\sigma^2/L^2 \tag{22}$$

If we want to be 90 or 99% certain that L will not be exceeded in a sample, the formulae become respectively,

$$n = 2.7\sigma^2/L^2 \tag{23}$$

and

$$n = 6.6\sigma^2/L^2 \tag{24}$$

To be able to use these formulae to find n, it is necessary to have an estimate of the population standard deviation σ, from previous samplings of the same or similar populations. If no information is available it is worth doing a small amount of pilot sampling to estimate σ. The estimate does not have to be very precise.

To estimate the size of sample that is required to measure the binomial proportion p of damaged items in a population with 95% probability of being right:

$$n = 4p\,(1 - p)/L^2 \tag{25}$$

This requires an advance estimate of p. Again the advance estimate need not be very precise if p is between 0.35 and 0.65 but should be fairly precise where p is nearer 0 or 1.

The binomial formula above is not the correct one to use to determine p by cluster sampling because p is then a continuous variable, with standard deviation,

$$\sigma = \sqrt{\frac{\Sigma\,(p_i - p)^2}{n - 1}} \tag{26}$$

The number of clusters in a sample is therefore determined from Equation 22 as,

$$n = 4\sigma^2/L^2$$

2. Stratified Samples

While the standard deviations and costs of sampling in all strata are the same, proportional allocation is the most efficient and economic form of allocation to use. But, if standard deviations and costs are not the same in all strata, the optimum allocation is obtained by taking n_h proportional to $N_h\sigma_h/C_h$ where σ_h is the standard deviation of the sampling units in the hth stratum and C_h is the cost of sampling in that stratum.[14] This rule says that larger samples than would be taken by proportional allocation should be taken in very variable strata and smaller than proportional samples in strata that are very expensive to sample.

3. Multiple-Stage Samples

In two-stage sampling as we have seen in the previous section the standard error of the mean per subunit is estimated by:

$$S_{\bar{y}} = \sqrt{\left(\frac{S_1^2}{n} + \frac{S_2^2}{n_1 n_2}\right)} \tag{27}$$

If the values of $S_1{}^2$ and $S_2{}^2$ are known, the equation can be solved for any chosen value of $S_{\bar{y}}$ to determine suitable primary, n_1, and subsample, n_2, sizes. In the example given (Table 1) the value of $S_{\bar{y}}$ was 0.57 when n_1 and n_2 were, respectively, 19 and 3. If the investigators who conducted the pilot survey (Table 1) wanted to know what effect reducing the subsample size from 3 to 2 would have on the accuracy of the survey estimate of $\bar{y}$ per subunit they needed only to solve

$$S_{\bar{y}} = \sqrt{\left(\frac{3.56}{19} + \frac{8.09}{(19)(2)}\right)} = 0.63 \tag{28}$$

The 95% confidence limits of $\bar{y}$ per sub unit would therefore be increased from $\bar{y} \pm 2(0.57)$ to $\bar{y} \pm 2(0.63)$. By likewise giving different values to n_1 they could test the effect of reducing or increasing the size of the primary sample.

The same methods can be used in three stage sampling schemes using the components of variance $S_1{}^2$, $S_2{}^2$, and $S_3{}^2$ when $S_{\bar{y}}$ becomes

$$S_{\bar{y}} = \sqrt{\left(\frac{S_1{}^2}{n_1} + \frac{S_2{}^2}{n_1 n_2} + \frac{S_3{}^2}{n_1 n_2 n_3}\right)} \tag{29}$$

Often, however, in national crop surveys primary divisions tend to be convenient administrative areas, like counties or districts. Usually these vary in size and have dissimilar areas under the crop that is the subject of the survey. When that is the case, sample sizes should be allocated in proportion to the contribution that the area of the crop planted in each county or district makes to the total area of the crop surveyed.

C. Methods of Drawing Random Samples From Crops

1. Simple Random Samples

Simple random samples are most easily drawn from crops in which the sampling units can be individually identified and numbered. With the population size, N, known, a simple random of sample of n units can be drawn with the aid of a table of random numbers or less scientifically, but as effectively, by putting N numbered tickets into a container, mixing them well, and drawing out n.

2. Cluster Samples

The method of drawing a random sample of clusters, instead of individuals, from a crop is basically the same except that in the first instance it is necessary to define the clusters which must be contiguous and not over-lapping. A population of 5000 1-m^2 quadrats would exist, for example, in a field of wheat that was 100 × 50 m. To draw a random sample of n clusters from that population of 5000 one would start by choosing and recording **in the order of choice** n random numbers between 1 and 100. These numbers will define 1 m units along the 100 m length of the field. The chosen values would then be paired in turn with another set of n random numbers between 1 and 50 defining 1-m units across the breadth of the field. The n pairs of values thus formed could be the coordinates that identified

1-m^2 clusters within the field. The pair of values 80,35 would signify the quadrat that lay 80 m along the length of the field and 35 m across the width. Exactly the same procedure can be used to choose a random sample of rice hills in a plot when the numbers of rows of hills across and along the length of the plot are known.

Random ordinates of length or breadth that are drawn more than once in a sample must not be rejected because it is well within the bounds of random probability to chose more than one unit in any row. Only when a pair of random values together form the coordinates of a unit that has already been drawn should they be rejected. In that case both values of the pair should be rejected and a new pair drawn.

3. Stratified Samples

Either of the methods described above can be used to draw random samples from the strata of stratified populations.

4. Diagonal Samples

To be able to draw a random sample of n plants from the total N on a plot's diagonal, one really needs to know the value of N. But to have to count the numbers of plants or population units on diagonals before choosing samples is quite impractical. The way out of this difficulty is to start sampling at a random corner of the plot, and to sample plants (or units) at random intervals along the diagonal until n have been sampled. The second diagonal may also have to be traversed to obtain all n units.

A surveyor using the above system could carry special tables of random numbers between 1 and N. A first reference to the tables would be made to choose a number between 1 and 4 to determine from which corner to start sampling. Successive references would be made thereafter to choose units.

Rennison[13] suggested using a system of this kind for rice rat damage surveys in Malaysia. To make the random number tables convenient to use each line comprised a set of n random numbers between 1 and N arranged in order from lowest to highest. A surveyor using the tables had thus only to choose a random line of the table each time he sampled a plot to identify the 25 hills (n = 25) to inspect. The value given to N in this case was 150; that being the total number of rice hills estimated to occur on the diagonal of a square rice field of 0.5 acre (0.2 ha). Each line of the table thus comprised 25 ordered random numbers between 1 and 150. Other values of N would be more suitable for sampling other crops or rice in other regions.

III. METHODS OF RELATING DAMAGE TO YIELD LOSS

Surveys of rodent damage in growing crops seldom directly measure actual yield losses. It is usually necessary to apply an adjustment based on either intuitive reasoning or a correction factor determined experimentally.

Generally speaking, it is easier to predict yield losses from damage in ripe crops than it is to predict the losses that will occur as a result of damage recorded during earlier growth stages. In the first case, the relationship can be determined by direct field observations of rat damage and yield in damaged and undamaged areas. In the second, the relationship has to be derived by simulating damage in experimental plots.

A. From Observed Damage in Mature Crops

Metcalfe and Thompson[10] showed that the percentage of damaged sugar canes had a simple, linear relationship with two estimates of yield loss, the percentage loss of sugar per ton extracted and the percentage loss of sugar extracted per acre.

Fulk and Akhtar,[7] investigating the relationship between rat damage and yield in fast-maturing rice, found that the effect of damage on yield was to an extent obscured by other factors that also affected yield. The main obscuring factor was tiller density which was not only correlated positively with yield but also with damage due to rats feeding preferentially in denser areas of rice. The multiple regression equation of yield (Y) on the density of tillers per m^2 (X_1) and on the percentage of rat-damaged tillers (X_2) was

$$Y = 44.24 + 0.83\ X_1 - 1.42\ X_2 \tag{30}$$

When damage was zero the potential yield ($\hat{Y}$) was estimated by the first part of the equation

$$\hat{Y} = 44.24 + 0.83\ (X_1) \tag{31}$$

Thus the percentage yield loss (Y_L) caused by rodent damage could be estimated from the observed yield (Y) as

$$Y_L = \frac{\hat{Y} - Y}{\hat{Y}} \times 100 \tag{32}$$

The advantage of establishing such a relationship is that, once established, it should obviate the need to measure damage again in the crop. However, the authors of both the examples given pointed out that the relationships established probably only applied to the varieties and locations studied. The relationships would have to be proved constant over a variety of conditions to be of real practical value.

Rennison[13] adopted a different approach for measuring yield loss in mature rice. He suggested that because rat damage resulted in fewer ripe tillers at harvest, ripe tiller loss should be measured at harvest thus:

If A = tipe tillers/hill in the undamaged hills of a sample of n hills and B = ripe tillers/hill in all n hills of the sample, the ripe tiller loss will be estimated by

$$L = A - B \tag{33}$$

and the percentage ripe tiller loss by

$$P_L = 100\ (A - B)/A \tag{34}$$

Provided that there is no significant compensating increase in grain weight in undamaged tillers of rat-damaged hills, the percentage ripe tiller loss should be equal to the percentage yield loss.

An advantage of the method is that it is independent of nonrodent sources of yield variation and so does not have to be proved before it can be put to practical use.

B. By Simulating Damage

Williams[18] simulated rat damage on immature coconuts of the age favored by rats and so succeeded in demonstrating that the resulting percentage copra loss, from the palms with damaged nuts, was only about half as great as the percentage nut loss.

Rice presents more problems because it is liable to rat attack from transplantation to maturity. In spite of this, however, interesting relationships between damage, crop age, and yield loss have been found. A large-scale experiment in the Philippines, for example, in which different percentages of rice tillers were cut by hand at fixed intervals between

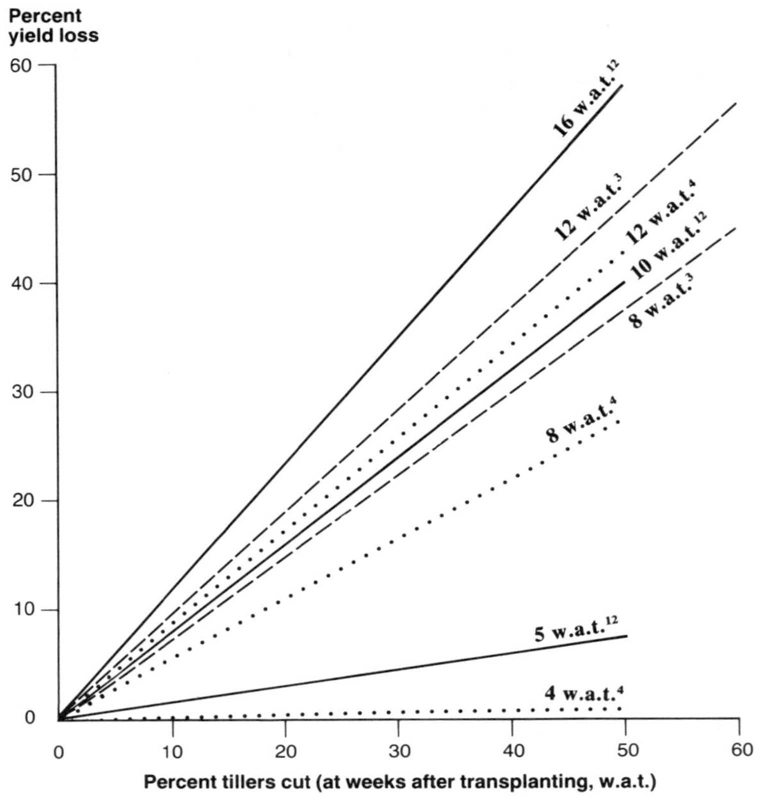

FIGURE 1. The linear regression lines (y = bB) relating percentage yield losses of rice (Y) to the percentages of tillers cut by hand (B) at different numbers of weeks after transplanting (w.a.t.) in simulated rat damage trials conducted independently in Malaysia[3,4] and Bangladesh.[12]

transplanting and harvest showed that the relationship between percentage yield loss (Y), the percentage of tillers cut (B), and time of cutting (A) measured in weeks after transplanting (w.a.t.) was[2]

$$Y = 0.0625\ AB - 0.1889\ B \tag{35}$$

A multiple regression of such a kind will have practical value if it can be shown to predict yield losses accurately over a wide range of conditions. The wider the range the greater will be the practical value. In that respect, published results of simulated rat damage trials conducted in Malaysia[3,4] and Bangladesh[12] are interesting. The trials were carried out quite independently in two countries but it is clear from a retrospective examination of the results that the rice response to damage was the same. This is first apparent from the slopes of the linear regressions of the form Y = bB (Figure 1) that can be calculated from the percentage tiller losses (Y) and percentages of cut tillers (B) at A = 4,5,8,10,12 and 16 w.a.t. Separate "Philippine type" regressions for Malaysia, Y = 0.1003 AB − 0.2431 B, and Bangladesh, Y = 0.0903 AB − 0.2301 B proved to be only insignificantly different ($F_{2,23}$ = 1.03; $P > 0.05$). The common regression therefore derived from the combined data,

$$Y = 0.0932\ AB - 0.2041\ B \tag{36}$$

Table 2
THE LOSSES OF RICE YIELD OBSERVED (Y) IN SIMULATED RAT DAMAGE TRIALS AND THE ESTIMATED LOSSES DERIVED FROM MULTIPLE EQUATIONS (36) $Y_1 = 0.0932$ AB − 0.2041 B AND (37) $Y_2 = 1.8865$ B Log_{10} A − 1.0996 B THAT FITTED THE DATA BEST

Time of cutting w.a.t. (A)	Percent tillers cut (B)	Percent yield loss recorded (Y)	Estimated yield losses from Eq 36 (Y_1)	Eq 37 (Y_2)	Deviations from observed values $Y-Y_1$	$Y-Y_2$	Slopes of regressions in Figure 1	Ref.
4	15	−12.8	2.5	0.5	−15.3	−13.3		4
4	30	0.0	5.1	1.1	−5.1	−1.1	0.02	
4	50	5.1	8.4	1.8	−3.3	3.3		
5	0	0.3	0.0	0.0	0.3	0.3		12
5	10	−2.3	2.6	2.2	−4.9	−4.5	0.15	
5	25	3.5	6.5	5.5	−3.0	−2.0		
5	50	8.5	13.1	11.0	−4.6	−2.5		
8	15	7.7	8.7	9.1	−1.0	−1.4		4
8	30	15.4	16.2	18.1	−0.8	−2.7	0.55	
8	50	28.2	27.1	30.2	1.1	−2.0		
8	20	12.7	10.8	12.1	1.9	0.6		3
8	40	29.8	21.7	24.2	8.1	5.6	0.75	
8	60	45.7	32.5	36.2	13.2	9.5		
10	0	−0.9	0.0	0.0	−0.9	−0.9		12
10	10	9.6	7.3	7.9	2.3	1.7	0.80	
10	25	24.7	18.2	19.7	6.5	5.0		
10	50	37.3	36.4	39.3	0.9	−2.0		
12	15	20.5	13.7	14.0	6.8	6.5		4
12	30	28.2	27.4	28.1	0.8	0.1	0.85	
12	50	38.5	45.7	46.8	−7.2	−8.3		
12	20	20.4	18.3	18.7	2.1	1.7		3
12	40	41.1	36.6	37.5	4.5	3.6	0.94	
12	60	53.6	54.9	56.2	−1.3	−2.6		
16	0	0.6	0.0	0.0	0.6	0.6		12
16	10	21.6	12.9	11.7	8.7	9.9	1.16	
16	25	31.1	32.2	29.3	−1.1	−1.8		
16	50	54.7	64.4	58.6	−9.7	−3.9		
Variance		330.1	295.3	309.0	35.3	24.5		
Percent of variance in Y		89.4	89.4	93.6	10.7	7.4		

accounted for 89% of the total variance in Y (Table 2). A slightly better fit (94%) was obtained with the regression

$$Y = 1.8865 \text{ B} \log_{10} \text{A} - 1.0996 \text{ B} \quad (37)$$

This latter regression was calculated because the slopes of the eight separate linear regressions (Table 2) clearly had a curved relationship with A rather than the linear relationship that is implicit in the "Philippine" regression (Equation 35). The equation for the curve (Figure 2) was

$$b = 1.90 \log_{10} \text{A} - 1.12 \quad (38)$$

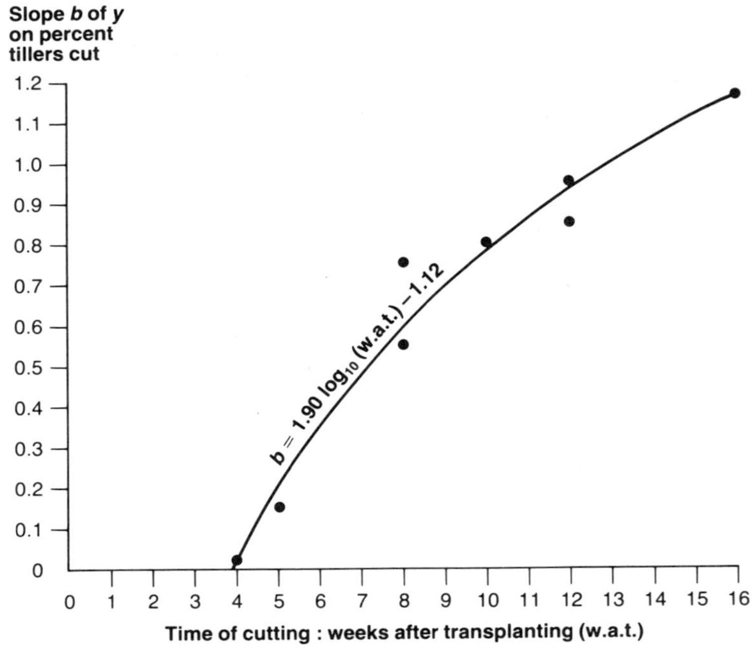

FIGURE 2. The slopes (Table 2) of the regressions in Figure 1 plotted against weeks after transplanting (A) showing the curvilinear relationship.

so that, since Y = bB in each data set, the multiple regression equation involving A becomes

$$Y = B\,(1.90 \log_{10} A - 1.12) \qquad (39)$$

or, more precisely, as given above (Equation 37) when calculated from the observed data (Table 2) with B $\log_{10}$ A and B as the two independent variables. The use of $\log_{10}$ A instead of A improved the fit most at A = 4 and 16 w.a.t.

All three equations (35, 36, 37) showed that the yield loss, Y, increased in relation to B as the rice crops matured. This was clearly demonstrated by plotting Y as a proportion of B against A w.a.t. in Figure 3. The increase from the lowest point at 4 w.a.t. was much more marked in the Malaysian/Bangladeshi experiments than in the Philippine; to the extent that Y became greater than B during the final 3 weeks of growth. There is no obvious explanation for this observation.

IV. SEASONAL DAMAGE CROP LOSS SURVEYS

A. Timing

Surveys to measure the damage or crop losses that have been caused by rats should be carried out nearly as possible to the time when the total effect of the damage can be most accurately measured. Generally that will be when the crop is ready for harvest. A problem with rice is that sampling of the ripe crop itself causes shedding of grain. To minimize this, and farmers' objections to it, the end-of-season surveys are probably best conducted 1 to 2 weeks before harvesting.[13]

B. Treatment of Data

When a damage survey has been completed the data should be tabulated (Table 3) and

Table 3
THE PERCENTAGE RIPE TILLER LOSSES RECORDED IN A PILOT RAT DAMAGE SURVEY CARRIED OUT IN FOUR DISTRICTS OF PULAU PINANG[4]

South west		Southern district		Central district						Northern district			
Sample unit no	Tiller loss (%)	Sample unit no	Tiller loss (%)	Sample unit no	Tiller loss (%)	Sample unit no	Tiller loss (%)	Sample unit no	Tiller loss (%)	Sample unit no	Tiller loss (%)	Sample unit no	Tiller loss (%)
	0		8.9		7.3		6.0		4.7		4.8		1.4
1	33.0	4	47.9	8	30.5	11	12.2	14	2.8	15	0	18	0
	5.3		8.6		8.8		8.9		2.4		2.2		0
	19.3		18.8		12.5		0				44.3		4.4
2	21.1	5	39.8	9	26.1	12	4.1			16	16.6	19	0.6
	6.2		5.9		8.1		5.7				0		2.5
	5.7		2.9		3.4		11.2				19.6		49.4
3	4.3	6	11.5	10	6.0	13	50.0			17	7.1	20	12.8
	1.1		11.7		13.4		5.1				17.4		3.6
Mean	10.7		17.3			10.9						10.4	
Variance	125.2		249.2			134.5						218.3	

Note: The numbers of units sampled in each district was proportional to the area of rice and in each unit 25-hill diagonal samples were taken in three plots.

Analysis of variance of transformed percentage [transformed percentage = $\log_{10}$(percentage + 1)]

Source	df	SS	MS	
Districts	3	1.06	0.353	1.67 Not significant
Within districts	53	11.17	0.211	
Areas within districts	15	4.86	0.324	1.15 $p<0.05$
Error	38	6.31	0.166	
Total	56	12.23		

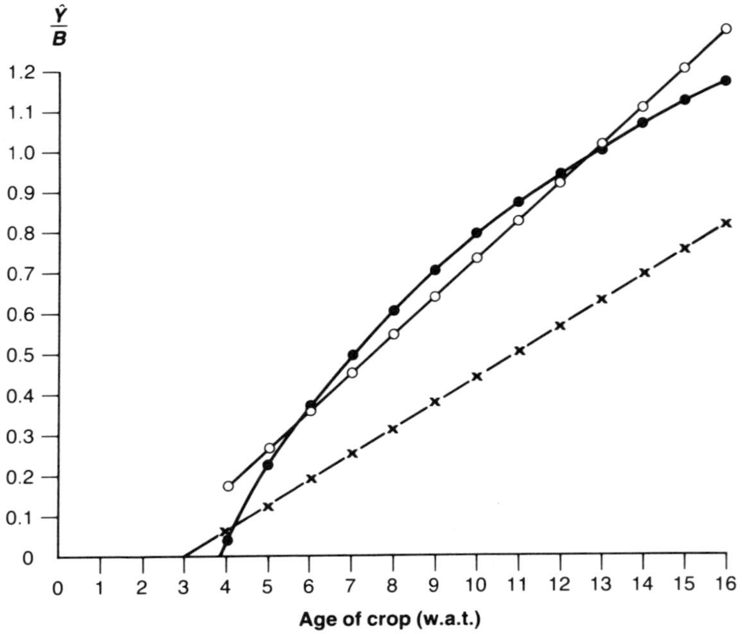

FIGURE 3. The estimated yield losses, (Y) derived from Equations 35 (x—x), 36 (○—○), and 37 (●—●) plotted as proportions of the percentages of tillers cut (B), against w.a.t. to show how the yield losses increased in relation to the tillers cut as the crops matured.

scrutinized for possible errors. At the same time a check should be carried out to see whether or not the data require transformation because very often damage/loss data that have been collected from a large area are not normally distributed but to some extent skewed. If the skewness is not very great it can probably be ignored but it is is significant, as in Figure 4, the data must be transformed before being analyzed by parametric methods.

1. Transformation

Rat damage data, like many biological data that reflect differences between population sizes, often have to be transformed for analysis. A simple test is available to determine which transformation, if any, is best suited to the data.[9] As described, the test entails plotting treatment means against treatment variances on double logarithmic paper. If a line through the plotted points from the origin of the graph has a slope of 1, a square root transformation is necessary. If the line has a slope of 2, a logarithmic transformation should be used, but if it has a slope of 4, a reciprocal transformation is more appropriate. With small calculators nowadays more readily available than double logarithmic paper it is usually easier to calculate the slope, b, from the uncorrected sums of squares and products thus

$$b = \frac{\Sigma\, XY}{\Sigma\, X^2} \tag{40}$$

using the logarithms of the variances (Y) and logarithms of the means (X). The test carried out using the logarithms of the district means and variances in Table 3 gave the result

$$b = \frac{9.71}{4.70} = 2.1 \tag{41}$$

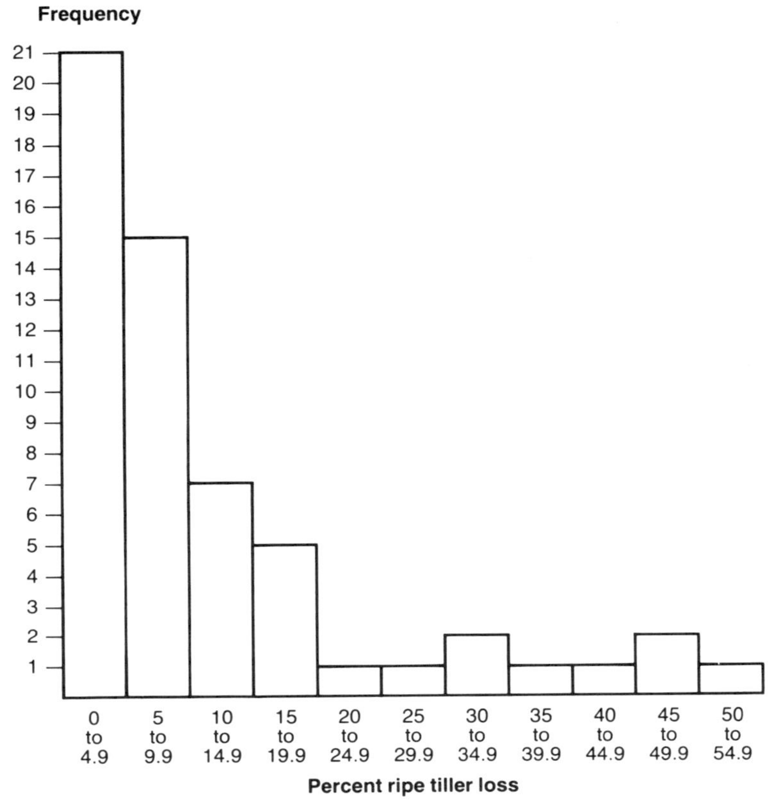

FIGURE 4. A frequency histogram of the percentage ripe tiller losses given in Table 3 showing the positively skewed distribution.

which indicated that a log transformation of the data was necessary. Because some values in the table were zeros, 1 was added to all values before transformation. The analysis of variance (anovar) shown below Table 3 was performed on the transformed data. It differed from an anovar of the untransformed values in that it showed a significant difference ($P = 0.05$) between sampling areas within districts.

2. Measures of Central Tendency

Transformation can cause difficulties when it comes to presenting results, because means, variances, and other parameters shown should be those pertaining to the transformed variables. However, in the interests of clarity it is often useful to show the back-transformed means. The back-transformed means of log-transformed data are, of course, geometric means and differ from the arithmetic means of log-normally distributed data. The arithmetic mean of the 57 values in Table 3, 11.7%, is not a measure of central tendency. There are 38 values larger than 11.7% but only 18 smaller. The geometric mean, 6.8%, (the antilog -1 of the mean of the 57 transformed variables) is, on the other hand, centrally placed between 29 larger and 28 smaller values.

Another often ignored measure of central tendency, which is useful when dealing with skewed data, is the median (M). It is,[14] when n is odd, the order statistic whose number is $(n + 1)/2$. In other words if the 57 variables in Table 3 are arranged in order of size, the median is the $(57 + 1)/2 = 29$th observation; in this case 6.2%. It is, thus, not very different in value from the geometric mean. The 95% confidence limits of M are[14]

$$M \pm \frac{2\sqrt{n}}{2} \tag{42}$$

which in the example are,

$$29 \pm 7.6 \tag{43}$$

and therefore the 21st and 37th ordered variables, respectively, 4.8 and 8.9%. These also are not very different in value from the 95% limits of the geometric mean. The latter are, in the transformed state,

$$0.89 \pm 2\sqrt{0.166/57} \tag{44}$$

which become, when back-transformed, 5.1 and 9.0%.

The median is, therefore, demonstrably a parameter which could be used much more often than it is as a measure of central tendency in skewed survey data.

C. Control Planning with Survey Data

The levels of rat damage recorded in a crop at harvest presage the levels that will recur in the next season if control is not applied. The results of an end-of-season survey can therefore be used predictively to plan a prophylactic control strategy.[5]

1. Establishing Priorities for Control

The simplest control strategy is to start control in the areas in which the heaviest damage occurred and to proceed progressively to less and less heavily damaged areas until a predetermined low threshold of damage is reached, until funds for control run out, or until all areas have been treated. Whether or not financial or other limits apply, it is useful to be able to estimate in advance the benefits that can be expected to accrue as control progresses. A method of doing this is demonstrated in Table 4 using the survey data given in Table 3 enhanced by further data from Buckle and Rowe.[4]

The accumulated percentages in the fifth and eighth columns of Table 4 are, respectively, the percentages of survey units that will be progressively treated and the accumulating percentages of the potential total damage that will thereby be alleviated.

The arithmetic mean ripe tiller losses were used in this example because they were positively skewed and there was greater economic benefit to be gained from treating preferentially the 'tail' of heavily damaged units. The skewness of the arithmetic mean losses is not obvious in Table 3. Nevertheless, the five (26.3%) sampling areas with the highest arithmetic means accounted for 51.6% of the total ripe tiller loss. By comparison, the five units with the highest geometric mean losses accounted for only 46% of the total recorded tiller loss.

2. Establishing an Economic Injury Level

Control will cease to be economically beneficial when the cost of it becomes equal to the value of the crop loss that will be prevented. The threshold, T% loss, at which the two become equal, named the Economic Injury Level,[5] has the value

$$T\% = 100\ [(S + C)]/P\ V \tag{45}$$

when S is the cost per unit area of surveys and surveillance (see next section), C is the cost per unit area of control, V is the value per unit area of the crop, and P is the proportion by which the crop loss is, on average, reduced by control.

Table 4
THE SAMPLING UNITS OF TABLE 3 ARRANGED IN THE DESCENDING ORDER OF MEAN RIPE TILLER LOSS IN WHICH THEY WILL BE TREATED IN A PROPHYLACTIC RAT CONTROL PROGRAM, THE INCREASING PERCENTAGES OF UNITS THAT WILL BE TREATED, AND DAMAGE ALLEVIATED AS CONTROL PROGRESSES

Sample unit	Mean ripe tiller loss (%)	Number of units (grouped)	Percentage of units	Accumulating percentage	Number of ripe tillers lost[a]	Percentage of total ripe tillers lost	Accumulating percentage
13	22.1	5	26.32	26.3	1551	51.55	51.6
20	21.9						
4	21.8						
5	21.5						
16	20.3						
9	15.6	4	21.05	47.4	803	26.69	78.7
8	15.5						
2	15.5						
17	14.7						
1	12.7	1	5.26	52.6	105	3.49	81.7
11	9.0	2	10.53	63.2	225	7.48	89.2
6	8.7						
10	7.6	1	5.26	68.4	118	3.92	93.1
3	3.7	3	15.79	84.2	138	4.59	97.7
14	3.3						
12	3.3						
19	2.5	2	10.23	94.7	64	2.13	99.8
15	2.3						
18	0.5	1	5.76	100	5	0.17	100
Totals:		19	100		3009	100	

Buckle et al.[5] found that, in the case of rice rat damage in Malaysia, T was less than 1% because control was very cheap. Under such circumstances control in all areas in Table 4 except the last will be economically beneficial but, as the table shows, the law of diminishing returns applies.

V. SURVEILLANCE

Prophylactic rat control planned and executed on the basis of an end-of-season damage survey does not guarantee protection under all circumstances.[5] Rat populations can increase spectacularly under favorable conditions. Populations that were perhaps too small to warrant prophylactic control at the beginning of the season, or even populations that were reduced by control can become large enough before the end to do serious economic damage. Protection against resurgence can only be provided by a program of regular surveillance. If not carefully structured, however, surveillance sampling can be uneconomically expensive or, if costs are too carefully pared, ineffective. A way of striking a balance is to adopt an industrial quality control technique which has been designed to provide a known degree of protection from damage for an economic amount of sampling. The most economical plans which are also the most suitable for rice surveillance are sequential.

A. Sequential Sampling

In sequential sampling, sample size is not predetermined but is variable and determined by results as they are obtained. The objective of the sampling is not to measure the mean level of the damage in the crop, as is the objective of the sampling plans considered in Section II, but to confirm that the damage has not reached a predetermined unacceptable level. The further the amount of damage really is above or below that threshold, the sooner will the fact become apparent in the sampling and the fewer will be the total number of items that will have to be sampled.

Figure 5 is a sequential sampling graph that was drawn up for rice surveillance in Malaysia.[13] The horizontal axis is the cumulative number of rice hills inspected in a sample and the vertical axis is the cumulative number found with rat-cut tillers. Sampling proceeds at a rate of five hills per plot and the results are recorded hill by hill on the graph. Each undamaged hill is recorded as a point one unit to the right and at the same level on the graph as the previously plotted point. A hill with one or more rat-cut tillers is plotted one unit to the right and one unit up the scale from the previously plotted point. Sampling and plotting continue until the line of points crosses one or other of the two parallel lines on the graph or, if neither is crossed, until 100 hills (20 plots) have been sampled. Crossing the upper line indicates that 15%* or more of the rice hills in the population from which the sample has been drawn are rat damaged and that remedial control measures must therefore be undertaken. Crossing the lower line indicates that 5%* or less of hills are damaged and that resampling will not be necessary for 3 to 4 weeks.

When the plotted points remain between the two threshold lines either control may be carried out or the area resampled after only two weeks.

1. The Parameters of a Sequential Scheme

To operate a sequential scheme four values must be decided:

1. The proportion, P_1, of rat-damaged items that must not be exceeded (which if reached will trigger remedial control).
2. The lower acceptable proportion, P_0, which can be taken as a warning level.
3. The probability, α, of failing to detect when the proportion of damaged items in the population is P_1 or greater.
4. The probability, β, of failing to detect when the proportion of damaged items is below P_0 or less.

The values that are given to these parameters are a matter of choice. However, neither of the two probabilities should exceed 0.10, although they can be differentially lower than 0.10. The parameters of the Malaysian graph (Figure 5) are $P_0 = 0.05$, $P_1 = 0.15$, $\alpha = 0.005$ and $\beta = 0.10$. A higher value was given to β than to α because to fail to detect the lower level was regarded as having less economically damaging consequences than failing to detect the upper.

* In Malaysia, the relationship between damaged hills and damaged tillers is expressed by the significant ($p < 0.001$) regression equation $y = 1.4 \times -1.4$, in which x and y are the logarithms of the percentages of, respectively, hills with damage and damaged tillers.[13]

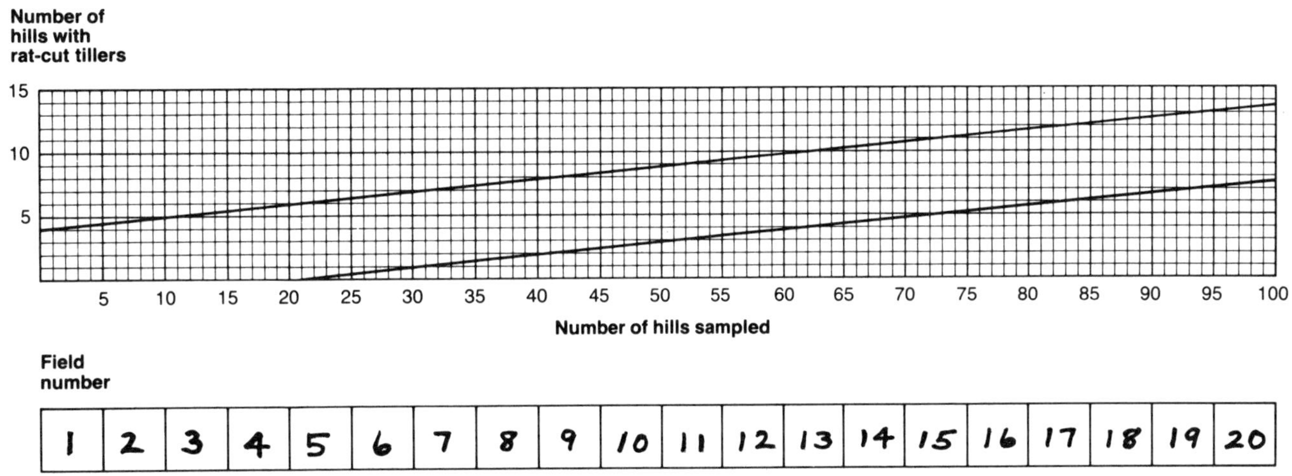

FIGURE 5. A sequential sampling graph for rice rat damage surveillance in Malaysia.[13]

2. Constructing a Sampling Graph

This method of constructing a graph is taken from Goulden.[8] Having assigned values to the parameters P_0, P_1, α, and β first calculate the following values:

$$B = \frac{\beta}{1-\alpha}; \quad A = \frac{1-\beta}{\alpha}; \quad r_1 = \frac{P_1}{P_0}; \quad r_2 = \frac{1-P_2}{1-P_0} \qquad (46, 47, 48, 49)$$

Then to construct the graph calculate:

$$h_0 = \frac{\log B}{\log r_1 - \log r_2} = \text{the point of origin of the lower parallel line on the vertical axis of the graph} \qquad (50)$$

$$h_1 = \frac{\log A}{\log r_1 - \log r_2} = \text{the point of origin of the upper parallel line on the vertical axis} \qquad (51)$$

$$S = \frac{-\log r_2}{-\log r_1 \log r_2} = \text{the slope of the parallel lines} \qquad (52)$$

$$E_S(n) = \frac{-\log B \log A}{-\log r_1 \log r_2} = \text{Maximum sample size (round to the nearest whole number)} \qquad (53)$$

$$h_0 + S[E_S(n)] = \text{height of the lower parallel line at the maximum sample size} \qquad (54)$$

$$h_1 + S[E_S(n)] = \text{height of the upper parallel line at the maximum sample size} \qquad (55)$$

Taking the parameters of Figure 5, $P_0 = 0.05$, $P_1 = 0.15$, $\alpha = 0.005$ and $\beta = 0.10$.

$$h_0 = -1.90$$

$$h_1 = 4.29$$

$$S = 0.09$$

$$E_S(n) = 98.0 \qquad (56)$$

3. Random or Nonrandom Selection of Plots?

The probabilities assigned to α and β in the sequential scheme assume the selection of both rice plots and hills within rice plots is random. If the selection is not random the values of those probabilities will not be as assigned. It can be argued that if plots are chosen preferentially from those with a higher than average risk of rat damage, because they will be the plots that will usually suffer economic damage first, crop protection will be increased. This is a reasonable argument as long as crop protection is the only purpose of the sampling and the data gathered will not be put to any other use. If the data are to be used, for example, in lieu of end-of-season survey data for control planning, the rules of random selection (Section II) should be maintained.

REFERENCES

1. **Alfonso, P. J., Fiedler, L. A., and Sumangil, J. P.,** Rodent ecology, population dynamics and behaviour, in *Rodent Biology and Control (With Special Reference to the Philippines),* Sanchez, F. F. and Benigno, E. A., Eds., National Crop Protection Centre, Laguna, Philippines, 1985, 152.
2. **Benigno, E. A.,** The assessment of in field damage to crops caused by rats, Ph.D. thesis, University of the Philippines, Los Bancos, 1979, 124.
3. **Buckle, A. P., Yong, Y. C., and Rowe, F. P.,** Yield response of the rice variety Improved Mashuri to simulated rat damage, *Malays. Agric. J.,* 52, 135, 1980.
4. **Buckle, A. P. and Rowe, F. P.,** Rice Field Rat Project, Malaysia, Technical Report 1977-80, Centre for Overseas Pest Research, Overseas Development Administration, College House, London, 1981, 99.
5. **Buckle, A. P., Huson, L. W., and Rowe, F. P.,** The frequency distribution of rat damage in rice fields and its implications for the cost benefit of control programmes, *Proc. Conf. Organisation and Practice of Vertebrate Pest Control,* ICI, Surrey, England, 1985, 289.
6. Food and Agriculture Organization, Vertebrate Pest Management in Asia and the Far East, FAO Regular Programme, FAO No. RAFFE 14, Food and Agriculture Organisation of the United Nations, Bangkok, 1973, 35.
7. **Fulk, G. W. and Akhtar, M. T.,** An investigation of rodent damage and yield reduction in rice, *Trop. Pest Manage.,* 27, 116, 1981.
8. **Goulden, C. H.,** *Methods of Statistical Analysis,* 2nd ed., Modern Asia, Ed, John Wiley & Sons, New York, 1959.
9. **Jeffers, J. N. R.,** *Experimental Design and Analysis in Forest Research,* Almqvist and Wicksell, Stockholm, Sweden, 1960.
10. **Metcalfe, J. R. and Thompson, G.,** Preliminary experiments in Jamaica with a method of determining loss of sugar resulting from rat damage to sugar cane, *Proc. Br. West Indies Sugar Technol.,* 276, 1966.
11. **Poché, R. M., Mian, Y., Haque, E., Mian, D., Mannan, A. I., and Sultana, P.,** Rodent damage to wheat in Bangladesh, Special report No. 1, Bangladesh Agricultural Research Institute, Vertebrate Pest Division, Joydebpur, 1979, 29.
12. **Poché, R. M., Haque, M. E., Mian, M. Y., Sultana, P., and Karim, M. A.,** Rice yield reduction by simulated rat damage in Bangladesh, *Trop. Pest Manage.,* 27, 242, 1981.
13. **Rennison, B. D.,** *Monitoring Rat Damage in West Malaysia,* A consultant's report and recommendations for the Malaysia/UK Rice Rat Control Project, Ministry of Agriculture, Fisheries and Food, Rodent Pests Department, Tolworth Laboratory, Surrey, U.K., 1979, 29.
14. **Snedecor, G. W. and Cochran, W. G.,** *Statistical Methods,* 6th ed., Iowa State University Press, Ames, Iowa, 1967.
15. **Swink, F. N., Sumangil, J. P., and Tolentino, D. O.,** *A method of estimating rat damage to growing rice,* Paper presented at the Regional Training Seminar on Field Rat Control and Research Manila, Philippines, March 1-15, 1974.
16. World Health Organization, Report on Seminar on Public Health and Economic Aspects of Rodent Control, World Health Organization, Regional Office for the Eastern Mediterranean, Alexandria, 1974.
17. **Williams, J. M.,** Rat damage assessement in cocoa, *Fiji Agric. J.,* 35, 15, 1973.
18. **Williams, J. M.,** The effect of artificial rat damage on coconut yields in Fiji, *PANS,* 20, 275, 1974.
19. **Williams, J. M.,** Rat damage to coconuts in Fiji. I. Assessment of damage, *PANS,* 20, 379, 1974.

Chapter 14

SELECTION OF BAITS AND THEIR DISTRIBUTION

M. Lund

TABLE OF CONTENTS

I. INTRODUCTION

The efficacy of a rodenticide is often more dependent on the quality of the bait base than on the active ingredient itself. In most situations the poison bait has to compete in attraction with a surplus of nonpoisonous food — if there was no food available for the rodents they would simply not be present. As the toxicant almost inevitably reduces palatability, the quality of the bait material should compensate for this.

As rodent pest species are either granivorous (sometimes omnivorous) or herbivorous, it is not possible to produce a bait which is equally attractive to all rodents. A thorough knowledge of the food habits and preferences of the particular rodent species in question is indispensable to achieve success. Whereas house mice nibble at almost any food type you present to them, *Rattus* species are usually much more reluctant to feed on an unknown type of food, or food having a strange taste or odor, and it may last several days before they overcome this shyness.[2] The more palatable the bait the less pronounced is the rat's reaction.[14] Previous experience or food habits may also strongly influence the behavior towards a poison bait, but it does not always imply that the food which has been available to the rodent also must be used as a bait base; rats tend to select a balanced diet and may be attracted to something different — rats at a fishing port are often very fond of a cereal bait. Rats living at a refuse dump with a variety of food presented regularly are more readily and quickly attracted to a new type of poison bait than rats living in a stable environment with few changes in their diet.

II. CHARACTERISTICS OF RODENT BAITS

Basically, baits must be of a quality suitable for human or animal consumption, and contaminated, dirty, old, or musty food must be avoided. This can be a problem, especially in developing countries with a shortage of food for human consumption. The following points must be considered when selecting baits for the control of commensal rodents:

1. *The bait must be attractive to the species in question.* For commensal rodents a high quality cereal seems to be ideal in most situations, and a particle size of 0.4 to 0.7 mm seems to be most attractive.[22] Larger particles, the size of or bigger than whole maize, are readily hoarded by rats, making it difficult to evaluate bait consumption and sometimes also making the bait available to nontarget animals; cereals with a particle size close to flour or dust are less palatable, possibly because the rodents prefer to hold the food in their front feet. In a study of the food preferences of the Norway rat, Barnett found that whole meal is preferred to wheat grains which are again preferred to flour.[1] Wet cereals are consumed in much larger quantities than dry food but they must obviously be used immediately after mixing. The more fresh the cereals are the more palatable, and if cut or crushed (especially maize) they rapidly lose attraction even in ordinary indoor conditions.[23] If the maize germ is removed the attraction keeps for a longer time. Other studies indicate that ground barley is highly palatable and keeps better than ground wheat.[17] In studies in the U.S. millet *(Panicum miliaceum)* and sweet rice *(Oryza sativa)* were the cereals preferred by Norway rats when compared with maize and rolled oats.[4] In Japan, Ikeda et al. found that apart from cereals, sweet potatoes were highly preferred by all rats independent of previous experience.[13] For the control of herbivorous field rodents (especially Microtines) a perishable bait may be useful. Not only because it is attractive to the rodents (sliced apple or carrot), but also because it, after a very short period, is rendered unattractive so that other animals are not endangered. Some quick-acting poisons and the second-generation anticoagulants may be used as the active ingredients in such perishable baits that only need to be ingested once by a rodent.

2. *The bait should, as far as is possible, not be attractive to nontarget animals.* Baits such as meat and fish should, as a rule, be avoided as they represent a serious risk to predators, and in warm climates they decay rapidly.
3. *The bait should be easily obtained.* Locally available cereals should always be the bait types of choice for the control of commensal rodents, especially in developing countries where the import of large quantities of prepared baits or plain bait base for rat control would be unwise from an economic point of view.
4. *The bait should be easily mixed with the toxicant.* Crushed cereals are usually easy to mix with the toxicant, especially when it is a powder concentrate. For oily or water concentrates whole grains may be a better bait type, but attention must be paid to the fact that whole grain with husks and hulls are not suitable as most rodents remove and do not consume this outer layer where most of the toxicant would be concentrated.

III. ADDITIVES

All experts agree that there is no alternative to a high-quality bait and that no additives can make a bad bait attractive. To counteract the taste of the toxicant a few possibilities, however, exist:

1. *Sugar,* at a concentration up to 5%, may increase the palatability of a cereal bait 2 to 3 times and maybe even more for liquid baits.[12] A disadvantage, however, is that liquids containing sugar are more easily attacked by bacteria and fungi, and insects, especially ants, are often attracted in large numbers. This can be avoided by using saccharin instead (0.1%).
2. *Vegetable oil* at concentrations of 3 to 10% may initially improve palatability of cereal baits, but if it turns rancid it is worse than nothing. To avoid this, mineral oil (pure medicinal-grade) can be used instead.[11] Oil is often added as a sticker to make the poison adhere to the bait particles, especially if whole grains are used. Cod liver oil, in contrast, seems to be a deterrent.[3]
3. *Water* may be used to increase bait palatability. A damp bait is preferred by many rodents, and wheat soaked in water overnight has been found to be an excellent bait base in rat control. Fresh bait has to be prepared each day and the method is primarily recommended when using quick-acting poisons like zinc phosphide or 1080.
4. Sometimes *pheromones* have been added to baits in order to attract rodents, but even if they might increase interest in the bait site they usually do not improve intake and the persistance of pheromones is generally too short to have a practical value.

Most other additives obviously *reduce* palatability but are still used for various reasons:

1. *Dyes* are used in dry as well as liquid baits as a warning agent to minimize the poisoning risk to humans and birds. Rodents are color-blind but they can still detect the dyes because of their taste or odor, and it is therefore always recommended to use as little as possible. The most common dyes are blue, red, green, and black, such as Chlorazol Sky Blue (0.05%), Alkali Fast Green, Aqua Green, Monastral Green, and especially for liquids, Methylene Blue (0.05%) and Nigrosene Black.[6,11,19]
2. In order to *inhibit the growth of mold* in baits fungistatic agents such as paranitrophenol (0.25%) or dehydroacetic acid (0.1%) may be added for use in sewers and other damp places, but palatability is often negatively influenced.[15]
3. It may also be emphasized that the *toxicant* itself is an additive reducing palatability, and for that reason alone the concentration must always be kept as low as possible.

Naturally, this is wise for other reasons too, i.e., to minimize the risk of accidental poisoning and to reduce costs.

4. Insects may often attack cereal baits and this usually reduces the attractiveness to rodents, but the use of *insecticides* to cope with the problem is no solution as they often lower palatability more than insect contamination.

IV. SPECIAL FORMULATIONS

The most common bait type for commensal rodents is the dry cereal bait, but in certain situations other types can be recommended.

A. Liquid or Water Baits

Liquid or water baits are very important tools, seldom used in many countries. They are used preferably where food is abundant, as in bakeries, food stores, grain elevators, etc., or in dry/arid climates where water may be difficult to obtain for the rodents. In spite of house mice being able to survive long periods without drinking, they drink readily when water is available, and Norway rats need a considerable amount of water each day. Sometimes liquid baits may even be recommended in humid conditions, where dry baits are quickly rendered unpalatable due to molding. Liquid baits are usually water solutions of the sodium salt of anticoagulants, but 1080 and fluoracetamide are often used, and recently calciferol liquid baits have been developed.

Advantages of liquid baits	**Drawbacks of liquid baits**
Easy to use	Evaporates and thus increases concentration and reduces palatability
Cannot be hoarded by the rodents	Attractive to many other animals
Attractive to most rodent species	Cannot be used in very cold environments
Can compete in a food surplus situation	Must be inspected often
	Spillage may easily occur

As molding is often seen, especially when sugar is added, the containers must be frequently and thoroughly cleaned.

B. Wax Blocks or Paraffin Blocks

Wax blocks or paraffin blocks are special formulations for use in damp conditions, such as sewers or humid tropical environments. They are made by mixing melted paraffin with the toxicant, some vegetable oil, and a suitable cereal. They may be quite small (less than 20 g) or larger (300 g) depending on the toxicant and the field of application. Small blocks containing a potent second-generation anticoagulant are used to control rats in oil- or coconut palms. They may be nailed or fixed by a cord to the stem, and the same method is used in the U.S. and elsewhere for the control of roof rats climbing poles, rafters, and fences. Small blocks may also be used for field rodent control where other bait types may quickly be destroyed by rain. Larger blocks, containing anticoagulants or calciferol are suitable for sewer rat control, as they keep well and do not have to be checked as often as normal baits.

C. Pellets

Ready-to-use poison baits are now often formulated as pellets, attractive to a number of different rodent species and usually less readily consumed by granivorous birds as normal cereal baits. They are always quite small to avoid releasing the hoarding behavior of rodents. Microtines, difficult to attract to ordinary baits, may be attracted to such pellets.[16] Paraffinized pellets are quite weatherproof without losing very much in palatability. Commercially available pellets are, however, usually considerably more expensive than cereal baits and prebaiting is naturally out of the question as nonpoisonous pellets are rarely available.

V. THE MIXING PROCESS

In many situations a freshly prepared cereal bait is the number one solution to a commensal rodent problem from an economic as well as an efficacy point of view. Small quantities (less than 2 kg) of poison bait can easily be mixed in a plastic bag or a small bucket, but larger amounts either have to be thoroughly mixed by a shovel on a concrete floor or in a cement-mixer turned by hand and supplied with a lid to keep the dust inside. The addition of a coloring dye facilitates the process by indicating when the toxicant or master-mix is evenly distributed in the cereal.

VI. SELECTION OF BAIT TYPE

When dealing with an unfamiliar rodent species or when having a choice between several local bait types, the most attractive one to the species in question can be selected by choice tests in the laboratory or in the field. In the laboratory a simple test could involve the following procedure:

1. Trap rats from the area in question and bring them to the laboratory.
2. Give them 2 weeks to acclimate to lab conditions on standard food.
3. Compare the candidate bait(s) to a standard well-known bait by placing them symmetrically in food containers in the front of the cage, alternating the position of the two containers from cage to cage.
4. Weigh the consumed amount once a day and change the position of the trays.
5. Run the test for at least 4 days.

In the field, weighed amounts of the bait candidates are placed in shallow trays or in bait boxes on rodent infested sites. In a bait box usually no more than two bait types can be compared at a time. The amount consumed is measured every day and the position of the trays is interchanged. Preferably, the procedure is repeated on ten sites.

VII. BAIT DISTRIBUTION AND BAITING TECHNIQUE

Bait distribution and baiting strategy is often basically different when dealing with field rodents or commensal rodents. The number of nontarget animals at risk may be smaller where commensal rodents are living, but these domestic animals may, on the other hand, be fairly vulnerable. Various methods to reduce the poisoning risk may be applicable to the urban or village situation, but rarely when dealing with rodents in vast areas of grassland or other field situations. Bad odor from rodent carcasses may be a serious problem in premises, but in the field secondary poisoning hazards are, in contrast, of primary importance.

A. Prebaiting

When using quick-acting poisons against commensal rodents, especially rats, it is of paramount importance that a lethal dose is ingested in one intake and, as the new object reaction and general suspicious behavior of rats to unfamiliar odors and taste is a serious obstacle to this, the technique called "prebaiting" has been introduced.[5] The idea is that by presenting the nonpoisonous bait base to the rodents for a period of 4 to 8 days they are accustomed to the site, the bait box or tray, and to the new type of food, so that they are induced to consume more when the placebo is replaced by the poison bait. The prebaiting period varies according to the species and situation, but as a general rule it should continue until the intake is roughly the same on 2 days running. Such a technique is time and labor consuming, but indispensable if a satisfactory result with compounds like zinc phosphide or 1080 is to be obtained against rat species. When prebaiting, an idea of the size of the rodent population is simultaneously obtained, so the amount of poison bait can be adjusted to the needs.

Commercially prepared bait, such as pellets impregnated with an acute toxicant, cannot be used as an identical nonpoisonous prebait will not be available.

When using the slow-acting anticoagulants or calciferol, not producing poisoning symptoms within 2 to 3 days, the prebaiting technique is usually not necessary, as these compounds act as their own prebait.

B. Saturation and Pulsed Baiting Techniques

A common baiting technique when using the less potent first-generation anticoagulants, such as warfarin, has been to make a surplus of the bait available to the rodents during the entire control period. Bait stations are visited and the bait replenished every 2 to 3 days until no more is taken. By this method control experts in the U.K. have found that, on an average, 3 weeks are needed for a complete control of *Rattus norvegicus* on farms.[21] As the same compound can give a total kill in a few days in the laboratory, and as the more potent second-generation anticoagulants, producing mortality after just 1 or 2 days feeding, cannot improve the results in the field, it has been suggested that marginal rats or subordinate family groups may not get access to the bait points until the established, dominant ones have been killed. As the amount of second-generation anticoagulant consumed by a certain rat after the second day is actually wasted, it has been proposed to introduce another technique instead: pulsed baiting.[8] By this technique small amounts of a second-generation anticoagulant are laid down at 4 to 7 day intervals in order to remove the established individuals that may keep the others away from the baiting points. The method is designed to optimize the probability of individual rodents finding the bait and consuming a lethal dose, while minimizing the possibility of excessive bait consumption; and, savings of 75% bait and 50% labor have been reported.[9]

C. Permanent Baiting or Maintenance Baiting

When a particular site (block, district, town) has been cleared of rodents it may, if vulnerable, be protected by permanent baiting: at strategic points weatherproof, durable bait boxes are placed and supplied with a surplus of an anticoagulant bait formulated to keep for at least 1 month. Paraffin-blocks or cereal baits in closed plastic bags are useful for this purpose. In certain countries almost "rat-free" areas or towns have been established by this method.[7,24]

D. Placement of Bait

Commensal rodents should be controlled whenever encountered, and baits placed where traces or fresh droppings are observed. The closer the poison bait can be placed to the rodent burrows or nests the greater the probability that the rodents will feed from it before they reach their normal food source. This may be difficult as far as house mice are concerned, as they often live very close to or in the middle of the food source (grain sacks, etc.). It may also be difficult to place the poison bait without the risk of contaminating human or animal food, and it should always be considered to use traps in such situations.

The larger the number of bait points the better and quicker the results. When dealing with mouse infestations the number of points should be significantly higher than with rat infestations and placed no more than 3 to 5 m apart. If possible, baits should be laid at floor level to avoid spillage on other food items, but when dealing with the acrobatic climber *Rattus rattus* this is not always feasible.

When baiting field rodents a cereal bait or small pellets are either placed at bait stations or, more frequently, broadcast by hand or distributed by aircraft. Myllymäki has reported that success is very much dependent on the bait density: with blanket spread (0.5% crimidin pellets) he achieved 91.5% reduction in the numbers of *Microtus*, with bait stations 5 m apart the reduction was 81.0%, 7 m apart 57.2%, and 14 m apart only 52.5%.[20]

The use of aircraft for baiting field rodents is particularly common in the U.S. (against ground squirrels, *Citellus*). An area of up to 2000 acres can be treated in 1 hr with good results, but it has to be done during the right season when only a fraction of the surface needs to be treated.[18] To reduce poisoning hazards, baits may be laid down in the rodent burrows or holes, but this is a very time-consuming practice. Underground rodents, like pocket gophers, in large areas of uniform grassland, may be baited by artificial burrow-builders.[25]

E. Bait Boxes

In an urban environment some protection of the poison bait is indispensable, unless placed inside rooms with no access for children and pets — here open trays may be used. Bait-containers or bait boxes serve several aims. Primarily they reduce poisoning hazards by giving access only to animals of the size of a rat (or smaller); but furthermore, they make the rodent feel more comfortable and protected thus inducing it to consume larger quantities and minimizing transporting behavior. Out-of-doors weatherproof bait-boxes protect even loose cereal baits for considerable periods of time.

A bait box is simply a container with a lid and a hole in each end, large enough for a rat to enter (6 to 7 cm). Wooden boxes are generally better accepted by rodents than metal ones. The size is not important, but 50 × 30 × 20 cm has been found suitable. A self-feeder hopper inside for holding the bait is convenient and reduces the risk that bait is dug out of the box. Instead of boxes, pipes (pitch fiber, bamboo, plastic) can be used as bait containers with a recommended length of about 60 cm and a diameter of no more than 10 cm. Existing materials in the environment may also be used for the protection of bait points. A plain bait may be used instead of a poison bait if the floor of the box is powdered by an anticoagulant dust. Sometimes a combination of dry baits and liquid baits can improve the effect, and if necessary, flea control can be achieved by dusting the box floor with an insecticide.

REFERENCES

1. **Barnett, S. A.,** Behaviour components in the feeding of wild and laboratory rats, *Behaviour,* 1, 24, 1956.
2. **Barnett, S. A. and Spencer, M. M.,** Feeding, social behaviour and interspecific competition in wild rats, *Behaviour,* 3, 229, 1953.
3. **Barnett, S. A. and Spencer, M. M.,** Responses of wild rats to offensive smells and tastes, *Br. J. Anim. Behav.,* 1(1), 1, 1953.
4. **Brooks, J. E. and Bowerman, A. M.,** Preferences of wild Norway rats for grains, seeds and legumes, *Pest Control,* 41(8), 13, 1973.
5. **Chitty, D.,** The study of the brown rat and its control by poison, in *Control of Rats and Mice,* Vol. I, Chitty, D., Ed., Clarendon Press, Oxford, 1953, 160.
6. **Drummond, D. C.,** Biology and control of domestic rodents, in *Vector Control in International Health,* World Health Organization, Geneva, 1972, 46.
7. **Drummond, D. C.,** Urban rat control: further experimental studies at Folkstone, *Environ. Health,* 85(12), 365, 1977.

8. **Dubock, A. C.,** Alternative strategies for safety and efficacy of rodenticides, Proc. 5th Br. Pest Control Conf., Stratford-upon-Avon, England, September 26 to 29, 1979, 15.
9. **Dubock, A. C.,** Pulsed baiting — a new technique for high potency, slow acting rodenticides, *Proc. Conf. Org. Pract. Vertebrate Pest Control 1982,* ICI Plant Protection Division, Surrey, England, 1984, 105.
10. **Howard, W. E. and Marsh, R. E.,** Rat control manual, *Pest Control,* 42(8), 24, 1974.
11. **Howard, W. E. and Marsh, R. E.,** The rat: its biology and control, Leaflet 2986, Division of Agricultural Science, University of California, Davis, Calif., 1976, 23.
12. **Howard, W. E., Marsh, R. E., and Palmateer, S. D.,** Rat acceptance of different sugar concentrates in water baits, *Int. Pest Control,* 11, 17, 1972.
13. **Ikeda, Y., Tabaru, Y., Yuyama, Y., and Naganuma, K.,** Studies on the food habits of rats. III. Feeding preferences of wild Norway rats in various habitats, *Botyu-Kagaku,* 41, 75, 1976.
14. **Krisnakumari, M. K.,** Effects of early food experiences on later food preferences in adult rats, *Pest Control,* 41(11), 22, 1973.
15. **Larthe, Y.,** The preservation of baits with special reference to the control of rats in sewers, *The Sanitarian,* 65, 276, 1957.
16. **Lund, M. and Lodal, J.,** Control of mice and voles, *Dan. Pest Infest. Lab. Annu. Rep. 1983,* 70, 1984.
17. **Lund, M. and Lodal, J.,** Palatability testing, *Dan. Pest Infest. Lab. Annu. Rep. 1984,* 71, 1985.
18. **Marsh, R. E.,** An aerial method of dispensing ground squirrel bait, *J. Range Manage.,* 21(6), 380, 1968.
19. **Marsh, R. E.,** Water, or liquid baits, *Pest Manage.,* 2(8), 13, 1983.
20. **Myllymäki, A.,** Efficacy of a number of toxic baits and baiting against the voles *Microtus agrestis* and *Arvicola terrestris,* Proc. 11th Vertebrate Pest Conf., Sacramento, Calif., March 6 to 8, 1984, 38.
21. **Rennison, B. D. and Hadler, M. R.,** Field trials of difenacoum against warfarin-resistant infestations of *Rattus norvegicus, J. Hyg. Cambridge,* 74, 449, 1975.
22. **Smythe, W. R.,** Criteria for rodent bait selection, Proc. 7th Vertebrate Pest Conf., Monterey, Calif., March 9 to 11, 1976, 211.
23. **Telle, H-J.,** Zur Frage der Köderbereitung bei Rattenbekämpfungsaktionen, *Prakt. Schädl. Bekämpf.,* Dec., 2, 1964.
24. **Telle, H-J.,** Die Problematik der Beibehaltung eines rattenfreien Zustandes, *Prakt. Schädl. Bekämpf.,* 18(5), 72, 1966.
25. **Ward, A. L. and Hansen, R. M.,** The burrow-builder and its use for control of pocket gophers, Special Science Report Wildlife No. 47, U.S. Department of Interior, Fish and Wildlife Service, Washington, D.C., 1960, 7.

Chapter 15

LARGE-SCALE EVALUATION OF RODENT CONTROL TECHNOLOGIES

C. G. J. Richards

TABLE OF CONTENTS

I. INTRODUCTION

A wide range of methods has been used to evaluate rodent control techniques in the field. In this chapter, an attempt is made to present the approaches which are most commonly used, to compare different methods, and to illustrate how they can be used in particular situations. Much of the text consists of examples, which have been chosen to highlight particular strengths of each approach. Attention is concentrated on large-scale testing and on methods suitable for the continual evaluation of management programs.

Large-scale evaluation is important because rodent pest management itself is usually most effective when implemented on a large scale. Results generated on a small scale are not directly applicable to the large scale, because of both immigration from untreated areas and the increasing importance of people management with increasing scale. It is therefore vital that the same attention is given to evaluating large-scale programs as to small-scale trials.

The literature is full of good reports on the success of small-scale trials. Data on large-scale operations are much more scarce. This is an area where further studies are important, for all types of control techniques. The following account is based almost entirely on chemical and traditional nonchemical control techniques such as trapping. Although the same evaluation methods could be used to evaluate biological control techniques, there are no examples of such studies in the literature. If biological control is to become more important in the future, large-scale evaluation is essential.

Evaluation on a large scale might include appraisal of people management techniques, as well as of the final effectiveness of the program in managing rodent damage. This, however, involves sociological approaches beyond the skill of this author. The following, therefore, concentrates only on evaluating the end product, the rodent control itself.

II. PRINCIPLES OF FIELD EVALUATION

Before considering particular methods in detail, it is useful to set out some guiding principles for the evaluation of rodent control. It is necessary to have the objectives of this type of evaluation clearly in mind, especially because much of the methodology used has been derived from other disciplines, such as animal ecology or agronomy. These disciplines, particularly animal ecology, have different objectives and constraints to rodent pest management.

The objective of all rodent control is to reduce rodent damage below the economic threshold level. Simply killing rodents is not important. Killing rodents does not control rodent populations. For instance, if some rodents are left behind they may breed or there may be rapid immigration from other areas. Nor is reducing the rodent's population size necessarily relevant. Not all rodents cause damage. To evaluate rodent control, it is necessary to estimate the reduction in rodent damage achieved. Moreover, it is necessary to determine the ratio between the costs of achieving that level of control and the resulting benefit in reduced damage.

In the case of crops, it is often easier to estimate rodent damage than rodent populations. Since it is not only easier but also closer to the final objective of the evaluation, damage assessment is therefore often the preferable approach in agricultural rodent control. Damage is, however, not so easy to measure in commensal situations. In such cases, some measure of rodent activity may be directly related to damage.

Whatever evaluation method is used, the first criterion must be that it should be independent of the control technique itself. For instance, intensive snap-trapping may itself reduce rodent numbers; repeated trapping on the same area is likely to show a reduction in the number of rodents trapped with time, even in the absence of control.

One of the major problems in rodent control is that damage tends to be heterogeneous — it varies widely between places and in the same place between one time and another. A

Table 1
PRINCIPLE OBJECTIVES AND CONSTRAINTS FOR THE EVALUATION OF RODENT CONTROL TECHNIQUES

- Measure cost-effective damage reduction
- Independent control and evaluation
- Simple and practicable methods
- Replication vs. plot size
- Sufficient precision
- Maximum sensitivity in range 75 to 100% damage reduction
- Use untreated reference areas
- Apply to chemical and nonchemical techniques

useful rodent control technique should reliably give a substantial reduction in damage, in spite of this heterogeneity. The consequence for evaluation is that an accurate assessment of one small trial is much less useful than an approximate assessment of many replicates. The information that a technique gave 80 to 90% damage reduction in seven replicates but only 10 to 30% in three others is more useful than that it gave 83.7% damage reduction in one replicate. This gives quite a different priority to accurate evaluation than that given in the less applied disciplines such as ecology, where accuracy is paramount. It is particularly important to remember this when adapting ecological methods to rodent control. Although some level of accuracy is obviously necessary, replication is more important than accuracy.

There is often a conflict between replication and plot size. Plots must always be sufficiently large to minimize the impact of immigration, and these large plots inevitably reduce the number of replicates which are feasible. In many cases very large plots can be a substitute for replication, provided they include a representative range of situations. For instance, proper evaluation of a city-wide rodent control program may be more useful than a series of small replicated trials. It should be understood, however, that subsamples within the large plots are not, statistically speaking, replicates — and so experimental variation cannot be measured statistically within large plots.

Changes with time are a problem with any biological testing. It is always useful, especially in agriculture, to extend evaluation to untreated reference areas as "control" plots.

The inevitable conclusion of the need for replication and large-scale testing is that effective evaluation methods must be simple to use. In any trial of a control method, the evaluation must not require so much effort that it detracts from the treatment itself. The method must, therefore, be practicable, given the particular situation and the time and resources available.

Table 1 summarizes a set of principles for the evaluation of rodent control techniques. This list is not comprehensive and the items are not necessarily in order of importance. They may, however, serve as a guide as we go on to consider a number of approaches in more detail.

III. ECONOMIC EVALUATION

A. General

The ultimate criterion of success in pest management is a favorable ratio of costs to benefits. This requires that the benefits be measured in terms of the reduction in some economically quantifiable type of damage and that the control effort be costed. The scarcity of this type of evaluation would suggest that it is difficult to accomplish. While this may be true where rodent control is primarily aimed at disease elimination, it is frequently not so in agricultural rodent control. In an ever more competitive world, economic evaluation is becoming ever more important. National priorities are set mainly by economic factors and without figures, however imprecise, rodent control will always be a low priority.

Table 2
RICE (PADI) YIELD AND THE PERCENTAGE OF RAT-CUT TILLERS IN EXCLOSURES AND "AT RISK" QUADRATS IN TREATED AND REFERENCE AREAS OF THE CENTRAL JAVA PROJECT[24]

		"At risk"	
Yield (g/hill)	**Exclosures**	**Treated**	**Ref.**
Mean	24.70	24.44	20.51
S.D.	5.90	6.70	6.41

Meaningful economic evaluation need not aim to be very precise but it should include all elements of costs and benefits. Benefits are estimated from the reduction in damage levels; this will be covered in more detail in the following section. Estimating costs is relatively straight forward. Costs involved in a large-scale program will include not only rodenticide bait, bait stations, traps, etc. but also items such as labor, training, public information, and transport. In large urban programs, some form of cost-accounting will often be imposed and administered by city treasury departments and this can be a useful source of information for the scientist.[2]

Three examples, one from a cereal crop, one from a tree crop, and one in stored products, have been selected to illustrate approaches to economic evaluation.

B. Rice Rat Control in Bukateja, Indonesia

A large-scale project was carried out in Indonesia to determine the cost-efficacy of the Government's recommendations for rat control.[6,24] Integrated control using brodifacoum baits and burrow fumigation was implemented over 1141 ha of rice fields. Rat damage and rice yield at harvest were estimated using an exclosure method. At the time of transplanting 25 2-m^2 plots were set up at random across the treated area. One half of each of these plots was protected from rodents by erecting a fence of polythene film, 30 cm above ground and dug 30 cm into the ground, which prevented access by rats. The other half of each plot was left "at risk" to rats.

At harvest, the rice in the exclosures and the "at risk" quadrats was collected and weighed; the former represented the yield in the absence of rodent control, the latter the average yield across the project area. Similar plots were set up on a reference area close to the treated area to allow comparison of the effectiveness of rat control within the project, with areas in which farmers continued to use traditional rodent control methods.

The results (Table 2) showed that there was no significant difference in rice yield inside the exclosures (24.7 g/hill) and in the "at risk" quadrats in the treated area (24.4 g/hill; $t = 0.18$, d.f. $= 72$, $p > 0.05$). Rice yield in the reference area was, however, 17% lower (20.5 g/hill; $t = 2.97$, d.f. $= 72$, $p < 0.01$). Average yields on the treated area, estimated by extension staff sampling randomly selected plots, were 3.76 t/ha. Farmers on the treated area harvested 17% less or 0.64 t/ha; using a world value of rice at $200/t, the financial gain on the treated area was $132/ha. The costs of the control program were recorded and expressed on an area basis (Table 3). The cost:benefit ratio of the project can therefore be estimated at 1:14 (Table 3) showing that the Government recommendations are highly cost-effective.

C. Control of Rats Damaging Coconuts in Fiji

Williams[25,26] carried out a detailed study of rat damage to coconuts while investigating

Table 3
COSTS AND BENEFITS OF THE RICE RAT MANAGEMENT SCHEME USED IN THE CENTRAL JAVA PROJECT IN 1982[24]

Item	Cost/ha ($)	Benefit/ha ($)
Klerat bait	3.42	
Bait stations	1.49	
Labor (1 day)	1.49	
Farmer training	0.74	
Surveillance	0.59	
Fumigation	1.67	
Total:	9.40	132

Cost:benefit ratio = 1:14

Table 4
COSTS AND BENEFITS OF RAT-CONTROL IN COCONUTS IN FIJI

A. **Costs**

Bait = 3 kg/ha × 3 applications/yr × $0.50 warfarin bait/kg
= $4.50
Labor = $3.72
Total: $8.22 ha/yr

B. **Cost-benefits**

Rat damage	$54/tonne[a]			$154/tonne[a]		
	Value of loss/ha	Cost of control	Cost:benefit ratio	Value of loss/ha	Cost of control	Cost:benefit ratio
(Loss nuts per palm per yr)						
4	$2.95	$8.22	Negative	$18.85	$8.22	1:2
10	$7.40	$8.22	Negative	$47.40	$8.22	1:6

[a] Value of copra.

Adapted from Williams, J. M., *PANS*, 21(1), 19, 1975.

the economics of control strategies. Initial studies showed that rat damage in Fiji was concentrated on immature nuts, a porportion of which would not in any case develop to maturity. Counts of rat-damaged coconuts compared with counts of harvestable nuts therefore over-estimated damage levels. A simple measure of rat damage could, however, be derived by counting the harvestable nuts every 2 weeks on trees with and without signs of rat damage. Williams[26] then compared the efficacy of banding each tree with that of poison baiting with warfarin. By comparing the yield of harvestable nuts on treated and untreated plots, it was revealed that banding was ineffective. Poison baiting, however, virtually eliminated rat damage for up to 3 months after treatment. Turning to cost-efficacy, the cost of baiting could be estimated by simply adding the costs of bait and labor. The benefit could be calculated from the yield increase and the value of the copra, but this was complicated by the wide variation in the value of copra. When the copra was of low value, the yield increase was less than the costs of control (Table 4). When copra values were high, control was cost-effective, particularly where rat damage levels were also high. The table thus provides a rough guide for the value of copra and the level of rat damage at which control is economically justifiable.

D. Stored Food Protection in Cuba

A project which aimed to measure the cost-effectiveness of reducing rodent damage to stored food in a number of selected warehouses in three provinces of Cuba[16] covered four phases:

1. Phase I. A pretreatment phase, in which rodent damage was assessed monthly for at least 3 months.
2. Phase II. A rodent control phase of 1 to 2 months, starting with chemical control and following up with improved storage practices and sanitation to prevent re-infestation.
3. Phase III. An interim maintenance phase of 3 months, in which rodent control was continued by permanent baiting; damage monitoring was continued but included products which had been attacked before poisoning had begun.
4. Phase IV. A long-term maintenance phase in which damage was monitored after the full effects of the poisoning had worked through.

Damage to food was recorded by identifying any packages which had been damaged by rodents and measuring the loss by weight. The amount lost was substantially more than the amount consumed by rodents because the entire contents of any damaged pack was usually condemned as unfit for human consumption and given free to pig farmers. For each warehouse, monthly returns were made on the amount of food leaving the store and the amount lost due to rodent damage. The value of the loss in dollars was calculated for each commodity. An example of the data recorded is shown in Table 5, for a specific warehouse during the pretreatment phase.

The costs of rodent control were measured during Phases II, III, and IV by recording the quantity of poison bait used, the man-hours of labor needed and the transport costs associated with control (Table 6). The cost of reducing the rodent population in Phase II was about five times as high as the cost of the maintenance phase. In calculating a cost:benefit ratio, Hernandez and Drummond[16] argued that the cost of Phase II was not relevant since it would eventually form a very small portion of the total cost of rodent control because the maintenance phase was intended to continue indefinitely. The cost was therefore taken to be the monthly cost of the maintenance phase. The benefit was correspondingly taken to be the difference between the level of rodent damage before treatment and that in the long-term maintenance phase. The resulting cost:benefit ratio (Table 6) showed that rodent control was highly cost-effective, even when the proportion of food damaged by rodents was very small.

IV. DAMAGE REDUCTION

A. General

Where it is not possible to do a full economic evaluation, measurements of rodent damage to crops are often used to evaluate control methods. They can be simpler to use than many measures of rodent populations. There are a number of such studies in monocultures such as rice[7,8] and sugar cane.[14] Methodology becomes more complex in polyculture systems, unless one crop clearly predominates in either area or susceptibility to rodent attack. Where several rodent-damaged crops are of equal importance in the treated area, damage reduction may be simply too complex a tool for evaluation.

Damage assessment for stored products is laborious[13] and probably not worthwhile unless a full economic evaluation is possible. Physical damage by urban rodents is also difficult

Table 5
EXAMPLE OF A WAREHOUSE DAMAGE REPORT SUMMARY FROM THE PRETREATMENT PHASE OF A STORED FOOD PROTECTION PROJECT IN CUBA16

Commodity	1 Mean time in store (months)	2 Total no. units moved from store	3 Total weight units moved from store (kg)	4 No. of units damaged	5 Weight of food eaten by rodents (kg)	6 Weight of food contaminated by rodents (kg)	7 Total loss (kg) (5 + 6)	8 Value ($/kg)	9 Total loss ($) (7 × 8)
Rice	3.5	5,242	262,100	55	96	2,654	2,750	0.20	550.00
Beans	1.0	1,328	66,400	0	—	—	0	0.31	0
Pasta	1.5	522	6,264	1	12	0	12	0.38	4.55
Wheat flour	2.5	542	34,146	9	15	552	567	0.18	101.65
Sugar	4.0	930	106,950	10	28	1,122	1,150	0.22	253.79
Boiled sweets	1.2	2,320	2,320	15	15	0	15	0.34	5.17
Total	—	15,998	615,353	109	220	4,328	4,528	—	915.16

Table 6
EXAMPLE OF A COST-BENEFIT ANALYSIS FROM A STORED FOOD PROTECTION PROJECT IN CUBA[16]

Phase	Mean weight of food leaving store/month (tonne)	Weight loss due to rodent damage (%)	Value of loss ($/month)	Cost of rodent control ($/month)	Cost:benefit ratio
I	205	0.73	308.27	0	
II	263	0.37	200.69	37.93	
III	194	0.15	60.00	10.69	
IV	229	0.00	0	10.69	1:29

Table 7
YIELD (g/m^2) OF FOUR VARIETIES OF DEEP-WATER RICE IN TWO AREAS OF TANGOUL DISTRICT, BANGLADESH[5]

Variety	No. fields sampled	Mean yield (g/m^2)		Yield increase treated/untreated (%)
		Untreated	Treated	
Injal degha	14	120.5	196.9	+63.4
Chamhara	16	150.0	336.1	+124.1
Channa baoalia	2	103.6	220.5	+112.8
Baron baoalia	2	133.1	188.8	+41.8

to assess. Dubock[11] describes experiments by Brooks, who approached the problem simply by listing the replacement cost of items damaged by rodents. The primary motivation for urban rodent control is usually public health, but there are few programs during which health parameters have been measured. A rare example is described by Abu Sabra,[1] where the incidence of murine typhus in hospitals was monitored during the first years of a city-wide rodent control program in Kuwait. Such methods, of course, are only of use in association with large-scale projects. Moreover, it is not possible to express public health benefits in terms of economics.

The principles of damage assessment as a means of evaluating rodent control can be illustrated by two further examples.

B. Control of Rats in Deep-Water Rice, Bangladesh

In the deep-water rice areas characteristic of the river deltas of Southeast Asia, rodents move extensively between fields and villages as the flood waters rise and fall. In such areas, rodents may cause extensive damage not only to rice crops but also to stored products and public health, as the floods compress widespread field populations into the higher ground on which the houses are built.[4,24] A series of studies was carried out in such an area of Bangladesh.[5] The approach in this situation was to attempt control over relatively large areas to prevent the migration of rats between villages and fields.

Two areas of 1 km^2 were selected to test this approach. One was treated with baits containing brodifacoum, the other was an untreated reference. Rat activity, measured by tracking tiles and burrow counts before and after baiting, fell by 85 to 90% after the first three bait applications. In measuring rice yield, it was necessary to take into account the wide variation in yield between varieties. Yield was therefore estimated by comparing samples of four different rice varieties collected from the treated and untreated areas. A sample of fields of each variety was selected at random, taking the same number of fields in both the treated and the untreated areas (Table 7). In each field, all of the rice panicles

Table 8
RAT DAMAGE TO RICE ON AREAS FOLLOWING THREE DIFFERENT RODENTICIDE TREATMENTS

Treatment	Area (ha)	No. fields sampled	Mean % tillers cut by rats
Brodifacoum bait			
Supervised — Village 1	518	20	0.11
Supervised — Village 2		20	0.41
Normal farmer practice			
Unsupervised		20	3.15[a]

[a] Indicates significantly different (p <0.05).

were clipped inside five randomly placed 1 m^2 quadrats. The panicles were dried, threshed, and weighed. Yield was then estimated as fresh rice by adding 12% as a standard moisture content. Results showed that yield was between 41 and 124% greater in the treated area.

C. Rice Rat Control in Indonesia

A different approach to measuring rodent damage to cereals was proposed by Rennison[22] and has been utilized in several rice rat projects.[8,24] This involves inspecting samples of rice hills and counting tillers which are ripe and those which have been cut by rats. The percentage of tillers cut by rats is a good measure of yield loss.[8]

One such project was set up to evaluate the efficacy of supervised applications by farmers of brodifacoum baits compared with normal farmer practices. Brodifacoum baits were applied around the time of transplanting the rice crop in two different villages, while a third village was a reference area. Damage assessment was carried out approximately 10 weeks later, 2 weeks before harvest. In 20 fields selected at random in each village, 25 rice hills were examined along a diagonal transect; the first and last hills were at each end of this transect, with the remaining 23 spaced at intervals of 5 hills. This transect aimed to take into account the patchy nature of damage within fields. The results, summarized in Table 8, showed that rat damage in the two brodifacoum-treated villages did not differ significantly but both had significantly less damage than the reference area.

V. TRAPPING

There is extensive literature on the estimation of rodent numbers, generated during the course of ecological studies.[9] Enumeration (i.e., the accurate census of a rodent population) is virtually impossible. Most methods require trapping, which is laborious and subject to behavioral variability within populations, leading not to true enumeration but to estimates of population size which can be widely inaccurate. For instance, it was found that a trapping method which had long been accepted as standard for the much studied *Microtus* species, sometimes underestimated population size by at least 50%.[3] Obtaining even an approximate estimate of rodent numbers is time-consuming and not practicable in the context of rodent pest management. Trapping really only provides a measure of rodent activity, which is related to population size.

A. Mark-Release-Recapture (MRR)

MRR techniques have been widely used in ecological studies to estimate population size, but their accuracy varies between species. For instance, *Rattus norvegicus* tend to divide into "trap-happy" and "trap-shy" individuals and MRR cannot be recommended.[12] For *Mus musculus* and for some microtine rodents *(Microtus* spp, *Clethrionomys* spp), however, the method has been considered to be of value.[12] In this case, trapping is carried out for 3 days before treatment and another 3 days after the treatment has had its full effect. Animals caught before treatment are marked, e.g., by fur clipping, before release. The number of animals caught, marked, and released in the first trapping is used as a pretreatment index of population size, the number of marked animals in the second trapping is a posttreatment index. Success is then estimated simply as:

$$\% \text{ mortality} = \frac{100\ (\text{pretreatment index} - \text{posttreatment index})}{\text{pretreatment index}}$$

B. Small Scale Removal Trapping

Removal trapping is frequently used to evaluate control success on a small scale. A wide variety of kill-traps or live-traps are employed. The major drawback of this approach is that it interferes with the control treatment itself. This effect is particularly severe in small-scale evaluations; untreated reference areas are essential here. It is suggested that, where there is a decrease of >30% in numbers caught on the reference area, the experiment should be rejected.[12] In large-scale programs, removal trapping should not affect the size of the treated population as a whole but the positioning of assessment plots may be critical.

Traps may be placed along line transects or on quadrats. The type, number, and spacing of traps will depend on the species involved. For small microtines, the European and Mediterranean Plant Protection Organization (EPPO) recommend one ''Longworth'' or ''Ugglan'' trap every 2 m along two transects 100 m long and 20 to 30 m apart, or alternatively, 3 traps at each station on a 15 × 15 m quadrat. These traps are run for 2 days before and after the treatment.

Estimation of control success should not be based on changes in the treated plot only. The Henderson-Tilton correction is recommended,[15] which takes into account the changes in catch numbers on the untreated plot:

$$\% \text{ control} = 100 \times \left(1 - \frac{t_2 \times u_1}{t_1 \times u_2}\right)$$

where letters indicate the number of animals trapped: t, on treated areas; u, on untreated areas; 1, before treatment; 2, after treatment.

C. Island-Wide Rodent Control Program in Taiwan

An example of the use of removal trapping to evaluate the effectiveness of a large-scale program in Taiwan is given by Ku.[18,19] A trapping method has been used to estimate rodent populations before and after area-wide baiting since 1976.

Between 1977 and 1981, one area was selected in each of 15 major agricultural districts. In each area, two 4-ha plots of mixed crops were selected at random and 400 live-traps set up on a 10 m grid on each plot. A slice of sweet potato was used as bait in all traps. Traps were prebaited (i.e., baited but not set) for 3 days then set for a further 5 days. All rodents were removed at daily inspections and their species, sex, and maturity recorded. The assessment was carried out in January each year and again in March; an island-wide warfarin baiting program took place in February each year.

Representative results are shown in Table 9. These show the wide variation in rodent density between areas. They also give an indication of the variation in the effectiveness of baiting between rodent species; such information can only be derived from trapping.

Table 9
RESULTS OF LIVE-TRAPPING ON SELECTED ASSESSMENT PLOTS BEFORE AND AFTER AN ISLAND-WIDE WARFARIN BAITING PROGRAM IN TAIWAN[18]

	Bandicota nemoriovaga			*Rattus losea*			*Rattus norvegicus*		
Region	Before baiting[a]	After baiting[a]	% Decrease	Before baiting[a]	After baiting[a]	% Decrease	Before baiting[a]	After baiting[a]	% Decrease
Northeast	10	6	40	105	46	56	2	0	100
North	36	17	53	242	96	60	10	8	20
Central	77	36	53	772	278	64	86	4	95
Southwest	180	138	33	435	104	76	168	28	83
Others	71	26	63	409	165	60	44	9	80
Total	374	223	40	1963	689	65	310	49	84

[a] Number of rodents captured.

VI. ACTIVITY ASSESSMENT

A. General

It was pointed out above that trapping is mainly an indication of population activity, broadly related to population size. There are many alternative ways of obtaining such an indication, similarly related to population size and frequently less laborious than trapping. These include such methods as burrow counts, tracking boards, and token baiting, of which there are many variants; a useful review of techniques is provided by Kaukeinen.[17] The essence of an activity assessment is that it should be simple, cheap, and easy to use; for this reason, the exact methods used are often specific to a particular set of conditions.

It is important that the activity being measured is being caused by rodents. For instance, foot-prints on tracking boards or gnawing on token baits may be due to insects, crabs, or other nonrodents. Possible sources of confusion must be identified and avoided by proper training of those who will be carrying out the work. Small changes to the method can be useful, e.g., slices of sweet potato with skin intact show rodent tooth-marks much more distinctly than do peeled slices. It is also important to ensure that the rodent species involved have been identified and this may require some preliminary trapping.

One advantage of a good activity assessment method is that it does not affect either the population or the control treatment. A common problem when using bait as an indicator of activity is making sure that the evaluation is independent of the control treatment. This is particularly difficult where an acute rodenticide such as zinc phosphide is used with pre-baiting. In such cases, any survivors of the treatment are likely to be poison shy; establishing that a bait, used to assess activity after the poison has been laid, will not be avoided by the rats because of association with the poison, is very difficult. For this reason, it is probably better to use a distinctly different bait for activity assessment, e.g., a root vegetable where rodents are being baited with cereals, or to use a method such as tracking in these cases.

B. Urban Rodent Control at Zhuo Xian, China

Evaluation of large-scale urban projects is complicated by the area to be covered and the number of people actively and passively involved. One such project took place in Zhuo Xian, a town of some 50,000 people in Hebei, China.[21,23] The abundance and distribution of rodents before and after a single application of brodifacoum bait was estimated using three independent assessment methods — tracking patches, token baiting, and trapping. Each was used during one night before the treatment and again after treatment; the interval between assessments was 6 days due to labor problems (14 days would normally be preferred for assessing the efficacy of an anticoagulant treatment). Samples of premises (i.e., a house, factory, or office) were selected for use in the assessments, representing the various types of buildings in the city.

For the tracking method, patches of wheat flour (20 $\times$ 20 cm, approximately) were put down in rodent runways and scored as positive (rodent signs visible) or negative (no rodent signs) the following morning. A total of 949 patches were put down.

Token baiting involved the use of unpoisoned wax blocks, each weighing 25 g. This assessment was carried out in 10% of all premises. The availability of labor would usually constrain this to the smallest number of premises which would give a statistically reliable sample; guidelines for determining the appropriate size of sample can be found in standard texts.[10] Twelve blocks were placed in each room or for each 45 m^2 in nonresidential premises. Blocks were scored the following morning as 0 (uneaten), 1 (partially eaten), or 2 (completely eaten by rodents).

To compare these assessment methods with trapping, 405 snap-traps were placed in a sample of premises and set for one night. The results of the assessment methods (Table 10) suggest that all three gave similar estimates of control success. It is not possible to determine

Table 10
SUMMARY RESULTS OF THREE ASSESSMENT METHODS USED TO EVALUATE AN URBAN RODENT CONTROL PROGRAM IN ZHUO XIAN, CHINA[23]

Method	Before treatment	After treatment	Estimated population reduction
1. Tracking patches			
No. patches with rodent signs	538	12	—
No. patches put down	664	510	—
Percentage positive	81.0%	2.3%	97.1%
2. Wax blocks			
Total blocks eaten	6,509	790	—
No. blocks put down	15,948	16,416	—
Percent eaten	40.8%	4.8%	88.1%
3. Trapping			
No. rodents caught	91	7	—
No. traps set	330	313	—
Percent trap success	30.3%	2.2%	92.6%

which of these estimates is closest to the correct result. None of them is precise and, unless it is possible to replicate the trial many times, accuracy to greater than ± 10% should probably not be claimed.

Further information can be derived from the token baiting data. Before poison treatment 91% of premises were infested by rodents, whereas only 10% were infested after treatment. Moreover, more than 50% of premises had total scores of more than 6 and 30% had scores of 12 or more before treatment. After treatment, only 10% of premises had scores of more than 6. This suggested that only one or two rodents were left behind in those houses which had any at all. Moreover, these rodents were probably confined to particular locations and could be cleared relatively easily.

C. Village Rodent Control at Mai Xa, Vietnam

Token baiting was used on a large scale during a recent project in Vietnam.[20,23] The project aimed to evaluate the effectiveness of placing bait (0.005% brodifacoum in 5 g wax blocks) in and around houses and in rice fields; the primary objective was increased rice yield but public health and stored products benefits were also expected. The cooperative of Mai Xa comprises 2393 people living in 663 houses and cultivating 202 ha of rice and 113 ha of mixed vegetable and tree crops. All bait was applied by six trained groups of farmers at rates refined from earlier trials (100 g per house, 0.5 kg/ha of rice, and 2 kg/ha in other ''garden'' crops). Two bait applications were made, with an interval of 10 days, around the time of land preparation.

Evaluation was carried out in areas baited by each of the six groups, using slices of sweet potato secured to the ground with bamboo sticks. In the rice-fields, a 120 m transect was selected for each group. These transects followed the bunds between rice fields and were placed such that they were always at least 20 m from houses and 50 m from the boundary of the treated area. Slices of sweet potato were placed at 5 m intervals along these transects (25 in all). A sample of ten houses was selected at random for each group from lists of householders' names. Houses at the boundary of the treated area were avoided. Five slices of potato were placed inside each house and a further five in the associated garden. For each assessment, the sweet potato was placed one afternoon and inspected the followig morning for signs of gnawing by rats; confusion with signs of crabs, snails, and other nonrodents was minimized by training. Three assessments were carried out, one on the day

Table 11
RESULTS OF A TOKEN BAITING ASSESSMENT METHOD IN A VILLAGE RODENT CONTROL PROGRAM IN VIETNAM[23]

(a) *Mai Xa* (treated area)

	Rice fields		Gardens		Houses		Total		Premises
	No. laid	No. eaten	No. laid	No. eaten	No. laid	No. eaten	No. laid	No. eaten	infested (%)
Pretreatment census (February 12)	210	63	350	60	350	77	910	200	77.1
Midterm census (February 20)	210	26	345	49	345	37	900	112	57.1
Posttreatment census (March 2)	210	0	350	5	350	4	910	9	8.6
% Reduction (February 12 to March 2)	100		91.7		94.8		95.5		88.9

(b) *Loc An* (untreated reference area)

	Rice fields		Gardens		Houses		Total		Premises
	No. laid	No. eaten	No. laid	No. eaten	No. laid	No. eaten	No. laid	No. eaten	infested (%)
Pretreatments census (February 14)	50	7	100	27	100	37	250	71	100
Posttreatment census (March 3)	50	23	100	23	100	19	250	65	70
% Change (February 14 to March 3)	329		−15		−49		−8		−30

Note: The number of pieces of sweet potato eaten by rodents during rodent activity assessments are shown for the treated village, Mai Xa (a) and the untreated reference village, Loc An (b). Each house with its associated garden was regarded as one premises and considered to be infested if any potato pieces were eaten by rodents in either house or garden.

before each of the bait applications and a third 10 days after the second bait application. For reference purposes, similar assessments were carried out at a nearby untreated village (Loc An) at the same time as the first and third assessments at Mai Xa.

This very simple assessment (Table 11) not only showed clearly that the rodenticide treatment was successful (>90% reduction in activity) but also gave valuable information about the distribution of rodent activity. For instance, it showed the very high level of rodent activity around houses before treatment. There was also a shift in rodent activity away from houses towards rice fields in the untreated area, which reflected the increase in food as the rice crop developed; this movement did not take place in the treated area. The same data can be used to construct a graph of overall rodent activity (Figure 1).

The primary objective of this project was to reduce rodent damage to rice. The activity assessment was therefore supplemented by a rice damage assessment 2 weeks before harvest on the treated and untreated areas, using methods similar to those described above (see Section IV.B). Of 800 hills examined in the treated area, only one was damaged by rodents, giving an estimated 0.032% yield loss or tiller damage, compared to 2.89% on the untreated area. This showed that the effect of the rodenticide treatment lasted at least 12 weeks. More useful information would have been derived had activity assessments been carried out at regular intervals (e.g., 4 or 8 weeks) after the treatment, to plot the recovery of the population.

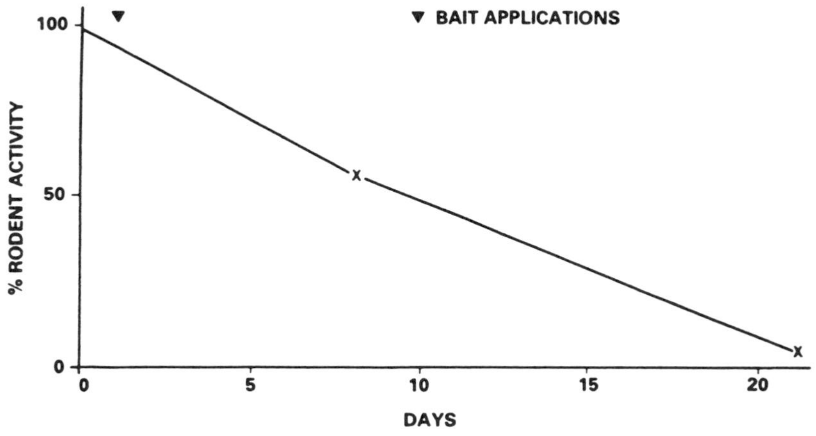

FIGURE 1. The decline in rodent activity following poison bait applications during a village rodent control program in Vietnam, as estimated using a token baiting assessment method.

REFERENCES

1. **Abu Sabra, I.,** Retrospective study of murine typhus in Kuwait, in *Proc. 1st Symp. Recent Advances in Rodent Control,* Helmy Mohammad, A. H., Zaghloul, T. M., and Zakaria, M., Eds., Kuwait, 1983, 149.
2. **Al Sanei, K., Zaghloul, T. M., Salit, A. M., and Omar, M. T.,** Cost-efficacy of country-wide programme in Kuwait, in *Control of Mammal Pests,* Richards, C. G. J. and Ku, T. Y., Eds., Taylor & Francis, London, 1986.
3. **Boonstra, R. and Krebs, C. J.,** Pitfall trapping of *Microtus townsendii, J. Mammal.,* 59, 136, 1978.
4. **Brooks, J. E., Htun, P. T., King, R. E., and Naing, H.,** A field trial to control rats and shrews in Rangoon, Burma, manuscript.
5. **Bruggers, R. L., Ed.,** Vertebrate damage control research in agriculture, Annual report of the Denver Wildlife Research Center, U.S. Fish & Wildlife Service, U.S. Agency for International Development, Denver, Co., 1983.
6. **Buckle, A. P.,** Food crop protection through integrated rodent control, Unpublished report TCP/INS/003, Food and Agriculture Organisation, Rome, 1982, 35.
7. **Buckle, A. P., Rowe, F. P., and Husin, A. R.,** Field trials of warfarin and brodifacoum wax block baits for the control of the rice field rat, *Rattus argentiventor,* in Peninsular Malyasia, *Trop. Pest Manage.,* 30(1), 51, 1984.
8. **Buckle, A. P., Yong, Y. C., and Husin, A. R.,** Damage by rats to rice in South-east Asia with special reference to an integrated management scheme proposed for Peninsular Malaysia, *Acta Zool. Fenn.,* 173, 139, 1985.
9. **Caughley, G.,** *Analysis of Vertebrate Populations,* John Wiley & Sons, New York, 1977, 234.
10. **Cochran, W. G.,** *Sampling Techniques* 2nd ed., John Wiley & Sons, New York, 1963, 413.
11. **Dubock, A. C.,** Pulsed baiting — a new technique for high potency, slow-acting rodenticides, in *Proc. Conf. Organization and Practice of Vertebrate Pest Control,* Dubock, A. C., Ed., ICI Plant Protection Division, Haslemere, England, 1984, 105.
12. **EPPO,** Guidelines for the development and biological evaluation of rodenticides, EPPO Bull., 5(1), 3, 1975.
13. **Greaves, J. H.,** Rodents: Loss determination by population assessment and estimation procedures, in *Post-Harvest Grain Loss Assessment Methods,* Harris, K. L. and Lindblad, C. J., Eds., ICI Plant Protection Division, Haslemere, England, 1976, 109.
14. **Hampson, S. J.,** A review of rodent damage to sugar cane with criteria for the use of rodenticides, in *Proc. Conf. Organisation and Practice of Vertebrate Pest Control,* Dubock, A. C., Ed., ICI Plant Protection Division, Haslemere, England, 1984, 227.
15. **Henderson, C. F. and Tilton, E. W.,** Test with acaricides against the Brown Mite, *J. Econ. Entomol.,* 48, 157, 1955.

16. **Hernandez, A. and Drummond, D. C.,** A study of rodent damage to food in some Cuban warehouses and the cost of preventing it, *J. Stored Prod. Res.*, 20(2), 83, 1984.
17. **Kaukeinen, D. E.,** Activity indices to determine trends in vertebrate pest populations, in *Proc. Conf. Organisation and Practice of Vertebrate Pest Control,* Dubock, A. C., Ed., ICI Plant Protection Division, Haslemere, England, 1984, 73.
18. **Ku, T. Y.,** Distribution and population fluctuations of field rodents and their control in Taiwan, in *Proc. Conf. Organisation and Practice of Vertebrate Pest Control,* Dubock, A. C., Ed., 1984, 487.
19. **Ku, T. Y.,** Population fluctuations of field rodents and their control in Taiwan, in *Control of Mammal Pests,* Richards, C. G. J. and Ku, T. Y., Eds., Taylor & Francis, London, 1986, 335.
20. **Le Truong, Bui Van Ich, Richards, C. G. J., and Taylor, K. D.,** Village-level rodent control in Vietnam, manuscript.
21. **Ma Lianquan, Li Jiuru, Wang Chengxin, Deng Zhi, Luterbacher, J. O., and Dubock, A. C.,** Urban rodent control in Zhuo Xian, Hebei Province, Peoples Republic of China, manuscript.
22. **Rennison, B. D.,** Monitoring rat damage in West Malaysia. A consultant's report and recommendations for the Malaysian/UK rice rat control project. Overseas Development Administration, London, 1980, 29.
23. **Richards, C. G. J.,** Cost-effective approaches to large-scale rodent control programmes, in *Proc. 2nd Symp. Recent Advances in Rodent Control,* Heleny Mohammad, A. H., Zaghloul, T. M., Salit, A. M., and Zakaria, M., Eds., Kuwait, 1986, 291.
24. **Richards, C. G. J. and Buckle, A. P.,** Towards integrated rodent pest management at the village level, in *Control of Mammal Pests,* Richards, C. G. J. and Ku, T. Y., Eds., Taylor & Francis, London, 1986, 293.
25. **Williams, J. M.,** Rat damage to coconuts in Fiji. I. Assessment of Damage, *PANS,* 20(4), 379, 1974.
26. **Williams, J. M.,** Rat damage to coconuts in Fiji. II. Efficiency and economics of damage reduction methods, *PANS,* 21(1), 19, 1975.

Chapter 16

RODENT PEST MANAGEMENT: THE PRINCIPLES

Walter E. Howard

TABLE OF CONTENTS

I. INTRODUCTION

There is no longer any biological, ecological, or economic reason for rodent health and pest problems to occur in villages, cities, and intensively cultivated agricultural areas. With adequate funding and adequate government support, it is now feasible to virtually eliminate small mouse- or rat-like pest rodents and to do it safely, inexpensively, simply, and in a more environmentally desirable manner than is possible with most current control programs.

There are many benefits to having a good understanding of the principles involved in rodent control before initiating control programs. Not only will knowledge of these principles make the control operation more effective and efficient, hence more economical, but it can help minimize any undesirable impact that the destruction of rodents may have upon the biotic community and on the general public.

The social values now placed on all wildlife, including rodents, cannot be ignored. Controlling rodents is no longer the exclusive province of landowners or elected public representatives: now the public not only wants but demands greater participation in environmental decisions. Hence, rodent control, as with all resource management, is difficult because it requires managing humans.

II. DEFINITION OF RODENT CONTROL

A rodent pest is defined as ''a rodent that is either native or introduced, wild or feral, which is currently troublesome locally or over a wide area to one or more people.''[1] At times rodents create health and safety hazards, they eat or contaminate our food, or they damage agricultural crops, private and public property, livestock production, forests and pastures, other resources, and even urban and rural structures. Or they may simply become a general nuisance. When any of these happens, the affected parties classify the rodents involved as pests and want them controlled. Even in the case of rodents, an important principle is that what may be a pest to one person may concurrently be desirable or of neutral value to someone else.[1]

The pestiferousness of a rodent depends upon one's relationship to the animal or species. Some people like to refer to the control of rodents as ''animal or wildlife damage control'' because they say the objective is to reduce or prevent damage and not necessarily to remove the animals. I prefer to call any action against rodents, or other vertebrate pests, ''pest control''. As Timm[8] points out, ''damage control'' is ambiguous, whereas ''pest control'' is more subjective and accommodates others who may not consider the animal a pest. To many people, rodents are pests whether or not they are doing damage. A homeowner usually will not tolerate a single rodent, whereas a farmer usually does not object to the presence of different rodent species unless they cause economic loss.[2]

When rodents such as beaver *(Castor canadensis)* are ''managed,'' the objective is usually to assist in the well being of that population, whereas beaver control implies action designed to primarily benefit humans or species other than beaver. The types of rodent problems requiring control measures may range as a continuum from hardly being a nuisance to being a serious competitor, health hazard, or even predator.

Pest rodents can readily adapt to new habitats created by humans. They are highly tolerant of environmental changes, are highly resilient, and usually recover promptly after being controlled.

III. WHY IS RODENT CONTROL NECESSARY?

Rodent pests must be controlled because humans have encroached upon their environment. Rodents become such serious pests to humans because not only are they highly reproductive, but, like all pest species, they have much resilience to environmental perturbations and can readily adapt to many of the environmental changes made by people. Most wildlife species do not do as well when people modify the environment.

The introduction of nonnative species of rodents by agricultural pursuits and urbanization has often resulted in an increase in the number of kinds of rodents present. Since agricultural systems can be made more productive than natural systems if the ecosystem is soundly managed and not left to the whims of nature, they may then support greater densities of some pest rodents than occurs in natural environments.

IV. ECOLOGICAL ASPECTS

One would think that in deserts and arid regions, the struggle for life is almost exclusively with the elements. That may be so from an evolutionary point of view, but many species of rodents, which have evolved an adaptation for xerophytic conditions, do not thrive if the communities are irrigated, thus stimulating dramatic increases in the biomass of the primary products. Even though these new habitats provide a dramatic increase in acceptable food and cover, the nonadaptable species of rodents disappear.

Rodent control can be an integral part of land harmony and is often essential in helping preserve specific forage species and our wildlife heritage. Even with the reintroduction of fauna and flora in restoration ecology programs aimed toward creating more pristine conditions, rodent control is frequently necessary. Rodent-control programs are not known to have destroyed the vitality of biotic communities within man-altered ecosystems.

It is very difficult to prove that an existing population of any rodent species is significantly important to the well being of that ecosystem, and it is even more difficult to show in agriculture or other man-modified habitats that any pest species has any beneficial aspects that will help counterbalance its pest status. It seems inconceivable that introduced rats (*Rattus* spp.) and mice (*Mus* spp.) could have any ecological value in urban settings.

In natural communities, the removal or artificial change in density of a native species of browsing, grazing, seed-eating, or predatory rodent does not precipitate chain reaction of other components of the community that disturbs the balance of nature. In nature, the animal-soil-vegetation complex is surprisingly stable and certainly not delicately balanced.[3,4] Of course, if nearly all of the rodent prey that certain predators require for food is removed, those predators will not survive in that locality. In many urban and suburban areas, rodents attract predators such as skunks, but usually those predators also become pests because of disease hazards, problems with pets, or other reasons, hence the pest rodent's value as prey is a moot point.

Rodent control ideally should be the successful translocation of ecology into a control policy. In agriculture we need to learn how to create tolerable densities of the pest species of rodents so that they largely become self-regulatory ecological units not requiring other measures.

A principle often overlooked is that even though most modifications of the environment are relatively irreversible, rodent control is unique in that most rodent populations, as with other vertebrate pest populations, promptly recover once control is discontinued. This is why they are pests. Unless reduced in density to the lower portion of the sigmoid (S-shaped) population growth curve when controlled, they quickly recover, and temporarily may even exceed the carrying capacity for that environment.

A rodent-control principle that surprises many is that control problems are much greater in diversified agriculture than in a monoculture. Most rodent species cannot survive in monoculture, and the species that do survive usually can be predicted. Even so, too often preventive control measures are not taken and some of the few surviving species may become more abundant than the total density of all the original rodents that occupied the area before it was changed to a monocultural crop.

V. MISCONCEPTIONS ABOUT BALANCE OF NATURE

To help understand the principles of rodent control, it is important to be aware of some of the common misconceptions about the balance of nature. Some key biological principles that are often misunderstood follow:

1. Nature is neither rational nor ethical. In environments that humans have modified, nature seldom "knows best"; hence, in disturbed environments people must usually manage or control rodents and other wildlife instead of leaving it to nature.
2. The balance of nature is not compassionate: it is worse than a battle field, for no holds are barred and animals can commit the most bizarre acts of cruelty, even against family members.
3. Naturally evolved rodent and plant communities are NOT delicately balanced; they are quite stable and have much resilience toward disturbances by humans.
4. Trying to save the lives of displaced or unwanted rodent pests by capturing them, to be released elsewhere, is not humane, for they will probably die unpleasantly. It is more humane to poison, shoot, or kill-trap them with a minimum of suffering.
5. Most control methodologies are more humane than nature's brutality. Of course, intentional torture of animals should not be tolerated.
6. A law of nature is that a sizeable proportion of all species must die prior to breeding if their populations are to remain healthy.
7. What is right or wrong about any environmental decision depends upon one's viewpoint. Ethically there are no right or wrong decisions about environmental issues.

VI. CULTURAL, SOCIETAL, AND RELIGIOUS ASPECTS

Rodent control is largely a people problem. Whether a rodent is "good" or "bad" is not determined ecologically but rather by each person's viewpoint.

Efficacious rodent control is possible for most pest species. The basic problem is in designing a program that can be executed successfully in the existing local cultural, societal, religious, and political climate. Some of the principles involved follow.

1. There is a need to incorporate, to a degree, the growing social attitudes and sanctity of animal rights movements. However, the prevailing emotional and extremist views against rodent control need to be replaced with a sensible environmental ethic that dignifies animal death.
2. Values placed on rodents and views opposing control are relative matters that depend upon how each person is affected.
3. Rodent-control programs should be based on applied ecological principles.
4. It is shortsighted to let economics be the sole factors setting policy as to whether, when, and how to control pest rodents.
5. The proposed control program must take into account all potential hazards and involve the safest, most selective, and efficacious rodent bait formulation available.
6. The primary objective of a control program is to accomplish the desired objective with maximum safety for nontarget species and minimum disturbance of that biotic community.
7. Since nature demands a high rate of premature deaths among rodent populations, perhaps we can humanize it in agricultural areas with modern rodent-control practices.

VII. HUMANENESS

How humane is rodent control? This is difficult to answer, because pain in animals can be measured only subjectively, and people tend to anthropomorphize and mistakenly give animals human characteristics such as sensitivity, feelings, and consciousness. It has already been shown how cruel nature is and that most lethal control methods of control probably cause less suffering than a natural death. In nature, animals are frequently very cruel to others, sometimes even to their own relatives. Without doubt, in disturbed environments, humans can and usually do ''husband'' rodent species under far better husbandry and animal-welfare conditions than nature can.

Instead of an extreme reverence for the life of all individual rodents and other animals, society needs to develop a death ethic, since it is a law of nature that large numbers of all animals must die (and usually unpleasantly) prematurely to enable a healthy population of each species to exist. Even though the intentional torture of animals is not tolerated, we must recognize the urgency for fairly high death rates in wild animal populations.

From an ethical and humane point of view, humans have an obligation, in environments they have modified, to manage (which includes some killing) the desirable animals and control those that become pests.

VIII. INTEGRATED CONTROL

Rodenticide-treated bait is frequently the most efficacious way of controlling pest rodents, but poisons should always be a last resort and used in the smallest possible amounts. This can be accomplished through various biological control and other integrated approaches.

Sanitation — Species like rats can be eliminated if their sources of food, water, or cover are removed.

Habitat manipulation — Sometimes it is feasible to eliminate rodent harborage by modifying the surrounding agricultural or other habitats. An important principle, however, is that whenever the vegetation is modified to provide habitat control of a pest species, it will probably have deleterious effects on other species, as most species of vertebrates suffer when a habitat is altered.

Rodent-proofing — Frequently, a small amount of concrete or mortar to seal holes and cracks in buildings, walks, and streets can solve a serious rodent problem, even if abundant garbage or other food is still available.

Chemosterilants — Pharmaceutical and chemical companies will not permit most of the potential rodent chemosterilants to be used in rodent control. Further, the practicality of using chemosterilants is limited, because most people will not tolerate even one rodent, whether or not it is sterile. There are situations where the use of chemosterilants is indicated to prevent a rapid recovery of a rodent population following conventional poisoning. It is also useful in parks and recreational areas, where the objective is not to eliminate the species but to prevent the rodents from being too abundant, and posing health and other problems.

Predation — Household predators, such as cats and dogs, may help keep pest rodent populations down to the amount of good harborage available; otherwise, rodents as prey have a much greater effect on their predators than the reverse. Natural predation helps sustain an optimal density of native prey, but predators can eliminate rodent pests in agriculture when their cover and harborage are destroyed through cultivation.

Epizootics — The intentional introduction of a specific contagious or infectious disease to control rodents is not recommended, even for local control purposes. Conventional control methods are safer and more effective. The one exception is the use of myxomatosis virus to control European rabbits *(Oryctolagus cuniculus)*.

Electricity — I have not seen any cost-effective electrocuting devices. Killing rodents by electrocution may be economically justified when used to protect rice research plots from rats.

Electromagnetic fields — Devices that create electromagnetic fields have proven ineffective in rodent control.

Ultrasonic devices — Based on research in many parts of the world, conventional control methods seem to be far more cost-effective than ultrasonics. From places of security, rodents soon learn that the ultrasounds will not harm them.

Payment of bounty — Bounty schemes are costly, usually create enforcement problems, and inherently do not provide sufficient financial incentive to encourage the reduction of rodent populations to the desired low density.

Other means of control — There is a vast array of control methods to use when needed, including dry and liquid baits, wax bait blocks, tracking powders, live and kill traps, glue boards, and various repellents.

IX. PERMANENT, PREVENTIVE CONTROL

Preventive control, which implies that pest rodent numbers are permanently maintained at very low densities, can usually be achieved with most species by placing bamboo, plastic, or other types of rodent bait boxes (bait stations) at those sites that provide favorable harborage for rodents to build their nests. However, rodent pest problems in pasture land and in forests are more difficult to control economically in any manner.

A good way to visualize preventive rodent control is to treat pest rodents like a serious disease. As with such diseases, the infestation should be eliminated and then carefully monitored to prevent any new outbreak. The primary objective of this type of preventive control is to reduce the pest rodent numbers to low levels during the first few weeks or months, usually by using anticoagulant rodenticides in bait stations. Depending upon the local cost of bait materials and labor, it will sometimes be desirable to risk some increase in secondary and direct hazards at the start of a preventive control program in order to reduce the populations of pest rodents more quickly. This can be done using second-generation anticoagulants or even acute rodenticides.

After a local population of pest rodents has been virtually eliminated, officials must then monitor key areas of favorable rodent harborage with nontoxic bait in order to promptly detect any reinvasion of rodents. With permanent preventive rodent control programs, very little rodenticide is used, except at the beginning, because no new infestation of pest rodents is ever permitted to become established and multiply. From the point of view of wildlife conservation, preventive control has many advantages.

It usually is not necessary to go to the great expense of censusing rodent populations when the objective is local elimination. With permanent preventive control, it is necessary to know only when new rodents have moved into the area. This can be determined using a monitoring system of bait stations baited with nontoxic cereals or grain-paraffin blocks placed at sites that provide the best harborage for rodents (but not in crops). The presence of one new rodent is a signal for action.

X. SAFETY

There is a tendency for many environmentalists to erroneously consider everything ''chemical'' to be unnatural and bad and everything ''biological'' to be intrinsically desirable. However, the manner in which a rodenticide is used to control pest rodents may contribute more to its potential environmental hazards than does its inherent toxicity. The hazard of a rodenticide to humans and nontarget animals is not directly correlated with toxicity. Potential use hazards, not toxicity, should be the main concern.

There are several criteria to consider when selecting a rodenticide:

1. It should be well accepted by the target species.

2. It should be poorly accepted by potentially vulnerable nontarget species and thus selective for the target species.
3. It should be safe and easy to handle.
4. It should be relatively slow acting to avoid bait shyness.
5. It should be painless and not cause violent death.
6. It should be noncumulative, except for second-generation anticoagulant rodenticides.
7. It should not translocate into vegetation unless it is to be used as a repellent at the cotyledon stage of seedlings.
8. It should be highly biodegradable in the environment.
9. It should have an effective antidote or practical means of treatment.
10. It should be economical.
11. It should be registerable.

In discussing methods of safeguarding nontarget species of wildlife when controlling rodents, Marsh[5,6] lists wildlife-pesticide education, the appropriate selection of rodenticide, bait type and formulation, rate and distribution of baits, restricting accessibility of nontarget species to bait, timing (season, time of day, weather conditions) of the control operation, kind of grain used, concentration of toxicant, artificial coloring of the baits to protect birds and increase safety to people, conditioned aversion of nontarget species, addition of emetics in baits, type of registration of bait, dilution baiting, and pulse baiting with second-generation anticoagulants. The use of rodent pheromones still requires more research.

When a bait box is put out for the first time in villages, cities, or an agricultural area, it should be baited with nontoxic bait; toxic bait is to be added only after rodents have removed the nontoxic bait. If toxic bait is used initially but the bait station is placed where rodents do not find it, the homeowner may erroneously conclude that the bait formula is not acceptable to the rodents. However, when toxic bait is added only after the rodents have been prebaited, acceptance of the toxic bait the first night is almost assured, and the public will then be enthusiastic about the bait formulation.

XI. COST-BENEFIT ASPECTS

There is a great need for more economic data about rodent control. Data concerning economic losses caused by different species of rodents in different areas are very limited. Most figures available are only educated guesses.

Why is there not more information concerning the economic losses from various rodent population to crops and in storage? Very simply, this type of data is rather expensive to obtain, and funds for this type of research have seldom been available, although a few studies have been reported and others are currently underway.

Research has been done to determine the economic threshold of rodent damage in attempts to learn when the damage to rice or other crops is severe enough to justify the expense of a control operation. However, if risk analyses were made, I think we would find that often it would be more expedient and less expensive to do preventive control, to prevent damage rather than trying to solve problems after the rodents have learned to feed on the crop or other material.

There are now established guidelines to indicate when rodent control should be undertaken or when it is justified economically and ecologically. Different people will view each situation differently, depending upon how the rodents affect them. But control decisions need to be based on sound ecological, economic, aesthetic, political, social, and administrative considerations.

XII. CONTROL EFFECTIVENESS

An important rodent-control principle frequently overlooked — although sometimes that is justified — is how to measure and report the effectiveness of a rodent-control program. It seems to be human nature to want to brag, or cry, about the number that were killed, when actually the true success of a control program can be measured only by the number and reproductive potential of the survivors. Some species of small rodents are so fecund that the population can recover very quickly even if more than 90% of the rodents are destroyed, whereas with species that breed but once a year and have small litters, this degree of control will have an effect for several years.

The number of animals destroyed in a rodent campaign seldom serves as a reliable measure of the efficaciousness of the operation, but it may provide some indication of the degree of infestation present before control. Control effectiveness can be measured by determining to what degree the damage has been abated. Control and damage are inseparable and must always be considered together.

XIII. GOVERNMENT SUPPORT

A minimum population density is required before a rapid numerical increase can occur. For this reason, a control program should be effective enough to bring pest rodent populations to much lower densities than is usually done. It is also important to recognize that, once a rodent population is brought to a low level, bait materials become much less costly and maintaining the low population is less hazardous. Further, when rodent populations become quite low, natural predators, which usually are not of much help, then seem to assist in keeping rodent populations low and the survival rate of immigrants also seems to be lower. In spite of these facts, however, if there is no current evidence of rodent damage, most farmers are not interested in controlling what seems to be an innocuous rodent population. For this reason, government supervision is essential when practicing preventive control.

Considering the importance of the function they are to fulfill, most rodent control staffs are undergraded and underpaid, even if they are well trained. To achieve the needed communications between control operations and the public, the government needs to establish respectability and status for the control workers.

The government should also play a role in bait quality control, as it is often neglected during the manufacturing of bait. Some branches of government must ensure that no inferior rodenticides or baits are used or sold to the public. Quality control of all rodent baits should be determined in a competent laboratory. Further, the government must be shown emphatically that using cheaper, but inferior, bait materials or toxicants usually does not, in the end, result in monetary savings, since bait acceptance usually will decline and much more bait will be required.

XIV. PLANNING A CONTROL PROGRAM

Some of the key points that should be considered before launching a rodent-control program include:

1. Verify that the claimed rodent damage actually exists and, when relevant, estimate its economic and ecological significance.
2. Identify the species involved. This may be difficult. For example, many rodent species feed on bark, foliage, buds, fruits, stems, branches, and roots of plants. It is also sometimes difficult to know what species removed seeds or ate seedlings that were at the cotyledon stage of growth.

3. The political and emotional factors involved must be examined closely, since the most difficult aspects of control programs are the people problems, not biological know-how. It often is not easy to find a control method that is completely acceptable both socially and politically for most rodent problems.
4. Define, in writing, the control objective and degree of expected approval and disapproval by others. Do not make economics the sole factor justifying the operation. Consider what alternative solutions are available, and carefully analyze the health, safety, and legal factors. Finally, state how selective the proposed control method is for the target species, how much the population must be reduced, and how long the operation must remain effective.
5. Before starting, establish priorities with precise objectives for each step, obtain all necessary permits, and establish effective communications with authorities and the public. Then carry out the program in proper sequence.
6. Monitor the effectiveness of the program.
7. Be alert for ways to improve the operation.

REFERENCES

1. **Howard, W. E.,** Means of improving the status of vertebrate pest control, in *Trans. 27th North Am. Wildlife and Natural Resources Conf.*, Wildlife Management Institute, Washington, D.C., 1962, 139.
2. **Howard, W. E.,** Vertebrate pest control, in *Proc. Vertebrate Pest Control Conf.*, National Pest Control Association, Dunn Loring, Va., 1962, 1.
3. **Howard, W. E.,** Principles of vertebrate animal control, in *Congress de la Protection des cultures Tropicales,* Chambre de Commerce et D'Industrie, Marseille, 1965, 627.
4. **Howard, W. E.,** Balance of nature: fiction and reality, in *Trans. 49th North Am. Wildlife and Natural Resources Conf.*, Wildlife Management Institute, Washington, D.C., 1984, 469.
5. **Marsh, R. E.,** Rodenticide selection and bait composition to minimize potential primary hazard to nontarget species when baiting field rodents, in *Proc. 1st Eastern Wildlife Damage Conf.*, Decker, D. J., Ed., Cornell University, Ithaca, New York, 1983, 155.
6. **Marsh, R. E.,** Methods of safeguarding nontarget wildlife when controlling rodents, in *Proc. Seminar on the Control of Forest Trees,* Kuo, P. C., Ed., Council of Agriculture, Forestry Service No. 2, Taiwan, Republic of China, 1985, 169.
7. **Pimentel, D.,** Energy flow in agricultural and natural ecosystems, *Intecol Bull.*, No. 11, International Association for Ecology, Athens, Ga., 1985, 51.
8. **Timm, R. M.,** Teaching vertebrate pest control: a challenge to wildlife professionals, in *Trans. 47th North Am. Wildlife and Natural Resources Conf.*, Wildlife Management Institute, Washington, D.C., 1982, 194.

Chapter 17

EXPLORING, SAMPLING, NEOPHOBIA, AND FEEDING

S. A. Barnett

TABLE OF CONTENTS

I. INTRODUCTION

The most menacing rodent pests owe their success largely to three features in which they resemble human beings: they can live in a great variety of environments, they tolerate a wide range of foods, and they are restlessly "inquisitive" and exploratory. The many species are, however, not uniform in their behavioral or ecological features, and few species have been thoroughly studied. An account of their feeding and exploratory behavior therefore runs the risk of understating their diversity. Most relevant experimental studies have been on *Rattus norvegicus,* the "common brown" or "Norway" rat. This species, though widely dispersed, is rare or absent in vast areas of the tropics and of the southern hemisphere. Worse still, much research (sometimes described as on "the rat") has been on its domestic varieties, which differ in many ways — not only tameness and color — from the wild type.[9] We have therefore to look for general principles inferred from narrowly based studies, and to apply them cautiously to the species in which we are interested. In the selection below, from the vast literature, I refer as often as I can to rodents of the wild type rather than domestic varieties. In doing so, I draw on the work of ecologists, ethologists, and psychologists. The contributions from these disparate disciplines do not, however, always use the same terminology. I have therefore been obliged to bring out inconsistencies in the use of certain key terms. Some recommendations on usage are made in an appendix.

The present review is concerned primarily with feeding; but an animal, in order to feed, either must know where food is (from previous movements about its range), or must search for food. We therefore begin with the analysis of movements in the living space, especially exploration.

II. EXPLORATION AND NEOPHILIA

A. Definition and Measurement

"Exploration" is not a precise term referring to a single kind of behavior: it refers to the fact that animals often move about when not impelled by immediate need, and in doing so store information about their surroundings. Movements outside the nest or burrow may be related to the need for food or water or for a mate, but satiation is not always followed by inactivity: an animal may regularly patrol its territory or home range, and it may also enter and move in a new area; such movements commonly occur *after* the immediate demands of survival and reproduction have been satisfied.[12,41,103]

The causes of an animal's movements can be clearly distinguished only by experiment. It is then necessary to measure "activity"; but this term too has several meanings. Figure 1 plots the time spent outside the nest by rodents of two species. For several hours around midday, these animals spend much time above ground. In the figure "activity" means "time spent outside the nest"; hence it covers all the kinds of behavior performed by *Reithrodontomys* when not in its burrow.

In the laboratory, where movements can be directly observed, other kinds of behavior have been measured and called activity. One much used method is to provide a solitary animal with a wheel or treadmill (Figure 2). The rate of running in the wheel reflects physiological changes such as those of the estrous cycle: the highest rates are during estrus. This measure, which is suitably named wheel-activity, is difficult to relate to anything that happens in nature. The presence of the wheel stimulates running: hence the distance covered, or energy used, does not reflect the "activity" that would be expended in more natural conditions. Another questionable but popular device is the "open field" (Figure 3A). An animal is abruptly placed on a flat surface, and its movements are recorded for a few minutes. The result varies with the species;[117] but again, the relationship of this behavior with what happens in nature is obscure.[1,32] A less unnatural situation is a maze, of which there are

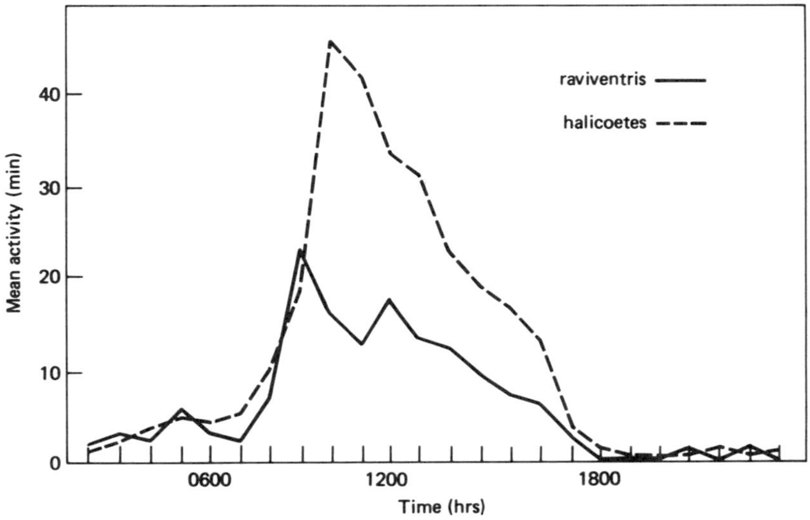

FIGURE 1. The circadian rhythm of "activity" of two diurnal species of harvest mice, *Reithrodontomys*. "Activity" here means time outside the nest. (After Fisler, G. F., *Univ. Calif. Berkeley Zool.*, 77, 1, 1965.)

FIGURE 2. Apparatus much used for measuring "activity" in the laboratory. The movements of an animal in the wheel or treadmill are recorded automatically. Access to a wheel provokes extra running.

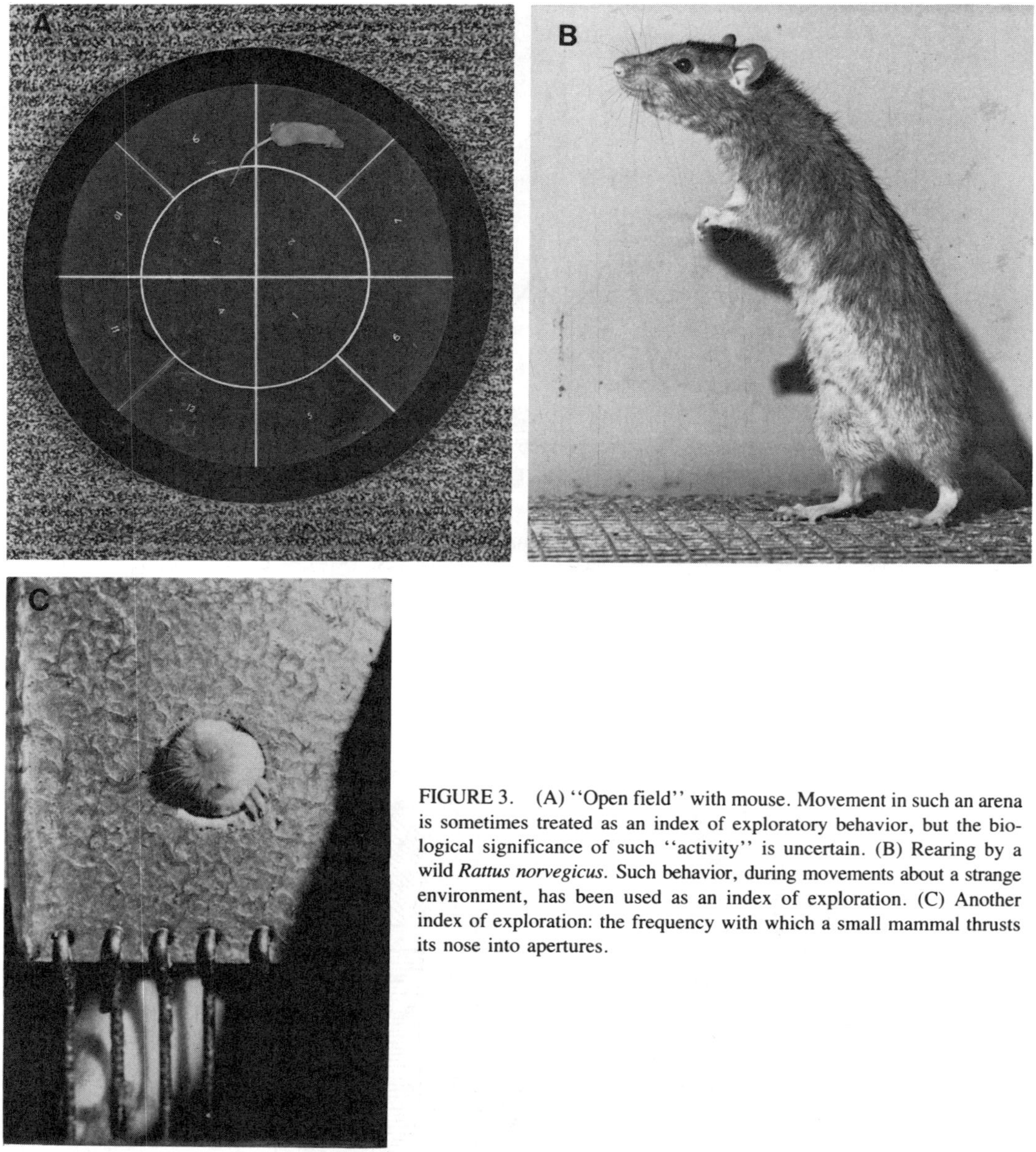

FIGURE 3. (A) "Open field" with mouse. Movement in such an arena is sometimes treated as an index of exploratory behavior, but the biological significance of such "activity" is uncertain. (B) Rearing by a wild *Rattus norvegicus*. Such behavior, during movements about a strange environment, has been used as an index of exploration. (C) Another index of exploration: the frequency with which a small mammal thrusts its nose into apertures.

many kinds. In typical maze experiments a tame rodent is again used, and is observed for only a few minutes. It is, however, possible to use a residential maze (Figure 4), in which an animal's movements can be recorded for many days.

Other measures which have been described as representing exploration or curiosity include rearing[76] (Figure 3B), and poking the nose into holes (Figure 3C).[49] If all these measures were positively correlated, we might be justified in saying that they all represent one process, "activity"; but they are not.[111] The findings shown in Figure 1, for example, tell us little about changes in rate of movement: much of the period of "activity" could be spent sitting and eating. Lack of correlation between measures of "activity" can be precisely recorded in a residential maze: Figure 5 shows how visits to the periphery of the maze are at a peak on the first day in the maze, but time spent outside the nest remains nearly constant. Similar observations have been made on wild house mice, *Mus musculus (vel M.m. domesticus)*.[19]

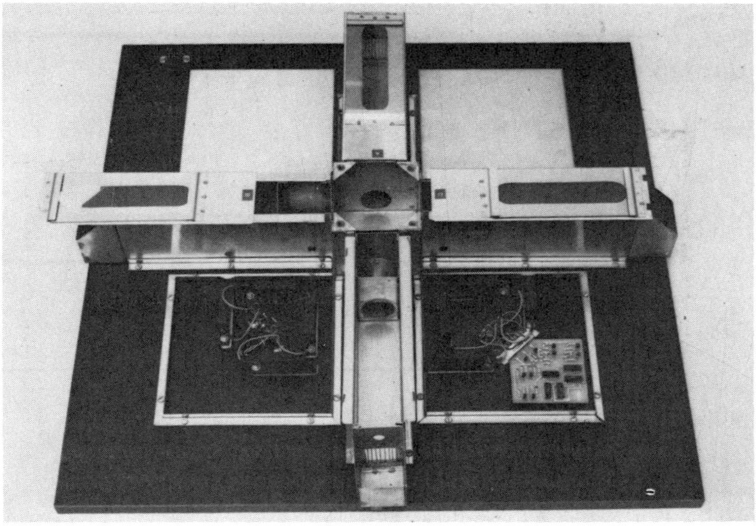

FIGURE 4. An artificial environment (a ''plus-maze'') in which an animal can live for many days, during which its movements are automatically recorded. Four arms lead from a central nest box. Food or water may be offered at the end of an arm.

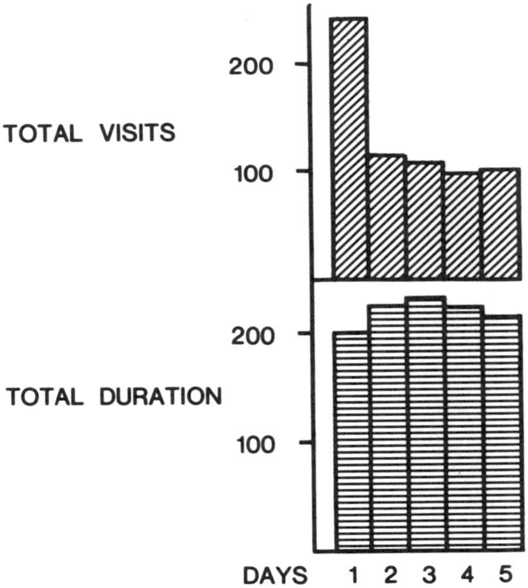

FIGURE 5. Two measures of ''activity'' in a ''plus-maze'' (Figure 4). Mean daily visits to the arms by domestic rats peak on day 1, when the animals are in a strange environment. Duration of stay in the arms (that is, time out of nest) does not correspond. (From Barnett, S. A. and Cowan, P. E., *Interdiscip. Sci. Rev.*, 1, 43, 1976.)

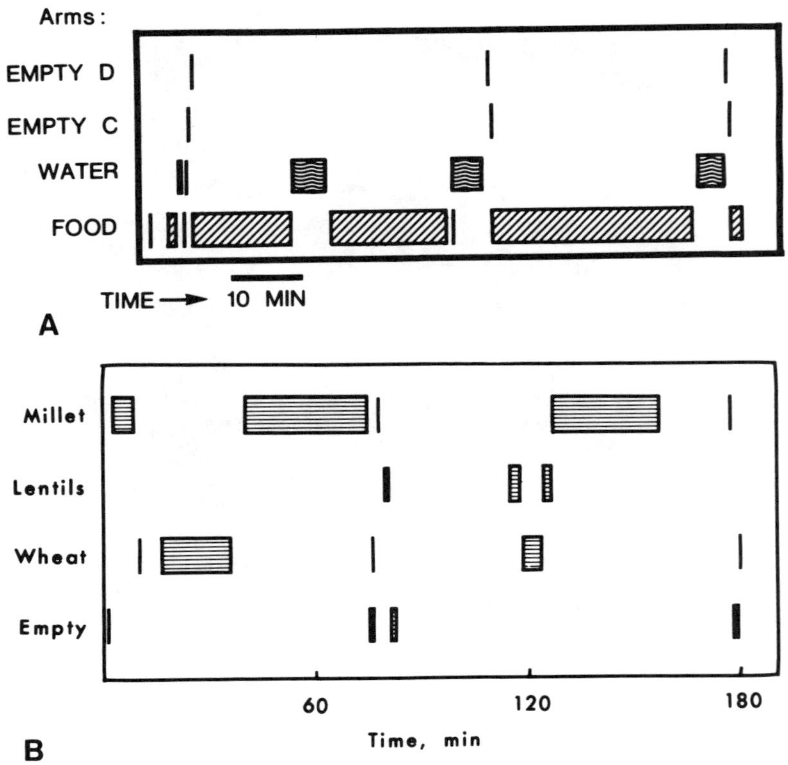

FIGURE 6. (A) Movements of one laboratory rat in a residential maze with four arms (Figure 4). The rat was released from the central nest box after 21 hr without food or water. Each block represents one visit to an arm. Long periods of eating or drinking were followed by brief visits to all arms, including the two empty ones. (From Barnett, S. A. and Cowan, P. E., *Interdiscip. Sci. Rev.*, 1, 43, 1976.) (B) As A, above, but the subject was a wild *Rattus norvegicus*, and three foods were offered. The patrols after each meal were accompanied by sampling of the less preferred foods. (From Barnett, S. A., Dickson, R. G., Marples, T. G., and Radha, E., *Behav. Proc.*, 3, 29, 1978.)

Hence, if we use terms such as exploration or activity, we should do so with caution: when measurements have been made, what has been measured should be precisely stated, and not equated with other, quite different measures.

B. Causes of Exploration

The causes of an animal's movements may be conveniently put under three headings: (1) the presence of a structured environment of which parts have not recently been visited, or have not been visited at all; (2) the presence of novel features in the familiar environment; and (3) an internal deficit, such as an empty stomach. We are now concerned with the first. The response to new objects is discussed in Section III, and the effects of a deficit belong with feeding behavior (Section IV.A).

A small mammal regularly patrols its home range. To distinguish such movements from a search for anything specific, behavior may again be studied in an artificial environment such as that of Figure 4. An animal may be fasted for 21 hr while confined to its nest; it is then released into an environment that is already familiar, and its movements are recorded for 3 hr (Figure 6).[14,40] Most of that time is taken up with eating or drinking; but each "meal" is followed by brief patrols of all parts of the environment, including those previously

FIGURE 7. Y-maze in which the two goal arms offer different visual stimuli: an example of a "maze" with only one choice point, in which spontaneous alternation may be observed.

experienced as empty. Such patrolling persists even when a female's movements in the living space are at a minimum, around the time of birth of a litter.[17]

During patrolling there is a marked tendency to enter the parts of the environment which have been less recently experienced. The preference for the less familiar has been extensively analyzed in experiments on domestic rats.[9] The simplest case is seen in spontaneous alternation: the animal is put in the stem of a "maze" shaped like a T or a Y (Figure 7), and walks into one of the arms; if it is then picked up and replaced in the stem, it usually next walks into the other arm. Provided that both arms are empty of incentives, such as food, alternation continues. If, however, one arm leads to a dead end, and the other to a spacious area that contains objects, the latter is regularly preferred: the opportunity to move around a complex environment is rewarding, just as food is to a fasted animal.

The tendency to approach unfamiliar, rather than familiar, places or objects is named neophilia. Wild-type rodents of which the neophilic behavior has been recorded include five species of *Rattus,*[4,40,42] *Mus musculus,*[19,26] and the golden hamster, *Mesocricetus auratus.*[105] In the experiments on rats and the house mouse, the animals were habituated to an artificial environment for some days; an additional part of the environment was then opened, and there was an immediate high rate of visits to the new place.

C. Inhibitors of Exploration

To examine the role of such behavior in nature, we must take account not only of what provokes exploration but also what limits or prevents it. Restrictions on free exploration are of at least five kinds: (1) the animal's internal state; (2) social interactions; (3) the topography of nest or burrow in relation to food and other environmental resources; (4) the performance of alternative activities such as nest building; and (5) other properties of the environment, especially the temperature.

(1) In a familiar environment, after a fast during the hours of daylight, feeding presumably has priority over generalized, neophilic exploration. Where a substantial source of food is accessible, movements are likely to be along a direct path between nest and food. Free exploration is ''released'' only when homeostatic needs are satisfied. There are, however, no full accounts of such behavior sequences in natural conditions.

(2) The most important social restrictions on exploring are probably territorial: movements are often limited by conspecific holders of neighboring territories. Correspondingly, rodent populations quickly flow into regions emptied by the death or removal of the occupants.[65,74] Such immigrations presumably begin with exploration.

(3) Movements are also restricted by the need to return to a central point. As an example, Sims and Wolfe[108] describe experiments on homing by wild house mice in Mississippi. Those that lived in fields were able to home from 100 m, but fewer than half from 200 m; a few succeeded from 250 m. Newsome et al.[84] have recorded similar behavior by this species in Australia: many of their mice, however, found their way home when they had been moved far outside their usual range. These two papers should be consulted for other work on homing by rodents.

The range of individual movements is influenced by the topography of food and shelter. As an example, Nass[83] used telemetry to record the movements of Polynesian rats, *R. exulans,* in Hawaii. Some lived in sugar cane fields, and their movements were almost confined to the fields. Other rats lived in nearby waste lands (evidently a less favorable environment), moved much longer distances, and made many visits to the cane fields. It would be interesting to know more about the differences between the two populations, and the interactions between them. In particular, did the rats from the waste lands explore less than the others?

(4) It is easy to guess that the longer the distance from nest to food, the less is the time or energy for free exploration. In the laboratory the ''competition'' of exploring with other activities can be analyzed quantitatively. Brant and Kavanau[27] give an extreme example from studies of the canyon mouse, *Peromyscus crinitus,* in a very complex artificial environment. The mice gradually learned their way around by making increasingly distant excursions from their nests; but, when an activity wheel was made available, much time was spent running in the wheel, and exploration almost stopped.

A wheel, however, has no obvious counterpart in a natural environment. When we turn to a normal activity, such as nest building, we find different relationships. Caged wild house mice were kept in a warm environment and given nest material, and the weight and quality of the nests they made was recorded. When the mice were also allowed to move around a larger cage than usual, their nests were smaller and less well made.[15] In this case, exploration evidently had priority over the alternative activity; but, if the environmental temperature had been lower, the result would probably have been different.

(5) The influence of physical demands imposed by the environment is clearly seen in the effect of temperature on behavior. Small mammals usually move about their living space

less as the ambient temperature declines.[16,64] Perrigo and Bronson,[89] in experiments on wild house mice, confirm this finding and describe an extreme situation in which their cold-exposed mice were obliged to spend so much energy on foraging that movement not directly related to feeding ceased.

The intensity of exploratory and other activities is, then, a resultant of a number of competing demands, not all of them easy to identify. Nonetheless, in the absence of severe debility an animal's movements are always likely to include a component not directly related either to homeostatic need or to reproduction: that is, either a patrolling of a familiar environment or an investigation of a completely new one.

D. Adaptability

The wandering tendency of rodents is one aspect of a general restlessness. The readiness to enter very confined places allows rodents to make use of seemingly impassable gaps. When proofing a structure against house mice, one has to block all apertures of more than 5 mm. Another familiar feature is their apparently indiscriminate gnawing of objects. *Rattus norvegicus* can chew its way through a concrete floor.

Rodents not only gnaw objects and materials but also handle them. The role of handling in natural conditions has been little studied, but it is certainly used in manipulating food and in nest-building. Occasionally, the manual dexterity of rodents has a disconcerting outcome. A standard form of wire trap has a lobster pot entrance and a trap door through which the captive animal falls into the main compartment. Both *R. norvegicus* and *R. rattus* can escape from such a trap by hooking a claw round the trap door and then forcing a way out through the entrance.[8]

Advantage has been taken of the handiness of laboratory rodents in traditional experiments on habit formation: they can be readily trained to press a lever or to perform some similar movement for a reward of food or water, or to avoid pain. Such findings rarely throw much light on the activities of free-living rodents; but the experiments show that, as rodents move restlessly about and sniff, gnaw, and manipulate objects, they continually put themselves in positions in which they can learn more about their environment. Any pattern of movements that pays off can then rapidly become a habit.

This, however, is not the whole function of "unrewarded" exploration: an important additional effect has been revealed in experiments, nearly all of which have been on domestic rats in mazes. If a rat is put in an unfamiliar environment (especially during the hours of darkness) its first response is to move about but to disregard the presence of special incentives such as food or shelter. Even if the animal has been fasted, the ranging movements have, at first, priority over eating. Observing such movements provides no evidence that anything is being learned, but in fact topographical information is being stored in the central nervous system. This is latent learning. Later, in a region previously explored, an animal can quickly learn to run from one point to another without error, whereas an animal without previous acquaintance with the environment takes much longer. Ellen et al.[45] have recently given a particularly clear illustration of such learning.

This ability helps animals not only to find food but also to avoid being eaten. Free-living rodents are liable to be attacked by predators. They are more likely to succumb to predation if they are in unfamiliar surroundings, and therefore unable to take cover quickly.

Movements about the living space have yet another, more subtle function: especially in early life they contribute to the development of "intelligence" — that is, the general ability to solve problems and to adapt behavior to experience. If young mammals are to develop the capacity quickly to solve such problems as moving by the shortest route between two points, or discriminating visual stimuli, they must have diversity of experience early in life.[9]

E. Sampling and Foraging

A surprisingly neglected accompaniment of rodents' restlessness is food sampling. Yet

FIGURE 8. An experiment on food sampling. A young *Rattus norvegicus* eats wheat flour in preference to alternatives; but, after a meal, it samples the other foods.

the behavior is easily observed, even in a small cage (Figure 8).[5] After a fast, an animal is offered several foods, one of which is known to be preferred to the others. A meal is made of the preferred food; water is drunk; and then there is movement about the cage during which the other foods are sniffed and small amounts are eaten. The pattern of such movements in an artificial environment is shown in Figure 6: the brief visits to the periphery after a meal are accompanied by consumption of small amounts of alternative foods.[14] The tendency of domestic rats to vary their sources of food when they have the opportunity to do so has also been analyzed by Olton and his colleagues.[86,87]

Such behavior is difficult to study in free-living populations, but Vickery[115] has described sampling by three species of rodents in a Canadian forest. A choice of foods, all acceptable to the rodents, was offered at bait stations. Preference was mainly for the food with the highest energy value, but the alternative foods were always sampled.

Sampling may be regarded as a component of the foraging "strategy" of omnivores. The foraging patterns most studied, however, have been those of predators.[72] Predation often requires movement over a wide range; but, unlike omnivores, predators are usually specialized feeders, and the nutritional value of their prey is fairly uniform. In addition, their food consists of discrete items that can be counted. Problems nonetheless arise because prey animals may be scattered, or in groups of various sizes; the prey themselves vary in size; and there are changes in these and other features with the seasons. Hence, for a predator, the best or optimal pattern of search for food varies with the kinds of prey eaten and with many features of the surrounding conditions.

Reichman[98] has applied the methods of foraging analysis to feeding by five sympatric species of desert rodents. Each species depends mostly on seeds, which are analogous to prey; but each also sometimes eats other plant food and insects. The seeds are evidently selected for their energy value. This process is, however, not simple, for selection occurs in two stages: first when the cheek pouches are filled, and second when the pouches are emptied and the contents picked over under cover. The first stage reflects not merely the need to maximize energy intake, but also (presumably) the need to collect food quickly and so to minimize danger from predators. Reichman and Rebar,[99] in a study of kangaroo rats, *Dipodomys spectabilis*, find yet a further elaboration. The stored seeds become moldy and, at a certain stage of mild moldiness, are nutritionally superior to fresh seeds. The rodents, when they have a choice, select the slightly moldy seeds. Such complexities regularly arise when one tries to analyze behavior in terms of cost benefit, or optimality.[96]

A further complication is variation among species. Reichman[98] gives evidence that the energy cost of the gait of one of his rodents, the quadrupedal pocket mouse, *Perognathus amplus*, does not allow it to travel long distances in the search for seeds in high density; in contrast, kangaroo rats, *Dipodomys merriami*, which are bipedal, can afford to range more widely.

Reichman's findings illustrate a method, so far little used, of studying the ecology and ethology of rodents. When the method comes to be applied to the still more complex feeding behavior of omnivores, a new dimension will be given to studies of foraging. Whether it will also help to make poison baiting more effective remains to be found out.

III. NEOPHOBIA

The preceding account represents rodents as incessantly "curious" and investigatory. Yet the pest species, at least, are notoriously "wary" — difficult to trap and apparently too "intelligent" to be easily poisoned. This seeming anomaly was first systematically studied during research, during the second world war, on poison baiting in the field.[36] If unfamiliar objects, such as boxes or even food items, are put on runways regularly used by wild *Rattus norvegicus*, the result is often a striking disappearance of the rats. This, *the avoidance of an unfamiliar object in a familiar place*, was initially named "new object reaction"; but when it was studied in the laboratory it was renamed neophobia.[6] One consequence of neophobia is the difficulty of trapping commensal rodents. An even more important effect is refusal of food previously eaten. For example, Chitty,[37] in one of his experiments, used food consumption as an index of neophobia: wild *R. norvegicus* were regularly eating about 2.0 kg of wheat on the floor of a barn; when, on the twelfth day of observations, the wheat was put out on tin trays, consumption for the next 24 hr was almost zero. Sometimes the effect is even greater: when food is offered in a strange container, or a familiar container is moved a short distance, it may be refused for several days.[41]

Food can itself be a "new object" or source of neophobia: a heap of edible grains or flour may, when first encountered, be avoided. Since about 1965 there has been much interest in such responses to novel foods, and many experiments have been done, mainly on domestic rats. As one result, there has been some confusion in the use of the term neophobia. It is common for an animal, even a domestic rat, to eat rather little of an unfamiliar mixture: the first approach to it is tentative. In some writings "neophobia" is restricted to such behavior, that is, to eating or drinking less on first presentation than later.[43] The ambiguity due to different definitions has misled some workers. For instance, Miller and Holzman[80] equate neophobia and flavor neophobia. They point out correctly that the hesitation of a rodent facing a new flavor usually has only a small effect on consumption. (A similar finding on wild *Rattus norvegicus* has been reported by Bull.[30]) They then refer to the work, cited above, on the avoidance of *strange objects*. This neophobic response, as we see further below (Section IV.D), helps to protect rodent pests from poisons; but these authors go on to state that the influence of flavor neophobia is too small to have this "alleged" protective function. No such action of flavor neophobia has, however, been proposed. The misunderstanding arises from the confusion of neophobia in its primary sense with the response to novel flavors.

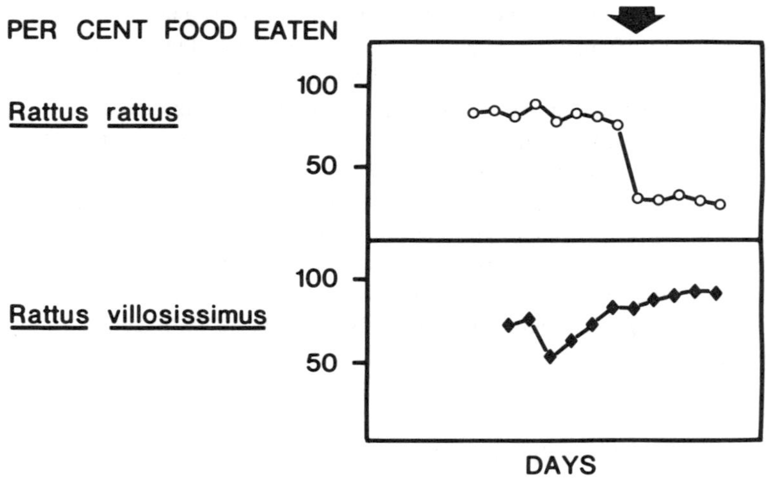

FIGURE 9. Neophobia of *Rattus rattus* contrasted with the absence of neophobia of *R. villosissimus*. Two foods were supplied, each in an identical container. The arrow marks the day on which the preferred food was offered in a different container. *R. rattus*, a commensal species, then largely rejected the formerly preferred food (upper graph), but *R. villosissimus*, which is not commensal, did not (lower graph).

When a neophobic animal is exposed continuously or repeatedly to a new food or other object, avoidance as a rule gradually declines. This is *habituation* of the neophobic response: the response is replaced by increasingly close and prolonged approach. Such a change is sometimes called loss of neophobia, but this is misleading: after one such experience, other completely novel objects are still avoided as before.

For a direct measure of neophobia (as originally defined) it is necessary to record movements, or their absence. For example, after an animal has spent several days in an artificial environment (Figure 4) a strange object is put in one arm of the maze. If the animal is wild-type *R. norvegicus* or *R. rattus*, visits to that arm typically stop for some hours or even for a whole day (Figure 8).[40,42]

It is not, however, possible to predict at all closely the duration of the neophobic response, for there is always much individual variation. The causes of this variation are not fully known, but those that are known or suspected may be put in three groups: (1) social; (2) other environmental; and (3) genetical. (1) As an example of the first, Summerlin and Wolfe[112] give evidence that the social status of cotton rats, *Sigmodon hispidus*, influences the response to traps: animals rated as "dominant" are described as more exploratory and less neophobic than those of lower status. (2) The inanimate environment may have an effect: neophobia may be attenuated or absent in some conditions. Boice[24] finds that *R. norvegicus* in land-fill refuse tips, where conditions are in a continual flux, are easily trapped, hence are evidently not neophobic. This difference from the typical is presumably environmentally, not genetically, caused. The animals have become habituated to change. (3) Genetical variation may be within or between species. The clearest examples of genetical differences within species come from the effects of domestication. All laboratory rats are of the species, *R. norvegicus*, yet their behavior retains only a trace of neophobia: within less than a minute they approach a novel object and investigate it.[6,40,81] The genus, *Rattus*, also illustrates variation between species. Cowan[40] tested two Australian species, *R. villosissimus* and *R. fuscipes*, for neophobia. He recorded consumption of two foods by caged animals and, after several days, changed the container used for the preferred food. These two species showed no neophobia (measured in this case by food consumption); but *R. rattus*, in identical conditions, was markedly neophobic (Figure 9).

The two certainly neophobic species, *R. norvegicus* and *R. rattus*, are pests. Their neophobia makes them difficult to trap and, as we see further below, also difficult to poison. The other two species of *Rattus* mentioned above are not commensal: *R. fuscipes* has not been recorded as a pest; and *R. villosissimus* affects human communities only sporadically, when it breaks out in vast "plagues" in grazing areas.[35] An obvious hypothesis is that the neophobia of the commensal species is a result of natural selection in conditions in which they are under continual human predation.[7] Such a guess about the past is, strictly speaking, untestable; but it implies a testable proposition, namely, that other commensal species, such as the Indian mole rat, *Bandicota bengalensis*, should be neophobic. The obverse has been tested on the Australian swamp rat, *R. lutreolus*, another independent species, and its response to strange objects has, as expected, proved to be predominantly approach, not avoidance.[4]

It might therefore be expected that the house mouse, *Mus musculus*, an outstandingly successful commensal, would be neophobic; but it is not. Wild mice combine the usual patrolling of the home range with an unpredictable or "random" component in their movements; and they do not (except, perhaps, momentarily) avoid new objects in a familiar place.[3,19] These findings confirm earlier observations of free-living populations. The "mercurial" movements of house mice are perhaps an alternative "strategy" to the neophobia of commensal rats.[110]

To sum up, the movement patterns of, probably, all mammals include a general neophilia: the typical response to both a new place and a new object is approach and familiarization. But in association with human populations uncontrolled "curiosity" is hazardous. Correspondingly, at least two of the most prominent commensal species are neophobic: in a familiar area, instead of approaching new objects, they avoid them. House mice, however, are not neophobic to any significant extent, and the erratic character of their behavior is perhaps an alternative means of avoiding human predation. An obvious hypothesis follows: all rodents well established as commensals should be either "mercurial" in their movements, or neophobic. Plenty of species, not yet studied, remain on which to test this proposal.

IV. FEEDING

A. The Patterns of Feeding

The dentition of rodents fits them especially for eating seeds and other hard foods; but, as we have seen, there is much variation among species; and individuals within species can live on a wide range of diets. Some rodents, such as voles, *Microtus*, feed mainly on the soft parts of grasses, but they can also live on roots or grain.[46,97] Other common European genera, such as *Apodemus* and *Clethrionomys*, are similarly adaptable.[60] The diet of the most studied species of squirrels (Sciuridae) consists largely of the hard parts of plants, but includes soft structures and even insects;[107] and the Indian palm squirrels, *Funambulus pennanti* and *F. palmarum*, which are serious pests, eat large quantities of flowers and fruits.[18] Gerbils, such as *Meriones*, at certain seasons eat a high proportion of insects.[93] The grasshopper mouse, *Onychomys torridus*, lives mainly on insects or other animal food.[66] As for *Rattus*, some species eat snails or slugs;[78,118] in certain conditions half their food consists of insects, especially termites,[63] and rats are also effective predators on smaller mammals, birds, and even fish.[2,39] The species of this genus, are however, not all equally versatile feeders. Yabe,[118] in Japan, contrasts the extreme omnivory of *R. norvegicus* with the tendency of *R. rattus* to eat principally fruit and seeds, and relates this difference to their different biotopes.

Whatever their diet, most rodents move about and feed during the dark period (Figure 10). If, within that period, food and water are readily available, rodents take meals, perhaps at intervals of 2 or 3 hr. Eating is likely to be interrupted by brief bouts of "prandial" drinking, but in any case a meal ends with a long drink (Figure 6). These statements are

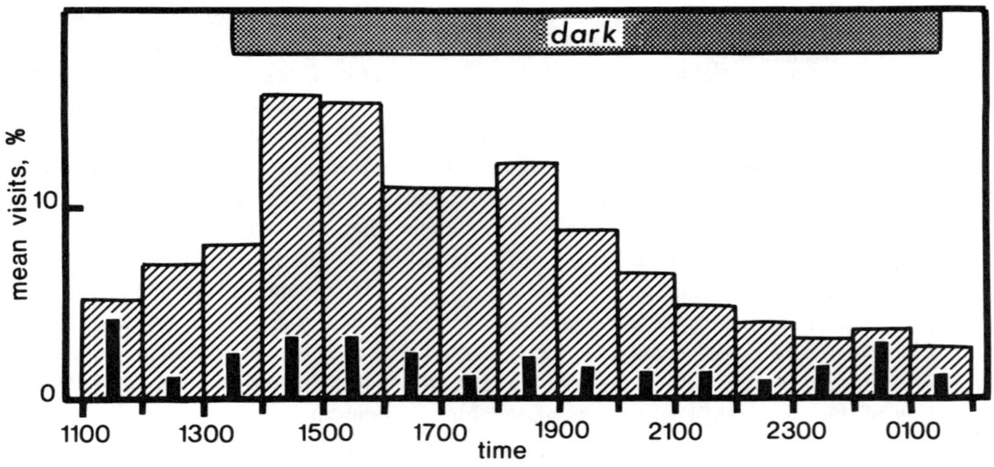

FIGURE 10. Circadian rhythm of "activity" of 8 *Rattus rattus* in an artificial environment (Figure 4). Mean visits to the periphery of the environment show two peaks during the dark period. (Black bars represent standard errors of the means.)

based mainly on studies of domestic rats.[9] No doubt rodents that need little or no water have a different feeding pattern.

Feeding rhythms can, however, be greatly altered to correspond with the times at which food is available (Figure 11). If food is available at only one time of day, rats eat an exceptionally large meal at that time and continue to flourish. Accordingly, free-living rodents can adapt their behavior rapidly to changes in their habitat. Cottam[39] describes a colony of *R. norvegicus* which lived near streams stocked with fish: when, during the day, food was thrown in the water, the waiting rats ate both food and fish. (This species swims well.) Such adaptability allows rodents to take advantage of seasonal increases in food supplies. In rice-growing regions Indian mole rats burrow at the edges of the fields, but at the harvest they quickly move into the areas where the grain is threshed. Many species behave in this way. Such movements have seldom been systematically studied; yet they are important for planning control measures.

For control, an intimate knowledge of the signs left by pest species is also desirable. In a stable environment, rodents typically establish regular routes between nests and food. A nest may be in a burrow; but some species, notably *R. rattus*, nest mainly above ground. Each species leaves characteristic traces (Figure 12), but these are rarely described in detail.

Feeding is usually under cover: portable food items are taken into concealment, often to the burrow or nest (Figure 13). Such items are sometimes hoarded. Hoarding, and consumption of the stored food in winter, is a familiar feature of the behavior of squirrels; but the scope of hoarding by other rodents is incompletely known. In a pioneering study of wild *R. norvegicus* in a large paddock, Calhoun[31] observed hoarding of food in burrows especially by rats which had come under mild attack by other rats. Such behavior could reduce conflict in a colony, and improve the nutrition of some of its members, but whether it does so in fact has still to be found out. Hoarding by *Bandicota bengalensis* has been studied quantitatively. The ecological niche of this species in the Indian subcontinent, in both fields and cities, resembles that of *R. norvegicus* in colder regions. In the paddy fields of West Bengal it has been reckoned to store at least 5.7% of the total rice production;[101] similar findings have been reported from Pakistan and Bangladesh.[51,91] Another very successful pest in the same region, the Indian gerbil, *Tatera indica*, hoards much food in its burrows, especially, it seems, after a period on short commons.[75] The mole rat of Sri Lanka, *Gunomys gracilis*,

FIGURE 11. The circadian rhythm of feeding can be drastically adapted to unusual circumstances. In a park in Calcutta, *Bandicota bengalensis* (usually a nocturnal species) feeds in brilliant sunlight from scraps provided by passers-by.

is also described as storing large quantities of rice in its burrows.[48] Such observations illustrate one of the many reasons why estimates of losses due to pests are unreliable; damage to crops or stored food may not be a simple function of the amounts eaten.

The observations by Calhoun, mentioned above, remind us that feeding, and movements generally, may be socially influenced. Rodents rub their bodies on surfaces as they move about, and leave odor trails consisting of skin and other secretions. Some, such as gerbils (Gerbillinae), have conspicuous sternal glands. The odors have been supposed to contribute to territorial defense, but experiments have usually shown them to attract conspecifics, not to drive them away.[70]

Rats, as they follow the trails of other rats, may be influenced in their choice of food. Galef and his colleagues[54] have shown wild *R. norvegicus* to eat foods recently eaten by

FIGURE 12. Entrance to burrow of Indian mole rat, *Bandicota bengalensis*, in a paddy field: a species-typical appearance.

FIGURE 13. Rodents usually take food under cover before feeding. Here a large male *Rattus norvegicus* drags a lump of liver toward its nest.

other rats rather than alternatives otherwise similarly palatable. The relevant stimuli are evidently odors carried by conspecifics, but the exact nature of the cues has not been established. Such influences may be especially important for the development of food habits early in life (Section IV.F). Further experiments, on other species, are likely to reveal more examples of social effects.

To sum up, the factors that can affect feeding patterns include (1) the topography of sources of food and water, (2) daily and seasonal changes in the food supply, (3) the available shelter, (4) the presence of hoarded food, and (5) social interactions. The primary determinants of food selection are, however, dietary needs and the avoidance of ill effects. We now turn to them.

B. Energy, Protein, and Water Intake

Rodents, like other animals, eat enough food to balance energy output. Adults in favorable conditions maintain a steady body weight or a slow rate of growth. Once again, the most exact and detailed information comes from experiments on domestic rats:[9,25] the calorigenic value of food can be decreased by adding indigestible fiber, or increased by adding sugar; in each case consumption is quickly adjusted. Wild *R. norvegicus* have similar abilities: Maller[79] gave them first a diet containing 10% oil, then one with 60%: intake was reduced in accord with the higher energy value. Domestic rats, in the same conditions, took longer to adjust their eating rate. These findings suggest that, when oil is used to make a bait more palatable, as little as possible should be added: if a bait is to contain poison, the more the pests eat, the better.

Evidence of the ability to adapt feeding to energy need has been recorded among free-living rodents. Gorecki and Gebeczynska[60] studied two European species, the bank vole, *Clethrionomys glareolus*, and a field mouse, *Apodemus flavicollis*. Both are versatile feeders and alter their choice of food with the season: in cold weather they increase their consumption of seeds with a high energy content. Vickery[115] describes analogous observations on three species of Canadian rodents.

An animal that feeds on a narrow range of foods, such as grasses, must adapt its intake to energy need but not to other dietary requirements: if it eats enough, all its dietary needs will be satisfied. But if an omnivore's diet were determined solely by energy intake, it might, even in natural conditions, eat an unbalanced diet. It could even go short of protein. Early experiments (on domestic rats) failed, however, to show a consistent ability to select a diet containing enough protein: it seemed that protein consumption, and hence growth, was a by-product of eating for energy. Maller,[79] however, made wild *R. norvegicus* deficient in protein, and found evidence of a preference for protein-rich food; and later meticulous experiments by Booth[25] and others have given convincing evidence that even domestic rats can adjust their choice of foods so that they avoid amino acid deficiency.

A further factor that can influence feeding behavior is the need for water, but rodents vary greatly in their water consumption. Some, such as *R. norvegicus*, eat wet in preference to dry foods, and are attracted to places where water is plentiful. For them at least, poisons in aqueous solution are sometimes useful. At the other extreme are the desert rodents which can survive for a long time on dry seeds alone: these include kangaroo rats, *Dipodomys*, pocket mice, *Perognathus*, and jerboas, *Jaculus*. But even these need water, and drink it readily when they can.[59]

C. Special "Hungers"

The ability to adjust food selection to need extends to inorganic salts and vitamins.[9,102] Rats given a choice can select foods or solutions that satisfy their need for salt, NaCl. If very severe salt deficiency is induced by removing both adrenal glands, rats drink salt solutions that they would otherwise reject. One effect of salt deficiency is evidently — in human terms — development of a liking for strong salty tastes. This is not a case of learning to choose a favorable diet. But rats less severely deprived of salt gradually alter their food habits and so increase their salt intake to an appropriate level. In this case a new habit is developed as a result of experiencing the internal effects of ingesting unusually large amounts of salt. There is also evidence of a similar capacity to regulate calcium intake.

The most remarkable findings in this field concern vitamins. Experiments, again, mainly on domestic rats, have revealed relationships of great subtlety between dietary needs and behavior. The most detailed information concerns thiamin (vitamin B1). Thiamin-deficient rats, offered several foods, can acquire the habit of eating the one mixture that compensates for the deficiency. They do this most readily if each food has a distinct flavor. Evidently, the favorable physiological action of the vitamin becomes associated with eating the food that contains it. Such adaptive feeding is aided by exploration and sampling: thiamin-deficient rats tend to reject the foods to which they are accustomed and to select novel diets; as a result, their chances of eating a food with plenty of the vitamin are enhanced. There is also evidence of regulation of two other vitamins of the ''B'' group, riboflavin and pyridoxin.

Except by inference, we know virtually nothing of the role of these abilities in natural conditions. They have, however, a possible implication for pest control: when rodents are to be poisoned, the mixture should provide a complete diet. This is presumably especially important when poisons such as anticoagulants are used: these are effective only if they are regularly eaten over a period of some days. There seems, however, to be no information on the effect of the nutritional quality of baits.

D. Acquired Aversions

For pest control, the most important aspect of food selection is the ability to learn to reject a toxic mixture. This is the obverse of learning to eat a favorable food, for whenever one food is preferred another must be at least partly rejected. We are now concerned with a phenomenon, originally called bait-shyness or poison-shyness, observed during research on the effects of trying to destroy rodent populations by poisoning.[37,44] These studies, with those on neophobia, for the first time made possible effective methods of poison baiting founded on systematic knowledge of behavior.

The first major findings were on free-living populations of *Rattus norvegicus* in England. Since only one species was present, it was possible to estimate populations by putting down wheat grains soaked in water until consumption had reached an asymptote: this gave a minimum figure for the number of adult rats present.[11,37] In 58 experiments, described by Chitty and Southern,[37] zinc phosphide or alphanaphthylthiourea was used, without prebaiting, in a bait base quite different from the wheat used for census. Most of the poisonings were failures: more than 15% of the rats survived, and so the remaining populations were at a density which allowed a high rate of increase by breeding. Since the survivors usually refused the bait if it was offered again, they were more difficult to kill than before. The researches of this group, reported in detail, remain a model for workers on rodent and other pests.

Probably, all mammals can develop aversions to toxins. Rodents that (in the laboratory) have acquired aversions for mixtures containing a poison (usually zinc phosphide) include *R. rattus*,[13,23] *R. meltada*, and *R. cutchicus*,[94] two gerbils, *Tatera indica* and *Meriones hurrianae*,[92] (Figure 14), the spiny mouse, *Acomys cahirinus*,[47] and two cavies, *Cavia porcellus* and *C. aparea*.[28,29,69] A notable feature of these findings is the long delay (sometimes 2 hr) between ingestion of a poison and the development of ill effects: that is, between an animal's action and the impact of the resulting aversive stimulus. In experiments of traditional design, delays of more than a few minutes, or even seconds, lead to no learning. Learning after long delays depends on the character of the stimuli involved; and the cues to which an animal responds in this way vary with the species.[56-58,100]

For details of what happens in the field, we return to *R. norvegicus*. If poison bait is put on a runway, the first response is usually neophobic: the strange food is avoided. Habituation

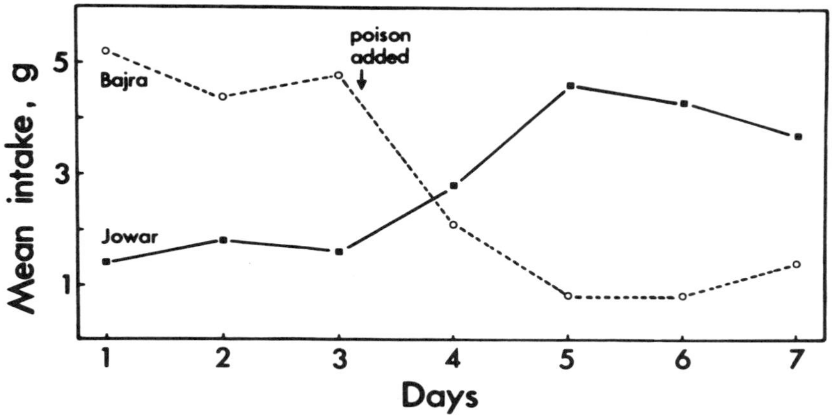

FIGURE 14. "Bait shyness" or learned aversion. Twelve caged desert gerbils, *Meriones hurrianae*, were fed on two foods. On Day 4, zinc phosphide was mixed with the preferred food (bajra). Thereafter, little bajra was eaten. (After Prakash, I. and Jain, A. P., *Ann. Appl. Biol.*, 69, 169, 1971.)

may take hours or days, but eventually the new food is sampled. There may then be an interval, during which the new food is not eaten, long enough for the poison to act. If the sample is small, the result is likely to be illness followed by recovery. During illness, nothing is eaten. When eating resumes, the mixture that has caused illness is refused.

It might be thought that bait shyness could be avoided by using very toxic substances. Zinc phosphide, the most used acute poison, has a LD_{50} for *R. norvegicus* of about 41 mg/kg body weight. Sodium fluoroacetate (1080) has a LD_{50} of 3.8 mg/kg or less (and is very dangerous to human beings and domestic animals). Yet failures due to shyness occur even to this poison.[20] The procedure that allows a high level of success against rat infestations is not to use highly toxic substances, but to train the rats to eat the bait. This is the now familiar method of prebaiting: the bait base is put down for several days without poison, so that the rodents become accustomed to eating it. Poison is then added; most of the rodents now make a full meal of the poison bait, and die (see Chapter 18).[114] The increasing readiness, with time, of animals to eat an initially unfamiliar food has led to the theory of "learned safety" proposed by Kalat and Rozin:[71] when eating a sample of food is *not* followed by illness, the animal behaves as if it had learned that the food is safe.

Aversions to poisonous mixtures are persistent and may be lifelong. Rzoska[104] gives 374 days as the longest duration of "shyness" recorded among domestic rats. Howard et al.,[67] in experiments on deer mice, *Peromyscus maniculatus*, found an aversion to sodium fluoroacetate to last 8 months, and similar or longer durations were found by Prakash and Jain[92] in their experiments on gerbils.

Although the expression "poison shyness" has been used, any component of the poisonous mixture may become repellent. Saccharin, which to us has a sweet taste, is readily accepted by rodents; but if it is a component of a poison bait survivors of the poisoning later reject it even in another mixture.

In the learned aversions of rodents (and of other mammals) the gustatory sense is the most important.[82] In experiments by Hankins et al.[61] domestic rats, made anosmic by applying zinc phosphate, $ZnSO_4$, to the nasal mucosa, were still able to develop a food-illness aversion. Similarly, Barnett et al.[13] showed *R. rattus* to retain a learned aversion when they had been made anosmic. Such aversions also generalize to substances which, for human beings, have similar tastes. An aversion to sodium chloride, NaCl, developed by domestic rats, is accompanied by rejection also of lithium chloride, LiCl; to us, both taste salty. The same applies to sweet-tasting substances, such as saccharin and glycine.[82]

Acquired aversions can, however, be mediated by odor. In experiments on domestic rats, Palmerino et al.[88] associated taste and odor with poisoning, and showed that rats could develop an aversion to an odor alone. Similarly, Mongolian gerbils, *Meriones unguiculatus*, studied by Pettijohn,[90] developed an aversion to the odor of acetic acid when it had been associated with illness. Some mammals can even learn to avoid a visible stimulus. Braveman[28,29] trained guinea-pigs, *Cavia porcellus*, to avoid a colored saccharin solution and, as usual, found taste to be the important cue; but the animals were also able to develop an aversion to the appearance of the mixture.

It is sometimes difficult to identify the cue to which animals are responding. Nachman et al.[82] review experiments in which caged domestic rats were made ill after drinking from a particular container and duly learned to avoid the container. The solutions offered in alternative containers did not differ in taste or odor. The rats may have been responding to (1) the appearance of the containers, (2) their positions in the cage, (3) the patterns of movement required for drinking, or (4) some combination of these.

The preceding paragraphs concern responses to the inanimate environment, but there is also evidence of socially influenced aversions, analogous to the socially determined preferences mentioned above. Galef and Clark[52] trained wild *R. norvegicus* to eat one food and to avoid an alternative, toxic mixture. The rats were breeding in small enclosures. When the young began to feed for themselves, the two foods were still offered, but neither was poisoned. Yet the young ate only the food that had not been poisoned. This was evidently because they followed adults to the feeding sites.

Another kind of socially caused aversion has been described by Lavin et al.[77] Adult domestic rats were exposed to a sick rat shortly after drinking a solution with an unfamiliar taste. The rats so treated developed an aversion to the new taste. Evidently a poisoned rat possesses some feature, perhaps an odor, which influences the behavior of conspecifics: the latter behave *as if* they had themselves been made ill by the unfamiliar solution, although they have not been poisoned. If, as is likely, wild rats also behave in this way, it is hardly surprising that, with all their other protective behavior patterns, they are so difficult to kill, or that they have such a reputation (however unjustified) for sagacity.

E. Questions of Taste

Choice of food does not always reflect dietary need or the avoidance of toxicity: sometimes the preferences of rodents resemble those of human beings who say they like or dislike some taste. The sparse information we have suggests that the palatability of food can be enhanced by increasing (1) its state of division, (2) its water content, (3) its oiliness, and (4) its sweetness.[9]

(1) *Rattus norvegicus* and *R. rattus*, given a choice between whole cereal grains and the same cereal as a flour, take the latter rather than the grains.[34,73] We do not know how widespread this preference is. Where it exists, the ability to make such discriminations has perhaps an application in poison baiting: *R. rattus* that have acquired an aversion to millet or wheat flour are said not to reject the same cereals offered as whole grains.[22] Such a finding could be applied in operations that require the successive use of two poison baitings (see Chapter 18).

(2) The need for water varies greatly among rodents. At one extreme, as we have seen, are the desert-living species which can subsist largely on dry seeds.[59] At the other are those, such as *R. norvegicus*, which thrive especially well near ponds and streams and drink as much water by weight as dry food; for them, wet baits are better than dry.

(3) Food mixtures can also be made more attractive by adding oils, especially those which to us have little taste. Wild *R. norvegicus* in England[9] and *R. rattus* in India[23] prefer cereals with added arachis oil. In Sweden, Hansson[62] has made similar observations on field voles, *Microtus agrestis*, bank voles, *Clethrionomys glareolus*, and wood mice, *Apodemus sylvaticus*.

(4) The method most used to make a food more attractive is to sweeten it with a sugar or saccharin.[21,68,116] Among the rodents that favor sweet mixtures are the five species mentioned in the previous paragraph. A sweet taste is not only attractive in itself: it may also disguise the repellent taste of another substance. Takahada et al.[113] made warfarin baits more acceptable to domestic rats by adding sucrose. A complicating factor is variation in the effect of sweetness with the state of the animal: if domestic rats (and probably other rodents) are fasted, their preference for highly sugary mixtures increases.[38]

Some tastes and odors are repellent. Unless there is no alternative, wild *R. norvegicus* reject grain mixed with aniseed oil or butyric acid; domestic rats, however, accept both.[9] This is one of many examples of how findings on laboratory varieties can be misleading when information is needed on the wild type. Inorganic ions may also be aversive. Ohnishi et al.[85] studied the responses of domestic mice to three salts of warfarin: the lithium salt was most repellent, that of potassium, the least; the sodium salt was intermediate. These authors relate their findings to neural physiology and to the chemical status of the three metals. Such a method could lead eventually to a systematic account of the responses of mammals to chemical stimuli.

Although knowledge of food preferences can help in the selection of baits, there is as yet no comprehensive account of the food preferences of commensal rodents. Much work has been fruitlessly directed toward finding substances which, by their odor, attract rodents to baits. Some well designed experiments, described by Bull,[30] are representative of findings in this area. The attempt was made, by using attractants, to control the choice of feeding sites by wild *Rattus norvegicus* in a spacious enclosure. The rats showed consistent preferences, but for positions, not odors or tastes.

F. The Ontogeny of Preferences

The preceding section may seem to imply that there are responses to tastes and odors which are independent of experience: that is, that they develop in each individual in any environment that permits survival. Such supposedly fixed features of behavior are, even today, sometimes said to be innate, or instinctive. To assume such fixity must, however, always be a presumption: development cannot be observed in all possible environments. It is more appropriate to refer to them as species-typical behavior patterns.[10] Preferences for sweet or oily mixtures are examples.

Preferences, even when they are apparently common to all members of a species, may still depend on individual experience. As we have seen, sweetness can become aversive. There is also some evidence that saccharin solutions are attractive to rats only when they have been previously associated with the presence of food in the stomach.[109] The environmental influences on the development of preferences are indeed much more complex than appears at first sight. Galef and his colleagues[52-55] have studied the feeding behavior of young domestic rats after their mothers had received different foods during lactation. According to their findings, at weaning the young are likely to choose the food eaten by the mother. They also give evidence that young rats can learn aversions with their mother's milk.

Capretta[33] has reviewed other experiments on domestic rats in which early experience of flavors influenced later preferences. In some experiments sucklings were obliged to ingest garlic with their milk by giving lactating females garlic solution to drink. Their young, tested later, drank more garlic solution than did controls which had received only garlic-free milk. The scope of such influences among free-living, wild-type rodents has still to be discovered.

V. CONCLUSIONS

Few species of rodents, even those most important economically, have been thoroughly studied, and our knowledge is distorted by the concentration of research on domestic varieties. Some tentative generalizations may, however, be made on their feeding and related behaviour.

1. All species are neophilic: that is, when their homeostatic needs are satisfied they move about, and approach especially the less familiar parts of their range; they also approach strange objects. While doing so, they store information about the environment, especially the whereabouts of food and shelter.
2. During such exploring, all possible foods are sampled.
3. If sampling of a new food has a favorable effect, that food comes to be preferred to alternatives; hence there is a tendency to select a balanced diet — an important ability for omnivores such as rats and mice.
4. Sampling also allows learning to avoid foods with ill effects, such as poison baits: the resulting learned aversions, or poison shyness, must be allowed for in designing methods of poisoning.
5. Social interactions, especially between parents and young, also influence food preferences.
6. Exploratory movements are limited by the need to return to the nest, but homing may occur from outside the usual range.
7. Exploration is also usually restricted by the territorial behavior of neighboring conspecifics.

A behavioral feature of at least two commensal species of rats is neophobia, that is, the avoidance of novel objects (including food items) in a familiar place. This behavior protects the animals from traps, and reinforces the effects of sampling dangerous foods. Whether other commensals are neophobic is not known. The house mouse is not neophobic.

Every rodent pest has an array of behavioral features which adapt it, often in unexpected ways, for life in human communities. Feeding behavior is usually adaptable to a wide range of foods, from grain to insects, and to a wide range of habitats. Successful management requires an intimate knowledge of the mode of life of each species.

APPENDIX: NOTES ON WORDS

ACTIVITY: A word with many meanings. In behavioral science these include: (1) the energy expended by an animal in a given time; (2) passages along a runway; (3) time spent out of a nest or lair; (4) carrying food or bedding; (5) gnawing; and (6) responding to novelty.

AVERSION, ACQUIRED: Refusal, or a decline in the amount consumed, of a food when illness has followed eating the food. "Food" in this context includes fluids.

CIRCADIAN RHYTHM: A cycle of behavioral or physiological changes recurring about every 24 hr.

COMMENSAL: Of members of different species: feeding in the same place, on the same food. In the present context, refers to species that live in human communities, and live on our food.

EXPLORATORY BEHAVIOR: An imprecise term referring to apparently unrewarded movements about an animal's living space. See latent learning.

HABITUATION: Decline in performance of an act or behavior pattern as a result of repeated exposure to a stimulus that evokes the act. In the present context, refers to a decline in the avoidance of a new object, or in the refusal of new food.

HOME RANGE: The whole region visited by an animal during a specified period. Should be distinguished from territory.

LATENT LEARNING: Storage of information (in the central nervous system) leading to more rapid solving of problems, or superior habit formation, as a result of exploration.

NEW OBJECT REACTION: See neophobia.

NEOPHILIA: Approaching unfamiliar, in preference to familiar, objects or places.
NEOPHOBIA: The avoidance of an unfamiliar object in a familiar environment. Should not be restricted to avoidance of foods, flavors, or odors. Contrast acquired aversions (''shyness''), in which avoidance is a result of past experience of what is avoided.
''SHYNESS'' (of baits or poisons): See aversion and neophobia.
TERRITORY: A region occupied by an individual, pair, or group to the exclusion of other conspecifics. Sometimes restricted to *defended* regions.

REFERENCES

1. **Archer, J.,** Tests for emotionality in rats and mice: a review, *Anim. Behav.*, 21, 205, 1973.
2. **Bagnall-Oakley, R. P.,** *Trans. Norfolk Norw. Nat. Soc.*, 18, 17, 1955.
3. **Bammer, G., Barnett, S. A., and Marples, T. G.,** Movement patterns of wild house mice in a residential maze, *Psychol. Rec.*, 37, 545, 1987.
4. **Bammer, G., Barnett, S. A., and Marples, T. G.,** Responses to novelty by the Australian swamp rat, *Rattus lutreolus, Aust. Mammal,* in press.
5. **Barnett, S. A.,** Behaviour components in the feeding of wild and laboratory rats, *Behaviour,* 9, 24, 1956.
6. **Barnett, S. A.,** Experiments on ''neophobia'' in wild and laboratory rats, *Br. J. Psychol.*, 49, 195, 1958.
7. **Barnett, S. A.,** Exploratory behaviour, *Br. J. Psychol.*, 49, 289, 1958.
8. **Barnett, S. A.,** Laboratory methods for behaviour studies of wild rats, *J. Anim. Tech. Assoc.*, 9, 6, 1958.
9. **Barnett, S. A.,** *The Rat: A Study in Behavior,* Chicago University Press, Chicago, Ill., 1981.
10. **Barnett, S. A.,** *Modern Ethology: The Science of Animal Behavior,* Oxford University Press, New York, 1981.
11. **Barnett, S. A., Bathard, A. H., and Spencer, M. M.,** Rat populations and control in two English villages, *Ann. App. Biol.*, 38, 444, 1951.
12. **Barnett, S. A. and Cowan, P. E.,** Activity, exploration, curiosity and fear, *Interdiscip. Sci. Rev.*, 1, 43, 1976.
13. **Barnett, S. A., Cowan, P. E., Radford, G. G., and Prakash, I.,** Peripheral anosmia and the discrimination of poisoned food by *Rattus rattus* L., *Behav. Biol.*, 13, 183, 1975.
14. **Barnett, S. A., Dickson, R. G., Marples, T. G., and Radha, E.,** Sequences of feeding, sampling and exploration by wild and laboratory rats, *Behav. Proc.*, 3, 29, 1978.
15. **Barnett, S. A. and Hocking, W. E.,** Are nests built for fun? Effects of alternative activities on nest-building by wild house mice, *Behav. Neural. Biol.*, 31, 73, 1981.
16. **Barnett, S. A., Hocking, W. E., and Wolfe, J. L.,** Effects of cold on activity and exploration by wild house mice in a residential maze, *J. Comp. Physiol.*, 123, 91, 1978.
17. **Barnett, S. A. and McEwan, I. M.,** Movements of virgin, pregnant and lactating mice in a residential maze, *Physiol. Behav.*, 10, 741, 1973.
18. **Barnett, S. A. and Prakash, I.,** *Rodents of Economic Importance in India,* Heinemann-Educational, London, 1975.
19. **Barnett, S. A. and Smart, J. L.,** The movements of wild and domestic house mice in an artificial environment, *Behav. Biol.*, 15, 85, 1975.
20. **Barnett, S. A. and Spencer, M. M.,** Sodium fluoracetate as a rat poison, *J. Hyg. Cambridge,* 47, 426, 1949.
21. **Bhardwaj, D. and Khan, J. A.,** Sugar preferences of ''black rats'', *Rattus rattus* L., *J. Linn. Soc. Zool.*, 64, 41, 1978.
22. **Bhardwaj, D. and Khan, J. A.,** Effect of texture of food on bait-shy behaviour in wild rats *(Rattus rattus), Appl. Anim. Ethol.*, 5, 361, 1979.
23. **Bhardwaj, D. and Khan, J. A.,** Responses of roof rat, *Rattus rattus* L., to non-oily and oily foods after poisoning in oily foods, *Proc. Indian Acad. Sci. B.*, 88, 125, 1979.
24. **Boice, R.,** Laboratorizing the wild rat *(Rattus norvegicus), Behav. Res. Methods Instrum.*, 3, 177, 1971.
25. **Booth, D. A.,** Approaches to feeding control, in *Appetite and Food Intake, Life Sciences Research Report 2,* Silverstone, T., Ed., Dahlem Konferenzen, W. Berlin, 1976, 417.
26. **Brant, D. H. and Kavanau, J. L.,** ''Unrewarded'' exploration and learning of complex mazes by wild and domestic mice, *Nature (London),* 204, 267, 1964.
27. **Brant, D. H. and Kavanau, J. L.,** Exploration and movement patterns of the canyon mouse *Peromyscus crinitus* in an extensive laboratory enclosure, *Ecology,* 46, 452, 1965.

28. **Braveman, N. S.,** Poison-based avoidance learning with flavored or colored water in guinea pigs, *Learn. Motiv.*, 5, 182, 1974.
29. **Braveman, N. S.,** Relative salience of gustatory and visual cues in the formation of poison-based food aversions by guinea pigs *(Cavia porcellus), Behav. Biol.*, 14, 189, 1975.
30. **Bull, J. O.,** The influence of attractants and repellents on the feeding behaviour of *Rattus norvegicus*, in *Proc. 5th Vertebrate Pest Conference*, Marsh, R. E., Ed., Fresno, Calif., 1972, 150.
31. **Calhoun, J. B.,** A method for self-control of population growth among mammals living in the wild, *Science, N.Y.*, 109, 333, 1949.
32. **Candland, D. K. and Nagy, Z. M.,** The open field: some comparative data, *Ann. N.Y. Acad. Sci.*, 159, 831, 1969.
33. **Capretta, P. J.,** Establishment of food preferences by exposure to ingestive stimuli early in life, in *Learning Mechanisms in Food Selection*, Barker, L. M., Best, M. R., and Domjan, M., Eds., Baylor University Press, Waco, Texas, 1977, 99.
34. **Carlson, A. J. and Hoelzel, F.,** Influence of texture of food on its acceptance by rats, *Science, N.Y.*, 109, 63, 1949.
35. **Carstairs, J. L.,** Population dynamics and movements of *Rattus villosissimus* (Waite) during the 1966-69 plague at Brunette Downs, N. T., *Aust. Wildl. Res.*, 3, 1, 1976.
36. **Chitty, D. and Shorten, M.,** Techniques for the study of the Norway rat *(Rattus norvegicus), J. Mammal.*, 27, 63, 1946.
37. **Chitty, D. and Southern, H. N., Eds.,** *The Control of Rats and Mice*, 3 Vols., Clarendon Press, Oxford, 1954.
38. **Collier, G. and Bolles, R.,** Hunger, thirst, and their interaction as determinants of sucrose consumption, *J. Comp. Physiol. Psychol.*, 66, 633, 1968.
39. **Cottam, C.,** Aquatic habits of the Norway rat, *J. Mammal.*, 29, 299, 1948.
40. **Cowan, P. E.,** Systematic patrolling and orderly behaviour of rats during recovery from deprivation, *Anim. Behav.*, 25, 171, 1977.
41. **Cowan, P. E.,** Exploration in small mammals, in *Exploration in Animals and Humans*, Archer, J. and Burke, L. I. A., Eds., Van Nostrand Rheinhold, Wokingham, England, 1983, 147.
42. **Cowan, P. E. and Barnett, S. A.,** The new-object and new-place reactions of *Rattus rattus* L., *Zool. J. Linn. Soc.*, 56, 219, 1975.
43. **Domjan, M.,** Attenuation and enhancement of neophobia for edible substances, in *Learning Mechanisms in Food Selection*, Barker, L. M., Best, M. R., and Domjan, M., Eds., Baylor University Press, Waco, Texas, 1977.
44. **Doty, R. E.,** Rat control on a Hawaiian sugar cane plantation, *Hawaiian Plant. Rec.*, 49, 71, 1945.
45. **Ellen, P., Soteres, B. J., and Wages, C.,** Problem solving in the rat, *Anim. Learn. Behav.*, 12, 232, 1984.
46. **Elton, C.,** *Voles, Mice and Lemmings*, Clarendon Press, Oxford, 1942.
47. **Etscorn, F.,** Illness-induced aversion learning in a desert species of rodent *(Acomys cahirinus), Physiol. Psychol.*, 5, 336, 1977.
48. **Fernando, H. E., Kawamoto, N., and Perera, N.,** The biology and control of the rice field mole rat of Ceylon, *Gunomys gracilis, FAO Plant Prot. Bull.*, 15, 32, 1967.
49. **File, S. E. and Wardill, A. G.,** Validity of head-dipping as a measure of exploration in a modified hole-board, *Psychopharmacologia*, 44, 53, 1975.
50. **Fisler, G. F.,** Adaptations and speciation in harvest mice of the marshes of San Francisco Bay, *Univ. Calif. Berkeley Publ. Zool.*, 77, 1, 1965.
51. **Fulk, G. W.,** Food hoarding of *Bandicota bengalensis, Mammalia*, 41, 539, 1977.
52. **Galef, B. G. and Clark, M. M.,** Social factors in the poison avoidance and feeding behavior of wild and domesticated rat pups, *J. Comp. Physiol. Psychol.*, 75, 341, 1971.
53. **Galef, B. G. and Henderson, P. W.,** Mother's milk: a determinant of the feeding preferences of weaning rat pups, *J. Comp. Physiol. Psychol.*, 78, 213, 1972.
54. **Galef, B. G., Kennett, D. J., and Wigmore, S. W.,** Transfer of information concerning distant food in rats, *Anim. Learn. Behav.*, 12, 292, 1984.
55. **Galef, B. G. and Sherry, D. F.,** Mother's milk: a medium for transmission of cues reflecting the flavor of mother's diet, *J. Comp. Physiol. Psychol.*, 83, 374, 1973.
56. **Garcia, J. and Levine, M. S.,** Learning paradigms and the structure of the organism, in *Neural Mechanisms of Learning and Memory*, Rosenzweig, M. R. and Bennett, E. L., Eds., Massachusetts Institute of Technology, Cambridge, 1976, 193.
57. **Garcia, J., McGowan, B. K., and Green, K. F.,** Biological constraints on conditioning, in *Classical Conditioning II: Current Research and Theory*, Black, A. H. and Prokasy, W. F., Eds., Appleton-Century-Crofts, New York, 1972.
58. **Garcia, J. and Rusiniak, K. W.,** What the nose learns from the mouth, in *Chemical Signals*, Muller-Schwarze, D. and Silverstein, R. M., Eds., Plenum Press, N.Y., 1980, 141.

59. **Ghobrial, L. I. and Nour, T. A.,** The physiological adaptations of desert rodents, in *Rodents in Desert Environments,* Prakash, I. and Ghosh, P. K., Eds., W. Junk, The Hague, 1975, 413.
60. **Gorecki, A. and Gebczynska, Z.,** Food conditions for small rodents in a deciduous forest, *Acta Theriol.,* 6, 275, 1962.
61. **Hankins, W. G., Garcia, J., and Rusiniak, K. W.,** Dissociation of odor and taste in baitshyness, *Behav. Biol.,* 8, 407, 1973.
62. **Hansson, L.,** Fatty substances as attractants for *Microtus agrestis* and other small rodents, *Oikos,* 24, 417, 1973.
63. **Harrison, J. L.,** The natural food of some rats and other mammals, *Bull. Raffles Mus.,* 25, 157, 1954.
64. **Hart, J. S.,** Rodents, in *Comparative Physiology of Thermoregulation II,* Whittow, G. C., Ed., Academic Press, London, 1971, 2.
65. **Healey, M. C.,** Aggression and self-regulation of population size in deermice, *Ecology,* 48, 377, 1967.
66. **Horner, B. E., Taylor, J. M., and Padykula, H. A.,** Food habits and gastric morphology of the grasshopper mouse, *J. Mammal.,* 45, 513, 1965.
67. **Howard, W. E., Marsh, R. E., and Cole, R. E.,** Duration of associative memory to toxic bait in deer mice, *J. Wildl. Manage.,* 41, 484, 1977.
68. **Howard, W. E., Marsh, R. E., and Palmateer, S. D.,** Rat acceptance of different sugar concentrations in water baits, *Int. Pest Control.,* 14, 17, 1972.
69. **Jacobs, W. W. and Labows, J. N.,** Conditioned aversion, bitter taste and the avoidance of natural toxicants in wild guinea pigs, *Physiol. Behav.,* 22, 173, 1979.
70. **Johnson, R. P.,** Scent marking in mammals, *Anim. Behav.,* 21, 521, 1973.
71. **Kalat, J. W. and Rozin, P.,** "Learned safety" as a mechanism in long-delay taste-aversion learning in rats, *J. Comp. Physiol. Psychol.,* 83, 198, 1973.
72. **Kamil, A. C. and Sargent, T. D., Eds.,** *Foraging Behavior,* Garland, New York, 1981.
73. **Khan, J. A.,** Laboratory experiments on the food preferences of the black rat *(Rattus rattus* L.), *Zool. J. Linn. Soc.,* 54, 167, 1974.
74. **Krebs, C. J.,** Demographic changes in fluctuating populations of *Microtus californicus, Ecol. Monogr.,* 36, 239, 1966.
75. **Kumari, P. V. and Khan, J. A.,** Food hoarding by Indian gerbil, *Tatera indica indica* (Hardwicke), *Proc. Indian Acad. Sci. B.,* 88, 131, 1979.
76. **Lát, J.,** Spontaneous exploratory reactions as a tool for psychopharmacological studies, *Proc. 2nd Int. Pharmacol. Meeting,* 47, 1965.
77. **Lavin, M. J., Freise, B., and Coombes, S.,** Transferred flavor aversions in adult rats, *Behav. Neural. Biol.,* 28, 15, 1980.
78. **Lim, B. L.,** Land molluscs as food of Malayan rodents and insectivores, *J. Zool. London,* 148, 554, 1966.
79. **Maller, O.,** Specific appetite, in *The Chemical Senses and Nutrition,* Kare, M. R. and Maller, O., Eds., Johns Hopkins Press, Baltimore, Md., 1967.
80. **Miller, R. R. and Holzman, A. D.,** Neophobias and conditioned taste aversions in rats following exposure to novel flavors, *Anim. Learn. Behav.,* 9, 89, 1981.
81. **Mitchell, D.,** Experiments on neophobia in wild and laboratory rats, *J. Comp. Physiol. Psychol,* 90, 190, 1976.
82. **Nachman, M., Rauschenburger, J., and Ashe, J. H.,** Stimulus characteristics in food aversion learning, in *Food Aversion Learning,* Milgram, N. W., Krames, L., and Alloway, T. M., Eds., Plenum Press, New York, 1977, 105.
83. **Nass, R. D.,** Movements and home ranges of Polynesian rats in Hawaiian sugarcane, *Pac. Sci.,* 31, 135, 1977.
84. **Newsome, A. E., Cowan, P. E., and Ives, P. M.,** Homing by wild house-mice, *Aust. Wildl. Res.,* 9, 421, 1982.
85. **Ohnishi, M., Sugawara, R., and Kusano, T.,** Effects of cation constituents on the gustatory responses of mice to aqueous solutions of Warfarin alkali salts, *Rep. Spec. Res. Project Trop. Agric. Resources,* 3, 107, 1984.
86. **Olton, D. S. and Schlosberg, P.,** Food-searching strategies in young rats: win-shift predominates over win-stay, *J. Comp. Physiol. Psychol.,* 92, 609, 1978.
87. **Olton, D. S., Walker, J. A., Gage, F. H., and Johnson, C. T.,** Choice behavior of rats searching for food, *Learn. Motiv.,* 8, 315, 1977.
88. **Palmerino, C. C., Rusiniak, K. W., and Garcia, J.,** Flavor-illness aversions: the peculiar roles of odor and taste in memory for poison, *Science, N.Y.,* 208, 753, 1980.
89. **Perrigo, G. and Bronson, F. H.,** Behavioral and physiological responses of female house mice to foraging variation, *Physiol. Behav.,* 34, 437, 1985.
90. **Pettijohn, T. F.,** Conditioned olfactory aversion in the male Mongolian gerbil, *Physiol. Psychol.,* 7, 199, 1979.

91. **Poche, R. M., Mian, M. Y., Haque, M. E., and Sultana, P.,** Rodent damage and burrowing in Bangladesh, *J. Wild. Manage.*, 46, 139, 1982.
92. **Prakash, I. and Jain, A. P.,** Bait shyness of two gerbils, *Tatera indica indica* Hardwicke and *Meriones hurrianae* Jerdon, *Ann. App. Biol.* 69, 169, 1971.
93. **Prakash, I. and Kumbkarni, C. G.,** Eco-toxicology and control of Indian Desert gerbille, *Meriones hurrianae, J. Bombay Nat. Hist. Soc.*, 59, 1, 1962.
94. **Prakash, I., Rana, B. D., and Jain, A. P.,** Bait shyness in three species of *Rattus, Z. Angew, Zool.*, 62, 89, 1975.
95. **Prakash, I., Taneja, G. C., and Purohit, K. G.,** Eco-toxicology and control of Indian desert gerbil, *Meriones hurrianae* (Jerdon). VIII. Relative number in relation to ecological factors, *J. Bombay Nat. Hist. Soc.*, 68, 86, 1971.
96. **Pyke, G. H., Pulliam, H. R., and Charnov, E. L.,** Optimal foraging, *Q. Rev. Biol.*, 52, 137, 1977.
97. **Ranson, R. M.,** The field vole *(Microtus)* as a laboratory animal, *J. Anim. Ecol.*, 3, 70, 1934.
98. **Reichman, O. J.,** Factors influencing foraging in desert rodents, in *Foraging Behavior,* Kamil, A. C. and Sargent, T. D., Eds., Garland, New York, 1981, 195.
99. **Reichman, O. J. and Rebar, C.,** Seed preferences of desert rodents based on levels of mouldiness, *Anim. Behav.*, 33, 726, 1985.
100. **Revusky, S.,** Interference with progress by the scientific establishment: examples from flavor aversion learning, in *Food Aversion Learning,* Milgram, N. W., Krames, L., and Alloway, T. M., Eds., Plenum Press, New York, 1977, 53.
101. **Roy, S. K.,** Pre-harvest loss of rice due to field rodents, *Econ. Pol. Week.*, 9, 66, 1974.
102. **Rozin, P.,** The selection of foods by rats, humans and other animals, in *Advances in the Study of Behavior,* Vol. 6, Rosenblatt, J. S., Hinde, R. A., Shaw, E., and Beer, C., Eds., Academic Press, New York, 1976, 21.
103. **Russell, P. A.,** Psychological studies of exploration in animals, in *Exploration in Animals and Humans,* Archer, J. and Burke, L. I. A., Eds., Van Nostrand Rheinhold, Wokingham, England, 1983.
104. **Rzoska, J.,** The behaviour of white rats towards poison baits, in *The Control of Rats and Mice,* Vol. 2, Chitty, D., Ed., Clarendon, Oxford, 1954, 374.
105. **Schneider, G. E. and Gross, C. G.,** Curiosity in the hamster, *J. Comp. Physiol. Psychol.*, 49, 150, 1965.
106. **Shorten, M.,** The reaction of the brown rat towards changes in its environment, in *The Control of Rats and Mice,* Vol. 2, Chitty, D., Ed., Clarendon, Oxford, 1954, 307.
107. **Shorten, M.,** *Squirrels,* Collins, London, 1954.
108. **Sims, R. A. and Wolfe, J. L.,** Homing behavior of the house mouse *(Mus musculus* L.), *Miss. Acad. Sci.*, 21, 89, 1976.
109. **Smith, M. P. and Capretta, P. J.,** Effects of drive level and experience on the reward value of saccharine solutions, *J. Comp. Physiol. Psychol.*, 49, 553, 1956.
110. **Southern, H. N.,** *The Control of Rats and Mice,* Vol. 3, Clarendon, Oxford, 1954.
111. **Strong, P. N.,** Activity in the white rat as a function of apparatus and hunger, *J. Comp. Physiol. Psychol.*, 50, 596, 1957.
112. **Summerlin, C. T. and Wolfe, J. L.,** Social influences on trap response of the cotton rat, *Sigmodon hispidus, Ecology,* 54, 1156, 1973.
113. **Takada, Y., Kusano, T., and Sugawara, R.,** Behavioral and physiological approach to the evaluation of rodenticidal effectiveness of sodium Warfarin, *Rec. Spec. Res. Project. Trop. Agric. Res.*, 2, 91, 1983.
114. **Thompson, H. V.,** Studies of the behaviour of the common brown rat, *Rattus norvegicus,* Berkenhout I, *Bull. Anim. Behav.*, 6, 1948.
115. **Vickery, W. L.,** Optimal diet models and rodent food consumption, *Anim. Behav.*, 32, 340, 1984.
116. **Wagner, M. W.,** Comparative rodent preferences for artificial sweeteners, *J. Comp. Physiol. Psychol.*, 75, 483, 1971.
117. **Wilson, R. C., Vacek, T., Lanier, D. L., and Dewsbury, D. A.,** Open-field behavior in muroid rodents, *Behav. Biol.*, 17, 495, 1976.
118. **Yabe, T.,** The relation of food habits to the ecological distributions of the Norway rat *(Rattus norvegicus)* and the roof rat *(R. rattus), Jap. J. Ecol.*, 29, 235, 1979.

Chapter 18

BAIT SHYNESS AND POISON AVERSION

Ishwar Prakash

TABLE OF CONTENTS

I. INTRODUCTION

The availability of anticoagulant rodenticides, particularly that of the second generation, has generated ample hope that, by their use, the rodent populations can be reduced below the threshold level, thereby substantially reducing the losses incurred by these pests. However, in developing countries, the use of anticoagulant rodenticides for the control of field rodents is still not in vogue due to a number of reasons: their nonavailability, poor knowledge about their efficacy, and relatively higher cost of operation. The acute toxicant, zinc phosphide, still remains the most used and popular rodenticide in most parts of the world. In a control operation, zinc phosphide baits are recommended for distribution in the fields on three consecutive days.[15,28,56] It has, however, been revealed by our experimental results that even after a single exposure of sublethal dose of zinc phosphide, a wide variety of rodent species not only develop poison aversion but also become bait shy.[37] This ability of rodents to discriminate between harmful and harmless foods is a protective shield or a "learned safety" for them to escape the adverse effects of poisonous foods.[10,26] For the purpose of enhancing control success, on the basis of our findings, it is now recommended that zinc phosphide baits should be laid in the field for rodent control (1) after 2 days of prebaiting (2) and only on a single day. The residual population should be tackled with aluminum phosphide fumigation of burrows.[16,35] In recent years, however, the use of aluminum phosphide has been restricted in India to trained rodent control operators and the hazardous fumigant cannot be used by illiterate farmers. As a consequence we are left with almost no alternative to kill the residual population after a single-day zinc phosphide operation for the control of field rodents. With a view to enhance the acceptability of zinc phosphide on subsequent exposures and to mitigate the shyness behavior of rodents, a number of experiments have been carried out, unfortunately, only with a partial success. The synthesized information on shyness behavior exhibited by Indian rodent species as induced by exposure to acute rodenticides is presented in this communication.

II. BAIT SHYNESS AND POISON AVERSION

When poison baits are exposed to the rodent populations, some of them consume a lethal quantity and succumb, whereas those which ingest only a sublethal dose of poison survive the poisoning. These survivors suffer a physiological disorder/gastro intestinal distress and are able to detect the poison and even the bait material after recovering from the illness. Further, they refuse to feed on the same for a considerable period. This phenomenon of rejecting the bait on subsequent exposures is termed as *bait shyness* and the aversion thus developed towards the poison is called *poison aversion*. Most of the predominant rodents of economic importance in India develop bait shyness even after a single exposure to the acute toxicant, zinc phosphide, the most-used rodenticide in India. There is no difference in the exhibition of this behavior in the members of Squiridae, Cricetidae, and Muridae: the five-striped squirrel, *Funambulus pennanti;*[31] the hairy-footed gerbil, *Gerbillus gleadowi;*[40] the Indian gerbil, *Tatera i. indica;*[37] *T. i. cuvieri;*[52] the desert gerbil, *Meriones hurrianae;*[37] the rock rat, *Rattus cutchicus;*[39] the metad, *Rattus meltada pallidior;*[39] the house rat, *Rattus rattus*[39] (Figure 1); the house mouse, *Mus musculus;*[41] Indian field mouse, *Mus booduga;*[42] brown mouse, *Mus platythrix;*[42] the bandicoots, *Bandicota bengalensis;*[53,56] and *B. indica.*[55] The development of bait shyness due to consumption of other rodenticides has also been reported. Rzoska[47] found that *Rattus norvegicus* exhibited distinct aversion to red squill, barium carbonate, and arsenic oxide baits, the shyness persisted for more than a year with red squill. Likewise ANTU and Norbormide, which are specific rodenticides for the *Rattus* group, could develop aversion in *Rattus norvegicus.*[17,19,47] Even one of the most potent rodenticides developed so far, a tasteless and odorless chemical, the compound 1080 (Sodium

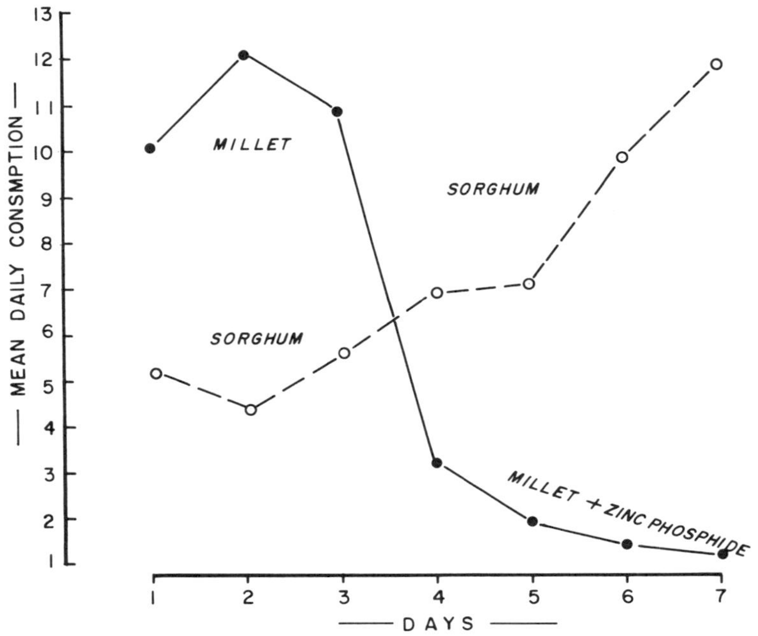

FIGURE 1. Poison shyness of *Rattus rattus rufescens*. Addition of sublethal dose of zinc phosphide to the preferred food changed the food preference.

monofluoroacetate) was memorized by *Peromyscus maniculatus* for 8 months after sublethal ingestion.[22] Another acute rodenticide RH-787 did not cause bait shyness among most of the rodent species. But *Rattus rattus* exhibited bait shyness for 42 days after exposure to a sublethal dose of RH-787 for 4 days, whereas among *Tatera indica* and *Mus platythrix* shyness was developed for 110 and 120 days after a single day exposure to its sublethal dose.[55] The chronic or multiple dose rodenticides which are used in very low concentrations normally do not induce bait shyness among rodents. However, Armour and Barnett[1] reported aversion by *Rattus norvegicus* to dicoumarol treated baits.

It was revealed that the intensity of shyness differs from poison to poison and from rodent species to species. Bhardwaj and Prakash[9] found it to be of low order with RH-787 as compared to zinc phosphide, all conditions of administration being the same. Moreover, it was evidenced that the magnitude of bait shyness as judged by its persistence, is greater when the intake of poisoned food is in a quick succession (continuously for 3 to 4 days) as compared to if it is ingested intermittently, say at an interval of 6 days.[9] It is quite likely that behaviorally the experience of illness or discomfort caused by ingestion of a sublethal dose of poison at intervals is not deeply imprinted and is memorized for a very short duration by *Rattus rattus* as compared to that when poison is accepted by them, though in declining proportion, in quick repeat succession.[5,9] The illness due to poison associated with some bases are remembered by *Rattus norvegicus* more clearly than other species.[48] In our experiments,[8] when *Rattus rattus* was poisoned with RH-787 for 4 days in wheat flour + sugar, the shyness towards the base continued only for 28 days, but with millet flour + sugar, it continued for 42 days. Rzoska[48] concluded that the nature of bait influences the degree and duration of memory. In this context, it was also observed that mixing of oil with the bait induces a severe shyness among *Rattus rattus* as compared to that due to the addition of sugar, which is more tasteful.[7] It appears that the stronger the taste of a food or a drink, the greater the aversion.[3,20] Social factors also have an important role in poison avoidance and feeding behavior. Galef and Clark[18] reported that rat pups born to colony members learn cues associated with food by following their parents. They do not eat the diet the adults avoid and continue to avoid that diet even after being isolated from the adults in a new environment. Early diet experience in life was found to have some influence in adult diet selection among *Rattus norveqicus*[27] and *Mus booduga*.[53] It is interesting to report in this context that 12.5% of *Rattus meltada* did not consume poison bait even on the first day of zinc phosphide exposure,[39] which may be contributed to their previous experience of feeding upon the poison. Experiments to investigate the role of odor and taste in food-illness aversion with *Rattus norveqicus*[3,20] and *Rattus rattus* indicated that once the avoidance has developed, the discrimination does not depend on the olfaction (olfactory senses were obliterated by making the rats anosmic by treatment of nasal mucous membrane by zinc sulphate), but the taste organs are probably responsible for it. Kalat[24] studied taste salience and indicated that the rats tend to form an aversion to the novelty of the taste (or unfamiliarity), and not on the "strength" of the taste. The residual population of rodents which survive poisoning may change its behaviour.[43] The tendency to sample new food is increased, free living populations may leave their burrows and make new ones elsewhere.[4] Revusky and Garcia[43] held that learning is a single, general process which, like gravity, is more or less the same wherever it occurs, except for parametric perturbations. On the contrary, Rozin and Kalat[46] and Shettleworth[49] used food aversion learning as the primary example to support the contention that learning is a diverse category including several specialized mechanisms each adapted to particular ecological situations and evolutionary pressures. Kalat,[25] in his exhaustive review concluded that "it makes sense to adopt a working hypothesis that learning differs from one situation or species to another". Aversions can be developed without exposing rodents to the poison baits, i.e., neophobia or new food reaction or new object reaction (see Chapter 17).[2] Such behavioral implications can also reduce the efficacy of a control operation if not considered adequately. When a new food is placed in the familiar environment of rodents they sample only a very small quantity (neophobia), hence a lethal dose cannot be ingested;[33,50,51] although, once strange foods are contacted by one rat others quickly follow and cease their avoidance behavior.[13] A prebaiting of 2 to 3 days is, therefore, suggested before undertaking poison baiting. When bait stations are kept in a habitat, rodents exhibit "container" neophobia.[10] Rodents usually acclimatize[11] to such objects in 4 to 5 days, therefore bait stations should be placed ahead for this period, prior to control operations.

III. PERSISTENCE OF SHYNESS

The bait shyness and poison aversion phenomena, due to ingestion of various poisons, persist for varying durations in different species and with different rodenticides. The period for which rodents can associate the memory of illness caused by the ingestion of a sublethal dose of zinc phosphide vary from 10 to 15 days in *Gerbillus gleadowi* to 135 days in *Rattus meltada* (Table 1), but the bait shyness persisted per 240 days in *Peromyscus maniculatus* due to sodium monofluoroacetate[22] and for 374 days in *Rattus norvegicus* due to red squill[48] poisoning.

The persistence of bait shyness may also depend on the bait and duration of baiting period. The house mouse, *Mus musculus* exhibited bait shyness for only 15 days when zinc phosphide was exposed for a single day. The shyness increased to 30 days when poison was exposed for 4 consecutive days.[41] The persistence of shyness in rodents, which are not prebaited prior to poison baiting, is of a higher magnitude than in those which are exposed to the toxicant after prebaiting.[11] The intensity of bait shyness is more severe among certain species than others. For example, some species (*Rattus rattus, Rattus cutchicus*,[39] *Meriones hurrianae, Tatera indica*[37]) completely stopped feeding on the preferred bait in which poison was mixed on 4th day after 3 days of prebaiting. However, others consumed smaller amount

Table 1
RESISTANCE OF BAIT SHYNESS AMONG RODENTS INDUCED BY VARIOUS RODENTICIDES

Rodenticide	Rodent species	Persistence of shyness (days)	Ref.
ANTU	*Rattus norvegicus*	180	17
Red Squill	*Rattus norvegicus*	374	48
Barium Carbonate	*Rattus norvegicus*	60	48
Arsenic Oxide	*Rattus norvegicus*	60	48
Norbormide	*Rattus norvegicus*	—	19
Compound 1080	*Peromyscus maniculatus*	240	22
Strychnine	*Rattus norvegicus*	—	23
Sulphate	*Rattus rattus* and Pocket gophers	—	23
Dicoumarol	*Rattus norvegicus*	—	1
RH-787	*Tatera indica*	120	52
	Rattus rattus	42	9
	Mus platythrix	120	9
Zinc Phosphide	*Funambulus pennanti*	30	31
	Gerbillus gleadowi	10—15	40
	Tatera indica	115	37
	Meriones hurrianae	35	37
	Rattus rattus	75	39
	Rattus cutchicus	75	39
	Rattus meltada	135	39
	Mus musculus	20	41
	Mus booduga	95	42
	Mus platythrix	75	42
	Mus platythrix	110	54
	Bandicota bengalensis	30	53
	Bandicota indica	105	55

of preferred bait even after the exposure to the zinc phosphide. Though, in both the cases the amount consumed was significantly less than that consumed prior to feeding on the poison.

IV. MITIGATION OF SHYNESS BEHAVIOR

Attempts have been made to overcome bait shyness in order to control the rodents surviving a control operation. Rzoska[47] concluded that bait shy rodents reject the bait material experienced earlier and not the rodenticide. This was, however, not confirmed by Prakash and Ojha,[38] indicating that shifting of zinc phosphide to a new bait in the case of Indian gerbil, *Tatera indica* did not reduce the shyness behavior towards the plain bait. Howard[21] suggested the use of durable baits and prebaiting of bait shy rats surviving a 1080 poisoning. Texture of food material can also play a role in overcoming bait shyness. Bait shy *Rattus rattus* do not avoid husked lentil and husked green gram after being poisoned with 0.02% zinc phosphide in whole lentil and green gram.[6] Extent of the bait shyness can be reduced to certain extent if during the successive baiting, poison is mixed without oil, if given with oil in the first baiting.[7] Can bait shyness be retarded by changing one or two of the components (baits, poison, additives) in subsequent baiting? Zinc phosphide induced bait shyness among *Tatera indica* and *Meriones hurrianae* could not be retarded even when the poison was given only once in a week or when the bait was changed in every subsequent poisoning.[14] Changing of both oil and bait was not enough to inhibit the Indian gerbils to develop bait shyness. Bhardwaj and Prakash[8] have inferred that if two additives of the poison bait, viz., sugar and groundnut oil, are mixed in a new food of which *Rattus rattus* has had no experience, the shyness continues, being more severe with the latter. However, Bhardwaj and Khan[6] examined a question of practical importance concerning the development of shyness to textural variants of the same food (e.g., grains in various states of division). Assuming that if textural variation is not important, then aversion developed for wheat flour would also result in the avoidance of whole wheat. However, their experiments revealed that poisoning in flour did not result in rejection of the corresponding whole grain.[6] The authors conjecture that possibly, due to the presence of outer seed structures (pericarp and testa), the taste of grains is perceived distinctly from flour and can be used successively in poisoning operations.

It has been clearly worked out that the persistence of shyness in rodents, which are not prebaited prior to poison baiting, is of a higher magnitude than in rodents that are exposed to zinc phosphide after 3 days of prebaiting.[5,11] The conspecific odor of body and fractions of urine significantly enhance the bait consumption.[29] When the findings were explored to mitigate bait shyness, it was revealed that 0.4% urine mixed in ordinary pearl millet mitigates bait shyness and poison aversion induced due to zinc phosphide poisoning to a significant extent in *Meriones hurrianae, Tatera indica* (Table 2), and *Mus musculus.*[30,32,36,41] Both *Tatera indica* and *Meriones hurrianae* preferred urine of the same sex indicating that the active ingredient in the urine has no sex attraction. These findings have opened a new line of thinking in mitigating bait shyness and poison aversion. Studies are in progress to identify the active fraction of urine and sebum of the ventral scent marking gland responsible for enhancing the bait consumption, so that it can be added in the poison baits during second control operation using Zinc phosphide. Likewise, Robbins[44,45] found that efficacy of prebaiting in *Peromyscus maniculatus* and *Mus musculus* may be greatly increased if the prebait is treated with a nontoxic flavor which mimics the flavor of subsequently used toxin. It also increased the effectiveness of a toxic bait by inhibiting the formation of illness-induced taste aversion.

Brodifacoum — a new single dose anticoagulant rodenticide, which was found very effective against a number of rodent pests,[34] has provided encouraging results to control bait shy *Rattus rattus*. The group of poison shy rats, which were earlier exposed to different dosages of Zinc phosphide, died within 2 to 6 days after feeding on 0.005% brodifacoum, followed by coumatetralyl which require 3 to 8 days.[12] The indigenously available warfarin also killed all bait shy rats in relatively longer period. The availability of such potent anticoagulants is likely to provide an alternative rodenticide to tackle bait shy rodents, besides their major use as the sole control agent in crop fields, residential premises, and godowns.

Even after acquiring considerable knowledge about the phenomena of bait shyness and poison aversion, we know very little about how to overcome them with a view to enhance success of the control operations. It would be interesting to know more about bait shyness and poison aversion in free living rodent populations. The changes in the behavior of bait shy rodents should be studied in order to exploit the results for further enhancing the control success in successive operations. The possible means of increasing efficacy of prebaiting for reducing or eliminating poison aversion and the role of attractants to mitigate shyness should form part of further studies in the field of rodent pest management.

Table 2
EFFECT OF ADDITION OF URINE IN POISON BAIT ON THE SHYNESS BEHAVIOR OF TWO GERBILS

	1st day	2nd day	3rd day	4th day[a]	5th day[b]	6th day[c]
				Meriones hurrianae	**Female urine**	
Millet	4.26 ± 0.59	3.04 ± 0.68	4.81*** ± 0.65	1.45*** ± 0.37	A 2.94* ± 0.40	5.35** ± 0.76
Sorghum	2.10 ± 0.55	2.18 ± 0.68	1.49 ± 0.39	2.78 ± 0.22	A 2.66 ± 0.46	3.85 ± 1.07
					B 1.15 ± 0.31	
					B 2.40*** ± 0.34	
				Tatera indica	**Male urine**	
Millet	4.23 ± 0.76	4.11 ± 0.81	4.70** ± 0.86	2.13 ± 0.35	A 2.40 ± 0.41	4.67** ± 0.89
					A 3.32* ± 0.78	
Sorghum	1.41 ± 0.51	2.85 ± 1.04	1.83 ± 0.68	3.80** ± 0.72	B 1.46 ± 0.89	2.15 ± 0.67
					B 3.52** ± 0.95	
				Tatera indica	**Female urine**	
Millet	3.68 ± 0.90	2.80 ± 0.79	4.70** ± 1.20	1.29* ± 0.48	A 2.30 ± 0.33	2.83** ± 0.66
					A 4.58** ± 0.77	1.09 ± 0.46
Sorghum	1.81 ± 0.59	1.72 ± 0.55	2.07 ± 0.77	2.50 ± 0.77	B 1.08 ± 0.37	
					B 2.25* ± 0.62	

Note: Mean daily intake, 100 g/100 g body weight, Mean ± S.E. Student's *t* test, level of significance between the consumption of millet and Sorghum: *, P <0.05; **, P <0.01, ***, P <0.001.

[a] 0.05% Zn_3P_2 ± 1% oil was mixed in millet.
[b] 0.05% Zn_3P_2 ± 1% oil was mixed with millet in both groups, 0.4% urine was also added to millet in group A.
[c] 0.4% urine was added to millet.

REFERENCES

1. **Armour, C. J. and Barnett, S. A.,** The action of dicocemaroe on laboratory and wild rats and its effect on feeding behaviour, *J. Hyg. London,* 48, 158, 1950.
2. **Barnett, S. A.,** Behaviour components in the feeding of wild and laboratory rats, *Behaviour,* 9, 24, 1956.
3. **Barnett, S. A., Cowan, P. E., Radford, G. C., and Prakash, I.,** Peripheral anosmia and discrimination of poisoned food by *Rattus rattus* L., *Behav. Biol.,* 13, 183, 1975.
4. **Barnett, S. A. and Prakash, I.,** *Rodents of Economic Importance in India,* Arnold-Heinemann, New Delhi, 1975, 175.
5. **Bhardwaj, D. and Khan, J. A.,** Mitigating poison and bait shyness by wild rats *(Rattus rattus* L.). Effect of poisoning at short intervals, *Indian J. Exp. Biol.,* 15, 624, 1977.
6. **Bhardwaj, D. and Khan, J. A.,** Effect of texture on the food preferences of baitshy wild rats *(Rattus rattus* L.). II, *Proc. Indian Acad. Sci.,* 87B, 77, 1978; as cited in I., Bhardwaj, D. and Khan, J. A., *Appl. Anim. Ethol.,* 5, 361, 1979.
7. **Bhardwaj, D. and Khan, J. A.,** Responses of roof rat, *Rattus rattus* L., to non-oily and oily foods after poisoning in oily foods, *Proc. Indian Acad. Sci.,* 88B, 125, 1979.
8. **Bhardwaj, D. and Prakash, I.,** Association of bait shyness with the components of poisoned food in *Rattus rattus, Z. Angew. Zool.,* 66, 329, 1979.
9. **Bhardwaj, D. and Prakash, I.,** Intensity of poison aversion and bait shyness in *Rattus rattus* due to intermittent and successive exposure to RH-787, *Z. Angew. Zool.,* 66, 427, 1979.
10. **Bhardwaj, D. and Prakash, I.,** Discrimination between harmless and harmful stimulate by the house rat *Rattus rattus rufescens* (Gray), *Indian J. Exp. Biol.,* 20, 302, 1982.
11. **Bhardwaj, D. and Prakash, I.,** Poison aversion in *Rattus rattus* through acclimatization, *Indian J. Exp. Biol.,* 20, 396, 1982.
12. **Bhardwaj, D. and Prakash, I.,** Efficacy of three anticoagulant rodenticides for the control of poison shy *Rattus rattus, Proc. 11th Vert. Pest Conf.,* Clark, D. O., Ed., University of California, Davis, Calif., 1984, 82.
13. **Calhoun, J. B.,** *The Ecology and Sociology of Norway Rat,* U.S. Public Health Service Publication, New York, New York, 1963, 1008.
14. **Cowan, P. E.,** Poison and bait shyness in two species of gerbil, *Meriones hurrianae* and *Tatera indica, Z. Angew. Zool.,* 65, 57, 1978.
15. **Deoras, P. J., Renapurkar, D. M., and Chaturvedi, G. C.,** Importance of baiting and trapping, *Pesticides,* 3, 41, 1969.
16. **Fitzwater, W. D. and Prakash, I.,** *Handbook of Vertebrate Pest Control,* Indian Council of Agricultural Research, New Delhi, 1978, 92.
17. **Gaines, T. B. and Haynes, W. J.,** Bait shyness to ANTU in wild Norway rats, *Public Health Rep.,* 67, 306, 1952.
18. **Galef, B. G., Jr. and Clark, M. M.,** Social factors in the poison avoidance and feeding behaviour of wild and domestic rat pups, *J. Comp. Physiol. Psychol.,* 75, 341, 1971.
19. **Greaves, J. H.,** Some laboratory observations on the toxicity and acceptability of norbormide to wild rats *(Rattus norveqicus)* and on feeding behaviour associated with sublethal dosing, *J. Hyg. Cambridge,* 64, 275, 1966.
20. **Hankins, W. G., Garcia, J., and Rusiniak, K. W.,** Dissociation of odour and taste in bait shyness, *Behav. Biol.,* 8, 407, 1973.
21. **Howard, W. E.,** How to overcome bait shyness in rodents, *Pest Control,* 27, 9, 1959.
22. **Howard, W. E., Marsh, R. E., and Cole, R. E.,** Duration of associative memory to toxic bait in deer mice, *J. Wildl. Manage.,* 41, 484, 1977.
23. **Howard, W. E., Palmateer, S. D., and Nachman, M.,** Aversion to strychnine sulphate by Norway rats, roof rats and pocket goptress, *Toxicol. Appl. Pharmacol.,* 12, 229, 1968.
24. **Kalat, J. W.,** Taste salience depends on novelty, not concentration, in taste-aversion learning in the rat, *J. Comp. Physiol. Psychol.,* 86, 47, 1974.
25. **Kalat, J. W.,** Biological significance of food aversion learning, in *Food Aversion Learning,* Milgram, N. W., Kramer, L., and Alloway, T. M., Eds., Plenum Press, New York, 1977, 73.
26. **Kalat, J. W. and Rozin, P.,** Learned safety as a mechanism in long delay taste aversion learning in rats, *J. Comp. Physiol. Psychol.,* 83, 198, 1973.
27. **Krishnakumari, M. K.,** Effects of early food experiences on later food preferences in adult rats, *Pest Control,* 41, 36, 1973.
28. **Krishnakumari, M. K. and Majumder, S. K.,** An integrated approach to rural rodent control programme, *Proc. Indian Rodent Symp.,* USAID, Calcutta, 1966, 253.
29. **Kumari, S. and Prakash, I.,** Conspecific odour as a phagostimulant for the Indian gerbil, *Tatera indica indica* Hardwiicke, *Indian J. Exp. Biol.,* 17, 981, 1979.

30. **Kumari, S. and Prakash, I.,** Role of conspecific urine in enhancing food consumption and masking of shyness behaviour in the desert gerbil, *M. hurrianae* (Jerdon), *Indian J. Exp. Biol.*, 18, 730, 1980.
31. **Kumari, S. and Prakash, I.,** Discrimination of poisoned food by *Funambulus pennanti* and persistence of learning behaviour, *Indian J. Rodentol.*, 2, 11, 1981.
32. **Kumari, S. and Prakash, I.,** Relative efficacy of male and female conspecific urine in masking shyness behaviour in Indian gerbil, *Tatera indica, Proc. Indian Acad. Sci.*, 93, 431, 1984.
33. **Mathur, R. P. and Prakash, I.,** New Food reaction among desert rodents, *Saugt. Mitteil.*, 28, 28, 1980.
34. **Mathur, R. P. and Prakash, I.,** Evaluation of brodifacoum against *T. indica, M. hurrianae* and *R. rattus, J. Hyg., Cambridge,* 87, 179, 1981.
35. **Prakash, I.,** *Rodent Pest Management, Principles and Practices,* Central Arid Zone Research Institute Monograph, Jodhpur 4, 28, 1976.
36. **Prakash, I.,** Efficacy of conspecific urine in masking the shyness behaviour in two gerbils, *Proc. 2nd Symp. on Recent Advances in Rodent Control,* Mohammad, A. H. H., Zaghloul, T. M., Salit, A. M., Zakaria, M., Eds., Ministry of Public Health, Kuwait, 131.
37. **Prakash, I. and Jain, A. P.,** Bait shyness of two gerbils, *Tatera indica indica* Hardwicke and *Meriones hurrianae* Jerdon, *Ann. Appl. Biol.*, 69, 169, 1971.
38. **Prakash, I. and Ojha, P.,** The phenomenon of poison bait aversion and its possible attenuation in the Indian gerbil, *Tatera indica indica* Hardwicke, *Z. Angew. Zool.*, 65, 385, 1978.
39. **Prakash, I., Rana, B. D., and Jain, A. P.,** Bait shyness in three species of *Rattus, Z. Angew. Zool.*, 62, 89, 1975.
40. **Rana, B. D., Prakash, I., and Jain, A. P.,** Bait shyness and poison aversion in the Hairy footed gerbil, *Gerbillus gleodowi* Murray, *Proc. All India Rodent Seminar,* Rodent Control Project, Sidhpur, 1975, 58.
41. **Rao, A. M. K. M. and Prakash, I.,** Bait shyness in house mouse, *Mus musculus bactrianus* Blyth, due to two acute poisons, *Indian J. Exp. Biol.*, 18, 1490, 1980.
42. **Rao, A. M. K. M. and Rajabai, B. S.,** Bait shyness in *Mus booduga* and *Mus platythrix,* 1986, unpublished manuscript.
43. **Revusky, S. H. and Garcia, J.,** Learned associations over long delays, in *The Psychology of Learning and Motivation,* Bower, G. H. and Spence, J. T., Eds., Academic Press, New York, 1970, 1.
44. **Robbins, R. J.,** Poison based taste aversion learning in deer mice *(Peromyscus maniculatus bairdi), J. Comp. Physiol. Psychol.*, 92, 642, 1978.
45. **Robbins, R. J.,** Taste aversion learning and its implications for rodent control, *Proc. 9th Vert. Pest Conf.* Clark, J. P., Ed., University of California, Fresno, Calif. 1980, 114.
46. **Rozin, P. and Kalat, J. W.,** Specific hungers and poison avoidance as adaptive specialisations of learning, *Psych. Rev.*, 78, 459, 1971.
47. **Rzoska, J.,** Bait shyness, a study in rat behaviour, *Br. J. Anim. Behav.*, 1, 128, 1953.
48. **Rzoska, J.,** The behaviour of white rats towards poison baits, in *The Control of Rats and Mice,* Vol. 2, Chitty, D., Ed., Clarendon Press, Oxford, 1954, 374.
49. **Shettleworth, S. J.,** Constraints on learning, in *Advances in the Study of Behaviour,* Lehrman, D. S., Hinde, R. A., and Shaw, E., Eds., Academic Press, New York, 1972, 1.
50. **Shorten, M.,** The reaction of the brown rat towards changes in its environment, in *Control of Rats and Mice,* Chitty, D., Ed., Clarendon Press, Oxford, 1954, 307.
51. **Sridhara, S.,** New food reaction in four Indian wild rodents, *Z. Angew. Zool.*, 64, 449, 1977.
52. **Sridhara, S.,** Rodenticide induced bait aversion and neophobia in *Tatera indica cuvieri, Z. Angew. Zool.*, 70, 429, 1983.
53. **Sridhara, S. and Srihari, K.,** Variability in the development of poison shyness of *Bandicota bengalensis bengalensis* Gray towards two acute poisons, *Z. Angew. Zool.*, 65, 179, 1978.
54. **Sridhara, S. and Srihara, K.,** Zinc phosphide and vacor induced bait shyness and its persistence in *Mus platythrix, Proc. Indian Acad. Sci.*, 89, 221, 1980.
55. **Sridhara, S. and Srihari, K.,** Bait shyness towards zinc phosphide and vacor in the larger bandicoot rat, *Bandicota indica indica, Indian J. Exp. Biol.*, 18, 1029, 1980.
56. **Srivastava, A. S.,** Evolution, development and standardisation of technique for the eradication of field rats, *Proc. Indian Rodent Symp.*, USAID, Calcutta, 1966, 283.

Chapter 19

NONANTICOAGULANT RODENTICIDES

M. Lund

TABLE OF CONTENTS

I. INTRODUCTION

The efficacy of a rodenticide is not only dependent on the toxicity to the species in question, but certainly also on palatability and mode of action. Like most other herbivorous and granivorous mammals, rodents too, usually, have an innate response to unfamiliar food items, thus reducing the risk of being accidentally poisoned by certain plant species. It is often believed by people lacking practical experience in rodent control that the stronger and faster the action of a certain compound the more efficient it is. This is not true, especially not as far as *Rattus*-species are concerned. An established rat in a stable environment typically shows a highly suspicious behavior towards unfamiliar objects (new object reaction), including baits with an unusual taste or smell.[45] It may last hours or even days before the rat ventures to nibble the bait, and unrestricted feeding may not take place until several days after the introduction of the bait. If nonfatal symptoms of poisoning appear shortly after the ingestion of the first tiny sample the rat is warned, and after recovery it rejects the bait in question. This reaction is called ''bait-shyness'' and it may last for several months or even the entire lifespan of the rat. The ability of the rat to associate between cause and effect is remarkable, but limited in time. If symptoms of poisoning are delayed for more than 6 to 7 hr the rat seems not able to associate, and bait-shyness is consequently not likely to occur. This is the primary reason why slow-acting rodenticides are, in the long run as a rule, much more effective against rats than quick-acting compounds.

Some improvement of the quick-acting compounds can be achieved by the so-called ''prebaiting'' method in which the plain bait base is offered to the rodents for some time prior to the application of poison bait. When the rodents have familiarized themselves with the baiting sites, the bait-containers, and the bait type they may be induced to consume a larger amount of the poison, resulting in a higher mortality. The period of prebaiting varies according to population density, type of infestation, and local conditions, but as a rule prebaiting should be continued until the total take is roughly the same on two consecutive days. The method is predominantly applied in rat control, sometimes also when dealing with other commensal species, but for practical reasons rarely in the field.

Prior to 1950 all rodenticides were nonanticoagulants, most of them quick-acting ones, but after the introduction of warfarin and the other anticoagulants their importance was very much reduced. After the emergence of resistance to anticoagulants around 1960 the interest for nonanticoagulants was revived, and a number of new compounds have been investigated during the last 25 years. Surprisingly few, however, were marketed, or they were withdrawn after a short period due to various shortcomings or unexpected side-effects. Other, older compounds, have been reconsidered on the basis of present knowledge, and recommended against by a World Health Organization (WHO) Expert Committee.[48] In the following, the various nonanticoagulant rodenticides are listed and described in alphabetic order. Those *not* recommended by WHO, as well as those withdrawn by the manufacturers, are left out.

II. NONANTICOAGULANT RODENTICIDES

A. Alpha-Chloralose (1,2-0-(2,2,2-trichloroethylidene)-D-glucofuranose)

Mode of action — Alpha-chloralose is a narcotic drug retarding metabolic processes and causing fatal body-heat loss in small-sized mammals and birds at low environmental temperatures. Symptoms of poisoning occur within an hour after intake.

Toxicity — Toxicity is low, LD_{50} in *Rattus norvegicus* is 205 to 365 mg/kg.[14]

Field of application — As laboratory tests have shown that the environmental temperature must be below 16°C to achieve satisfactory mortality in small rodents, the compound is used preferably against *Mus musculus* in temperate regions and against infestations in cold stores.

Hazard to man and nontarget animals — In spite of the recommended concentration in prepared baits being fairly high, 4%, alpha-chloralose is considered one of the safest rodenticides, also as far as secondary poisoning is concerned.[9] It is not recommended for control of field rodents due to the risk to small granivorous birds.

Special recommendations — Prebaiting may be recommended in situations where alternative food is abundant. Rats are not effectively controlled by this compound, as they do not consume a lethal dose before poisoning symptoms occur. Bait-shyness is readily induced.

B. Alpha-Chlorohydrin (3-chloro-1,2-propanediol)

Mode of action — Alpha-chlorohydrin is primarily a chemosterilant, but a high proportion of the rodent population is killed through glomerulonephritis and renal insufficiency.

Toxicity — LD_{50} values on *R. norvegicus* vary from 26 mg/kg to 116 to 172 mg/kg.[20,27] The lesser bandicoot rat (*Bandicota bengalensis*) has been found to be susceptible to alpha-chlorohydrin, the LD_{50} being 50 mg/kg.[6]

Field of application — Alpha-chlorohydrin is only used against rats as 1% prepared baits and only in special situations, i.e., where anticoagulant-resistance has been found or where alternative rodenticides have failed for other reasons.

Hazard to man and nontarget animals — It seems to be fairly safe in use, but information is scarce.

Special recommendations — Very attractive baits must be used as the compound is unpalatable to rodents. For the same reason a long period of prebaiting must be carried out, in spite of poisoning symptoms being delayed for 1 to 2 days. As sublethal doses may induce male sterility, a high mortality is not essential for long-term efficiency.

C. Bromethalin (EL-614) (N-methyl-s,r-dinitro-N-(2,4,6-tribromophenyl)-6-(trifluoromethyl)-benzenamine)

Mode of action — Bromethalin is the most recently developed nonanticoagulant rodenticide. It works by uncoupling oxidative phosphorylation in the mitochondria in the central nervous system leading to diminished membrane transport and increased cerebrospinal fluid pressure.[7] It is a fairly quick-acting compound, death usually occurring within 24 hr after intake.

Toxicity — It is a highly toxic rodenticide, the LD_{50} for *R. norvegicus* being 2 mg/kg, for *Mus musculus* 5 mg/kg.

Field of application — Laboratory and field trials indicate a high degree of efficiency, including a good acceptance by *R. norvegicus* and *M. musculus.*[24] The recommended concentration in prepared baits is 0.005 to 0.01%. Small wax-blocks containing 0.02 to 0.03% have shown promising results against *R. tiomanicus, R. argentiventer, R. exulans,* and *R. rattus* in oil palm plantations in Malaysia.[13]

Hazard to man and nontarget animals — The susceptibility of domestic animals, including cats and dogs, is close to that of *R. norvegicus* and the same is true for the monkey. Secondary poisoning is, however not likely to occur.[24]

Special recommendations — Prebaiting is, due to the relatively slow action, not considered necessary.[11] A system of "pulsed baiting", supplementing bait stations once a week, may be recommended in order to bring down the costs.[24]

D. Calciferol (24-methyl-9,10-secocholesta-5,7,10(19),22-tetraene-3-B-01)

Mode of action — Calciferol is a fat-soluble vitamin in two forms: ergocalciferol (vitamin D_2) and cholecalciferol (vitamin D_3). In high doses calciferol promotes the absorption of calcium from the gut and its subsequent deposition in lungs, cardiovascular system, and kidneys. Death seems to be due to circulatory blockage, heart and renal failure. This process is very slow, and symptoms of poisoning usually do not occur until 2 to 3 days after intake.[17]

Toxicity — The two forms of calciferol are equally toxic to most mammals, whereas cholecalciferol is considered to be significantly more potent to birds. The acute LD_{50} of ergocalciferol to *R. norvegicus* is 40 mg/kg, to *M. musculus* 15.7 mg/kg, and if ingested through 3 days 8 mg/kg and 11.5 mg/kg, respectively. Cholecalciferol seems to be equally toxic to *R. norvegicus* and *M. musculus,* i.e., 42.5 mg/kg to 43.6 mg/kg.[34] *R. rattus* seems to be more susceptible to calciferol than *R. norvegicus.*[3] The cotton rat (*Sigmodon hispidus*) has, in laboratory tests, shown the same degree of susceptibility to ergocalciferol as *R. norvegicus.*[15]

Field of application — The recommended concentration is 0.1% in prepared baits (1 mg calciferol is equivalent to 40.000 I.U.). It is effective against all rodent species but its high toxicity usually limits its use to commensal rodents. Mixed with vegetable oil, paraffin, and a suitable cereal calciferol is recommended for sewer rat control. In the U.K. calciferol is often used in combination with warfarin (0.025%) against *R. norvegicus* and with canary seed as the bait base against *M. musculus,* but the advantage of this combination is somewhat obscure.[28,39,42]

Hazard to man and nontarget animals — No antidote is available and great care must be taken to avoid poisoning of other mammals. Secondary poisoning is not likely to occur and the compound is readily decomposed in the field.

Special recommendations — As rats usually feed for 2 to 3 days on calciferol-baits before symptoms occur, prebaiting is *not* necessary. The stability of calciferol has been questioned, but studies indicate that baits stored in closed containers away from light remain effective for up to 10 months.[32] Bait laid in dry conditions in food stores, warehouses, etc., keeps its toxicity for at least 3 months; in damp and warm conditions, however, toxicity may decrease in 3 weeks, but the change in palatability of the cereal is more important in such circumstances.[30]

E. Crimidin (2-chloro-4-(dimethylamino)-6-methylpyrimidine)

Mode of action — Crimidin is a very fast acting rodenticide inducing poisoning symptoms within 15 to 45 min after intake.[26] It is an antagonist to vitamin B_6 and affects the central nervous system, resulting in death from respiratory failure.

Toxicity — It is a highly toxic compound, the LD_{50} to *R. norvegicus* being 2 mg/kg. The susceptibility of most mammal and bird species seems to be in the same level.

Field of application — Crimidin is used at a concentration of 0.1% in dry baits, especially for the control of mice in and around premises. Its rapid action renders it possible to prevent the poisoned mice from entering the houses, thus preventing odor problems. Crimidin is not recommended for the control of *Rattus*-species, as bait-shyness is readily induced, and lethal doses are seldom consumed. Field trials of *Microtus*-species with a pellet containing 0.5% crimidin has shown promising results in West Germany, Holland, Denmark, and Finland, but this formulation is not marketed.[25,29,37,44] The spiny mouse (*Acomys cahirinus*) has been found easily controlled by ordinary crimidin-baits in Egypt.[2]

Hazard to man and nontarget animals — All mammals are susceptible to crimidin. Birds are relatively tolerant but cases of primary and secondary poisoning of birds have been encountered in large-scale operations. Crimidin is rapidly broken down in the organism and also relatively easily in the field, and it shows no cumulative effect. An antidote, pyridoxin or vitamin B_6, is available.[26] Crimidin has disappeared from the market in most European countries.

F. Fluoracetamide (1081)

Mode of action — Fluoracetamide is closely related to sodium fluoracetate, i.e., it interferes with the normal acetate metabolism in the ''Krebs cycle''. Death is either due to cardiac failure or respiratory paralysis, preceded by convulsive seizures, but the action is somewhat slower than that of sodiumfluoracetate, warning symptoms occurring 1 to $3^1/_2$ hr after intake.[4]

Toxicity — It is less toxic than 1080, the LD_{50} in *R. norvegicus* being 13 to 16 mg/kg, but as it is slower in action many times the lethal dose may be ingested before warning symptoms occur.[4] LD_{50} for *Apodemus agrarius* is 27.7 mg/kg, but other species seem to be more susceptible, i.e., the LD_{50} for ground squirrels (*Citellus spp.*) is 0.46 mg/kg and for *Meriones spp.* 1.7 mg/kg.[10]

Field of application — Fluoracetamide is considered one of the most efficient, acute poisons and it can be used at a concentration of 1 to 2% in baits against all rodent species, but its high toxicity limits its use to the three commensal species. It should only be available to trained pest control operators and only be used in areas where access by unauthorized persons and pet animals can be prevented. It is the second most widely used rodenticide in China.[10]

Hazard to man and nontarget animals — Fluoracetamide is highly toxic to man and domestic animals, especially the dog (LD_{50} appr. 0.5 mg/kg). Secondary poisoning of predators or scavengers is very likely to occur, as a single mouse may contain enough poison to kill an average dog. Acetylamine is claimed to be an effective antidote.[47]

Special recommendations — Prebaiting is recommended in most situations. Prepared wax-blocks are particularly useful for sewer rat control, but the poison can also be formulated as a liquid or dry bait.

G. Gliftor (1,3-difluoro-2-propanol)

Gliftor is a mixture of two related compounds: 70 to 80% glycerol-α-difluorohydrin and 20 to 30% glycerol-α-chlor-γ-fluorohydrine and it is used widely in the U.S.S.R. and China for the control of field rodents, but very little information on mode of action and hazards involved is available. Acetylamine is claimed to be an antidote. It is usually recommended at a concentration of 0.25 to 1% in baits and at 0.05% as a ground spray. For field rodents only 7.5 to 10 g of active ingredient (a.i.) is needed per ha. Treated fields must not be used for grazing until after 30 days. No toxic effect is found in grassland after 113 days.[47] It exhibits a high degree of absorption to leaf, stems, and roots of plants. It has been found potentially promising for control of *Microtus* in Finland, but is not on the market.[37] The LD_{50} to ground squirrels (*Citellus*) is 4.5 mg/kg, and to *Meriones* is 10.0 mg/kg.[10] Poultry is considered to be rather tolerant to gliftor.

H. Norbormide (Shoxin) (5-(α-hydroxy-α-2-pyridylbenzyl)-7-(α-2-pyridyl-benzylidene)-norborn-5-ene-2,3-dicarboximide)

Mode of action — This very fast acting compound introduced in the early 1960s, causes peripheral, irreversible vasoconstriction leading to ischemia and ultimately death.[41] The physiological mechanism has never been fully understood, and no explanation can be given to its apparent selectivity to *Rattus*-species. Symptoms of poisoning may occur within 15 min after intake and death is seen from 40 min to 4 hr after the intake of a lethal dose. It seems essential that the lethal dose is consumed in a very narrow period of time. Sublethal doses may lead to blindness.

Toxicity — Norbormide is particularly toxic to *R. norvegicus*, less so to *R. rattus*, and apparently not at all to other mammals.[41] The median lethal dose to *R. norvegicus* is 9.0 mg/kg, and the LD_{95} around 17 mg/kg.[16] Recently it has been reported suitable for the control of other *Rattus*-species in China, but no data are given.[10]

Field of application — The recommended concentration in prepared baits is 0.5%, which in laboratory trials is reported to give results comparable to 2.5% zinc phosphide.[16] Field experience has revealed a number of shortcomings, especially unpalatability to a considerable proportion of the rat population and a great individual variation in susceptibility.[46] Field trials against warfarin resistant rat populations on farms in the U.K. showed a poorer performance of 0.5 and 1.0% norbormide than 2.0 and 5.0 % zinc phosphide.[40] Attempts to improve the palatability of norbormide by encapsulation have so far not been successful. Even if norbormide could be rendered more acceptable its use would be limited due to its specificity.

Hazard to man and nontarget animals — Norbormide is the only known rodenticide with a high degree of specificity. It is considered the safest compound available in rodent control.

Special recommendations — Due to its very rapid action and profound unpalatability prebaiting is a must if reasonable efficiency should be achieved. It is also essential that the most attractive bait-types are used. Bait-shyness is very likely to occur.

I. Red Squill

Mode of action — Red squill is one of the oldest known rat poisons and has been used since medieval times. It is an extract of the bulb of the red form of the Mediterranean plant *Urginea maritima,* containing varying amounts of the glycoside scilliroside (see later). This cardiac glycoside affects the central nervous system causing violent spasms and convulsions leading to death from heart paralysis. The action is fairly quick, symptoms occurring 2 to 3 hr after intake, but death is delayed for 24 hr, sometimes much more.

Toxicity — The varying content of scilliroside in the extract has caused problems in the evaluation of its efficacy and toxicity, but a method of fortification and standardization has been developed so that a minimum LD_{50} of around 500 mg/kg can be claimed for *R. norvegicus,* females being more susceptible than males.[12]

Field of application — Red squill is only used for the control of *Rattus*-species. The prolonged poisoning symptoms preceding death, combined with the relatively poor efficacy, have excluded this rodenticide from being used in numerous countries.

Hazard to man and nontarget animals — Red squill has a distinct bitter taste rendering it unpalatable to most mammals except rodents. Furthermore, it induces vomiting in mammals capable of that (not rodents). This, in combination with its low toxicity, places red squill among the safest rodenticides. It must be noted, however, that the extract is extremely irritating to the skin.

Special recommendations — Baits containing up to 10% of the extract are used for the control of *R. norvegicus* and prebaiting is necessary in most situations. Survivors from sublethal doses become permanently bait-shy.[12]

J. Scilliroside (6-acetyloxy)-3-(D-glucopyranosyloxy)-8,14-dihydroxybuta-4,20,22-trienolide)

Mode of action — Scilliroside is the pure glycoside derived from red squill (see above), now also being artificially synthesized. Its effect is much less variable than that of the red squill extract. Symptoms of poisoning appear after half an hour, but death may be delayed for 6 days.

Toxicity — The LD_{50} to *R. norvegicus* is reported to be 0.7 mg/kg, females being more susceptible than males.[31]

Field of application — The recommended concentration in rat baits is 0.015% and formulated as a tracking powder the content is 0.05% scilliroside. For the lesser bandicoot rat (*Bandicota bengalensis*) the optimal concentration in baits seems to be 0.05%, but a high degree of unpalatability is found.[5] Against *R. rattus* baits containing 0.075% scilliroside are reported to be most efficient.[38] Testing scilliroside, as well as a number of other cardiac glycosides, Meehan concludes that it is unpalatable to rats and mice.[35]

Hazard to man and nontarget animals — Scilliroside presents a potential hazard to children and pets and should be used with great care.

Special recommendations — Prebaiting should be used whenever possible.

K. Sodium Monofluoracetate (1080)

Mode of action — Sodium monofluoracetate is the sodium salt of fluoracetic acid which is the main toxic component of the South-African plant *Dichapetalym cymosum* and is similar in action to Fluoracetamide.[33] It is metabolized to fluorocitric acid, which blocks the citric acid cycle, causing death due to cardiac failure or respiratory paralysis. It is somewhat faster in action than fluoracetamide.

Toxicity — Sodium monofluoracetate is as a rule more toxic than fluoracetamide, LD_{50} in *R. norvegicus* being around 1 to 5 mg/kg. LD_{50} to ground squirrels (*Citellus*) is 0.22 mg/kg, to *Meriones* 0.65 mg/kg, and to *Apodemus agrarius* 6.25 mg/kg.[10] In *M. musculus* LD_{50} is reported to be 6 to 10 mg/kg.[12]

Field of application — This poison is efficient against all rodents, but due to the risk to nontarget animals it should be restricted in use to certain situations and should only be available to trained personnel. It is particularly recommended for sewer rat control, where it has been found more effective than zinc phosphide.[4] It has also been used against field rodents in China.[10]

Hazard to man and nontarget animals — 1080 is highly toxic to all mammals and birds and should be administered with great care. Dogs are particularly susceptible, LD_{50} being around 0.1 mg/kg. In large-scale operations against field rodents in China numerous cases of secondary poisoning of wildlife and domestic predators have been reported — the population of weasels, foxes, and hawks being drastically reduced.[10]

Special recommendations — As symptoms of poisoning occur within 30 min to $1^1/_2$ hr, bait-shyness is very likely to be induced. Consequently, prebaiting is recommended whenever possible. The recommended concentration in dry baits is 0.25% and in liquids 0.4%. The latter is an efficient means of control where food is plentiful and where immediate action has to be taken.

L. Strychnine

Mode of action — Strychnine is found in the seeds of the Indian tree *Strychnos nuxvomica* where it is one of a number of different alkaloids. It has a long history as a rodenticide, being used first in Germany in the 16th century. It affects the central nervous system leading to paralysis in a few minutes after intake and to death within half an hour.

Toxicity — Strychnine and its salts (especially strychnine sulphate) have a bitter taste and are highly toxic to all mammals, less so to birds. The LD_{50} to *R. norvegicus* is 5 to 6 mg/kg.[12]

Field of application — Strychnine is not recommended for the control of rats as the bitter taste and rapid onset of poisoning symptoms act as warning signals. It is primarily used for house mouse control, and in the U.S. also for control of pocket gophers and ground squirrels (*Spermophilus beecheyi*).[22,43]

Hazard to man and nontarget animals — Strychnine is highly toxic to all mammals and there is a great risk of secondary poisoning to animals feeding on strychnine-killed rodents. Some predators are particularly susceptible to strychnine, for instance the LD_{50} to mink is only 0.6 mg/kg.[1]

Special recommendations — Strychnine should only be used by trained control operators and only in situations where other rodenticides have failed.

M. Zinc Phosphide (Zn_3P_2)

Mode of action — Zinc phosphide has been used for rodent control since World War I, and it was the predominant rodenticide until the introduction of warfarin around 1950. The action of zinc phosphide is due to the evolution of phosphine (PH_3) in the stomach and intestines, causing heart failure and liver or kidney damage. It is a quick-acting compound, symptoms of poisoning appearing within 25 min after intake of high doses, and death usually takes place within 24 hr. With the low doses it may take 2 to 3 days before the rodent succumbs. Zinc phosphide is a blackish powder with a characteristic garlic-like odor and is stable for several years under normal storage conditions, but breaks down rather rapidly in dilute acids and in wet baits. Dry baits protected from rainfall are effective for extended periods.[18]

Toxicity — The LD_{50} to *R. norvegicus* is around 40 mg/kg, to ground squirrels (*Citellus*) about the same (36 mg/kg), and to *Meriones* only 12 mg/kg.[10,12]

Field of application — Zinc phosphide baits containing from 1 to 5% a.i. are recommended for the control of commensal rodents as well as field rodents. For *M. musculus* 1% baits and for *R. norvegicus* 2.5% baits are found most effective. Laboratory tests indicate that 4% would be the concentration of choice for the control of cotton rats (*Sigmodon hispidus*).[15] In Burma, 2.5% zinc phosphide baits are recommended as suitable alternatives to the older anticoagulants as far as *Bandicota bengalensis* is concerned.[23] In the U.S. ground squirrels and meadow voles are successfully controlled by 1 to 2% baits.[8,36,43] Zinc phosphide is widely used in China for the control of commensal as well as field rodents, but bait shyness is reported as a problem.[47] In contrast, Myllymäki, on the basis of trials against *Microtus* in Finland, doubts that field vole control by zinc phosphide can ever be reasonable because of the aversion reaction of these rodents to the odor and taste.[37]

Hazard to man and nontarget animals — Other mammals and especially birds are very susceptible to zinc phosphide (the LD_{50} of pheasants is 9 mg/kg) and secondary poisoning is very likely to occur even if vomiting is induced.[19] Predators and scavengers (cats and dogs) consuming the stomach and intestines of the prey are particularly at risk. According to U.S. Fish and Wildlife Service the greatest hazard is as a primary toxicant to gallinacious birds and water fowl.[21] Particular precautions must be taken when preparing wet baits, but inhalation of zinc phosphide dust also has to be avoided. A respirator should be worn when preparing baits within doors.

Special recommendations — Prebaiting is necessary when controlling rats, and bait-boxes are recommended, especially in indoor situations. To give all flock members ready access to the poison bait in a short period of time a high number of bait stations is essential. Remaining poison bait and dead rodents must, as far as possible, be collected. Zinc phosphide is sometimes used as a tracking powder (10%), especially in small bait stations for the control of house mice. For the control of meadow voles (*Microtus californicus* and *M. montanus*) in the U.S., teaspoon quantities of 1% bait are scattered on runways and close to burrows, or 2% bait is broadcast by hand at the rate of 5 to 10 kg/ha.[8]

REFERENCES

1. **Anthony, R. M., Lindsey, G. D., and Evans, J.,** Hazards to golden-mantled ground squirrels and associated secondary hazard potential from strychnine baiting for forest pocket gophers, Proc. 11th Vertebrate Pest Conf., Sacramento, Calif., March 6 to 8, 1984, 25.
2. **Arafa, M. S., Nasr, N. T., Salil, A. H., and Khalil, M. S.,** A comparative study of slow and quick-acting rodenticides in a confined area in Egypt, *Int. Pest Control.*, 5, 17, 1974.
3. **Bai, K., Krishnakumari, M. K., and Majumder, S. K.,** Toxicity of calciferol, warfarin and their combinations to *Rattus norvegicus* (albino) and *R. rattus*, *Pest. Sci.*, 9, 44, 1978.
4. **Bentley, E. W., Hammond, L. E., Bathard, A. H., and Greaves, J. H.,** Sodium fluoracetate and fluoracetamide as "direct" poisons for the control of rats in sewers, *J. Hyg. Cambridge*, 59, 413, 1961.
5. **Brooks, J. E. and Htun, P. T.,** Laboratory evaluation of scilliroside used as a rodenticide against the lesser bandicoot rat, *Bandicota bengalensis*, *J. Hyg. Cambridge*, 85, 227, 1980.
6. **Brooks, J. E., Thun, P. T., Naing, H., Walton, D. W., and Tun, M. M.,** Laboratory evaluation of rodenticides for use against South-east Asian commensal small mammals, *WHO/VBC 79.720*, World Health Organization, Geneva, 1979, 15.

7. **Cherry, L. D., Gunnoe, M. D. and Van Lier, R. B. L.,** The metabolism of bromethalin and its effect on oxidative phosphorylation and cerebrospinal fluid pressure, *Toxicologist,* 2, 108, 1982.
8. **Clark, J. P.,** Vole control in crops, Proc. 11th Vertebrate Pest Conf., Sacramento, Calif., March 6 to 8, 1984, 5.
9. **Cornwall, P. B. and Bull, J. O.,** Alphakil — a new rodenticide, *Pest Control,* 35(8), 31, 1967.
10. **Deng, Z. and Wang, C.-X.,** Rodent control in China, Proc. 11th Vertebrate Pest Conf., Sacramento, Calif., March 6 to 8, 1984, 47.
11. **Dreikorn, B. A., O'Doherty, G. O. P., Clinton, A. J., and Kramer, K. E.,** EL-614, A novel rodenticide, Proc. Br. Crop Prot. Conf. Pests and Diseases, Brighton, November, 1979, 491.
12. **Freeman, R. B.,** Properties of the poisons used in rodent control, in *Control of Rats and Mice,* Vol. 1, Chitty, D., Ed., Clarendon Press, Oxford, 1954, Chap. 2.
13. **Gai, S. K., Juan, H. K., Mansor, M., and Guse, L. R.,** Evaluation of bromethalin for the control of rats in oil palms, Proc. Int. Conf. Plant Prot. in the Tropics, Kuala Lumpur, Malaysia, 1985, 1.
14. **Giban, J.,** Note sur la toxicite du chloralose alpha a l'egard des rats, *Ann. Epiphyties,* 17(4), 525, 1966.
15. **Gill, J. E. and Redfern, R.,** Laboratory trials of seven rodenticides for use against the cotton rat *(Sigmodon hispidus), J. Hyg. Cambridge,* 85, 443, 1980.
16. **Greaves, J. H.,** Some laboratory observations on the toxicity and acceptability of norbormide to wild *Rattus norvegicus* and on feeding behaviour associated with sublethal dosing. *J. Hyg. Cambridge,* 64, 275, 1966.
17. **Greaves, J. H., Redfern, R., and King, R. E.,** Some properties of calciferol as a rodenticide, *J. Hyg. Cambridge,* 73, 341, 1974.
18. **Guerrant, G. O. and Miles, J. W.,** Determination of zinc phosphide and its stability in rodent baits, *J. Assoc. Off. Anal. Chem.,* 52(3), 662, 1969.
19. **Hayne, D. W.,** Zinc phosphide: its toxicity to pheasants and effect of weathering upon its toxicity to mice, *Mich. Agric. Exp. Sta. Q. Bull.,* 33(4), 412, 1951.
20. **Helal, T. Y.,** Chemosterilant and rodenticidal effects of 3-chloro-1,2-propanediol against the albino laboratory rat and the Nile field rat, *Int. Pest Control,* 24(1), 20, 1982.
21. **Hood, G. A.,** Zinc phosphide — a new look at an old rodenticide for field rodents, Proc. 5th Vertebrate Pest Conf., Fresno, Calif., March 7 to 9, 1972, 85.
22. **Howard, W. E., Palmateer, S. D., and Nachman, M.,** Aversion to strychnine sulfate by Norway rats, roof rats and pocket gophers, *Toxic. Appl. Pharm.,* 12, 229, 1968.
23. **Htun, P. T., Muller, W. W., and Naing, H.,** Paired rodent control field trial with zinc phosphide and brodifacoum in Rangoon, Burma., *WHO/VBC/84.902,* World Health Organization, Geneva, 1984, 14.
24. **Jackson, W. B., Spaulding, S. R., Van Lier, R. B. L., and Dreikorn, B. A.,** Bromethalin — a promising new rodenticide, Proc. 10th Vertebrate Pest Conf., Monterey, Calif., March 23 to 25, 1982, 10.
25. **Jobsen, J. A., and Brink, A. E.,** Personal Communication, 1979.
26. **Karlog, O. and Knudsen, E.,** Vitamin B_6 as an antidote against the rodenticide "Castrix" (2-chloro-4-methyl-6-dimethyl-aminopyrimidin), *Nature,* 200, 790, 1963.
27. **Kassa, H. and Jackson, W. B.,** Bait acceptance and efficacy of alphachlorohydrin in the Norway rat, *Rattus norvegicus, Int. Pest Control,* 26(1), 7, 1984.
28. **Lund, M.,** Calciferol as a rodenticide, *Int. Pest Control,* 16(6), 10, 1974.
29. **Lund, M. and Lodal, J.,** Crimidin pellets, *Danish Pest Inf. Lab. Ann. Rep. 1979,* 1980, 78.
30. **Macnicoll, A. D. and Simpson, W. J.,** The stability of ergocalciferol in rodenticidal baits, *J. Hyg. Cambridge,* 90, 253, 1983.
31. **Maddock, D. R. and Schoof, H. F.,** New red squill derivates: Laboratory and field studies on stabilised scilliroside against Norway rats, *Pest Control,* 38(8), 32, 1970.
32. Ministry of Agriculture, Fisheries, and Food, Sorexa CR-supplementary information, *Technical Circular 31,* 1974.
33. **Marais, J. S. C.,** Monofluoracetic acid, the toxic principle of "giftblaar" *Dichapetalum cymosum,* Engl. Onderstepoort, *J. Vet. Sci. Anim. Ind.,* 20, 67, 1944.
34. **Marshall, E. F.,** Cholecalciferol. A unique toxicant for rodent control, Proc. 11th Vertebrate Pest Conf., Sacramento, Calif., March 6 to 8, 1984, 95.
35. **Meehan, A. P.,** The rodenticidal activity of some cardiac glycosides, *Int. Pest Control,* (3), 1975. 12.
36. **Merson, M. H. and Byers, R. E.,** Pellet size and effectiveness of rodenticides against orchard *Microtus,* Proc. 11th Vertebrate Pest Conf., Sacramento, Calif., March 6 to 8, 1984, 33.
37. **Myllymäki, A.,** Efficacy of a number of toxic baits and baiting against the voles *Microtus agrestis* and *Arvicola terrestris, Proc. 11th Vertebrate Pest Conf.,* Calif., 1984, 38.
38. **Pahwa, R., Doharey, R. B., and Varma, B. K.,** Laboratory evaluation of Silmurin (1% technical powder) against black rat *(Rattus rattus* (L.)), *Int. Pest Control.,* 24(1), 18, 1982.
39. **Rennison, B. D.,** Field trials of calciferol against warfarin resistant infestations of the Norway rat *(Rattus norvegicus* Berk.). *J. Hyg. Cambridge,* 73, 361, 1974.

40. **Rennison, B. D., Hammond, L. E., and Jones, G. L.,** A comparative trial of norbormide and zinc phosphide against *Rattus norvegicus* on farms, *J. Hyg. Cambridge,* 66, 147, 1968.
41. **Roskowski, A. P., Poos, G. I., and Mohrbacker, R. J.,** Selective rat toxicant, *Science,* 144, 412, 1964.
42. **Rowe, F. P., Smith, F. J., and Swinney, T.,** Field trials of calciferol combined with warfarin against wild house-mice *(Mus musculus* L.), *J. Hyg. Cambridge,* 73, 353, 1974.
43. **Salmon, T. P. and Schmidt, R. H.,** An introductory overview to California ground squirrel control., Proc. 11th Vertebrate Pest Conf., Sacramento, Calif., March 6 to 8, 1984, 32.
44. **Schutz, W.,** Feldmausbekämpfung in Weser-Ems, *Gesunde Pflanzen,* 27(10), 201, 1975.
45. **Shorten, M.,** The reaction of the brown rat towards changes in its environment, in *Control of Rats and Mice,* Vol. 2, Chitty, D., Ed., Clarendon Press, Oxford, 1954, Chap. 5.
46. **Telle, H-J.,** Some field observations on the acceptability and effectiveness of norbormide to *Rattus norvegicus, WHO/VBC/67.39,* World Health Organization, Geneva, 1967, 10.
47. **Wang, Z.,** Acute rodenticides in the control of rodent pest in China. A review., Proc. 11th Vertebrate Pest Conf., Sacramento, Calif., March 6 to 8, 1984, 54.
48. World Health Organization, Safe use of pesticides, 20th Report of the WHO Expert Committee on Insecticides, World Health Organization Technical Report Series, No. 513, WHO, Geneva, 1973, 38.

Chapter 20

ANTICOAGULANT RODENTICIDES

M. Lund

TABLE OF CONTENTS

I. INTRODUCTION

Like many other great discoveries the so-called indirect anticoagulants were detected by some coincidence. In 1942 a disease occurred among cattle in Wisconsin, causing high mortality through substantial internal bleedings. It was soon established that the diseased cattle had been feeding on moist and molded sweet-clover hay, and a closer investigation revealed that the compound coumarin, a common substance in green plants, characterized by its pleasant, sweet smell, had been converted into a derivate, the hydroxycoumarin, dicoumarol. When administered to healthy cattle this compound interferes with the normal clotting process of the blood and damages the smaller blood vessels, leading to fatal bleedings in a few days. Apart from the obvious veterinary importance of this discovery, it was used in human medicine too, for the prevention and treatment of thrombosis. Years later, however, a small British pharmaceutical company Ward, Blenkinson, and Co., suggested that dicoumarol might also have rodenticidal properties. Subsequent trials carried out by Armour and Barnett[2] definitively confirmed this idea and started the era of the anticoagulant rodenticides.

A long series of anticoagulants have since been synthesized, most of them biologically much more potent than dicoumarol, and they all belong to one of the two groups: the hydroxycoumarines and the indandiones. They are quite different in chemical structure but basically similar in their physiological effect. A great deal of effort has been carried out to reveal the actual mechanism and the site of action of these compounds, mainly for medical reasons but to some degree also stimulated by the detection of resistant rodent populations. A brief review of the present knowledge relevant to the topic of rodent control follows.

In the liver cells a constant cyclic process is going on, the biologically inactive vitamin K_1-2,3-epoxide being reduced by a microsomal enzyme into the biologically active vitamin which again is converted by an oxidase into the vitamin K_1-2,3-epoxide. Vitamin K_1 is essential for the synthesis of prothrombin and other factors of paramount importance to the clotting process.[59]

Hydroxycoumarines and indandiones probably act as antagonists to the vitamin K_1-epoxide reductase, causing a gradual depletion of the vitamin and consequently of the prothrombin complex from the organism resulting in an increase in clotting time until the point where no clotting occurs. This is a fairly slow process rendering the anticoagulants one of their most important characteristics, i.e., that symptons of poisoning usually do not occur until 3 to 4 days after ingestion. This delay in action is imperative for the efficiency of a rodenticide as it counteracts the initial fear in some rodent species of feeding from an unfamiliar food source (see previous chapter).

As many mammals and man can survive for prolonged periods with a dramatically increased clotting time, if not wounded or put under stress,[41] it is believed that the anticoagulant also has a damaging impact on the smaller blood vessels, increasing the risk of "spontaneous" hemorrhages. The hydroxycoumarines and indandiones first introduced were characterized by their relatively moderate toxicity, the acute LD_{50} values ranging from about 10 to 50 mg/kg, and by the fact that repeated intake increased toxicity unproportionally. The newer anticoagulants (also called the "second-generation", "single-dose", or "acute" anticoagulants) are much more potent, with acute LD_{50} values ranging from 0.22 to 3.0 mg/kg, but they do not show a similar marked difference between chronic and acute doses. Being introduced on the market around 1950 the anticoagulants, in a remarkably short time, replaced the acute, fast acting, rodenticides in the U.S., Europe, and many other parts of the world, at least as far as the control of commensal rodents was concerned. This was true in spite of the fact that most of the early anticoagulants are more effective against *Rattus norvegicus* than against *R. rattus* and *Mus musculus*. Where *R. norvegicus* is considered the most important species the introduction of these new, slow acting, rodenticides was almost a revolution in the sense that the efficiency of practical control campaigns was very much improved. It was now possible to reduce rat populations to a minimum not only on single farms but in entire districts or towns, and for the first time the concept of "rat-free areas" was introduced.[20,74] After 8 to 10 years, however, the heavy selection pressure exerted by the systemic and regular use of the anticoagulants led to the development of resistant rodent populations.

In spite of the massive application of anticoagulant rodenticides in many countries for more than 35 years their safety record is remarkably good as far as man is concerned — so far no fatal, accidental poisoning has been documented.[23] This is not only due to the very low concentrations used in prepared baits (0.001 to 0.05%), but also to the fact that a very effective antidote is available: during the first 3 to 4 days after intake massive doses of vitamin K_1 can completely neutralize the adverse effects of the anticoagulant. Sometimes, however, the treatment has to be very much prolonged, dependent on the biological half-life of the particular anticoagulant.[45] Primary poisoning of nontarget animals, especially domestic animals such as dogs and pigs, happens however, fairly frequently, in spite of rather large quantities of prepared bait being necessary for producing lethal poisoning in larger animals. Secondary poisoning risk was considered relatively low with the older anticoagulants, even if experimental evidence existed that predators (dogs and mink) can be killed by being fed anticoagulant-killed rodents.[22] After the introduction of the more toxic second-generation anticoagulants for the control of field rodents the primary as well as the secondary poisoning risk of nontarget animals has increased.

Workers in pharmaceutical industries or in pest control companies, handling anticoagulant concentrates, must take all precautions to avoid inhaling dust or being contaminated on the skin by oil-soluble compounds. The anticoagulants possess several of the characteristics of an ideal rodenticide: slow action, high palatability, low risk, and easy application. In contrast to quick-acting rodenticides, prebaiting is not necessary, as the slow-acting anticoagulants act as their own prebait, allowing the rodents to become confident with the bait before symptoms occur.

The fact that most anticoagulants produce a 100% mortality in *R. norvegicus* after less than 6 days feeding in the laboratory must not lead to the assumption that the same effect can be achieved after a similar feeding period in the field. Highly infested farmsteads, refuse dumps, etc., have to be treated for much longer periods, often for at least 3 weeks, before rat activity stops completely. This seems partly to be due to the social structure of the rat communities, being divided into separate family groups that do not allow members of other groups to approach the bait sites. When the established territory owners have been removed by the anticoagulant, neighboring groups might get access to the bait and cause an increase in consumption. Usually this phenomenon is taken into consideration by the ordinary baiting technique where a *surplus* of the anticoagulant bait is placed at several sites and replenished a couple of times until feeding stops. The newer anticoagulants, especially brodifacoum, that can produce complete kill after a single dose, may, with advantage, be applied by the "pulsed baiting technique",[63] where very small quantities of bait are laid down at as many sites as possible and replenished at 1-week intervals until feeding stops.

In the following the various, commercially available, anticoagulant rodenticides are described in some detail, divided into the two main categories: indandiones and hydroxycoumarines, and listed in alphabetic order.

II. INDANDIONES

A. Chlorophacinone (2-(α-4-chlorophenyl-α-phenylacetyl)-1,3-indandione)

Chlorophacinone is a French, oil-soluble anticoagulant, synthesized in the mid-sixties and used for the control of commensal as well as field rodents, especially *Ondathra zibethica* and *Microtus arvalis* in Europe for many years.[33,57] The ordinary concentrations are 0.005%

in baits and 0.2% in tracking dusts. It is somewhat less toxic to *R. norvegicus* than warfarin, the acute LD_{50} being around 20.5 mg/kg, but it has a stronger initial effect on rats and mice.[46] For the control of house mouse populations a prolonged feeding period is needed, and a complete kill not readily achieved. It has been found effective against deer mice (*Peromyscus maniculatus*) in the laboratory, giving complete kill after a 2-day feeding period on 0.02% bait.[54] In fruit orchards in the U.S. pine voles (*Microtus pinetorum*) have been successfully controlled by either a ground-spray at a rate of 0.2 kg/ha[13] or by wax-grain pellets in two applications or 11 kg/ha.[13] Chlorophacinone is not effective against warfarin-resistant rodents.

B. Diphacinone (2-diphenylacetyl-1,3-indandione)

Diphacinone is an old anticoagulant, produced and primarily used in the U.S. as a 0.005% dry or liquid bait. The acute LD_{50} to *R. norvegicus* is reported to be 22.7 mg/kg in females and 43.3 mg/kg in males;[44] but still it is claimed to be more toxic than warfarin to rats, mice, and dogs,[24] or comparable to warfarin in the case of *R. norvegicus*,[36] in spite of the palatability being somewhat lower. Bentley and Larthe[6] concluded, from laboratory as well as field tests, that it was more effective against *R. rattus* than other anticoagulants at that time. House mouse populations may be effectively controlled by a 0.0125% bait, but prolonged baiting periods are necessary.[67] It is reported to be less effective than chlorophacinone against *Peromyscus maniculatus* in conifer-forest regenerations.[39,54] Diphacinone is not effective against warfarin-resistant rodents.

C. Pival or Pindone (2-pivalyl-1,3-indandione)

Pival is one of the earliest synthesized anticoagulants, introduced on the market in the U.S. in 1953[42] as a bait containing 250 ppm active ingredient (a.i.). A sodium salt is water soluble (Pivalyn) and used at a concentration of 0.005% as liquid baits, well accepted by, among others, *Mus musculus*.[66] No LD values are published, but the compound seems to be comparable to warfarin in its effect on *R. rattus* and *M. musculus,* whereas it is considered less effective against *R. norvegicus*.[7,36] Palatability is apparently lower than that of warfarin, diphacinone, and chlorophacinone.[7,70] It has been claimed to possess some insecticidal and fungicidal properties, but this has not been substantiated by Bentley et al. or Crabtree et al.[7,18] Pival is not effective against warfarin-resistant rodents.

D. PMP or Valone (2-isovaleryl-1,3-indandione)

PMP is produced in the U.S. and mainly used there. It is available as a dry bait (0.055%) as well as liquid bait and a tracking dust. It is considered less effective than warfarin to *R. norvegicus*.[6] PMP is not effective against warfarin-resistant rodents.

III. HYDROXYCOUMARINES

A. Bromadiolone (3-(3-(4′ bromobiphenyl-4-yl)-3-hydroxy-1-phenylpropyl)-4-hydroxycoumarine

Bromadiolone was synthesized and marketed by a French company in the mid-seventies, and has since been widely used against commensal and field rodents in many countries. The LD_{50} values in different species can be seen in Table 1.[27]

In baits the concentration is usually 50 ppm and even at higher concentrations it is considered more palatable than most other anticoagulants.[56,61] A complete kill can be obtained in *R. norvegicus* after 24-hr feeding, whereas 5 days are needed in *R. rattus*[61] and a much longer period in *M. musculus*.[28,53] Many field rodent species are reported to be effectively controlled by bromadiolone, e.g., gophers (*Thomomys* spp.), ground squirrels (*Spermophilus* spp.), wood rats (*Neotoma* ssp.), and gerbils (*Meriones* ssp.).[53] The important African species

Table 1
LD_{50} VALUES FOR BROMADIOLONE[27]

Species	LD_{50} value (mg/kg)
Rat	0.65 (acute)
Rat	0.06 to 0.14 × 5 (chronic)
Mouse	0.99
Rabbit	c.1.0
Pig	c.3.0
Dog	c.10
Cat	c.25
Chicken	c.5.0
Guinea pig	2.8

Table 2
LD_{50} VALUES FOR BRODIFACOUM[27]

Species	LD_{50} value (mg/kg)
Rat	0.27 (acute)
Rat	0.05 to 0.08 × 5 (chronic)
Mouse	0.4
Rabbit	0.2
Pig	c.10
Dog	3.5
Cat	c.25
Chicken	c.10 to 20
Sheep	c.10
Opossum	0.17

Mastomys natalensis and the Indian field mouse (*Mus booduga*) are also found to be very susceptible to this compound.[3,25] Bromadiolone wax-blocks are suitable for sewer rat control and for permanent baiting in humid conditions. Coconut damage caused by *R. rattus* in India has been significantly reduced by the use of such formulations.[40] Reports from Switzerland[34] indicate that hazards to wildlife do exist when using bromadiolone against field rodents indiscriminately, even if the secondary poisoning risk to carnivores and raptors seems to be of minor importance.[27,51]

In spite of bromadiolone belonging to the group of more potent ''second-generation'' anticoagulants, serious resistance problems have been encountered with *R. norvegicus* and *M. musculus* in the U.K. and Denmark.[47,68]

B. Brodifacoum (3-(3-(4′-bromobiphenyl-4-yl)-1,2,3,4-tetrahydronaphth-1-yl)-4-hydroxycoumarine)

Brodifacoum is one of the newest and most potent of the so-called ''second-generation'' anticoagulants, synthesized in 1977 in the U.K. and marketed in several countries for the control of a wide range of rodent pest species, comensal as well as field rodents. Brodifacoum is available as a 0.005% pellet for rat and house mouse control, a smaller 0.001% pellet for field rodent control, and furthermore, as 20 g wax-blocks for sewer rat control, humid conditions in general, permanent baiting stations, and for the control of certain field rodents.

The toxicity of brodifacoum appears in Table 2.[27] Brodifacoum is the only anticoagulant found to produce a 100% mortality in most rodent species after only 24 hr feeding (''single-dose'').[9,15,21,26,48,55,58,75] The only species showing such a low susceptibility that practical control is unlikely to be achieved are *Acomys cahirinus*, the spiny mouse, and the gerbil *Meriones shawi*.[50,52] Successful control of *Microtus*-species has been obtained in orchards in the U.S. by hand-baiting with small brodifacoum-pellets (0.001%) at the rate of 29 kg/ha.[14] Damage to rice-fields in Malaysia caused by *R. argentiventer* has been significantly reduced by the use of 5 g-brodifacoum wax blocks (0.005%), applied once a week for 8 weeks.[10]

In spite of the extremely high toxicity of the technical compound the practical use of the low-concentration brodifacoum baits seems not to have resulted in a serious increase in the risk to domestic animals or wildlife compared to that of the other anticoagulants. Dogs are, however, quite susceptible to this compound and living in the same environment as the commensal rodents they are particularly exposed to primary poisoning.[21] The average pig, sheep, or chicken must consume considerable quantities of prepared bait to be endangered;[21,49] but as there is a great variation in susceptibility, and as brodifacoum and several other anticoagulants have been found to have a very long biological half-life (brodifacoum: 150 to 200 days), sublethal doses may accumulate and become fatal over a long period.[27] A great variation in susceptibility of birds has also been demonstrated by Godfrey,[27] ranging from an LD_{50} of 0.75 mg/kg for Canadian Geese (*Branta canadensis*) to 10 mg/kg in the Pheasant (*Phasanius colchicus*) and the Harrier Hawk (*Circus approximans*), and to more than 20 mg/kg in the Paradise Duck (*Tadorna variegata*). The secondary poisoning hazards to Barn Owls (*Tyto alba*), feeding partially on brodifacoum-killed rodents, was found negligible in a telemetry study in New Jersey.[37]

Up to 1984 brodifacoum was reported effective against most warfarin-resistant rat and mouse infestations, but recently laboratory and field evidence has been presented of increased tolerance to this potent compound in *R. norvegicus* as well as *M. musculus* in the U.K. and Canada.[32,50,72]

C. Coumachlor (Tomorin) (3-α-p-chlorphenyl-β-acetylethyl)-4-hydroxycoumarin)

Coumachlor is a German anticoagulant introduced shortly after warfarin. It has a very low toxicity with an acute LD_{50} to *R. norvegicus* of around 900 to 1200 mg/kg, and accordingly it is considered less effective than warfarin.[6,62] It is used mainly as a tracking dust (1%), and the original 0.025 to 0.5% bait is now almost out of use. Coumachlor is not effective against warfarin-resistant rodents.

D. Coumatetralyl (3-α-tetralyl-4-hydroxycoumarin)

Coumatetralyl is a German product used for many years and in many countries for the control of commensal rodents. It is formulated as a dry bait (0.0375%), a liquid bait based on the sodium-salt, and a 0.75% tracking dust. The acute LD_{50} to *R. norvegicus* is 16.5 mg/kg, females being less susceptible than males.[38] In spite of its lower toxicity than warfarin it is reported to be a little more effective against *R. norvegicus*, apparently due to a higher palatability.[47] The susceptibility of the two other major commensal species, *R. rattus* and *M. musculus*, is reported to be close to that of warfarin,[16,17,70,71] but coumatetralyl seems to be more effective against *Tatera indica*.[30]

Some specificity is claimed, the chronic toxicity being 5 × 0.3 mg/kg to *R. norvegicus* in contrast to more than 8 × 50 mg/kg to the chicken.[38] More recent tests, however, seem to indicate that chickens are also endangered in practice.[49] The risk to fish has exeptionally been investigated, and no symptoms were provoked until very high concentrations has been reached, e.g., around 1 g a.i. per liter of water.[38]

Shortly after the detection of rat populations being resistant to warfarin, coumatetralyl was introduced with considerable success for a few years, but then serious resistant problems were met in the U.K. as well as Denmark.[47,70]

E. Difenacoum (3-(3-p-diphenyl-1,2,3,4-hydronaphth-1-yl)-4-hydroxycoumarin)

Difenacoum was synthesized in the U.K. and marketed in 1975 as a 0.005% pelleted

Table 3
LD_{50} VALUES FOR DIFENACOUM[12]

Species	LD_{50} value (mg/kg)
Rat	1.8 (acute)
Rat	0.15 × 5 (chronic)
Mouse	0.8
Rabbit	2.0
Pig	c.80
Dog	c.50
Cat	c.100
Chicken	c.50
Guinea pig	c.50
Sheep	c.100

bait, later also as a wax block. It is found less palatable than warfarin,[35] but much more toxic to most rodent species. Some specificity is claimed (Table 3).

Feeding for 24-hr on a 0.005% bait does not lead to a complete mortality in *R. norvegicus, R. rattus, M. musculus, Tatera indica,* or *Nesokia indica.*[30,48] *Bandicota bangalensis* seems, however, to be very susceptible.[9,17] Studies indicate that efficient field control can be achieved in *Mastomys natalensis, Sigmodon hispidus,*[26] *Microtus ssp.*[48] *R. norvegicus, R. rattus, R. exulans* and *M. musculus.*[9,68]

Difenacoum is still efficient against many populations of warfarin-resistant rats, but in the U.K. serious resistance problems to this compound have also been encountered in recent years.[31,60,68] In Denmark, difenacoum is more effective against warfarin-resistant rats than bromadiolone, in spite of its lower toxicity.[50] *M. musculus* populations have also shown a high degree of resistance to difenacoum in the U.K., some individuals surviving 21 days on 0.005% bait, consuming up to 147 mg/kg.[68]

F. Fumarin (Coumafuryl) (3-(α-acetonylfurfuryl)-4-hydroxycoumarin)

Fumarin was produced and introduced on the market by a German company in 1952. It is used in baits at the same concentration as warfarin (0.025 to 0.05%). A water-soluble sodium salt exists for the preparation of liquid baits. It is considered almost equal in toxicity to warfarin as far as *R. norvegicus* is concerned and slightly less efficient against *M. musculus.*[6] No acute LD_{50} values are found in the literature, but the chronic LD_{50} in *R. norvegicus* is reported to be 1.4 mg/kg × 5.[73] Cats and dogs seem to be almost as susceptible as rats, dogs being killed by 2 mg/kg × 5 and cats by 10 mg/kg × 4.[73] Fumarin is not effective against warfarin-resistant rodents.

G. Warfarin (3-(α-acetonylbenzyl)-4-hydroxycoumarin)

Warfarin was the first anticoagulant introduced as a rodenticide shortly after the World War II (see I). The syllable WARF comes from the first letters in Wisconsin Alumni Research Foundation that introduced the rodenticide. It rapidly became almost synonymous with a rodenticide and was used indiscriminately against most rodent species in many countries for several years. After the detection of resistance around 1960 other types of rodenticides again had a chance until the new, more potent anticoagulants were developed. Warfarin is, however, still used widely, especially for the control of *R. norvegicus* in areas where resistance has not been discovered. It is generally applied as the S-isomer, which is more than ten times as toxic as the R-isomer. Information on the acute LD_{50} values differs considerably, but it now seems established that it is around 10 to 12 mg/kg (for the S-isomer) for *R. norvegicus*; the chronic LD_{50} is 0.75 mg/kg × 5.[4] Warfarin is formulated as dry baits (0.005

Table 4
NUMBER OF DAYS REQUIRED TO ACHIEVE LD_{50} or LD_{99}, RESPECTIVELY, BY FEEDING BAIT CONTAINING 250 PPM WARFARIN

	Feeding period	
Species	LD_{50}	LD_{99}
Nesokia indica	1.9	3797.0
Acomys cahirinus	5.4	239.3
Mus musculus	4.8	29.5
Mastomys natalensis	4.8	26.0
Bandicota bengalensis	1.4	25.0
Rattus rattus	3.6	21.0
Tatera indica	5.8	19.2
Rattus argentiventer	3.2	15.1
Sigmodon hispidus	3.7	8.1
Arvicanthis niloticus	3.8	6.0
Rattus norvegicus[a]	1.7	5.8

[a] 50 ppm

After Greaves, J. H., *Acta Zool. Fenn.*, 173, 173, 1985.

to 0.05%) as well as liquid baits based on the sodium salt. Tracking dusts at a concentration of 0.5 to 1.0% are also available. Palatability of baits may vary according to the content of certain impurities in the technical compound, especially phenolic ketones.[8,43]

Warfarin is primarily efficient against *R. norvegicus,*[5] whereas most other species show a remarkable variation in susceptibility as can be seen in Table 4.

The LD_{99} values indicate, and this has been substantiated by field trials, that warfarin may also be recommended for the control of *Arvicanthis niloticus, Sigmodon hispidus,* and perhaps *Rattus argentiventer.*[11] In prolonged treatments with higher concentration baits or tracking dust, satisfactory control may be achieved in *M. musculus* populations,[65] male mice being more susceptible than females.[64,69]

Warfarin is sometimes combined with an antibacterial agent, sulfaquinoxaline, in order to reduce the bacterial production of vitamin K in the gut of the rat,[19] but the advantage of this combination has not been proved.[1] In the U.K. the anticoagulant may also be combined with calciferol for the control of warfarin-resistant rats.

Warfarin is considered one of the safest anticoagulants, also as far as domestic animals are concerned, and especially chickens are at a low risk.[49] Serious resistance problems in connection with the three major commensal rodents exist in Europe, the U.S., and a few other areas.

REFERENCES

1. **Anonymous,** The efficiency of warfarin with sulfaquinoxaline against rats and mice, Inf. Control Lab., Tolworth, Surbiton, England, Techn. Circular No. 7, 1964, 2.
2. **Armour, C. J. and Barnett, S. A.,** The action of dicoumarol on laboratory and wild rats and its effect on feeding behaviour, *J. Hyg. (Cambridge),* 48, 158, 1950.

3. **Balasubramanyam, M., Christopher, M. J., and Purushotham, K. R.,** Laboratory trials of three anticoagulant rodenticides for use against the Indian field mouse, *(Mus booduga Gray), J. Hyg. Cambridge,* 93, 575, 1984.
4. **Becker, K.,** Können Haustiere mit Cumarin vergiftet werden?, *Prakt. Schädl. Bek.,* 23(9), 128, 1971.
5. **Bentley, E. W., Hammond, L. E., and Taylor, E. J.,** The comparative toxicity of 0.025% and 0.005% warfarin to *Rattus norvegicus, Plant Pathol.,* 4(4), 120, 1955.
6. **Bentley, E. W. and Larthe, Y.,** The comparative rodenticidal efficiency of five anticoagulants, *J. Hyg. Cambridge,* 57(2), 135, 1959.
7. **Bentley, E. W. and Rowe, M.,** Pival, an anticoagulant rodenticide, *J. Hyg. Cambridge,* 54(1), 20, 1956.
8. **Bowerman, A. M. and Brooks, J. E.,** What makes a palatable warfarin?, *Pest Control,* 40(2), 28, 1972.
9. **Brooks, J. E., Than Htun, O., Walton, D. W., and Maung Tun, M.,** Laboratory evaluations of rodenticides for use against south-east Asian commensal small mammals, WHO/VBC/79, 720, World Health Organization, Geneva, 1979, 15.
10. **Buckle, A. P., Rowe, F. P., and Rahman, A.,** Field trials of warfarin and brodifacoum wax block baits for the control of the Rice field rat, *Rattus argentiventer,* in peninsular Malaysia, *Trop. Pest. Manage.,* 30(1), 51, 1984.
11. **Buckle, A. P., Rowe, F. P., and Young, J. C.,** Laboratory evaluation of 0.025% warfarin against *Rattus argentiventer, Trop. Pest Manage.,* 26(2), 162, 1980.
12. **Bull, J. O.,** Laboratory and field investigations with difenacoum, a promising new rodenticide, Proc. 7th Vertebr. Pest Conf., Monterey, Calif., March 9 to 11, 1976, 72.
13. **Byers, R. E.,** Effect of hand baits and ground sprays on pine vole activity, *Hortic Sci.,* 10(2), 122, 1975.
14. **Byers, R. E. and Merson, M. H.,** Current improvements in baiting pine and meadow voles, Proc. 10th Vertebr. Pest Conf., Monterey, Calif., March 23 to 25, 139, 1982.
15. **Chmela, J. and Rupes, V.,** Testing the efficacy of the rodenticide brodifacoum on Norway rats and house mice, *Vešt. Cš. Špolec. Zool.,* 47, 6, 1983.
16. **Chopra, G., Ahmad, N., and Parshad, V. R.,** Efficacy of Racumin (0.75% coumatetralyl) against *Rattus rattus, National Program Rodent Pest Manage. Rodent Newslett.,* 7(4), 21, 1983.
17. **Chopra, G. and Parshad. V. R.,** Evaluation of coumatetralyl against two predominant murid species, *J. Hyg. Cambridge,* 94, 327, 1985.
18. **Crabtree, D. G. and Robinson, W. H.,** Pivalyl, the new insecticidal rodenticide, *Pest Control* 21(7), 1953.
19. **Derse, P.,** Anti-K factor in anticoagulant rodenticides, *Soap and Chem. Spec.,* March, 82, 1963.
20. **Drummond, D. C.,** Rat free towns?, Proc. 2nd Br. Pest Control Conf., Brighton, England, April 3 to 5, 1967, 19.
21. **Dubock, A. C.,** The development and practical use of the novel anticoagulant rodenticide brodifacoum, *Plant Prot. Bull. (Taiwan R.O.C.),* 22, 223, 1980.
22. **Evans, J. and Ward, A. L.,** Secondary poisoning associated with anticoagulant-killed nutria, *J. Am. Vet. Med. Assoc.,* 151(7), 856, 1967.
23. **Fristedt, B. and Sterner, N.,** Warfarin intoxication from percutaneous absorption, *Arch. Environ. Health,* 11, 205, 1965.
24. **Gates, R. L.,** Diphacin, new anticoagulant rodenticide, *Pest Control,* 25(8), 14, 1957.
25. **Gill, J. E. and Redfern, R.,** Laboratory tests of seven rodenticides for the control of *Mastomys natalensis, J. Hyg. Cambridge,* 83, 345, 1979.
26. **Gill, J. E. and Redfern, R.,** Laboratory trials of seven rodenticides for use against the cotton rat *(Sigmodon hispidus), J. Hyg. Cambridge,* 85, 443, 1980.
27. **Godfrey, M. E. R.,** Non-target and secondary poisoning hazards of "second-generation" anticoagulants, *Acta Zool. Fenn.,* 173, 209, 1985.
28. **Grand, M.,** Experimental results on a new anticoagulant rodenticide — bromadiolone, *Phytiatrie-Phytopharmacie,* 25, 69, 1976.
29. **Greaves, J. H.,** The present status of resistance to anticoagulants, *Acta Zool. Fenn.,* 173, 173, 1985.
30. **Greaves, J. H. and Rehman, A. B.,** The susceptibility of *Tatera indica, Nesokia indica* and *Bandicota bengalensis* to three anticoagulant rodenticides, *J. Hyg. Cambridge,* 78, 75, 1977.
31. **Greaves, J. H., Shepherd, D. S., and Gill, J. E.,** An investigation of difenacoum resistance in Norway rat populations in Hampshire, *Ann. Appl. Biol.,* 100, 581, 1982.
32. **Greaves, J. H., Shepherd, D. S., and Quy, R.,** Field trials of second-generation anticoagulants against difenacoum-resistant Norway rat populations, *J. Hyg. Cambridge,* 89, 1982, 295.
33. **Grolleau, G.,** Comparaison de la toxicitéde divers raticides anticoagulants a l'egard du campagnol des champs, *Microtus arvalis* (Pallas), *Ann. Zool. Ecol. Anim.,* 3(2), 261, 1971.
34. **Grolleau, G.,** Le rodenticide anticoagulant Bromadiolone est il dangereux pour les animaux prédateurs et en particulier les rapaces?, *Def. Veget.,* 2119, 14, 1983.
35. **Hadler, M. R., Redfern, R., and Rowe, F. P.,** Laboratory evaluation of difenacoum as a rodenticide, *J. Hyg. Cambridge,* 74, 441, 1975.

36. **Hayes, W. J. and Gaines, T. B.,** Laboratory studies of five anticoagulant rodenticides, *Public Health Rep.*, 74(2), 105, 1959.
37. **Hegdal, P. L. and Blaskiewicz, R. W.,** Evaluation of the potential hazard to Barn owls of Talon (Brodifacoum) used to control rats and house mice, *Environ. Toxicol. Chem.*, 3, 167, 1984.
38. **Herman, G. and Hombrecher, S.,** Ratten-und Mäusebekämpfung mit Racumin 57-Präparaten, *Pflanzenschutz Nachr. Bayer,* 15(2), 93, 1962.
39. **Howard, W. E., Marsh, R. E., and Cole, R. E.,** A diphacinone bait for deer mouse control, *J. Forest.*, 68(4), 220, 1970.
40. **Jhala, R. C., Shah, A. H., Patel, R. C., and Patel, R. L.,** Effectiveness of bromadiolone and warfarin wax cakes against rats damaging coconut palms in South Guajarat, *Central Arid Zone Res. Inst., Jodhpur, Rodent Control Newslett.*, 8(1-4), 5, 1984.
41. **Jaques, L. B.,** Dicoumarol drugs and the problem of hemorrhage, *Can. Med. Assoc. J.*, 81, 848, 1959.
42. **Kilgore, L. B., Ford, J. H., and Wolfe, W. C.,** Insecticidal properties of 1,3-indandiones, *Industr. Engng. Chem.*, 34, 494, 1942.
43. **Krishnamurthy, K., Ramasivan, T., and Singh, D. P.,** Studies on rodents and their control. VII. Effect of impurities in warfarin on its acceptability and mortality to black rats *(Rattus rattus), Bull. Grain Technol.*, 9(4), 252, 1971.
44. **Kusano, T.,** The toxicity of diphacinone (2-diphenylacetyl-1,3-indandione) to laboratory rats and mice, *Jap. J. Sanit. Zool.*, 24(3), 207, 1974.
45. **Lipton, R. A. and Klass, E. M.,** Human ingestion of a ''superwarfarin'' rodenticide resulting in a prolonged anticoagulant effect, *JAMA,* 252(2), 3004, 1984.
46. **Lund, M.,** The toxicity of chlorophacinone and warfarin to house mice *(Mus musculus), J. Hyg. Cambridge,* 69, 69, 1971.
47. **Lund, M.,** The resistance to anticoagulants, *Danish Pest Inf. Lab. Annu. Rep. 1971,* Lyngby, Denmark, 1972, 57.
48. **Lund, M.,** Comparative effect of the three rodenticides warfarin, difenacoum and brodifacoum on eight rodent species in short feeding periods, *J. Hyg. Cambridge,* 87, 101, 1981.
49. **Lund, M.,** Hens, eggs and anticoagulants, *Int. Pest Control,* 5, 126, 1981.
50. **Lund, M.,** Resistance to the second-generation anticoagulant rodenticides, Proc. 11th Vertebr. Pest Conf., Sacramento, Calif., March 6 to 8, 1984, 89.
51. **Lund, M. and Rasmussen, A. M.,** Secondary poisoning hazards to Stone Marten *(Martes foina)* fed bromadiolone poisoned mice, *Nord. Vet. Med.*, 38, 241, 1986.
52. **Mahmoud, W. and Redfern, R.,** The response of the Egyptian spiny mouse, *(Acomys cahirinus)* and two other species of commensal rodents to anticoagulant rodenticides, *J. Hyg. Cambridge,* 86, 334, 1981.
53. **Marsh, R. E.,** Bromadiolone, a new anticoagulant rodenticide, *EPPO Bull.*, 7, 502, 1977.
54. **Marsh, R. E., Howard, W. E., and Cole, R. E.,** The toxicity of chlorophacinone and diphacinone to deer mice, *J. Wildl. Manage.*, 41(2), 298, 1977.
55. **Mathur, R. P. and Prakash, I.,** Evaluation of brodifacoum against *Tatera indica, Meriones hurrianae* and *R. ratus, J. Hyg. Cambridge,* 87, 179, 1981.
56. **Meehan, A. P.,** Rodenticidal activity of bromadiolone, Proc. 8th Vertebr. Pest Conf., Sacramento, Calif., 1978, 122.
57. **Moens, R. and Ghesquiere, A.,** La destruction des rats musques dans les polders ''De Moeren'' au moyen d'appats traités a la chlorophacinone, *Rev. Agric.*, 22(8), 1089, 1969.
58. **Myllymäki, A.,** Efficacy of a number of toxic baits and baiting against the voles, *Microtus argrestis* and *Arvicola terrestris,* Proc. 11th Vertebr. Pest Conf., Sacramento, Calif., March 6 to 8, 1984, 38.
59. **Park, B. K., Scott, A. K., Wilson, A. C., Haynes, B. P., and Brechenridge, A. M.,** Plasma disposition of vitamin K_1 in relation to anticoagulant poisoning, *Br. J. Clin. Pharmacol.*, 18, 655, 1984.
60. **Redfern, R. and Gill, J. E.,** The development and use of a test to identify resistance to the anticoagulant difenacoum in the Norway rat *(Rattus norvegicus), J. Hyg. Cambridge,* 81, 427, 1978.
61. **Redfern, R. and Gill, J. E.,** Laboratory evaluation of bromadiolone as a rodenticide for use against warfarin-resistant and non-resistant rats and mice, *J. Hyg. Cambridge,* 77, 419, 1980.
62. **Reiff, M. and Wiesmann, R.,** Untersuchungen über ein neues Rodenticid mit kumulativer Wirkung auf Basis der Cumarin-Derivate, *Acta Trop.*, 8(2), 98, 1951.
63. **Richards, C. G. J. and Huson, L. W.,** Towards the optimal use of anticoagulant rodenticides, *Acta Zool. Fenn.*, 173, 155, 1985.
64. **Roll, R.,** Über die Wirkung eines Cumarinpräparates (Warfarin) auf Hausmäuse *(Mus musculus), Z. Angew. Zool.*, 53(13), 278, 1966.
65. **Rowe, F. P.,** The control of house mice *(Mus musculus)* with a warfarin dust surrounding water baits, *The Sanitarian,* Dec. 4, 1957.
66. **Rowe, F. P.,** The toxicity and acceptability of the sodium salt of pindone, an anticoagulant rodenticide, to the house mouse *(Mus musculus), J. Hyg. Cambridge,* 59(3), 335, 1961.

67. **Rowe, F. P. and Neve, M.,** Two field trials of diphacinone against the house mouse *(Mus musculus), Extrait de Parasitica,* 17(4), 170, 1961.
68. **Rowe, F. P., Plant, C. J., and Bradfield, A.,** Trials of the anticoagulant rodenticides bromadiolone and difenacoum against the house mouse *(Mus musculus), J. Hyg. Cambridge,* 87, 177, 1981.
69. **Rowe, F. P. and Redfern, R.,** The effect of sex and age on the response to warfarin in a non-inbred strain of mice, *J. Hyg. Cambridge,* 65, 55, 1967.
70. **Rowe, F. P. and Redfern, R.,** Laboratory studies on the toxicity of anticoagulant rodenticides to wild house mice, *Mus musculus* L., *Ann. Appl. Biol.,* 61, 303, 1968.
71. **Rowe, F. P. and Redfern, R.,** Comparative toxicity of the two anticoagulants, coumatetralyl and warfarin, to wild house mice *(Mus musculus), Ann. Appl. Biol.,* 62, 355, 1968.
72. **Siddiqi, Z. and Blaine, W. D.,** Anticoagulant resistance in house mice in Toronto, Canada, *Pest Manage.,* 10, 1982.
73. **Steiniger, F.,** Über die Wirksamkeit des Fumarins, eines neues Antikuagulants zur Rattenbekämpfung und seine Verträglichkeit für Haustiere, *Nachr. Blatt Deutsch. Pfl. Schutzd.,* 5(11), 1, 1953.
74. **Telle, H-J.,** Permanente Rattenbekämpfung oder rattenfreier Zustand?, *Gesundheitswes. Desinfekt.,* 60, 151, 1967.
75. **Tongtavee, K.,** Efficacy of different rodenticides on some species of rats in Thailand, Proc. 9th Vertebr. Pest Conf., Fresno, Calif., March 4 to 6, 1980, 143.

Chapter 21

CHEMOSTERILANTS FOR RODENT CONTROL

Rex E. Marsh

TABLE OF CONTENTS

I. INTRODUCTION

The theoretical and applied feasibility of chemosterilants for rodent control has been the subject of numerous papers, yet progress in this area has been less than overwhelming.[13,56,58,65] Since the chemosterilant approach to rodent control is based on sound biological prinicples, it offers a highly acceptable means of biological control.[41] The extensive research directed toward oral contraceptives for humans has assisted the search for effective candidate chemical agents.

For the purpose of this chapter, a chemosterilant is defined broadly as a chemical that can cause permanent or temporary sterility or infertility in either or both sexes or, through some other physiological means, reduce the number of offspring or alter the fecundity of the offspring produced. This definition encompasses antifertility agents, a term frequently used interchangeably with chemosterilants when discussing them in the context of rodent control. Chemosterilants also encompass a variety of other descriptive terms that denote specific modes of action, such as spermatocides, embryocides, and gametocides.

There are many ways in which chemical agents can interfere with the physiological events of reproduction. "Interference with reproduction may take the form of direct or indirect damage to the developing or mature gametes prior to copulation, or after this event but before fertilization. The union of sperm and ovum may be prevented or the fertilized ovum may be hindered during the process of implantation into the uterus. Finally, the implanted embryo may be the focus of chemical attack in various ways at different stages of its development."[45] A rodent population can also be suppressed by chemical neonatal interference with the sexual development of the offspring, causing irreversible sterility, or by other chemically induced factors in parent or offspring that cause a high rate of early mortality, such as reduced lactation or abnormalities in the young.

II. THE INITIAL CONCEPT

Credit for the concept of sterilization of insects goes to Dr. E. F. Knipling, who proposed it in 1938.[78] Some two decades later, after many years of research, the concept was applied to the eradication of the screwworm, *Cochliomyia hominivorax* (Cqr.), a serious pest of lifestock in southeastern U.S. Since that success, this approach has been used for the control of a number of insects, some more successfully than others.

The concept of controlling or regulating reproduction in rodents and other wild vertebrates was discussed by Knipling,[56] Davis,[13] and Wetherbee.[83] In the U.S. in 1957 to 1958, Davis conducted one of the earliest studies of this approach to rodent control using triethylenemelamine (TEM) on a Norway rat (*Rattus norvegicus*) population in Baltimore, Md. The results, though promising, were inconclusive, and the studies were terminated for other reasons.[13] This approach to resolving pest rodent problems received considerable attention during the 1960s and early 1970.[4,10,41-44,53,63,64,66,73,74,77,79] Since the mid 1970s, chemosterilants for rodent control have received much less attention, although some research has continued. Recent studies predominantly involve two active ingredients — diethylstilbestrol (DES) and alpha-chlorohydrin.[3,19,24,27-30,40,51,52]

Most rodent control-related chemosterilant research has centered on rat (*Rattus*) species with relatively little research on house mice (*Mus musculus*). Of the field rodents, at least four different species of voles (*Microtus* spp.) have been studied, and a variety of other rodents and lagomorphs, including ground squirrels, prairie dogs, tree squirrels, beaver, jackrabbits, and European rabbits, have been the subjects of chemosterilant research.

III. ACCEPTANCE BY THOSE OPPOSED TO KILLING PEST VERTEBRATES

Reproductive inhibitors of chemosterilants as a means of rodent control are more compatible than toxic rodenticides with certain religious philosophies and human lifestyles, as well as more acceptable to environmental and protectionist groups. They are also much less controversial from a ''humane'' or ''animal rights'' point of view. In general, chemosterilants are more acceptable to those with strong convictions against killing animals. Opponents of pesticides view nonlethal chemosterilants for vertebrate pest control with considerable less alarm, just as they do the chemical repellents.

IV. CHARACTERISTICS OF THE IDEAL CHEMOSTERILANT

Sufficient research and field studies have taken place to permit a good characterization of the ideal chemosterilant. The importance of some of these characteristics may vary, depending on the pest rodent to be controlled. Some chemosterilants, because of their mode of action, hold much greater promise for certain rodent species than for others.

Some general characteristics of an ideal rodent chemosterilant include the following:

1. Orally effectice, preferably in a single feeding.
2. Effective in very low doses (not exceeding 10 mg/kg).
3. Permanent or long-lasting sterility (preferably) lasting 6 months or longer, or at least through the major breeding period of the pest species).
4. Effective for both sexes or preferably for females if only one sex.
5. Rodent-specific or genus-specific.
6. Relatively inexpensive.
7. There should be a wide margin between the chemosterilant effects and lethal doses. (If high in specificity, this may be unimportant or the narrow margin be of value.)
8. Well accepted (i.e., highly palatable) in baits at effective concentrations.
9. Biodegradable after a few days in the environment.
10. If not highly specific, rapid elimination from the body of the primary target to avoid secondary effects.
11. No acquired tolerance nor genetic or behavioral resistance.
12. Free of behavioral modification (such as altering libido, aggression, or territoriality).
13. Free from producing discomfort or ill feelings that could suppress consumption (i.e., bait shyness) on repeat or subsequent feedings.
14. Humane (i.e., produces no stressful symptoms).
15. Easy to formulate into various kinds of baits.
16. Sufficiently stable when prepared in baits (i.e., adequate shelf life).
17. Not translocated into plants (or only at a very low level), thus permitting use on crops.

An impressive number of compounds have reproductive-inhibiting or chemosterilant qualities. When these compounds are judged on the basis of their ideal characteristics, the list narrows considerably. No known compound meets all of the ideal criteria; however, the list serves as a guide when selecting for chemicals with useful attributes. Many formulating and operational methods are available to compensate for any lack of desirable characteristics with a chemosterilant in order to keep it efficacious and safe.

V. CURRENT RODENT CHEMOSTERILANTS

To this author's knowledge, there are only two chemosterilants marketed for rodent control, and both of these are probably best defined as toxicants/chemosterilants. Although others may be in an experimental stage, they are not being used on any regular or practical basis for rodent control. Glyzophrol® by Rowa Ltd., of Ireland, was the first. It seems to have been first used in Europe. Little has been published on this chemosterilant or its effectiveness in rodent control practices. The Austrian patent application (1968) indicates the active ingredient is butandiol-bis-methane sulphonic acid ester, possibility in combination with other active ingredients.[50,70] Glyzophrol, an alkylating agent, interferes with embryo development, and sterility apparently is most pronounced in the sexually immature Norway rats of both sexes.[70] When ingested in amounts above the sterility level, mortality occurs in both males and females. The product can be considered a toxic rodenticide as well as a chemosterilant, depending on the amount consumed.

Alpha-chlorohydrin was the second chemosterilant marketed for rodent control. It is registered and sold under the trade name Epibloc® by Gametrics, Ltd., of the U.S. and is currently registered in 12 countries and licensed in some 29 others for eventual sale.[20] At low doses the product causes temporary sterility; at somewhat higher doses it produces epididymal lesions, causing permanent sterility in adult males.[19] At still higher doses alpha-chlorohydrin is lethal to both males and femeles of susceptible species.

Despite this worldwide attention, neither Glyzophrol nor Epibloc (alpha-chlorohydrin) has come into significant use in rodent control, possibly because their characteristics and potential uses are not sufficiently understood by many practitioners to maximize their effectiveness. A great deal of effort will be needed to educate the user on how any specific chemosterilant can be effectively applied for rodent control, just as it was necessary with warfarin, the first anticoagulant to come on the market. Anticoagulant rodenticides brought about a whole new concept in rodent control — that of required multiple feedings. Chemosterilants, likewise, involve a totally new rodent control concept, and potential users must fully understand the concepts of how to use them effectively.

VI. OUTLOOK FOR FUTURE DEVELOPMENT

The proliferation of new antifertility compounds and the development of potential chemosterilants directed at humans in the last 15 years provide extraordinary possibilities for future chemosterilants for rodent control. The possibilities are overwhelming as they include many compounds that have been rejected for human use for a variety of reasons, including minor adverse side effects, toxicity, irreversible sterility, sterility of offspring, etc., which would not rule them out for rodent control, in fact, in some instances might be advantageous. A considerable amount is known about the inhibitory effect of these compounds on reproduction of rodents, as rodents are extensively used in the early evaluation of these compounds.

The interest in rodent chemosterilant research by pharmaceutical companies, governmental, and university researchers has dwindled since the 1960s and early 1970s, with relatively little research known to be currently under way. Little interest is being expressed by research and development (R & D) departments of manufacturers of traditional toxic rodenticides. Pesticide companies in all but a few instances have been less than enthusiastic over chemosterilants. The reason for this is probably linked to several factors:

1. They are unfamiliar with chemosterilants and their related physiological activities and lack in-house expertise in this area.
2. It is difficult to determine how much of the existing toxic rodenticide market an effective chemosterilant would capture, or whether it would expand the total market.
3. The problems associated with registration of vertebrate chemosterilants are relatively unknown because there are few previous chemical reproductive inhibitors or chemosterilants that have been pursued to registration.

Also, the difficulties and costs of future registrations of rodent chemosterilants are unpredictable, at least in the U.S. Therefore, future strides in rodent chemosterilant research and development will probably have to be supported in a major way by governmental organizations.

In the 1960s when extensive emphasis was placed on the "pill" as an antifertility or contraceptive agent for humans, a number of pharmaceutical companies (Syntex; Parke, Davis and Co.; Warner-Lambert; and Upjohn Co.) also became very interested in the potential use of their compounds for regulating or inhibiting fertility in domestic animals and as chemosterilants for rodent control. For the most part, this interest in chemosterilants for rodent control was short-lived. There appeared to be a reluctance on the part of some other pharmaceutical companies to want their proprietary compounds explored for use for domestic animals or vertebrate pests.

VII. CHEMOSTERILANTS REVIEWED

A variety of potential reproductive inhibitors for pest rodent control have been discussed previously.[62,65] Both estrogens and androgens can be reproductive inhibitors.[76] Synthetic estrogens have received much attention as potential rodent chemosterilants and, as a group, appear to hold the greatest promise when efforts are directed primarily at the females. Generally speaking, estrogens have a greater physiological effect on females than males, and the degree of efficacy (i.e., duration of inhibited reproduction) is for the most part dose-dependent.

Mestranol (17 α ethynylestradiol-3-methyl ether), a proprietary compound of Syntex Corporation, was the first potential rodent chemosterilant to receive much attention in the U.S.[44,63,71]

Other synthetic steroids, including quinestrol, 8-aza quinestrol (an analog of quinestrol), chlorotrianisene (CTA), and diethylstilbestrol (DES), were explored as rodent chemosterilants with some success.[10,28-30,31,67,71] The estrogenic compounds SC20775 and SC24674 (Searle) and AY20, 121 (Ayerst Laboratories) have also been evaluated with varied results.[5,6,25,62]

One estrogen, 17 α-hexa-1′,3′-dynyl oestra-1,3,5 (10)-trien-17 β-ol (coded BDH 10131), without doubt holds the greatest promise as a rodent chemosterilant. This potent synthetic estrogen has been evaluated in both the laboratory,[53] and in the field with surprisingly good results.[59,73,74] With a single exposure to bait, adult females will remain infertile for an exceptionally long period (about 10 months). In our laboratories some female Norway rats remained infertile for 11 months after treatment without need for repeat ingestions. After the 11th month, the study was terminated, thus the possible effect might have been longer. This is impressive, considering that in the wild the average Norway rat has a lifespan of about 1 year.

A matter of particular concern when evaluating hormonal compounds is that their efficacy in the rodent community may hinge on the fact that they do not unfavorably upset social hierarchies, territoriality, or other behavioral traits. In the case of males, they should not reduce libido. A chemosterilant that would result in a high percentage of submissive or subdued individuals, especially males, may prove self-defeating by a realignment of social dominance, whereas the nontreated or unaffected might become the dominant individuals. Information on the effects of chemosterilants on the social aspects of the rodent community is very limited for any of the prospective rodent chemosterilants, and their effects on social hierarchies offer some exciting future studies.

Several nonsteroid compounds, such as clomiphene and transclomiphene, have been evaluated for possible rodent control.[65] The antifertility activity of a diphenyldihydronaphthalene derivative (U11100A) has been studied in rats.[16,36] This nonsteroidal compound was evaluated for the control of the European rabbit, *Oryctolagus cuniculus*.[31,32] A nonsteroidal estrogen antagonist, CI-628 (1-[2[p-[α-(p-methoxyphenyl)-β-nitrostyryl]phenoxy] ethyl] pyrrolidine monocitrate), has received a fair amount of attention.[37,38] In the opinions of some, it is accepted in baits better than most other potential chemosterilants.[39,61,66] This short-lived estrogen antagonist disrupts implantation and, therefore, must be consumed by the female rats just prior to copulation or within a few days post-insemination. To have an effect on a rat population, it must be available nearly continuously, and such traits make CI-628 (Parke, Davis and Co.) a potentially useful additive for incorporation into anticoagulant baits to prevent the development of anticoagulant resistance in rodent populations, or to prevent already-occurring genetic resistance from increasing. Female rodents surviving the anticoagulant and continuing to feed on the bait would be nonproductive.

Since the gestation period in rodents is short and offspring are born relatively undeveloped, treatment with steroids and some nonsteroids just before or soon after birth can cause irreversible sterility. Mestranol and DES, for example, when received via the mother's milk within the first few days of life, can produce sterility in both sexes.[28,55,63]

Certain steroidal and nonsteroidal compounds can inhibit implantation, act as abortifacients, or interrupt postcopulatory events in other ways. If these are the only mode of sterilant action, such compounds often must be available to the female rodents immediately after insemination or early during gestation, which requires precise timing of application. Such compounds will be of greater potential value for those species that have one or two distinct annual peak breeding periods or preferably a species that breeds only once a year.

Long-term exposure to the steroid cyproterone acetate has produced sterility in male rats by acting as an anti-androgen. Cyproterone acetate inhibits the action of androgen (i.e., testosterone) and the resulting sterilization produced in rats affects spermatozoa in the epididymis.[46] Johnson and Tait[47] reported on some investigations with cryproterone acetate as a reproductive inhibitor for the possible control of pest gray squirrels (*Sciurus carolinensis*) with some promising results, although there was a time lag for inhibition and results were relatively short-lived (about 4 weeks). In the light of existing knowledge of endocrine control and the duration of spermatogenesis, the hormonal approach of inducing male sterility may not be very practical from a control point of view, at least in mature rodents.

The alkylating agents seem to offer the greatest possibility for sterilizing male rodents, although their physiological action is by no means limited to males. Skinner[77] reported on ethyleneimine and methanesulfonate alkylating agents that were under study as antispermatogenic compounds. Alkylating agents are reported to be quite diverse in their biological activities, which increases the likelihood of finding effective compounds with considerable specificity.[77]

Nitrofurans (e.g., nitrofurazone, nitrofurantoin, and related compounds) have been reported to interfere with spermatogenesis. Furadantin was studied in bandicoot rats (*Bandicota bengalensis*) in India and its use, along with colchicine, was indicated for possible control of field populations.[79,80]

Table 1 provides examples of potential chemosterilants, the pest rodent species, and the researchers involved. In no way is the table to be considered all-inclusive and is intended to illustrate the diversity of the test species and potential chemosterilants explored.

Glyzophrol, mentioned earlier as one of the two rodent chemosterilants marketed, is an alkylating chemical but little else is known of this product.[70] Alpha-chlorohydrin (3 chloro-1,2-propanediol), the second rodent chemosterilant to be marketed, belongs to a different group of chemicals, is quite different in many respects, and much has been published on this chemosterilant. Alpha-chlorohydrin (Epibloc®) has been investigated, both in the laboratory and under natural conditions, as a sterilant of male rats.[2-4,10,11,22,51,52,67] In mature male Norway rats, this compound causes permanent epididymal lesions that block sperm passage, creating irreversible sterility in a single feeding of treated baits.[17,21] Its lesion-

Table 1
EXAMPLES OF SOME OF THE PEST RODENT SPECIES AND POTENTIAL CHEMOSTERILANTS WHICH HAVE BEEN RESEARCHED FOR THEIR CONTROL

Species	Common name	Chemosterilant(s) explored[a]	Ref.
Rattus norvegicus	Norway rat	BDH 10131	59, 73, 74
R. norvegicus	Norway rat	Quinestrol	10
R. norvegicus	Norway rat	Quinestrol, mestranol	71
R. norvegicus	Norway rat	Mestranol	44, 63
R. norvegicus	Norway rat	TEM	13
R. norvegicus	Norway rat	Alkylating agents (several)	77
R. norvegicus	Norway rat	SC-20775	25
R. exulans	Polynesian rat	SC-20775	25
R. exulans	Polynesian rat	SC24674, CI-628	5
Mus musculus	House mouse	DES	49
Bandicota bengalensis	Lesser bandicoot rat	Furadantin, colchicine	79, 80
Microtus californicus	California vole	Mestranol	44
Microtus pennsylvanicus	Meadow vole	Mestranol, DES	81
M. guentheri	Levant vole	DES	27-30
M. townsendii	Townsend's vole	Testosterone, mestranol	82
Sciurus carolinensis	Gray squirrel	Cyproterone acetate	47
Spermophilus richardsoni	Richardson's ground squirrel	Mestranol	34, 35
S. richardsoni	Richardson's ground squirrel	Mestranol, DES	1
Cynomys ludovicianus	Black-tailed prairie dog	DES	24, 72
Castor canadensis	Beaver	SC-24674, quinestrol	33
Lepus californicus	Black-tailed jackrabbit	Mestranol	64
Oryctolagus cuniculus	European rabbit	Mestranol, DES, U11100A, BDH 10131	31, 32

[a] Excluding alpha-chlorohydrin (see Table 2).

producing action is dose-related and is limited to certain rodent species, whereas temporary sterility is produced in a greater variety of rodents. During early studies, alpha-chlorohydrin was found to be relatively toxic to Norway rats and, when prepared in a highly preferred bait, produced significant rat mortality under field conditions.[67] The marketing of this product now capitalizes on this lethal aspect since reduction of the rat population contributes significantly to the effectiveness of a male sterilant. In some situations, alpha-chlorohydrin as a toxic rodent bait is moderately effective without relying on any sterilant effect; thus, the compound may be valuable for the most susceptible species even if no sterility occurs. Epibloc® has been tested on more species of rodents than any other rodent chemosterilant.[19] A list is provided in Table 2 along with references for each.

The effect of alpha-chlorohydrin have been studied in other rodents also, but the intent of that research was not directed towards the control of those species: guinea pigs, *Cavia cobaya;*[21] gerbils, *Meriones hurrianae;*[14] hedgehogs, *Erinaceus* sp.;[15] and hamsters, *Mesocricetus auratus.*[60]

VIII. CHEMOSTERILANTS VS. TOXICANTS

A lack of adequate specific materials and methods presently prevents us from using traditional rodent control methods in all situations in the best ecological manner. Destruction of the rats or other rodents by artificial means, such as trapping, poisoning, etc., may have only a temporary population effect where the habitat remains essentially unchanged, as is often the case with agricultural and forestry rodent problems. Frequently, such populations

Table 2
EXAMPLES OF SOME TOXICANT AND CHEMOSTERILANT EVALUATIONS OF ALPHA-CHLOROHYDRIN (EPIBLOC®) FOR A NUMBER OF PEST RODENT SPECIES IN BOTH THE LABORATORY FIELD

Species	Common name	Country	Evaluation		Ref.
			Sterility[a]	Toxicity	
Rattus norvegicus	Norway rat	Many	Yes	Yes	4, 23, 69
R. rattus	Roof rat	U.S.	—[b]	—[c]	54
R. exulans	Polynesian rat	New Zealand	—[b]	Yes	12, 54
R. losea	Lesser ricefield rat	Taiwan	Yes	Yes	19
R. argentiventer	Ricefield rat	Philippines	No	—[c]	4, 54
Microtus montebelli	Japanese field vole	Japan	Yes	Yes	19
M. californicus	California vole	U.S.	—[d]	Yes	68
Apodemus agrarius	Striped field mouse	Taiwan	Yes	—[c]	19
Arvicanthis niloticus	Nile field rat	Egypt	Yes	Yes	40
Mus musculus	House mouse	U.S.	No	Yes	17
Bandicota bengalensis	Lesser bandicoot rat	Burma	—[e]	Yes	19
Sigmomys alstoni	Cotton rat	Venezuela	—[d]	Yes	18
Holochilus brasilensis	Marsh rat	Venezuela	—[d]	Yes	18
Gerbillus gerbillus	Pygmy gerbil	Niger	—[d]	Yes	8
Mastomys natalensis	Multimammate rat	Nigeria	—[d]	Yes	9

Note: The ability of alpha-chlorohyrin (Epibloc®) to produce temporary or permanent sterility varies with the species.

[a] Sterility refers to permanent sterility.
[b] Results not uniformly positive.
[c] No precise data presented, although the chemical is known to be toxic in various degrees to all rodents.
[d] Not evaluated or data too limited for a conclusion.
[e] Unknown.

then quickly recover to levels that equal or even exceed the densities that existed before control. This population resurgence (compensatory replacement), has been demonstrated in the field many times.[7,75] If reproduction of rodents can be economically suppressed, such a chemosterilant procedure may be far superior to conventional attempts at regulating population densities by increasing one or more mortality factors.

Populations of rats and other prolific rodent species have steep growth curves as is represented by the classic Sigmoid growth curve. To artificially destroy great numbers of rodents with poison or other methods over a short period of time frequently merely pushes the population down from the carrying-capacity plateau onto the steep slope of the growth curve. When this happens, it is of little value unless there are some more permanent means of preventing or slowing the recovery.[41] This is where chemosterilants can play a major role.

Chemosterilants that produce only temporary sterility or are short-term in effect will necessarily have to be used with great regularity to avoid nullification by the compensation principle. On the other hand, individuals with long-term infertility or permanently sterilized will contribute nothing to reproduction and will remain in the population to compete for physical needs of space, food, and shelter in addition to competing for mates, territorial dominance, or social status. Chemosterilants alone will not reduce rodent populations immediately, but without normal reproduction the population will gradually decline through natural mortality. The merits of sterilization of rodents and the advantages over the traditional poisoning or trapping has been discussed in other publications in some depth.[57,58,65]

A frequent question asked by the layman concerning chemosterilants is, if you must bait the rodents to achieve control, then why not use poison to get rid of them? From a pure biological point of view as pointed out earlier, if a control program using poisons does not reduce the number of rodents to zero or a very low level, the success of the control will be short-lived and the population will recover quickly unless their habitat can be altered sufficiently to make it inhospitable. This is not feasible in many situations, e.g., sewer systems and agricultural crops. This is why poison baits must be repeatedly used in order to maintain a relatively low rodent population.

When a void is created in suitable rodent habitat as the result of poisoning or trapping of a substantial portion of the rodent population, these voids are often filled by immigrant rodents from surrounding areas. With sterilization, instant voids will not occur as the sterile rodents will presumably defend their territories from intruders. But, of course, once the sterile animals begin to die off, some immigrants may begin to encroach upon the area.

Although not usually intentional, lethal rodenticides are often used when the rodent population is already high and appreciable damage is occurring or has already been done. To be effective, chemosterilants require a totally different strategy that necessitates they be applied well in advance of the time damage is apt to occur. Thus, chemosterilants can be used as a preventive measure in an integrated program, which, in most situations, is a biologically sound approach to rodent control.

Depending on the characteristics of the chemosterilant, it may be more selective and present fewer potential hazards to humans, pets, domestic livestock, and nontarget wildlife than some toxic rodenticides now used. Environmentally and biologically, chemosterilants are more comparable to natural events.

Chemosterilants are subtle and slow-acting. Therefore, it is sometimes difficult for the layman to appreciate their advantages. Unlike acute rodenticides, where results are often seen within 24 to 48 hr, the benefits of chemosterilants cannot be demonstrated in a short time period. As with most new approaches to control, the public must be adequately informed through educational programs, as was necessary when the slow-acting anticoagulants were first used.

In the use of lethal rodenticides, the initial satisfaction of seeing dead rodents is at least psychologically convincing that control is being achieved. In some parts of the world, the success of a rodent control program is mistakenly based on the number of carcasses counted rather than the propensity of the survivors to repopulate or continue to do damage. This is one reason why anticoagulant rodenticides are not well received in some regions of the world. The lack of visual evidence of results will unfortunately be a detriment to the promotion of chemosterilants as an effective approach to rodent control.

More easily understood and accepted than the use of chemosterilants alone are the integrated programs that initially use traditional poisons to reduce rodent populations. This is followed by chemosterilants to keep the population down or to prevent it from rebounding rapidly. The modeling of alternating chemosterilants and poisons as a rat control strategy has received some attention and deserves more.[28]

Permanent or long-term sterility from a single feeding of an antifertility agent would be superior to temporary sterility or some postcopulatory effects. Compounds that can (or will) be consumed in sufficient amounts in one feeding are superior to those that require several days of feeding. The minimum acceptable antifertility effect is probably about 45 days, although this period would be inadequate in a number of pest situations because of the cost of materials and labor for repeat applications. As many pest rodents breed frequently, have short gestation periods, and the young mature rapidly, reproductive inhibitors of short duration would have little effect on the population unless repeat exposures were part of the control strategy. Such repeat exposures may be necessary over the approximate reproductive span of some species in order to take full advantage of natural mortality.

Repeat exposures present other problems because certain potential antifertility agents — such as mestranol, quinestrol and others — produce bait shyness or an aversion that is more severe in each subsequent exposure, thus reducing the overall effectiveness on the feeding population the longer they are repeatedly used on a regular basis.

IX. STERILIZATION OF ONE SEX VS. BOTH SEXES

Although Knipling[56] postulated the use of sterile males in vertebrate pest, he also pointed out that his theoretical models would not be valid for polygamous species. This includes most rodent species. Knipling and McGuire[58] and Knipling[57] later published theoretical appraisals of the role of sterilization of both sexes for suppressing rat populations. In the instance of rats, such as the Norway rat, sterile male rats can have a greater influence in reducing the reproductive potential of a population than would the elimination of the same number of males by poison or traps, because female rats of this and some other species go through a pseudopregnancy after mating with a sterile male. The physiological events of pseudopregnancy temporarily disrupts the normal estrous cycle, which essentially prevents the female from mating again for about 14 days. The occurrence of pseudopregnancies and competition by sterile males for receptive mates lends some support to the sterile-male approach for rat control. The extent to which pseudopregnancies played a role in the effectiveness of alpha-chlorohydrin as a male chemosterilant has not been fully evaluated under field conditions.

It is evident that the effectiveness of a male sterilant such as alpha-chlorohydrin can be effective.[2-4] Johnson[48] developed probabilistic analyses of the effect of alpha-chlorohydrin as a male sterilant on rat populations. The lower the population density (i.e., rats per-unit-space), the greater the reproductive inhibitory effect. It is speculated that this relates to space, time, and social dominance and the mathematical probability that with a high percentage of sterile males, few nonsterile male rats (especially those low in the social hierarchy) will have an opportunity to copulate with a sufficiently high percentage of receptive females in a given area. Every female that is unbred with a fertile male for even one estrus cycle means diminished reproduction.

In the search for effective rodent chemosterilants, the major emphasis has been placed on the females rather than the males inasmuch as the females contribute more to the population growth. Even better would be compounds effective for both sexes. With polygamous rodents that breed several times a year, sterile or infertile females would have a greater inhibiting effect on reproduction than would an equal number of sterile males. However, if a high percentage of both sexes were sterilized, the results would be substantially greater than if only the females were affected.[58]

X. WHERE CHEMOSTERILANTS WILL PLAY MAJOR ROLES

Chemosterilants will find their greatest use in agricultural and forested areas where a few rodents may be tolerable but where high numbers cause serious economic damage. Chemosterilants are perceived to be useful for rodent control on poultry and hog farms, in dairies and on horse farms. They should be valuable for rodent control in zoos, around pet breeding facilities, fur breeding farms, and for protecting certain types of aquaculture where rodents are a problem. The influence of chemosterilants on rodent outbreaks (plagues), as have occurred with house mice and various *Microtus* species, could be substantial, but this approach has yet to be adequately explored on a wide scale. In all of these situations, chemosterilants will play their greatest role as preventive measures to keep a population from increasing. If the rodent population has no natural low point in its annual cycle, then it may be necessary to use the chemosterilant following a reduction of the population with poison baits. The use of chemosterilants and poisons in an integrated effort holds significant potential and broadens their scope of use.

Chemosterilants may be useful in suppressing rodent populations out-of-doors that are a source of diseases, such as bubonic plague, murine typhus, and others. Apart from the fact that chemosterilants can be important in regulating populations of disease-bearing rodents, it is also fortunate that nonlethal sterilants permit this to be done without instantly causing rodent deaths. This means that, in contrast to lethal rodenticides, there will be no sudden influx of vectors such as fleas, mites, and ticks abandoning large numbers of poisoned rodents. A release of numerous ectoparasites into the environment increases the risk that vectors will spread rodent-borne diseases, hence normally necessitates pesticide treatment for ectoparasite control.

There are numerous situations in which chemosterilants, as perceived, will not be useful, and it is important that these are recognized. Where rats or other rodents are an immediate threat to public health, or if in a dwelling, we may not be able to tolerate a single rodent — sterile or not. Chemosterilants will be of no value in food-handling or processing establishments, or in hospitals or other structures where a high level of sanitation is necessary, because then even sterile individuals are not acceptable. A population of rodents, even if sterile, will still continue to cause damage, inflict bites, transmit disease, and contaminate foodstuffs. In general, chemosterilants will be more effective in outdoor situations rather than in buildings where all rodents need to be removed.

Effective use of chemosterilants requires considerable knowledge of the biology of the pest rodent species, especially its reproductive characteristics and its peak periods of reproduction. The most appropriate time of applying chemosterilants is more critical than with traditional control measures; therefore, chemosterilants are most efficient and effective if used by well-trained individuals or professional pest control operators.

XI. TYPES OF FORMULATIONS AND METHODS OF USE

Methods of exposing chemosterilants to rodent populations are similar to those used with toxic rodenticides. Cereal-based baits lend themselves to the widest range of application methods, although water baits and baits of high-moisture content have some potential values. Those chemosterilants that are less well accepted in baits may require prebaiting. Products that require several successive feedings to be effective may be exposed in bait boxes, as are many anticoagulant rodenticides. Certain chemosterilants (such as CI 628) may prove to be a useful additive to anticoagulant baits to aid in combating development of anticoagulant resistance in rodents. In some agriculture and sewer systems, paraffin-type bait blocks may be the formulation of choice. Aerosols may be effective in treating rodent burrow systems. Chemosterilant tracking powders have been shown effective, and current research suggests the treatment of nesting materials may be an effective approach for some species. Foliar sprays applied to food items of rodents, such as the bark of forest or fruit trees, and seed cones, forage, or even fruit or vegetable crops, although not yet explored, may be a useful means of exposure to rodents if the chemosterilant is relatively specific and such uses are not potentially hazardous to nontarget species.

APPENDIX: CHEMOSTERILANTS MENTIONED

1. AY 20,121 = 3-(2-propynyloxy)estra-1,3,5(10)-trien 17β-ol
2. 8 aza quinestrol, an analog of quinestrol
3. BDH 10131 = 17 α-hexa-1′,3′-dynyl oestra-1,3,5(10)-trien 17 β-ol
4. chorotrianisene (CTA) = TACE®
5. CI-628 = 1-[2[p-[α-(p-methoxyphenyl)-β-nitrostyryl] phenoxy] ethyl] pyrrolidine monocitrate

6. clomiphene
7. colchicine
8. cyproterone acetate
9. diethylstilbestrol (DES)
10. Epibloc® = alpha-chlorohydrin = 3 chloro-1,2-propanediol = U5897
11. ethyleneinine, an alkylating agent
12. furadantin, a nitrofuran compound
13. Glyzophrol® = butandiol-bis-methane sulphonic acid ester
14. mestranol = 17 α ethynylestradiol-3-methyl ether
15. methanesulfonate, an alkylating agent
16. nitrofurans (e.g., nitrofurazone, nitrofurantoin)
17. quinestrol = 17 α-ethynylestradiol-3-cyclopentyl ether
18. SC20775 = 17 α-ethynyl-11 β-methylestradiol-3-cyclopentyl ether
19. SC24674, an experimental synthetic estrogen
20. testosterone
21. transclomiphene
22. triethylenemelamine (TEM)
23. U11100A, a diphenyldihydronaphthalene derivative

REFERENCES

1. **Alsager, D. E.,** Experimental population suppression of Richardson's ground squirrels *Spermophilus richardsonii* in Alberta, in *Proc. 5th Vertebr. Pest Conf.*, Marsh, R. E., Ed., University of California, Davis, Calif., 1972, 93.
2. **Anderson, M. D.,** Influence of male sterility by an alpha-chlorohydrin and behavioral factors on fecundity of a rat colony, *Sci. Biol. J.*, 2, 154, 1976.
3. **Andrews, R. V. and Belknap, R. W.,** Efficacy of alpha-chlorohydrin in sewer rat control, *J. Hyg. Cambridge*, 91, 359, 1983.
4. **Andrews, R. V., Belknap, R. W., and Keenan, E. J.,** Demographic and endocrine responses of Norway rats to antifertility control. *J. Wildl. Manage.*, 38, 868, 1974.
5. **Averill, R. L. W.,** Chemosterilants for *Rattus exulans*, in *Proc. Symp. Ecology and Control of Rodents in New Zealand Nature Reserves*, Dingwall, P. R., Atkinson, I. A. E., and Hay, C., Eds., Department of Lands and Survey, Wellington, New Zealand, 1976, 213.
6. **Banik, U. K., Revesz, C., and Herr, F.,** An orally active long-acting estrogen (AY-20, 121), *Steroids*, 16, 289, 1970.
7. **Batcheler, C. L.,** Compensatory response of artifically controlled mammal populations, in *Proc. New Zealand Ecol. Soc.*, 15, 25, 1968.
8. **Brei, W.,** The reaction to rodenticides on rodent species occurring in northeastern Nigeria, in *Proc. Symp. Plant Protection*, Munster, West Germany, Sect. 3, May 1978.
9. **Brei, W.,** L'efficacite des differentes rodenticides sur *Gerbillus gerbillus nigeriae* dans un essai fait a Bani-Bangou, Niger, Tech. Rept., Deutsche Gessellschaft fur Technische Zusammenarbeit, Eschborn, West Germany, 1979.
10. **Brooks, J. E. and Bowerman, A. M.,** Estrogenic steroid used to inhibit reproduction in wild Norway rats, *J. Wildl. Manage.*, 35, 444, 1971.
11. **Cooper, E. R. A., Jones, A. R., and Jackson, H.,** Effects of α-chlorohydrin and related compounds on the reproductive organs and fertility of the male rat, *J. Reprod. Fert.*, 38, 379, 1974.
12. **Cummins, J. M. and Wodzicki, K.,** Effects of alpha-chlorohydrin on the male reproductive tissues of the Polynesian rat, *Rattus exulans*, *New Zealand J. Zool.*, 7, 427, 1980.
13. **Davis, D. E.,** Principles for population control by gametocides, *Trans. N. Am. Wildl. Conf.*, 26, 160, 1961.
14. **Dixit, V. P. and Lohiya, N. K.,** Mechanism of action of alpha-chlorohydrin on the testis and caput epididymides of rat, gerbil *(Meriones hurrianae)*, bat and mouse, *Acta Anat. Basel*, 95, 50, 1976.
15. **Dixit, V. P. and Lohiya, N. K.,** Effects of alpha-chlorohydrin on the testis and sex accessory organs of male hedgehog: a biochemical study, *Zool. Pol.*, 26, 135, 1977.

16. **Duncan, G. W., Stanley, C., Lyster, S. C., Clark, J. J., and Lednicer, D.,** Antifertility activities of two diphenyldihydronaphthalene derivatives, *Soc. Expt. Biol. Med.*, 112, 439, 1963.
17. **Ericsson, R. J.,** Male antifertility compounds: U-5897 as a rat chemosterilant, *J. Reprod. Fert.*, 22, 213, 1970.
18. **Ericsson, R. J.,** EPIBLOC: laboratory and field tests on two species of rodents in Venezuela *(Sigmomys alstoni* and *Holochilus brasilensis),* Tech. Rept. EPI/77/RJE/01, Gametrics Ltd., Sausalito, California, 1977.
19. **Ericsson, R. J.,** Alpha-chlorohydrin (Epibloc®): a toxicant-sterilant as an alternative in rodent control, in *Proc. 10th Vertebr. Pest Conf.*, Marsh, R. E., Ed., University of California, Davis, 1982, 6.
20. **Ericsson, R. J.,** personal communication, 1985.
21. **Ericsson, R. J. and Baker, V. F.,** Male antifertility compounds: biological properties of U-5897 and U-15,646, *J. Reprod. Fertil.*, 21, 267, 1970.
22. **Ericsson, R. J., Downing, H. E., Marsh, R. E., and Howard, W. E.,** Bait acceptance by rats of microencapsulated male sterilant alpha-chlorohydrin, *J. Wildl. Manage.*, 35, 573, 1971.
23. **Field, R. J.,** Urban Norway rat control, *Pest Control* 40, 20, 1972.
24. **Garrett, M. G. and Franklin, W. L.,** Diethylstilbestrol as a temporary chemosterilant to control black-tailed prairie dog populations, *J. Range Manage.*, 36, 753, 1983.
25. **Garrison, M. V. and Johns, B. E.,** Antifertility effects of SC-20775 in Norway and Polynesian rats, *J. Wildl. Manage.*, 39, 26, 1975.
26. **Gentry, J. W.,** Evaluation of rat eradication programs, *Environ. Sci. Tech.*, 5, 704, 1971.
27. **German, A.,** Diethylstilbestrol as a reproduction inhibitor in the Levant vole, *Phytoparasitica,* 8, 163, 1980.
28. **German, A.,** Effect of diethylstilbestrol treatment of the Levant vole female during lactation on the reproduction of her female offspring, *Phytoparasitica,* 9, 171, 1981.
29. **German, A.,** Effect of external contact with diethylstilbestrol on the reproduction of female Levant voles, *Phytoparasitica,* 10, 3, 1982.
30. **German, A.,** Contact effect of diethylstilbestrol (DES) on the suppression of reproduction in the Levant vole, *Microtus guentheri, Acta Zool. Fenn.*, 173, 179, 1985.
31. **Godfrey, M. E.,** Techniques for the evaluation of potential rabbit *(Oryctolagus cuniculus)* chemosterilants, in *Am. Soc. Testing and Materials Spec. Tech. Publ. 625,* Philadelphia, Penn. 1977, 99.
32. **Godfrey, M. E. R.,** An evaluation of potential rabbit *(Oryctolagus cuniculus)* chemosterilants, Ph.D. thesis, University of California, Davis, 1976.
33. **Gordon, K. L. and Arner, D. H.,** Preliminary study using chemosterilants for control of nuisance beaver, in *Proc. 13th Annu. Conf. Southern Assoc. Game and Fish Commissioners,* 1976, 463.
34. **Goulet, L. A.,** Aspects of population dynamics and social behaviour in the Richardson's ground squirrel as modified by a chemosterilant (mestranol), Ph.D. thesis, Simon Fraser University, British Columbia, Canada, 1979.
35. **Goulet,L. A. and Sadleir, R. M. F. S.,** The effects of a chemosterilant (mestranol) on population and behavior in the Richardson's ground squirrel *(Spermophilus richardsonii)* in Alberta, in *Proc. 6th Vertebr. Pest Conf.*, Johnson, W. V., Ed., University of California, Davis, 1974, 90.
36. **Greenwald, G. S.,** Effects of a dehydronafhthalene derivative (U11100A) on pregnancy in the rat, *Fert. Steril.*, 16, 185, 1965.
37. **Gwynn, G. W.,** Effects of a chemosterilant on fecundity of wild Norway rats, *J. Wildl. Manage.*, 36, 550, 1972.
38. **Gwynn, G. W.,** Field trial of a chemosterilant in wild Norway rats, *J. Wildl. Manage.*, 36, 823, 1972.
39. **Gwynn, G. W. and Kurtz, S. M.,** Acceptability and efficiency of antifertility agents in wild Norway rats, *J. Wildl. Manage.*, 34, 514, 1970.
40. **Helal, T. Y.,** Chemosterilant and rodenticidal effects of 3-chlor,1,2 propanediol (Epibloc) against the albino laboratory rat and the Nile field rat, *Int. Pest Control,* 24, 20, 1982.
41. **Howard, W. E.,** Biological control of vertebrate pests, in *Proc. 3rd Vertebr. Pest Conf.*, Cummings, M., Ed., University of California, Davis, 1967, 137.
42. **Howard, W. E.,** Biocontrol and chemosterilants, in *Pest Control: Biological, Physical, and Selected Chemical Methods,* Kilgore, W. W. and Doutt, R. L., Eds., Academic Press, New York, 1967, 343.
43. **Howard, W. E.,** Controlling rodents biologically and with sound, in *Proc. Rodents as Factors in Disease and Economic Loss,* Asia-Pacific Interchange, Honolulu, Hawaii, 1968, 220.
44. **Howard, W. E. and Marsh, R. E.,** Mestranol as a reproductive inhibitor in rats and voles, *J. Wildl. Manage.*, 33, 403, 1969.
45. **Jackson, H.,** Antifertility substances, *Pharmacol. Rev.*, 11, 135, 1959.
46. **Jackson, H.,** Chemical methods of male contraception, *Am. Sci.*, 61, 188, 1973.
47. **Johnson, E. and Tait, A. J.,** Prospects for the chemical control of reproduction in the grey squirrel, *Mamm. Rev.*, 13, 167, 1983.
48. **Johnson, R. F.,** A probabilistic analysis of a class of male rat chemosterilants, *Biometrics,* 30, 341, 1974.

49. **Johnson, R. F. and Barrett, G. W.,** Effects of diethylstilbestrol on feral house mouse *(Mus musculus* L.) population dynamics under experimental field conditions, *J. Appl. Ecol.*, 12, 741, 1975.
50. **Kamarudin, K. A.,** The potential of glyzophrol as a rodent chemosterilant, M. S. thesis, Bowling Green State University, Bowling, Green, Ohio, 1977.
51. **Kassa, H. and Jackson, W. B.,** The effect of alpha-chlorohydrin on the fertility of male rats, *Ohio J. Sci.*, 81, 281, 1981.
52. **Kassa, H. and Jackson, W. B.,** Bait acceptance and chemosterilant efficacy of alpha-chlorohydrin in the Norway rat, *Rattus norvegicus, Int. Pest Control.*, 26, 7, 1984.
53. **Kendle, K. E., Lazarus, A. B., Rowe, R. P., Telford, J. M., and Vallance, D. K.,** Sterilization of rodent and other pests using a synthetic oestrogen, *Nature,* 244, 105, 1973.
54. **Kennelly, J. J., Garrison, J. J., and Johns, B. E.,** Laboratory studies of the effect of U-5897 on the reproduction of wild male rats, *J. Wildl. Manage.*, 34, 508, 1970.
55. **Kincl, F. A., Pi, A. F., Maquero, M., Lasso, L. H., Oriol, A., and Dorfman, R. I.,** Inhibition of sexual development in male and female rats treated with various steroids at the age of five days, *Acta Endocrinol.*, 49, 193, 1965.
56. **Knipling, E. F.,** Sterile-male method of population control, *Science,* 130, 902, 1959.
57. **Knipling, E. F.,** The basic principles of insect population suppression and management, in *Agriculture Handbook No. 512,* U.S. Department of Agriculture, Washington, D.C., 1979.
58. **Knipling, E. F. and McGuire, J. U.,** Potential role of sterilization — a theoretical appraisal, Tech. Bull. No. 1455, Agricultural Research Service, U.S. Department of Agriculture, Washington, D.C., 1972, 1.
59. **Lazarus, A. B. and Rowe, R. P.,** Reproduction in an island population of Norway rats, *Rattus norvegicus* (Berkenhout), treated with an oestrogenic steroid, *Agro-Ecosystems,* 8, 59, 1982.
60. **Lubicz-Nawrocki, C. M. and Chang, M. C.,** Effect of α-chlorohydrin on the fertilizing ability of hamster epididymal spermatozoa, *J. Reprod. Fert.*, 38, 65, 1974.
61. **Marsh, R. E.,** unpublished data, 1971.
62. **Marsh, R. E.,** Potential chemosterilants for controlling rats, *Beitr. Z. Angew. Zool.*, 9, 191, 1973.
63. **Marsh, R. E. and Howard, W. E.,** Evaluation of mestranol as a reproductive inhibitor of Norway rats in garbage dumps, *J. Wildl. Manage.*, 33, 133, 1969.
64. **Marsh, R. E. and Howard, W. E.,** Chemosterilants as an approach to rodent control, in *Proc. 4th Vertebr. Pest Conf.*, Dana, R. H., Ed., University of California, Davis, 1970, 55.
65. **Marsh, R. E. and Howard, W. E.,** Prospects of chemosterilants and genetic control of rodents, *Bull. WHO,* 48, 309, 1973.
66. **Marsh, R. E. and Howard, W. E.,** New perspectives in rodent and mammal control, in *Proc. 3rd Int. Biodegradation Symp.*, Sharpley, J. M. and Kaplan, A. M., Eds., Applied Science, London, 1976, 317.
67. **Marsh, R. E. and Howard, W. E.,** unpublished data, 1973.
68. **Marsh, R. E. and Tunberg, A. D.,** Research on meadow mouse *(Microtus californicus)* biology and its control relative to artichoke production, Progress Report, Artichoke Research Association, University of California, Davis, 1985.
69. **Meehan, A. P. and Hum, M. C.,** The rodenticidal and chemosterilant effect of U-5897 (alphachlorohydrin), *Int. Pest Control,* 21, 1979.
70. **Meehan, A. P.,** *Rats ad Mice, Their Biology and Control,* Rentokil Ltd., E. Grinstead, W., Sussex, England, 1984, 277.
71. **Mischler, T. W., Welaj, P., and Nemith, P.,** Biological evaluation of two estrogenic steroids as possible rodent chemosterilants, *J. Wildl. Manage.*, 35, 449, 1971.
72. **Pfieffer, D. G.,** Effects of diethylstilbestrol on reproduction in the black-tailed prairie dog, M. S. thesis, South Dakota State University, Brookings, S.D., 1972.
73. **Rowe, F. P. and Lazarus, A. B.,** The effects of an oestrogenic steroid on the reproduction of wild rats, *Rattus norvegicus* (Berk.), *Agro-Ecosystems,* 1, 57, 1974.
74. **Rowe, F. P. and Lazarus, A. B.,** Reproductive activity in a wild rat, *Rattus norvegicus,* (Berk.), population treated with an oestrogenic steroid, *Agro-Ecosystem,* 1, 227, 1974.
75. **Rowley, I.,** Studies on the resurgence of rabbit populations after poisoning, *CSIRO Wildl. Res.*, 13, 59, 1968.
76. **Saunders, F. J.,** Effects of sex steroids and related compounds on pregnancy and on development of the young, *Physiol. Rev.*, 48, 601, 1968.
77. **Skinner, W. A.,** Chemosterilants in rodent control, in *Proc. Rodents as Factors in Disease and Economic Loss,* Asia-Pacific Interchange, Honolulu, Hawaii, 1968, 192.
78. **Smith, E. H.,** Advances, problems, and the future of insect control, in *Proc. Symp. Sci. on the Aspects of Pest Control,* National Academy of Science, National Research Council, Washington, D. C., 1966, 41.
79. **Srivastava, A. S.,** Rats and their control, *Bull. Labdev. J. Sci. Tech.*, 4, 207, 1966.
80. **Srivastava, A. S.,** *Rodent Control for Increased Production,* Rotary Club (West), Kanpur, India, 1968, 58.

81. **Storm, G. L. and Sanderson, G. C.,** Effects of mestranol and diethylstilbestrol on captive voles, *J. Wildl. Manage.*, 34, 835, 1970.

82. **Taitt, M. J. and Krebs, C. J.,** Manipulation of female behaviour in field populations of *Microtus townsendii, J. Anim. Ecol.*, 51, 681, 1982.

83. **Wetherbee, D. K.,** Vertebrate pest control by biological means, WHO/EBL/31.65, World Health Organization, Geneva, 1, 1965.

Chapter 22

THE POTENTIAL FOR PHEROMONAL INVOLVEMENT IN RODENT CONTROL PROGRAMS

D. Michael Stoddart

TABLE OF CONTENTS

I. INTRODUCTION

In view of the considerable recently made advances in rodent olfactory biology, and the proven involvement of pheromones in social biology,[1,5] it is puzzling that there have been so few attempts to involve them in integrated pest management schemes. The situation is in stark contrast to that pertaining in the field of insect pest control, in which pheromone-induced manipulation of behavior now plays a major role.[3] The reasons for this difference between the two types of pests, and the lack of interest in the development of rodent studies stems more from oversight than from unsuccessful experimentation. The objective of this chapter is to examine the potential for pheromone involvement in rodent pest management studies. The term "pheromone" is here used not in the original sense as defined by Karlson and Luscher,[19] but in the now widely accepted manner to embrace all intraspecific odors.

Table 1 summarizes the main facets of rodent biology in which pheromones have been demonstrated to play a part. Research into pheromonal involvement in rodent pest management studies have been restricted almost exclusively to the effect of pheromones as sex attractants, but this appears to be because pheromones have been used to supplement an existing means of rodent control.

II. SPECIES, GROUP, AND INDIVIDUAL RECOGNITION

Although it is slightly beyond the scope of this chapter, rodents have been observed to be repelled by the odor of predators, though not all individuals react with equal alacrity to exposure. Stoddart[35] showed that anal gland secretion of the European weasal (*Mustela nivalis*) exerted a strong repellent effect upon the trap entry behavior of the short-tailed vole (*Mictrotus agrestis*), but to a far lesser extent upon that of the woodmouse (*Apodemus sylvaticus*). The effect was also found to be influenced by the sex and age of the rodent.[36] Sullivan[38] indicates that the montane vole (*Microtus montanus*) and northern pocket gopher (*Thomomys talpoides*) tend to avoid traps which were treated with the main components of stoat (*Mustela erminea*) anal gland secretion, 2, propylthietane and 2,4,5-trimethyl-3thiazoline. Sullivan et al.[37,39,40] have presented evidence showing that predator odors, both synthetic and natural, may be used to repel snowshoe hares (*Lepus americanus*) and black-tailed deer (*Odocoileus hemionus*) from stands of growing plants.

No studies appear to have been conducted on the potential for group ("clan") or individual recognition as an adjunct to other control studies.

III. BROADCAST OF ALARM SIGNALS

Although rodents have been shown to respond to stressful situations by the production of a fear odor,[28,42] this response has not been investigated with a view to its application in rodent pest management. Mackay-Sim and Laing[25] have shown that fear odor is released from the body surface of stressed rats and in the urine, but feces are not involved. The compound(s) responsible for this effect has/have not been identified, and it is not known for how long it(they) remains active.

IV. SOCIAL ORGANIZATION

There is an abundance of evidence that rodents indulge in scent marking of their habitats,[41] but generally not enough is known about the function of territory or range exclusivity to enable a coherent explanation for the phenomenon to be given. In a few cases involving large, visible rodents, the role of scent marks can be clearly understood. Muller-Schwarze[27] reports that experimental translocation of beaver anal gland scent marks from one scent

Table 1
THE MAIN FACETS OF RODENT BIOLOGY IN WHICH PHEROMONES HAVE BEEN DEMONSTRATED TO PLAY A PART

1. Species recognition
2. Group (i.e., "clan") recognition
3. Individual recognition
4. Broadcast of alarm signals
5. Defense
6. Social organization
   - Maintenance of social dominance
   - Inhibition and elicitation of aggression
   - Maintenance of territory
7. Reproduction
   - Sex attraction: pheromone produced by the female
   - Sex attraction: pheromone produced by the male
   - Induction of estrus
   - Maintenance of pregnancy
   - Maintenance of litter cohesion prior to weaning
   - Control of rate of growth and sexual maturation of the young

mound to another resulted in considerably increased levels of behaviors associated with scent marking. Strange conspecific scent marks are obliterated as soon as they are encountered. Dispersing beavers are repelled from establishing residence in unoccupied lodges if experimental scent mounds have been positioned nearby,[26] suggesting that scent marks in this context serve to provide to transients information about the residency status of an area. This aspect of repellence needs further examination to establish if short term protection to particularly vulnerable sites may be afforded by some component of anal gland secretion.

Territoriality is closely associated with social dominance and the aggressive behaviors necessary to support and maintain the social hierarchy. Fass et al.[9] have demonstrated that fighting in rats is mediated by certain testosterone-related anogenital substances whose source is, at present, obscure. A number of authors have reported that rodents introduced into strange colonies are set upon by the residents, but if they are first daubed with the new colony odor they are protected from attack (e.g., Barnett[2] and Steiner[34]).

Social status has been shown by a number of workers to be reflected in body odor, and to be subject to selective response by conspecifics (e.g., Huck and Banks[16] and Huck et al.[15] for the brown lemming *Lemmus trimucronatus* and Hoffmeyer[13] for the bank vole *Clethrionomys glareolus*). Scent marking and the saturation of the environment with an individual's personal odor is also related to social dominance in many species (e.g., Golden hamster *Mesocriecetus auratus* Johnston[18] and mouse *Mus musculus* Desjardins et al.[8]) with the dominant individuals marking at higher frequency than subordinates.

It would appear that further study should be directed towards the elicitation of aggression through odor cues. Christiansen[7] draws attention to the preputial gland as the source of pheromones signalling maleness and aggressiveness, and this would seem to be a logical target for increased investigation.

V. REPRODUCTION

Following the early observations of Bruce,[6] Whitten,[44] and van der Lee and Boot[22] the involvement of pheromones in rodent reproductive biology has received a great deal of attention. Most of the work has been of an endocrinological or physiological nature, although more recently attention has been paid to behavioral and ecological considerations. The ecological significance of this is well summarized by Bronson.[4] Of particular interest in the present context is work on sex attractants, for they may be seen to play a role in attracting otherwise recalcitrant rodents to piles of poisoned bait.

Gawienowski et al.[11,12] first showed that extracts from the preputial gland of female rats had an attractive effect on sexually active males, and that an equal effect occurred with male preputial gland extract upon sexually mature females. Dimethyl disulphide and dimethyl sulphite have been identified as being highly attractive to male rats.[10] Gawienowski and his co-workers[10-12] list a number of studies which show that male rodent body odor (most often urine odor) is strongly attractive to female rodents. In the rat the lipid fraction of preputial extract appears to be responsible for the attraction of the opposite sex. Female rats are attracted to alkyl acetates of propyl, pentyl, and decyl alcohol.[10,32]

It is interesting, and perhaps significant to note that dimethyl disulphide has also been identified as a sex attractant in the vaginal secretion of female golden hamsters.[33] Male hamsters make frantic attempts to find the source of artifically ducted vaginal secretion odor, and choose to remain for long periods in an airstream carrying the odor.[18] It is not known whether dimethyl disulphide occurs widely among the rodents as a sex attractant. Idris and Prakash[17] have shown that the urine of female Indian gerbils *(Tatera indica)* produced during estrus is significantly more attractive to males than urine from other phases of the estrous cycle. In this species the level of urine marking activity by females peaks sharply during proestrus and estrus.

The age of urine samples influences their attractive properties. Reiff[31] mentions that freshly voided urine, or centrifuged urine from which solids and bacteria have been removed, is unattractive. Whole urine that is allowed to ferment, however, is more attractive. Examining this in more detail Price[29] indicates that aging for 7 to 8 days results in its being investigated significantly more frequently than fresh. The effect is seen in wild as well as domestic Norway rats. He points out that rat urine is basically attractive to both strange and familiar conspecifics of both sexes, and quotes a number of references to substantiate his view. Castration of male rats results in a decrease in urine marking, which would thus seem to be an androgen-dependent behavior, but it has little effect on the propensity of males to investigate urine. Lydell and Doty[24] showed that urine age, among other factors, played a role in the attractiveness of female rat urine to males.

Although male rodents have been shown mutually to induce estrus[4] and to accelerate the onset of first estrus (puberty) in young females, further investigation of these phenomena would not seem to yield profitable results in the present context. Neither would further study of the "pregnancy block" pheromone first demonstrated by Bruce.[6] The reproductive effects noted at the start of this section can only be demonstrated in the laboratory under conditions never encountered in the wild. It is highly unlikely that further study of these phenomena will produce results which are applicable to rodent control programs either immediately, or after further development.

Maternal pheromones, by which female rodents maintain the cohesion of their litters prior to weaning, have been demonstrated most completely in rats.[23] Maternal pheromone is synthesized in the caecum by microorganismal action during the production of caecotrophe, or soft feces, for reingestion. It has been shown that as a result of prolactin-induced rise in food intake levels, lactating female rats do not reingest as much caecotrophe as they produce. The result is that the mother, her young, and the nest become scented with the odor. The precise source and chemical identity of maternal odor has not yet been identified, although quite a lot is known about its properties. Females are attractive to pups only when their gut acidity is low, but it is not known whether acidity influences odor synthesis, emission, or both. If prolactin is suppressed by antiserum or by ergot derivatives, maternal food intake is suppressed as is caecotrophe production and hence odor emission, but odor synthesis remains unaffected. Maternal odor attractivity serves two important biological functions. One is to inhibit the young from straying far from the nest, and hence from shelter, warmth, and food. The other is to provide the youngsters with an odor image of foods preferred by their mothers. Experience gained early in life establishes food preferences for later, though such preferences may be expressed only in the controlled conditions of the laboratory. It is clear that there exists an important potential here for further work on the synthesis of maternal pheromone, for if it could be perturbed by the action of some externally administered agent it might be able to disrupt the important bonds between mother and young.

The role of pheromones in the sexual development and maturation of young rodents, and in their growth rates has been studied quite intensively,[43] but studies do not indicate the likely development of an agent(s) which will have the potential for permanent inhibition of maturation in the field; or for the disruption of maturation to such an extent that it will influence reproduction in wild rodent populations.

Finally, a number of studies have indicated a general attractiveness of conspecific body odor in rodents, a phenomenon not necessarily directly associated with sex attraction. Working with the desert gerbil (*Meriones hurrianae*) Kumari and Prakash[21] showed that the addition of a small amount of female gerbil urine to millet increased food consumption significantly. The millet was eaten by males and females alike, and female consumption after treatment was higher than male consumption. Additionally, they reported that bait shyness, which occurred to millet treated with a sublethal dose of zinc phosphide, could be masked by the addition of urine. Even gerbils which had developed shyness to poisoned bait following two exposures to sublethal doses overcame their shyness and consumed more bait than they did on the initial day of testing. The same authors[20] showed with the Indian gerbil that consumption of grain among which male gerbils had been housed was far higher than of control grain. No attempt was made to determine the source of the attractive odor. If the grain was removed from the males 5 days before being offered as food to the test subject, no enhanced consumption was observed. This suggests that the attractive odor is fairly volatile. Purushotham et al.[30] demonstrated significantly enhanced consumption of millet by the Indian field mouse, (*Mus booduga*) and the field rat (*Rattus meltada*) when rodents of these species were allowed access to the millet stock and could spoil it with urine and feces. How valuable the fundamental principle underlying these observations may be to the development of pheromonally based rodent control studies remains yet to be evaluated, but it would seem a fruitful area for further research.

VI. BIOASSAYS

In the studies mentioned above a wide array of bioassays have been used to demonstrate the odorous effects under investigation. Generally they fall into three main groups, viz: (1) the time a test individual spends in one or other arm of a "Y" or "T" maze, or some other choice chamber; (2) the amount of food consumed per unit time when a test subject if offered a choice between food treated with an odorous additive and untreated food; and (3) the growth or development of a gland or organ structure when a test individual is exposed to an odor. There are problems associated with each of these bioassay techniques. In the first the individual's choice is frequently reported as a preference, when it could just as easily be the result of repulsion to another odor. Careful pairings of odorous stimuli and judicious use of controls are required to overcome this problem. In the second, care must be taken to ensure that both treated and untreated food supplies are subject to identical physical constraints, such as proximity to side of cage, subject to identical lighting, etc. In the third, it is necessary to know to what extent any increase (or decrease) in an organ system is linearly related to the level of odorous stimulation presented, otherwise comparability between studies is hard to achieve. Before undertaking olfactory studies in applied rodent biology, researchers should consider their experimental methodologies with the utmost care.

VII. CONCLUSIONS

A consideration of the involvement of pheromones in rodent biology suggests that there are three areas which have potential application in integrated rodent pest management studies.

The first concerns aggression promoting pheromones, and implicates the preputial gland and its exudates. Synthesis of compounds which markedly increase aggressive interactions may help to destabilize populations by causing increased stress and thus predisposing the population to the effects of fatal disease or other mortality agents. The second concerns urinary compounds of an attractant nature. This is a particularly fruitful field for further investigation as the addition of an attractive odor to a poisoned bait may mask bait shyness as has been demonstrated for the desert gerbil in India. A further benefit may accrue from the more rapid consumption of bait, by the reduction of the primary hazard. Integrated pest management schemes need to consider the effect of control programs on other species in the environment, to minimize the nontarget species effects of those programs. As has been shown, the preputial gland produces attractive exudates which require further analytical and synthetic study. Finally, there would appear to be some potential for further study of the structure and action of maternal pheromone. Any means to reduce reproductive success has a useful application in integrated pest management schemes.

It has been reported[14] that pheromones have not yet proven very efficacious in control programs, and at the time of writing this is indeed so. But old problems frequently respond to new solutions, and while it is unlikely that the application of a pheromonally based technique will ever prove to be stunningly effective in rodent control, the potential for a measure of success to be gained from pheromonal involvement is high. It is a potential which state rodent control agencies must realize with all speed.

REFERENCES

1. **Albone, E. W.,** *Mammalian Semiochemistry,* John Wiley & Son, New York, 1984.
2. **Barnett, S. A.,** *A Study of Behaviour,* Metheun & Co., London, 1963.
3. **Birch, M. C. and Haynes, K. F.,** *Insect Pheromones,* Edward Arnold, London, 1982.
4. **Bronson, F. H.,** The reproductive ecology of the house mouse, *Q. Rev. Biol.,* 54, 265, 1979.
5. **Brown, R. W. and MacDonald, D. W.,** *Social Odours in Mammals,* Eds., Oxford, 1985.
6. **Bruce, H. M.,** An exteroceptive block to pregnancy in the mouse, *Nature,* 184, 105, 1959.
7. **Christiansen, E.,** Pheromones in small rodents and their potential use in pest control, 7th Vertebr. Pest. Conf., Monterey, Calif., March 9 to 11, 1976, 185.
8. **Desjardins, C., Maruniak, J. A., and Bronson, F. H.,** Social rank in the house mouse: differentiation revealed by ultraviolet visualisation of urinary marking patterns, *Science,* 182, 939, 1973.
9. **Fass, B. J., Mason, R., and Stevens, D. A.,** Evidence that chemical signals promote fighting between intact male hooded rats, *Bull. Psychoneurol. Soc.,* 17, 156, 1981.
10. **Gawienowski, A. M.,** Chemical attractants of the rat preputial gland, in *Chemical Signals in Vertebrates,* Muller-Schwarze, D. and Mozell, M. M., Eds., Plenum Press, New York, 1977, 45.
11. **Gawienowski, A. M., Orsulak, P. J., Stacewicz-Sapuntzakis, M., and Joseph, B. M.,** Presence of sex pheronomes in preputial glands of male rats, *J. Endocrinol.,* 67, 283, 1975.
12. **Gawienowski, A. M., Orsulak, P. K., Stacewicz-Sapuntzakis, M., and Pratt, J.,** Attractant effect of female preputial gland extracts on the male rat, *Psychoneuroendocrinology,* 1, 411, 1976.
13. **Hoffmeyer, I.,** Responses of female bank voles *(Clethrionmys glareolus)* to dominant vs. subordinate conspecific males and to urine odors from dominant vs. subordinate males, *Behav. and Neur. Biol.,* 36, 178, 1982.
14. **Howard, W. E.,** Effectiveness of chemosterilants, pheromones and repellents in field rodent control, in *Proc. 1st Symp. on Rec. Adv. in Rodent Control, Kuwait,* Kuwait Ministry of Public Health, Kuwait City, 1982, 41.
15. **Huck, V. W., Banks, E. M., and Wang, S-C.,** Olfactory discrimination of social status in the brown lemming, *Behav. and Neur. Biol.,* 33, 364, 1981.
16. **Huck, V. W. and Banks, E. M.,** Differential attraction of females to dominant males: olfactory discrimination and mating preference in the brown lemming *(Lemmus trimucronatus), Behav. Ecol. Sociobol.,* 11, 217, 1982.
17. **Idris, M. D. and Prakash, I.,** Response of male Indian gerbil *Tatera indica* to urine from females, *Indian J. Exp. Biol.,* 21, 97, 1983.

18. **Johnston, R. E.,** Sexual attraction function of golden hamster vaginal secretion, *Behav. Biol.*, 12, 111, 1974.
19. **Karlson, P. and Luscher, M.,** "Pheromones" a new term for a class of biologically active substances, *Nature,* 183, 155, 1959.
20. **Kumari, S. and Prakash, I.,** Conspecific odour as a phago-stimulant for the Indian gerbil *Tatera indica* Hardwicke, *Indian J. Exp. Biol.*, 17, 981, 1979.
21. **Kumari, S. and Prakash, I.,** Role of conspecific urine in enhancing food consumption and masking shyness behaviour in the desert gerbil *Meriones hurrianae* (Jerdon), *Indian J. Exp. Biol.*, 18, 730, 1980.
22. **Lee van der, S. and Boot, L. M.,** Spontaneous pseudopregnancy in mice, *Acta Physiol. Pharmacol. Neerl.*, 4, 442, 1955.
23. **Leon, M.,** Chemical communication in mother-young interactions, in *Pheromones and Reproduction in Mammals,* Vandenbergh, J. G., Ed., Academic Press, New York, 1984, Chap. 2.
24. **Lydell, K. and Doty, R. L.,** Male rat preferences for female urine as a function of sexual experience, urine age and urine source, *Horm. and Behav.*, 3, 205, 1972.
25. **Mackay-Sim, A. and Laing, D. G.,** The sources of odors from stressed rats, Physiol. Behav., 27, 511, 1981.
26. **Muller-Schwarze, D. and Heckman, S.** The social role of scent marks in beaver *Castor canadensis, J. Chem. Ecol.*, 6, 81, 1980.
27. **Muller-Schwarze, D.,** Experimental modulation of behavior of free ranging mammals by semiochemicals, in *Chemical Signals in Vertebrates, III* Muller-Schwarze, D. and Silverstein, R. M., Eds., Plenum Press, New York, 1983, 235.
28. **Muller-Velten, H.,** Uber den Augstgeruch bei der Hansmans *(Mus musculus* L.), *Z. Vergl. Physiol.*, 52, 401, 1966.
29. **Price, E. O.,** Urine marking and the response to fresh vs. aged urine in wild and domestic Norway rats, *J. Chem. Ecol.*, 3, 9, 1977.
30. **Purushotham, K. R., Raghupathi, M., and Ramamurthi, R.,** Role of body odours on consumption of food in two rodent species *Geobios,* 10, 32, 1983.
31. **Reiff, M.,** Utersuchungen über naturliche und synthetische Geruchstoffe, die bei Ratten und Mausen ein stimulierende Wirkung auslosen, *Acta Tropica,* 13, 289, 1956.
32. **Stacewicz-Sapuntzakis, M. and Gawienowski, A. M.,** Rat olfactory response to aliphatic acetates, *J. Chem. Ecol.*, 3, 411, 1977.
33. **Singer, A. G., Agosta, W. C., O'Connell, R. J., Pfaffman, C., Bowen, D. V., and Field, F. H.,** Dimethyl disulphide: an attractant pheromone in hamster vaginal secretion, *Science,* 191, 948, 1976.
34. **Steiner, A. L.,** Self- and allo-grooming behavior in some ground squirrels (Scuiridae); a descriptive study, *Can. J. Zool.*, 51, 151, 1973.
35. **Stoddart, D. M.,** Effect of the odour of weasels *(Mustela nivalis* L.) on trapped samples of their prey, *Oceologia,* 22, 439, 1976.
36. **Stoddart, D. M.,** *The Ecology of Vertebrate Olfaction,* Chapman & Hall, London, 1980.
37. **Sullivan, T. P. and Crump, D. R.,** Influence of mustelid scent-gland compounds on suppression of feeding by snowshoe hares *(Lepus americanus), J. Chem. Ecol.*, 10, 1809, 1984.
38. **Sullivan, T. P.,** Influence of predator odours on the feeding behaviour of herbivores, in Abstr. 4th Int. Theriol. Congress, Edmonton, August 13 to 20, 1985.
39. **Sullivan, T. P., Nordstrom, L. O., and Sullivan, D. S.,** Use of predator odors as repellents to reduce feeding damage by herbivores. I. Snowshoe hares *(Lepus americanus), J. Chem. Ecol.*, 11, 903, 1985.
40. **Sullivan, T. P., Nordstrom, L. O., and Sullivan, D. S.,** Use of predator odors as repellents to reduce feeding damage by herbivores. II. Black-tailed deer *(Odocoileus heminous columbianus). J. Chem. Ecol.*, 11, 921, 1985.
41. **Thiessen, D. and Rice, M.,** Mammalian scent gland marking and social behavior, *Psychol. Bull.*, 83, 506, 1976.
42. **Valenta, J. G. and Risby, M. K.,** Discrimination of the odor of stressed rats, *Science,* 161, 599, 1968.
43. **Vandenbergh, J. G.,** Pheromonal regulation of puberty, in *Pheromones and Reproduction in Mammals,* Vandenbergh, J. G., Ed., Academic Press, New York, 1984, chap. 4.
44. **Whitten, W. K.,** Modifications of the oestrous cycle of the mouse by external stimuli associated with the male, *J. Endocrinol.*, 13, 399, 1956.

Chapter 23

BIOLOGICAL PRINCIPLES, DEVELOPMENT, AND PERSPECTIVES OF THE USE OF BACTERIA AND VIRUSES

V. A. Bykovskii and N. V. Kandybin

TABLE OF CONTENTS

I. INTRODUCTION

An important biological peculiarity of rodents is their high level of phylogenetic diversity resulting in great variations in their morpho-physiological adaptations and ecological success. These animals represent one of the orders of mammals and it is very difficult to find a sufficiently selective means of control which is safe for representatives of other relative groups and man. The idea of the use of microorganisms for this purpose, the specificity of which is determined by natural-historical reasons, arose and resulted in the investigation of the pathogen/host system and the solution of a number of applied problems.

II. BACTERIA

Bacteria have been used against harmful mammals for well over a century. This approach was initiated by the well-known biologists Louis Pasteur and I. I. Mechnikov. The former, in 1870, discovered the bacterium of chicken cholera (*Bacterium bipolaris avisepticum*). Pasteur pointed out the possibility of its use in Australia and France, and Mechnikov, according to his own experience, proposed it for controlling susliks in the south of Russia.

Because there were doubts about the ecological safety of the chicken cholera bacterium, the search for other pathogenic agents for rodent control was undertaken. In Russia in 1893 a prospective bacteria was isolated from a mouse carcass. In Germany the bacteria that were believed to be a source of epizootics in white mice was isolated in 1890.[18] In France a new pathogen found in vole carcasses was proposed for rodent control in 1893.[12] Quite recently some other species or strains of bacteria have been suggested for the same aim.[19,22,46] However, many of the mentioned microorganisms were not of practical use because of their low efficiency or insufficient selective effect. Nowadays Issatschenko bacteria and the strain 5170 are used in the U.S.S.R.

The taxonomy of these bacteria repeatedly changed. Now they belong to genus *Salmonella,* group "D" according to the Kaufmann-White classification schedule. *S. enteritidis* serovar forms the independent biovars: *S. enteritidis* var. *danycz* and var. *issatschenko.*[30,42] The morphological and cultural features of Danycz and Issatschenko bacteria are identical. They are Gram negative, nonsporeforming rods, 0.6 to 0.7 $\times$ 2.0 to 3.0 nm in size, and rounded at the corners. They are characterized by motility due to the presence of flagella. In broth they produce uniform turbidity and occasionally sediment. The bacteria are susceptible to high temperatures. Boiling kills them immediately. However, even at low temperatures ($-25°$), they maintain their viability.

As far as physiological properties are concernted, Issatschenko and Danycz bacteria are similar to other biovars of *S. enteritidis,* but they have stable specific features. Unlike other biovars of *S. enteritidis,* while growing on Shtern broth or d-tartrate, they do not change the color of the media. The differentiation of Danycz and Issatschenko bacteria is done with the help of l-lysin and lysozyme. In a media containing l-lysin, Danycz bacteria change the media's color, in contrast to the Issatschenko biovar. Under a lysozyme influence the 1-day old culture of Issatschenko bacteria form lysis zones on the media, but lysis does not occur with other biovars, including Danycz bacteria. Using lysis as a means of identification Issatschenko bacteria and the relative strain 5170 were distinguished as independent biovars. This isolation is of practical value since medical specialists have found the Issatschenko bacteria to be safer than Danycz for use with men and domestic animals.[42]

Antigen structure of the Issatschenko bacterium in the modern form has the following indices: somatic antigen 1,9,12, flagellated antigen phase 1 - g, m, phase 2 - (1,7).[18] Somatic- or O-antigens are involved in the composition of the bacterium cell wall and represent a very sophisticated complex, including proteins, lipids, and sugars. These antigens are highly toxic and are considered to be endotoxins. Their basic structure has glucosamine,

6-desoxi-, and 3,6 didesoxihexoses as the side chains. Immunodominant function is peculiar to sugars occupying the terminal position in chains. In particular antigen 9 specificity in Salmonella is determined by tivelosa (3-6-didesoxi-d-mannose). Immune O-antigen 12 complex is determined by the availability of two lateral chains, one of which is completed with ramnose, the other with glucose, and so on. Besides, serological specificity of polysaccharide agents depend on sugar number and their spacial position in the structure.

The basis of flagellated or H-antigens is the low-molecular protein, flagellin. It is believed that the amino acid composition of various H-antigens and their molecular organization are not similar. The corresponding differences were used for creating a Sallmomella classification schedule. In these bacteria H-antigens can exist in two phases: ''specific'' and ''nonspecific''. The former is typical for one serovar of the group, the latter has a wider significance.

Salmonella H-antigens have little influence on the course of macro- and microorganism relationships. O-antigens are of considerable significance in this process due to the similarity with host isoantigens.[21] The availability of heterogenic antigens, similar in their specificity in various species, prevents antigen development by the target organism at early infection and allows the pathogen to reproduce freely during this period.

Issatschenko bacteria pathogenecity, i.e., their potential ability to cause an infectious disease, was studied by many investigators, first of all, naturally, on rodents. Much attention was paid to the families Muridae and Cricetidae as comparatively many harmful species belong to these families. The experiments were carried out on susliks and dormice as well. The above mentioned rodents were introduced per os or given as a bait, fresh bacterial cultures in doses from one to hundreds of millions of microbial bodies. Then the experimental individuals were observed for 22 days. These investigations involved the representatives of 31 species (Table 1), Issatschenko bacteria being pathogenic for most of them.[26] Of these rodent species 14 were highly susceptible, their mortality in trials was not less than 80%. This group included house mice (genus *Mus*) and representatives of the subfamily Microtinae (genus *Clethrionomys, Lagurus, Arvicola, Microtus, Lasiopodomys*). The host list should be supplemented with the harvest mouse (*Micromys minutus*) and mygratory hamsters (*Cricetulus migratorius*).

In applied aspects the following figures are of interest. Of all species given in Table 1, 21 species cause appreciable damage in the territory of the U.S.S.R. Among them, 33.3% are highly susceptible to the bacteria, 57% are slightly susceptible, and 9.6% are entirely resistant. The attempts to use Issatschenko bacteria to control rodent species with low susceptibility were unsuccessful. The application of combined preparations containing a bacterial and chemical basis was short and not practiced on a large scale. Thus, the usefulness of Issatschenko bacteria application is limited to a group of highly susceptible species of harmful rodents.

Issatschenko bacteria pathogenicity for use against warm-blooded animals (domestic and wild) was the object of special investigations carried out by microbiologists and veterinarians. The data obtained are described in a number of works and we shall refer only to the most informative ones.[22,27,30,38,46,58,59]

Domestic animal's resistance to Issatschenko bacteria and strain 5170 was studied many times. Oral introduction of high doses did not cause any signs of disease in horses, sheep, pigs (mature and young ones), dogs, cats, adult rabbits, geese, ducks, or chickens. Cases of enteritis and a short-term rise of temperature were observed in calves and lambs. There were fatal cases in the trials with young rabbits and lambs as well, but it is suspected that the Issatschenko bacteria only complicated other pathological processes in these animals. However, prohibiting the use of Issatschenko bacteria to control rodents in large livestock complexes was considered necessary. The effect of these bacteria on wild animals was studied less. It was found that the bacteria is pathogenic for musk-rats (*Ondatra zibethica*) and house sparrows (*Passer domesticus*). It is still unclear what level of resistance the hare

Table 1
SUSCEPTIBILITY OF RODENTS TO *SALMONELLA ENTERITIDIS* VAR. *ISSATSCHENKO*

	Rodent species	Susceptibility
1.	*Citellus pygmaeus* Pall	+
2.	*Citellus major* Pall	+
3.	*Dyromys nitedula* Pall	0
	Alactagulus pygmaeus Pall	+
	Dipus sagitta Pall	+
6.	*Apodemus sylvaticus* L	+
7.	*Apodemus flavicollis* Melch	0
8.	*Apodemus agrarius* Pall	0
9.	*Mus musculus musculus* L	+ +
10.	*Mus musculus hortulanus* Nordm	+ +
11.	*Micromys minutus* Pall	+ +
12.	*Rattus rattus* L	+
13.	*Rattus rattoides* Hodg	+
14.	*Rattus norvegicus* Berk	+
15.	*Nesokia indica* Gray	+
16.	*Ellobius talpinus* Pall	+ +
17.	*Cricetulus migratorius* Pall	+ +
18.	*Mesocricetus raddei* Nehr	0
19.	*Mesocricetus brandti* Nehr	0
20.	*Meriones tamariscinus* Pall	+
21.	*Meriones unguiculatus* M.E.	+
22.	*Rhombomys opimus* Licht	+
23.	*Clethrionomys glareolus* Schreb	+ +
24.	*Lagurus lagurus* Pall	+ +
25.	*Arvicola terrestris* L	+ +
26.	*Microtus majori* Thom	+ +
27.	*Microtus gregalis* Pall	+ +
28.	*Microtus socialis* Pall	+ +
29.	*Microtus agrestis* L	+ +
30.	*Microtus arvalis* Pall	+ +
31.	*Lasiopodomys brandti* Radde	+ +

Note: (+ +), Highly susceptible; (+), low susceptibility; (0), nonsusceptible.

After Kandybin, N. V., *Scientific Grounds of Plant Protection*, Kolos Publishers, Moscow, 1984, 191.

(genus *Lepus*) has for the bacteria, especially the young hares, as well as common grain-eating birds such as the bunting (genus *Emberiza*) and the gray partridge (*Perdix perdix*), etc. The problems of Issatschenko bacteria circulation in agrobiocenosis have not been well studied. Nevertheless, the pellets of birds of prey are known to contain strains of various virulences (i.e., unequal in their degree of pathogenicity for susceptible rodents), 37.5% of which are the highly virulent strains.[31]

Issatschenko bacteria pathogenicity for man was indirectly studied for a long time. There was some information about occasional consumption of bacterial culture that usually caused no dangerous consequences.[3] To prove the safety of these bacteria, Professor Prokhorov swallowed a dose nearly equal to 10 billion and, as it was expected, the dose was harmless.[46] Quite recently, purposeful investigations with the use of modern techniques have been carried out by the specialists of the Ministry of Public Health of the U.S.S.R.[42] According to the available data, Issatschenko bacteria and strain 5170 are referred to as microbes conditionally pathogenic for man — they can cause a disease provided they are consumed with food products in which they are able to reproduce intensively (for example, milk). In this connection official permission to apply these bacteria in storages, greenhouses, and under field conditions was given, provided simple rules of safety were followed.

Issatschenko bacteria virulence is not stable and during storage virulence can be markedly weakened, especially while reinoculating. The best means to prevent or markedly slow this process is to keep bacteria in nutrient-deficient media at a negative temperature. If the virulence level decreases it can be restored by passage of the organism through a susceptible rodent.

Under field conditions Issatschenko bacteria usually dissociates from S- to R-form (low virulence for rodents) under the influence of isolation, drying, an unsuitable pH medium, enzymes, phage, etc. In our special investigations the tagged Issatschenko bacteria were introduced in large quantities to rodents inhabiting haystacks, ricks, under vegetable cover of pastures, and in seedlings of cultivated plants. Some time later the previously introduced bacteria from these substracts were isolated and their virulence was found to decrease gradually.

In Russia, the bacterial method of rodent control has not been used extensively. The liquid bacterial cultures were mixed with the bait-food before application. In the middle of the 1950s a method of cultivating bacteria on dense nutritive media of granular structure, which allows one to obtain a preparation with a high bacterial titer, was worked out.[46] It was called bactorodencide. Originally it was prepared on fibrin, whole and broken kernals of wheat, oats, corn, or bone sawdust. Then, the technology of production was modified[33] and at present bactorodencide is mainly prepared on grain media. As the technology of bactorodencide production was improved, accompanied by an increase in its quality and efficiency, the interest of specialists in this preparation became greater. Now the grain bactorodencide is used in the U.S.S.R. in large areas (in some years, 4 million ha) to control mice, *Mus*, and voles, *Microtus*.

The control efficiency is influenced by the technique of bactorodencide application. It is rather important that the preparation is found easily and consumed well by those rodents against which it is applied, and the bacteria in it do not lose their virulence before consumed by the rodent. It is known that under field conditions the bacterial preparation preservation is higher with batch or many layer arrangements than with diffused distribution. The same advantage, in comparison with an open arrangement, is achieved by the bactorodencide application into burrows and other shelters.[8] Generally this preparation causes diseases only in the individuals that directly consume it. The possibility of epizootic development through reinfection (i.e., as a result of the contact between sick and healthy individuals) has not been recognized as yet. Some authorities (Polyakov[45]) fully deny it. The work was carried out in open-air cages and in field conditions. *Microtus arvalis* in the open-air cages and the population of small muriform rodents (*Clethrionomys glareolus, Apodemus sylvaticus, Mus musculus*) in the field were observed. Besides tagged individuals, the technique especially developed for the life determination of rodent infection by Issatschenko bacteria was employed, the results of necroposy being taken into account too.[30] Under the experimental conditions an epizootic was successfully induced in a high density population and a considerable proportion of them (31 to 50%) where artificially infected by bacterial culture in a lethal dose. At the natural concentration of muriform rodents in straw stacks or on the forest plots the introduction of infection into their numerous population caused a short-term or weak epizootic, but it waned out as the host numbers decreased. Thus, with the help of Issatschenko bacterium it is possible to cause an epizootic; but, its development depends on the parasite and host condition and their relationship in the particular ecological situation. Under the natural conditions Issatschenko bacteria usually dissociate and there is "a set" of strains of various virulence in the epizootic hotbed. "Weak" strains among them are the cause of the formation of an immune part of the rodent population. The various host susceptibility favors it as well, that is connected with the population structure peculiarities and the phase of its many-years cycle.

Cannibalism is noted as one of the possible natural ways of spreading infection.

The significance of actively acquired immunity in rodents as the result of having a disease caused by bacteria of *S. enteritidis* serotype was ascertained from different positions. The population study of Norway rats (*Rattus norvegicus*) in the places of Danysz bacteria application showed the occurrence of immune individuals,[13] the number of which can reach up to 20% according to more precise estimates.[37] The existence of group immunity to various biovars of *Salmonella enteritidis* in Norway rats (*Rattus norvegicus*),[20] and in water voles (*Arvicola terrestris*)[22] was demonstrated by other experiments. The variance of susceptibility of voles to Issatschenko bacteria of different ages was noted many times: susceptibility decrease as age increases.[43] The period of acquired immunity preservation in small muriform rodents was determined in the experiments carried out by a number of investigators. As a result it was found that such immunity was not stable or long lasting. It disappears in 30 to 50 days.[30] The modern studies of immunity in rodents[5,30] have allowed one to obtain both in vitro and in vivo the material demonstrating the significance of inherited nonspecific factors such as lysozyme in the manifestation of protective reaction. This enzyme produced by the mucous membranes is able not only to lyse Issatschenko bacteria but to cause the induction of productive phage of strictly selective effect in S-forms of these *Salmonella*. Its appearance can be accompanied by lysis of such bacterial cells or their transformation in an R-form that is resistant to lysozyme or phage but of low virulence for rodents. Lysozyme is easily found in spleen, liver, or intestine of rodents; its content is species-specific and usually found in nonsusceptible forms in a great quantity. However, the population of susceptible species is heterogenous and according to this feature there is a possibility to make a selection for resistance when bactorodencide is applied often. Issatschenko bacteria and rodents are links of a single ecological parasite-host system, the existence of which needs the balanced relationships of its constituents. This regularity determines the necessity to alternate the preparations of various action mechanism to control rodents.[8]

The influence of a sublethal dose on rodent prolificacy should be noted as the possible consequence caused by Issatschenko bacteria. For example, in rats (Wistar line), after the pregnant females had been fed bacterodencide the number of young in the litter was less and they developed more slowly as compared with the control.[9]

Thus, the relationships of pathogen and rodent-host are very complex and many sided. In this connection the following peculiarities should be taken into consideration in order to apply bacterodencide effectively.

1. It is necessary to determine the species composition on all arable lands liable to treatment.
2. The preparation must be fresh and attractive for rodents; it should be placed in concentrated form so that it may be conserved longer and be eaten more intensively.
3. It is useful to repeat the treatment to fully satiate the rodent population with pathogen if the preparation is completely eaten.
4. The most favorable conditions for bacterodencide application are created in the places of abundance of small muriform rodents (stacks, ricks, grain storage, etc.); it is more effective in a cold season.

III. VIRUSES

Viruses occupy a special place in the world of microorganisms. They were first recognized as living bodies at the end of the 19th and beginning of the 20th centuries,[26] but it took more than 30 years to recognize their practical utility.[23] Later the knowledge of viruses made rapid progress. Due to their infectious role much attention was and is being paid to the problems of treatment and prophylaxis of man, animal, and plant diseases; the problems of virus application to control farm crop pests are also being studied. Among vertebrates, the European rabbit, *Oryctolagus cuniculus* became the first to be studied. Formerly this species was included in Rodentia but later it was distinguished as a phylogenetically distinct Lagomorpha order. The purposefully wide virus application against the European rabbit is still the single example of the control of harmful vertebrates, and the experience obtained may have a significant lesson applicable to the problem of limiting the harmful rodent numbers.

It is advisable to preface the account of corresponding developments with a short characteristic of vertebrate viruses. They have much in common with viruses of other groups but possess a number of specific features. The main characteristic is that they are parasites on the gene level.[17] To reproduce they need the synthesizing mechanism of a cell (ribosomes and membranes), its enzymes, and an energy-generating system. A virion, otherwise called a virus corpuscle, is unable to reproduce out of a cell and there are no metabolic processes in it. Biosynthesis of virus components and host-organism damage takes place after the penetration of the virus into the target cell. This complex and many stage process is completed with viral destruction, and liberation of nucleic acid that fulfils the genetic function of the virus. In this connection the presence of virus corpuscles in the infected cell is found only after the first daughter virions appeared. Such a way of reproduction, attended by the loss of physiological integrity by the parent virion, is unique and called disjunction. Its realization is opposed by the protective reactions of the host-organism. In warm-blooded animals practically any cell is able to produce a special protein called interferon as soon as virus penetrates it.[25] Normally interferon is not produced, and although not toxic to the organism it inhibits virus reproduction. The surface virion antigens (hemaglutinin, neuraminidase, and others) stimulate, in host-organisms, the accumulation of immunoglobulins having virus-neutralizing activity. In mammals they are represented by three classes of globular proteins, IgM, IgG, and IgA. High-molecular IgM appear at first both in the ontogeny and during the virus infection or the artificial immunization. Low-molecular IgG are formed later than others and able to pass through the placenta to provide a young organism with maternal antibodies for the first month of its postnatal period. IgA presence is sufficiently identified only in man as yet, but similar proteins are present in the serum of rabbits, rats, mice, and some other mammals.[41] Immunoglobulins are produced by the highly specialized plasmatic cells, B-lymphocytes, being transformed in them under the antigen influence. One plasmatic cell forms the antibodies of the same specificity and can become an ancestor of hundreds of relative cells during maturity. It increases the accuracy of antibody differentiation. They are able to be adsorbed on the surface of proper virions and thus to prohibit the virions attachment to a cell membrane and the following penetration into the cell. The same antibodies agglutinate virions, and in cooperation with T-lymphocytes and macrophages promote the liberation of organisms from the pathogen. This results in the development of immunity that is characterized by effective disease suppression, especially when it again meets the given or homologous virus. Such immunity is acquired not at once, but usually preserved for a long time although it is not transferred to a progeny.

In spite of different mechanisms of protection that a mammal organism possesses, viruses can overcome them and cause dangerous epizootics. If it is of great difficulty to prevent such diseases and prevention is not always successful, then could it not be a highly effective means of suppressing the harmful species numbers? In this connection let us address the unique experience of myxomatosis use. The history, content, and results of respective investigations were described in detail in 1965.[16] The reliable study became the source of valuable information and not a single modern review on verterbrate viruses or biological control methods with these animals can be done without reference to it.[1,2,17,54] This allows us to pay attention only to the results of the main events which occurred and to take into consideration the data of last years.

Myxomatosis is a virus disease affecting rabbits of genus *Sylvilagus* and *Oryctolagus*. In sick animals the tumors appear on the body as the result of the virus induced active multiplication of fibroblasts — the cells in the connective tissue of the skin. In a mammal skin other kinds of tumors induced by viruses may be found: papiloma embracing epidermal cells and histiocytomes arising as a result of the proliferation of histiocytes, but they are produced by pathogens referring to other genera and families of this kingdom.

Myxomatosis became known in the scientific literature in 1898. In Montevideo (Uruguay) the high lethal "spontaneous" disease of domestic rabbits was reported, its pathogen being a virus.[50] The ancestor of these animals is known to be *Oryctolagus cuniculus*. The infection was easily transferred from sick to healthy animals and after a short-term incubation period the development of bilateral conjunctivitis accompanied by abundant, first mucous, then purulent outflows was observed. The eyelids are severely swollen, then tumors appear which, depending on the severity of a disease, may either be found on the whole body or localized on the head, genital organs, and near the anal orifice. The typical bolster-like "crimples" of skin give a specific "lion" appearance to the head of such animals. If the process is progressive, rhinitis and pneumonia with acute inflammation of mucous membranes of the respiratory tract develop. The lethal outcome in mature rabbits without immunity comes in 10 to 14 days, in the young in 5 to 6 days of onset of the disease. The mortality rate is usually over 90%. At necropsy the accumulation of yellow, easily threading, transparent mucous in subcutaneous connective tissue is observed. From here we have the name of the disease myxomatosis ("myxa" in Greek means slime, "oma", tumor).

According to the modern classification, the viral agent belongs to the genus *Leporivirus*, family Poxviridae.[15] Virons of the pathogen are characterized by a comparatively large size (the corpuscle size, near 200 nm), rectangular form, and the presence of an outer membrane.[1] As to other features, myxomatosis virus is similar to other representatives of the family. For example, about 30 proteins are involved in the virion content, including enzymes that are necessary for successful reproduction that takes place, not in a cell nucleus as it occurs in many DNA-containing viruses of other families, but in the cytoplasm. At the same time myxomatosis virus is more susceptible to heating (at 55°C it inactivates in 25 min) and the action of ether. It is also known that myxomatosis virus is well preserved in glycerin (50%) and after lyophilization. It survives many months in the skin of an infected rabbit at usual temperatures. In antigen properties it is similar to Shope and other viruses of the genus (causing fibromata in rabbits and squirrels), but as a disease-producing agent it is specific.

In South America, where myxomatosis in domestic rabbits was first observed, the same pathogen attacks the individuals of native species, *Sylvilagus brasiliensis* under natural conditions. In them, however, the disease takes a light course and results in the formation of fibrous tumors in skin. Another pathogen, the Shope virus, induces fibromata in *S. floridanus* distributed in the eastern U.S. Domestic rabbits easily endure this disease: although fibromata arise they regress very soon, in contrast to *S. floridanus* in which they are preserved for a long time. Only newly born domestic rabbits get sick with a generalized form of infection resulting in their death.[14,44] The third virus of this group was discovered in California. In the endemic species of rabbits, *S. bachmani*, it causes local fibromata and in domestic rabbits, a lethal disease that is not accompanied by tumor formation.[47]

There are differences between Brazilian and Californian myxomatosis viruses that are revealed in a serological reaction while using soluble antigen. Close relationships between myxomatosis and Shope viruses are beyond doubt. Rabbits having had fibromatosis are not susceptible to myxomatosis. In this connection it was supposed that nonmalignant fibroma could be induced by myxodmatosis virus with weakened infectivity. The antigenic affinity of these two pathogens is confirmed by the serological reaction.[60] At present two more viruses of American origin are known: Colombian and Panamanian ones.[17]

The main ways of spreading myxomatosis under natural conditions remained unknown for a long time. It was believed that direct contacts between individuals were of main importance. This was confirmed by the ease with which rabbits were infected when placed in a cage with sick animals. These observations led to the idea of using the virus against susceptible rabbits where they caused great losses to farming.[4] The suggesion had an official response in Australia, since the rabbits formerly introduced there reproduced to such an extent that they began consuming enough fodder for 40 million sheep. In 1926 some large, sick rabbits were released in the semidesert regions of that country, but the disease did not cause an epizootic. Shortly thereafter it was found that the infection could be transmitted by mosquitoes and fleas, and later that its source in the South America was rabbits, *S. brasiliensis*. In Australia and some European countries the possibility of culturing the virus, and its pathogenic spectrum were investigated. At that time in experiments carried out on various warm-blooded animals, including marsupials, only rabbits were infected.[6] New data on high virus selectivity stimulated its use for practical purposes. In May, 1950 some sick rabbits were again released in the south-east of Australia, but now in more humid regions. After a period of 6 months the epizootic occurred. It began only on one of seven possible sites, but then it spread over the territory with a radius of up to 800 km. The rabbit number on this territory decreased by more than 99%. The epizootic ceased towards the winter, but the next summer after the sick rabbits had been released it appeared again and embraced wider territory. As a result the rabbit populations inhabiting the south-east of Australia were reduced by 80% and the next year income in farming was estimated to be 50 million pounds. However, the annual release of sick rabbits soon became useless. Ecological studies aimed at elucidating the cause of this loss of effectiveness showed that two main events had taken place: the appearance of weakened virus strains and the increase of rabbit resistance to myxomatosis.

The study of mosquitoes as a vector enabled the discovery that *Anopheles annulipes* is of the greatest significance in Australia. The mosquitoes live in rabbit burrows during the daytime, were found to be less connected with the river valleys, and were more widespread than *Culex annulirostris*. Nevertheless, epizootic intensity was higher under the liberal rainfall conditions that favor the reproduction of various species of mosquitoes. It is believed that mosquitoes play the part of mechanical carriers, as myxomatosis virus does not reproduce in an arthropoda organism. In this connection a mosquito needs a rather large virus quantity to infect rabbits and that is possible if the proboscis penetrates not into blood, but into a tumor. The existence of highly infectious virus, causing the quick death of host-organism, proved to be biologically less successful. When the rabbits are infected with this strain, their numbers decrease so fast that the possibility of virus spread, transmission, and even its existence is reduced. The absence of mosquitoes in winter also reduces its existence. While weakening the virus to great extent, this strain induces in rabbits only the appearance of local "knots" with low pathogen concentration, and its transmission occurs in quantities not great enough for contamination.

Thus, the strain of average mean virility gets the selective advantages.[40] The attenuated variants appeared a year after the first successful virus introduction in Australia and 3 to 4 years later they became predominant in the whole country. This process was carefully examined and as a result the direct definite relation between the mortality rate and the average disease duration in rabbits was stated. According to these data the strains isolated under the field conditions amounted to five groups, the third one uniting variants with average infectiousness was found more often (Figure 1). They caused diseases with lethal issue in most parts of sick animals (70 to 95%), but the residual part of the population restored the numbers, especially during the winter months when there were no vectors.[14]

Rabbits that have had the disease develop immunity for their subsequent life, and the immune females transmit short-term passive immunity to their young. The average life span in rabbits, *O. cuniculus* in Australia, however, is about a year. In this connection the importance of acquired immunity for the populations existence is not of great value. Its

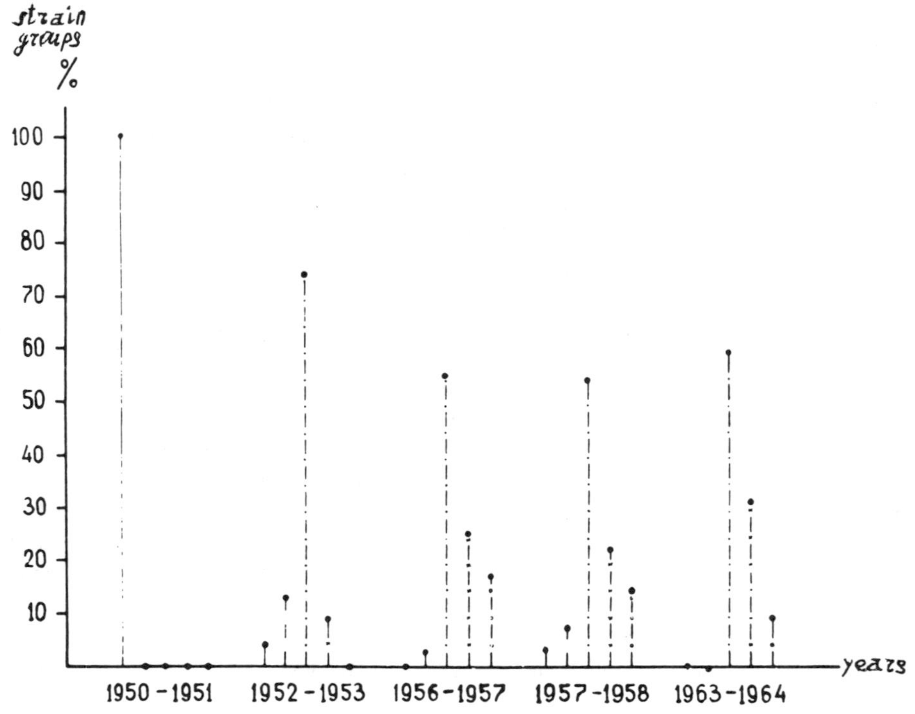

FIGURE 1. Indices of infectiousness of virus myxomatosis in different years after successful introduction in Australia. (1950-1951) Number of disease days <13, % mortality >99; (1952-1953) number of disease days = 14 to 16, % mortality = 95 to 99; (1956-1957) number of disease days = 12 to 16, % mortality = 70 to 95; (1957-1958) number of disease days = 29 to 50, % mortality = 50 to 70; (1963-1964) number of disease days >50. (After Fenner, F., *The Biology of Animal Viruses*, Vol. 2, Academic Press, New York, 1968, chap. 16.)

genetic irregularity is of considerable significance. The pathogen, capable of eradicating the populations of its host by more than 99%, became a powerful selection factor in the regions where the individuals contaminated with a highly infectious strain were released every year, the tolerance among rabbits increased quickly. Under field conditions during 7 years the rabbit mortality from equivalent doses of stable strain decreased from 90 to 25%.[53]

The subsequent attempts to use new highly infectious viral strains did not change the general scene: under field conditions they were quickly forced out by less active strains. Thus, the usefulness was not great, especially if one takes into consideration that such introduction promote the selection of individuals for resistance. This resistance stressed the advisability to use not only myxomatosis virus, but also the traditional means of rabbit control.

In 1952, on his own initiative a French physician-pensioner used myxomatosis for *O. cuniculus* eradication in his estate near Paris. The infection quickly spread throughout France in spite of the government efforts to suppress the epizootic — as rabbits are the subject of animal husbandry there, on the one hand, and the object of hunting, on the other. In particular, active control of mosquitoes was carried out, and domestic animals were vaccinated. Nevertheless, the natural population was destroyed by 90% and many farmers lost their domestic population.[52] Uncontrolled epizootics caused other consequences. The import of hares from Central Europe began and they became the source of spreading tularemia, a disease of bacterial origin dangerous for man.[39] At present the problems and usefulness of *S. floridanus* introduction from America are being discussed. The sharp *O. cuniculus* population decrease enabled the successful completion of studies on reforestation with no construction of expensive fences. The marked decrease in the occurrence of Marseilles fever, caused by *Rickettsia conorii,* is possibly connected with the break of the infectious chain under the myxomatosis influence.

In 1953 this disease emerged in the south-east of England, mainly in the regions situated opposite the French coast. In the beginning the quick weakening or even the disappearance of the infection was expected, but it gradually involved England and Scotland. It was found that the main carriers were not mosquitoes, as in Australia, but fleas, *Spilopsyllus cuniculi.* In contrast to mosquitoes, fleas are active the whole year and myxomatosis in Great Britain did not become a seasonal phenomenon. Besides, it is supposed that the virus doses transmitted by fleas are more than those of other arthropods.[11] These peculiarities exerted influence on the many-years dynamics of the disease. The appearance of attenuated strains occurred more slowly than in Australia; moreover, irregular unexpected outbreaks of disease were noted in the regions where the infection did not appear for a year or more. This again attracted attention to the vectors. The idea of low mobility of fleas was not confirmed in experiments: during 3 days a considerable exchange of these ectoparasites took place between rabbits. It became also known that fleas living on sick rabbits did not lose their infectiousness for up to 105 days, even when they were not on a host, but in holes.[10] The ovaries of *S. cuniculi* females feeding on a sick rabbit mature more quickly. Birds, especially those that live on carcasses as well as hares, take part in the displacement of fleas over great distances.[48] The mentioned investigations helped to realize that *S. cuniculi* were imported to Australia. Besides, the phenomenon of irregular re-infestation coincided with sheep import to the regions infected by mites. The possibility of viral transmission by mites was then confirmed in the laboratory.

Myxomatosis spread had a considerable influence on the vegetable cover. As bush preservation improved, bushes occupied a wider area of lowlands. This had an effect on species composition, as well as on the bird and insect numbers. Some changes took place in the life of warm-blooded predators for whom rabbits were an adequate kind of food. Besides, during the wide epizootic, hares with a heavy form of myxomatosis were observed. These cases convincingly demonstrated the necessity to study a large quantity of individuals to make valid conclusions about the natural resistance of a given species to infection. More and more data on myxomatosis epizootic consequences are now available.[24,55]

At present myxomatosis in Great Britain reduces the numbers of natural rabbit population by nearly 50% and its restoration now develops more quickly but it threatens farm crops with resumption of damage.[56,57] In various parts of that country the resistance of *O. cuniculus* to myxomatosis was estimated and in one region individuals resistant to the action of a highly infectious strain were found.

When the epizootic spread to Middle and Eastern Europe, the disease started losing its effectiveness. This is connected with rather low numbers and discontinuous distribution of rabbits in that region, vaccine use, and improvement of hygiene conditions to protect domestic animals.[29] In the U.S.S.R., myxomatosis was first found on the territory of the Latvian S.S.R. in 1977 and soon it was eliminated.[51]

The application of myxomatosis virus in different countries to control rabbits (*O. cuniculus*) showed its limited possibilities. The disease effectiveness gradually reduces in connection with the appearance of weakened strains of pathogen and the increased resistance of the target population. This process is influenced by natural selection and is advantageous for each participant. Irresponsible virus use caused great damage. An epizootic outbreak could not be ceased. Although this virus is a highly specified pathogen, in many cases the consequences of its use were unexpected and effected various biocenotic links. To determine the possibility of virus introduction to a new country thorough investigations of many problems are required, from the determination of the spectrum of species specificity of the pathogen to forecasting the changes in the biosphere of a geographical region or landscape zone.

REFERENCES

1. **Andrewes, C. H.,** *Viruses of Vertebrate,* Balliere Tindall, London, 1964, chap. 12.
2. **Andrewes, C. H.,** *The Natural History of Viruses,* Weidenfeld and Nicolson, London, 1964, chap. 18.
3. **Antonovsky, A. I.,** Principles of biological control of rodents using bacteria, in *Proc. All-Union Res. Inst. Agric. Microbiol, 8,* Avaschnle, Leningrad, U.S.S.R., 1936, 5.
4. **Aragao, H. B.,** Myxoma of rabbits, *Mem. Inst. Osvaldo Cruz,* 20, 237, 1927.
5. **Basilova, G. I. and Pilipenko, V. G.,** Changes in lysozyme content of blood in rabbits when immunized with live associated vaccine against plague, tularemia, brucellosis and anthrax, *Problems of Highly Dangerous Infections, USSR,* 4(20), 153, 1971.
6. **Bull, L. B. and Dickinson, C. G.,** The specificity of the virus of rabbit myxomatosis, *J. Counc. Sci. and Ind. Res. Aust.,* 10, 291, 1937.
7. **Bykovsky, V. A.,** Mouse-like rodents, *Plant Prot.,* 8, 15, 1982.
8. **Bykovsky, V. A.,** Selective preparations for plant protection against field rodents, in *Proc. 8th Int. Congr. Plant Protection,* II, 1975, 86.
9. **Bykovsky, V. A. and Peghelman, S. G.,** Influence of toxicant and pathogen on fertility of rodents, in *Proc. 5th All-Union Rodent Conf.,* Nauka, Moscow, U.S.S.R., 1980, 338.
10. **Chapple, P. I. and Lewis, N. D.,** Mxomatosis and the rabbit flea, *Nature,* 207, 388, 1965.
11. **Cooce, B. D.,** Changes in the age, structure and size population of wild rabbits in South Australia following the introduction of Europe as rabbit fleas *Spilopsyllus cuniculi* as vector of myxomatosis, *Austr. Wildlife Res.,* 10, 105, 1983.
12. **Danysz, J.,** Emploi des cultures artificielles de microbes pathogenes a la destruction des rongeurs (campagnolls et mulots) en grande culture, *C. R. Acad. Sci.,* 117, 1893.
13. **Danysz, J.,** Un microbe pathogene pour les rats *(Mus decumsnus et Mus rattus)* et son application a la destruction de ces anumaux, *Ann. Inst. Pastuer,* 14, 196, 1900.
14. **Fenner, F.,** The pathogenesis and ecology of viral infections, in *The Biology of Animal Viruses,* Vol. 2, Academic Press, New York, 1968, chap. 16.
15. **Fenner, F.,** Classification et nomenclature des virus, *Ann. Microbiol.,* 127 A, 322, 1976.
16. **Fenner, F. and Ratcliff, E. N.,** *Myxomatosis,* Cambridge University Press, London, 1965.
17. **Franz, J. M. and Krieg, A.,** *Biologische Schädlingsbekämpfung,* Parey, Berlin, 1982, chap. 12.
18. **Golubeva, I. V., Kilesso, V. A., Kiseleva, B. S., Pryamukhina, N. S., Tatarinova, S. D., Khomenko, N. A., and Yuschenko, G. V.,** Genus Salmonella, in *Enterobacteria Manual,* Meditsina Publishers, Moscow, 1985, 121.
19. **Golubeva, M. P.,** *Streptococcus* as an agent of epizootic in mouse-like rodents, in *Rodent, Control in Pre-Caucasus Steppes,* Rostov-on-Don, 1935, 74.
20. **Golubeva, M. P. and Posenfeld, E. F.,** Group immunity to Danycz bacillus in Norway rats, in *Proc. All-Union Res. Inst. Agric. Microbiol.,* 8, 1, Vaschnil, Leningrad, U.S.S.R., 1936, 75.
21. **Gostev, V. S.,** *Biochemical Grounds of Medical Bacterilogy,* Acacemy of Medical Sciences, Moscow, U.S.S.R. 1951, chap. 1.
22. **Gritsenko, I. N.,** *Microflora of Small Mammals,* Nauka Publishers, Novosibirsk, 1971, 128.
23. **D'Herrelle, M. D.,** *The Bacteriophage and its Behavior,* Boilliere Tindall and Cox, London, 1926, part 1, chap. 1-4.
24. **Hodgkin, S. E.,** Scrub encroachment and its effect on soil fertility on Newborough warren, Anglessey, Wales, U.K., *Biol. Conserv.,* 29, 99, 1984.
25. **Isaacs, A. and Lindemann, J.,** Virus interference. 1. The interferon, 2. Some proportion of interferon, *Proc. R. Soc. London, Ser. B,* 147, 258(1), 268(2), 1957.
26. **Issatschenko, B. L.,** About a new bacillus pathogen to rats, *Rep. Minist. Agric. State Property,* 20, 310, 1898.
27. **Issatschienko, B. L. and Zlatogorov, S. I.,** Typhimurium group, in *Manual on Microorganisms,* Vol. 3, Petograd, 1918, chap. 2.
28. **Ivanovsky, D. I.,** About two diseases of tabacoo, *Agric. Forest.,* 169, 103, 1892.
29. **Jeřabek, J.,** Applicability of Shope fibroma virus replicated in cell cultures for immunoprophylaxis of rabbit myxomatosis, *Acta Vet.,* 49, 259, 1980.
30. **Kandybin, N. V.,** *Microbiological Method of Rodent Control,* VNIITEISKX Publishers, Moscow, 1974, 6.
31. **Kandybin, N. V.,** Ecological peculiarities of Issatschenko bacteria, in *Proc. All-Union Res. Inst. Agric. Microbiol.,* 47, Vaschnil, Leningrad, 1978, 138.
32. **Kandybin, N. V.,** Bacteria for plant protection, in *Scientific Grounds of Plant Protection,* Kolos Publishers, Moscow, 1984, 191.
33. **Kandybin, N. V., Serebsryakova, L. K. and Koreshkova, G. N.,** A new technology of bacterodencide production, in *Proc. All-Union Res. Inst. Agric. Microbiol.,* 48, Vaschnil, Leningrad, U.S.S.R., 1980, 180.

34. **Löffler, T.**, Über Epidemien unter den hygienischen Institute zu Greifswald gehaltenen Mausen und über die Bekämpfung der Feldmaus plage, *Centrallbl. Bacter., Parasitenk. Infection.*, 11, 129, 1892.
35. **Luria, S. E., Darnell, J. E., Baltimore, D., and Campbell, A.**, *General Virology*, 3rd ed. John Wiley & Son, New York, 1978, chap. 1.
36. **Merezhkovsky, S. S.**, A new bacillus for mouse control, *Arch. Vet. Sci.*, 7, 4, 1894.
37. **Merezkhovsky, S. S.**, On the use of bacterial cultures in rat control, in *Proc. Agric. Bacteriol. Lab.*, 2, 3, 1910.
38. **Miller, A. A.**, Streptococci of spontaneous epizootics of mice, in *Rodent Control in Pre-Caucasus Steppe*, Rostov on the Don, 1935, 90.
39. **Moutou, E. and Artois, M.**, Human influence on mammals populations through epidemic diseases in France, in Abstract of papers and posters 4th Int. Theriol. Congr. Suppl. list, session B, University of Alberta, Endmonton, Canada, 1985.
40. **Mykytowicz, R.**, An attenuated strain of the myxomatoses virus recovered from the field, *Nature*, 172, 448, 1953.
41. **Nezlin, R. S.**, *Structure and Biosynthesis of Antibodies*, Nauka Publishers, Moscow, 1972, chap. 3.
42. **Omelyanets, T. G.**, Sanitary and bacteriological evaluation of conditions of bacteriodencide production and application, in *Hygiene of Application and Toxicology of Pesticides and Medical Treatment of Poisonings*, Medicina, Moscow, 2, 1977, 52.
43. **Peghelman, S. G. and Bobovich, V. G.**, On the influence of physiological condition of rodents on virulence of bacteria, in *Proc. All-Union. Plant Protection Res. Inst.*, 12, Vaschnil, Leningrad, 161, 1955.
44. **Pinus, A. A.**, Viral tumors, in *Viral Diseases of Animals*, compiled by Orlov. D. M., Agricultural Literature Publishers, Moscow, 1963, 532.
45. **Polyakov, I. Ya.**, Bacteriological method of control, in *Harmful Rodents and their Control*, Agriculture Publishers, Leningrad-Mowcow, 1961, 250.
46. **Prokhorov, M. I.**, *Microbiological Control of Harmful Rodents*, Kolos Publishers, Leningrad, 1966, 3.
47. **Regnery, D. C. and Marschall, J. D.**, Studies on the epidemiology of myxomatosis in California. IV. The susceptibility of six leporid species to Californian myxoma virus and the relative infectivity of their tumors for mosquitoes, *Am. J. Epidemiol.*, 94, 508, 1971.
48. **Rottschild, M.**, Myxomatosis and the rabbit flea, *Nature*, 207, 1162, 1965.
49. **Ross, J. and Sanders, M. F.**, The development of genetic resistance to myxomatosis in wild rabbits *(Oryctolagus cuniculus)* in Britain, *J. Hyg.*, 92, 255, 1984.
50. **Sanarelli, G.**, Das Myxomatogene virus. Beitrag zum studium der krankheitserreger ausserhalb des Sichtbare, *Centrallbl. Bacter. Parasitenk. und Infektion.*, 23B, 865, 1898.
51. **Saulite, E. G.**, Myxomatosis of rabbits and its causing agent, in *Trans. Acad. Sci. Latv. SSR*, 6, 97, 1981.
52. **Siriez, H.**, *La myxomatose, Moyen de Lutte Biologighe contre le Lapin*, Nuisible, R., Ed., SEP, Paris, 1957, 87.
53. **Sobey, W. R.**, Selection for resistance to myxomatosis in domestic rabbits *Oryctolagus cuniculus., J. Hyg.*, 67, 743, 1969.
54. **Syurin, V. N. and Fomina, M. V.**, *Veterinary Virology*, Kolos Publishers, Moscow, 1975, 38.
55. **Thomas, J. A.**, The ecology and status of Thymelius acteon (Lepidoptera, Hesperiidae) in Britain (U.K.), *Ecol. Entomol.*, 8, 427, 1983.
56. **Tittensor, A. M. and Cowon, D. P.**, Extrinsic factors influencing rabbit populations in Britain, France and Spain, in Abstract of papers and posters, 4th Int. Theriol. Congr., Session 9, University of Alberta, Edmonton, Canada, 1985.
57. **Tittensor, A. M. and Trout, R. C.**, The effects of myxomatosis on rabbit populations in Britain, in Abstr. of papers and posters, 4th Int. Theriol. Congr., Session 9, University of Alberta, Edmonton, Canada, 1985.
58. **Trakhanov, D. F. and Kadirov, A. F.**, Studies on the susceptibility of piglets to Issatschenko bacteria, in *Proc. All-Union Inst. Veter. Sanit.*, 17, 152, 1972.
59. **Vashkov, V. I. and Gandelsman, B. I.**, Deratization, in *Data on Organization and Methods of Desinfection, Desinsection and Deratization*. Medghiz Publishers, Moscow, 1950, chap. 3.
60. **Zhdanov, V. M. and Gaidamovich, S. Ya.**, *Virology*, Meditsina Publishers, Moscow, 1966, chap. 10.

Chapter 24

NATURAL RESISTANCE OF PLANTS TO PEST RODENTS

Lennart Hansson

TABLE OF CONTENTS

I. INTRODUCTION

The natural resistance of plants is a mechanism commonly exploited in agriculture, horticulture, and forestry. It is mainly employed against microorganisms such as virus, bacteria, and fungi.[1] It has also been used practically against higher organisms, such as nematodes and insects.[2] Its employment includes selection for resistant varieties of crop species and breeding for resistance. Such breeding has not been performed against injurious vertebrates but there are many instances of variable plant susceptibility to grazing and browsing mammals, e.g., rodents. Such observations are usually referred to as demonstrations of natural defense and may have a great potential in practical pest management. However, the theoretical basis for natural defense has changed considerably during recent years. It will be briefly reviewed here together with actual observations of rodent deterrence. Finally, implications for rodent management are discussed.

II. THE CHANGING CONCEPT OF PLANT DEFENSE

The common image of plant defense relates probably to mechanical devices such as spines and hairs to deter grazing animals. Another older aspect of natural defense is a decreased content of nutritional substances, foremost protein, in exposed tissues, thereby affecting the preferences, or reproduction, of attacking animals. Chemical defense has also been well known but is now in the center of interest, and the various chemicals involved have been related to different plant strategies.

The defense chemicals, or "secondary compounds", have been divided into two classes, constituting, respectively, quantitative and qualitative defense. The former appears in exposed plants as vegetation dominants or long-lived species, e.g. trees, and the latter in rare or "fugitive" species, e.g., annuals.[3,4] The former is based on energetically expensive compounds appearing in great quantities within the plant cells, especially phenolics and terpenes.[5] Many of these compounds form complexes with proteins and therefore reduce digestibility when the cells are crushed. The latter group instead includes strongly poisonous substances in small amounts, e.g., alkaloids and cyanogenic glucosides, which may disrupt important metabolic processes or affect the nervous system of the animals. The latter types of substances are assumed to be cheaper to synthesize by the plants. They also contain, in contrast to phenolics and terpenes, nitrogen, and are assumed to be especially common in plants growing on productive soils, i.e., with a surplus of inorganic nitrogen.[6] Seeds also contain at times a qualitative defense with, e.g., uncommon amino acids and toxic lipids.[7,8] The proportions of phenolics and alkaloids in plant leaves also change during the growing season.[9]

All plants do not contain any obviously defensive chemicals. Plants growing from a protected meristem below the ground or inside the leaves, e.g., grasses, are assumed to rely more on rapid regrowth than chemical defense.[6,10] It has even been argued that grasses have evolved to be eaten, especially by mammals.[11]

The chemical defense does not necessarily remain at a constant concentration in the various plant parts, then being a constitute defense, but may change in response to animal attacks, i.e., an induced or facultative defense.[2] This applies to damage caused by both microbes and animals. The plant defense system should thus be fairly similar to the immunological system of animals. Such changes have been documented in Scandinavian birches against certain geometrid moths[12] and in North American birches and willows against snowshoe hares.[13] In the first case, the content of phenolics in the leaves increased after a defoliation (and even in response to larval excrements in the soil around the trees), and in the second case root-suckers with higher contents of phenolics developed after hard browsing by peak hare numbers. Furthermore, stems and branches in reach of hares and other browsing

Table 1
TWO EXAMPLES OF VOLE *(MICROTUS AGRESTIS)* DAMAGE TO PROVENANCE TRIALS WITH NORWAY SPRUCE *(PICEA ABIES)*[a]

	Trial 1[b]			Trial 2[c]		
Origin	Vole damaged	Undamaged	% damaged	Vole damaged	Undamaged	% damaged
Sweden	17	90	16	24	173	12
Bielorussia + Ukraine	11	52	17	4	21	16
Belgium + Germany	8	31	21	39	82	32
Sweden × Belgium/Germany	29	63	32	—	—	—

[a] Based on an inventory at Kil, Sweden (59°N) in 1984 after damage during the preceding winter.
[b] $X^2 = 8.08$; $p<0.05$.
[c] $X^2 = 19.49$; $p<0.001$.

vertebrates contain more defensive substances (mainly phenolics) than plant parts high up in the same trees.[14] Trees have even been found to sense gaseous compounds emitted by other trees attacked by animals and then rapidly to increase their levels of defensive chemicals.[15]

External factors may interfere with the defense system. Fertilization with nitrogen has been found to diminish the content of phenolics but also to weaken the lignification.[16,17] Nitrogen fertilization usually decreases resistance against mammalian herbivores in nutrient-limited woody plants[6] but may increase resistance in rapidly growing herbaceous species. The effect seems to depend on the actual plant carbon-nutrient balance.

Defense chemicals may also interfere directly with the reproduction of attacking animals. Thus, phenolic compounds function as reproductive inhibitors in the vole *Microtus montanus*.[18] Some substances retard the growth of young voles[19] and rats.[20] However, other compounds may also work as oestrogens, increasing reproduction in, e.g., small rodents.[21,22] In the latter case, originally defensive plant chemicals are probably secondarily used as cues for information on suitable environmental (food) conditions for the animals.

III. EVIDENCE OF PLANT DEFENSE AGAINST RODENTS AND OTHER MAMMALS

Genetically based resistance has been definitely demonstrated in Douglas fir against snowshoe hares and black-tailed deer.[23,24] The genetic factor was chiefly additive. In many other cases a genetic interpretation is strongly supported.

Most observations of varying preferences of small rodents within plant species concern woody plant. Varieties of apple trees were barked to very different extents by field voles[25] and central European provenances of Norway spruce introduced to Scandinavia were barked considerably more by this species than were indigeneous provenances.[26,27] An example from the latter tree is provided in Table 1. Separate Ponderosa pine trees have been attacked differently by pocket gophers[28] and squirrels.[29] In both cases the different palatability was related to the contents of monoterpene hydrocarbons, β-pinene + sabinene for the gophers and in the case of the squirrel predominantly α-pinene. Hares were found to select intermediate-sized birch twigs,[30] presumably due to especially high contents of phenolics in the most nutritious, youngest twigs.[31] They attacked Scots pine bark low in total phenol content.[32]

Bark of conifer branches, e.g., from Scots pine, grafted to an established root for rapid seed production is eaten to a considerably greater extent than bark of seedlings/young trees of the same size by voles[33] and hares.[34] Seedlings originated by cuttings from tree-form Radiata pines were attacked more by hares than were seedlings derived from Radiata pines kept juvenile by hedging.[35]

A qualitative poison was found and identified in crownvetch forage (a forb) when fed to voles.[36] There are also a few observations of differences in palatability of varieties of forage grasses. Thus both rabbits[37] and voles[38] discriminated between varieties of *Phalaris arundinacea,* obviously due to varying contents of alkaloids; while no such differences could be shown for *Festuca arundinacea* or *Lolium* × *Festuca* hybrids. Sheep were adversely affected by the alkaloids contents of a *Festuca* variety but voles were not. Terpenic substances from *Compositae* plants affected food choice, weight, and survival of *Clethrionomys* voles when given in an artificial diet.[39]

Fertilization, especially by nitrogen, increases the palatability of both woody and herbaceous plants to rodents and lagomorphs.[27,34,40,41] Both fertilization and spacing of Lodgepole pines increased the damage by snowshoe hares and squirrels.[42]

Deer selection of browse species has been generally related to the terpene contents of the foliage, e.g., in *Artemisia,*[43] *Juniperus,*[44,45] Douglas fir,[46] and Scots pine.[47] Oxygenated compounds showed greater effects than hydrocarbonic monoterpenes. The effects are due to antibiotic influences on rumen microorganisms.[48] These findings should not therefore be uncritically transferred to monogastric mammals such as rodents and lagomorphs. However, most findings are similar for the two groups of herbivores. Furthermore, deer prefer new growth of Douglas fir to old needles and nitrogen-fertilized trees were more fermentable than unfertilized trees.[48]

Both rodents[34] and deer[47,49] select plant parts with relatively low content of dry matter. This may be related to stage of growth of the plants and at present appears to be a disregarded factor in animal preferences.

IV. A CASE STUDY: LODGEPOLE PINE PROVENANCES IN SCANDINAVIA

The North American Lodgepole pine *(Pinus contorta)* has been introduced to Finland and Sweden as a fast-growing and productive species. It has been badly damaged by voles, especially the field vole. Seedlings emanating from seeds from various localities (provenances) in Canada and the U.S. have been planted randomly in field trials all over Sweden and in northern Finland for examinations of survival and growth.[50] Vole damage appears only in the northern part of this area,[51] but both during increase and peak years of the vole cycles.[52] A high proportion of the seedlings are then often killed by voles (e.g., 25 to 85% damaged stems in various trials in 1980 to 1982). Various provenances are damaged to very different extents[51,53] and generally northern provenances (from Yukon and north British Columbia) are less damaged than southern (e.g., coastal and the U.S.) provenances (Figure 1). Only the northern provenances are considered hardy enough for north Scandinavia and thus selection according to damage levels and productivity in forestry coincide. In Sweden there was a higher level of various phenolic substances in bark of damaged provenances some years after damage,[54] being possibly an induced affect, whereas in Finland[55] there were no clear differences in total phenol contents between damaged and undamaged provenances. Instead, certain resin acids (e.g., neoabietin and levopimar) were higher in the bark of less damaged provenances in Sweden[54] and they might have repelled the voles to some extent. However, this possible chemical defense was not perfect in view of the great injury still sustained.

V. THE POSSIBILITY OF USING NATURAL DEFENSE AGAINST RODENTS

Little practical work has been done to manage rodent populations or their damage by using defensive or resistant plant varieties. However, the possibilities will certainly vary between various objects of damage.

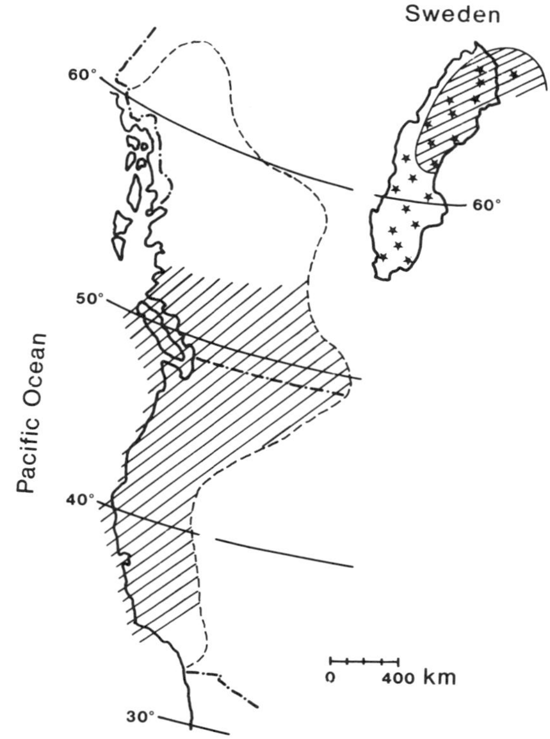

FIGURE 1. Vole damage to provenances of Lodgepole pine *(Pinus contorta)* introduced to Sweden and Finland. Provenance trials in these countries are shown by stars and areas with considerable vole damage (barking of stems) are hatched. The origin in North America totally and of the most damaged provenances is indicated by the broken line and the hatched area respectively.

Many or most natural plant defense chemicals effective against herbivorous mammals are also toxic to humans and their livestock.[2,36,38,56] Several severe diseases are caused by resistant plant species or varieties. Therefore, crop plants used as food for humans or livestock may generally be unsuitable for this type of manipulation. Grasses seem to have a too sporadical natural defense to be considered in this context. What remains are ornamental trees and herbaceous plants on one hand and forest trees on the other.

There are indications of a reverse relationship between natural resistance and biomass production, especially when nutrients are in short supply.[2,6] Biomass production is of little or no concern to ornamental plants but is vital for forestry plants. Central European spruces are planted in Scandinavia for their higher timber production and are attacked more by voles, which seems to fit this allocation scheme. However, in the case of Lodgepole pines the interests of rodents and humans differ and provenance selection related to vole attacks is possible. Selection for resistance against browsing mammals could also be made among superior phenotypes within a local race of Douglas fir.[24]

Various selection schemes will show a different economic ratio between rodent damage and production profit and only the local or regional outcome can decide when the balance is positive. This applies also to costs and benefits to be expected upon (nitrogen) fertilization. However, in the present context the emphasis generally has to be put on areas known to show a high level of tree damage from the rodents,[27,28] and where simple choices between provenances may be possible. For areas with low or occasional damage, large breeding programs may be needed to sort out especially repellent genotypes. It has also been suggested that various species, hybrids, or genotypes should be screened for especially defensive chemicals, e.g., salicin and its derivatives (phenolic substances) in willows and poplars against hares, rabbits, and opossums.[57]

REFERENCES

1. **Shaner, G.,** Genetic resistance for control of plant disease, in *Handbook of Pest Management in Agriculture,* Pimentel, D., Ed., CRC Press, Boca Raton, 1981, 395.
2. **Levin, D. A.,** The chemical defences of plants to pathogens and herbivores, *Annu. Rev. Ecol. Syst.,* 7, 121, 1976.
3. **Feeny, P.,** Plant apparency and chemical defence, in *Biochemical Interactions Between Plants and Insects,* Wallace, J. W. and Mansell, R. L., Eds., Plenum Press, New York, 1976, 1.
4. **Rhoades, D. F. and Cates, R. G.,** A general theory of plant antiherbivore chemistry, in *Biological Interactions Between Plants and Insects,* Wallace, J. W. and Mansell, R. L., Eds., Plenum Press, New York, 1976, 168.
5. **Matson, W. J.,** Herbivory in relation to plant nitrogen content, *Annu. Rev. Ecol. Syst.,* 11, 119, 1980.
6. **Bryant, J. P., Chapin, F. S., III, and Klein, D. R.,** Carbon/nutrient balance of boreal plant in relation to herbivory, *Oikos,* 40, 357, 1983.
7. **Bell, E. A.,** Toxic amino acids in the *Leguminosae,* in *Phytochemical Ecology,* Harborne, E. E., Ed., Academic Press, New York, 1972, 163.
8. **Janzen, D. H.,** Community structure of secondary compounds in plants, in *Chemistry in Evolution and Systematics,* Swain, T., Ed., Russak, New York, 1973, 529.
9. **Prudhomme, T. I.,** Carbon allocation to antiherbivore compounds in a deciduous and an evergreen subarctic shrub species, *Oikos,* 40, 344, 1983.
10. **Jung, H.-J. G., Batzli, G. O., and Seigler, D. S.,** Patterns in the phytochemistry of arctic plants, *Biochem. Syst. Ecol.,* 7, 203, 1979.
11. **Owen, D. F.,** How plants may benefit from the animals that eat them, *Oikos,* 35, 230, 1980.
12. **Haukioja, E.,** On the role of plant defences in the fluctuations of herbivore populations, *Oikos,* 35, 202, 1980.
13. **Bryant, J. P.,** Phytochemical deterrance of snowshoe hare browsing by adventitious shoots of four Alaskan trees, *Science,* 213, 889, 1981.
14. **Bryant, J. P. and Kuropat, P.,** Selection of winter forage by subarctic browsing vertebrates: the role of plant chemistry, *Annu. Rev. Ecol. Syst.,* 11, 161, 1980.
15. **Schultz, J. C.,** Coevolutionary implications of defensive communication among forest trees, Abstr. 3rd Eur. Ecol. Symp., August 22 to 26, 1983, Lund, Sweden.
16. **Király, Z.,** Plant disease resistance as influenced by biochemical effects of nutrients in fertilizers, *Fert. Use Plant Health,* 20, 233, 1976.
17. **Trolldenier, G. and Zehler, E.,** Relationship between plant nutrition and rice diseases, *Fert. Use Plant Health,* 20, 85, 1976.
18. **Berger, P., Sanders, E. H., Gardner, P. H., and Negus, N. C.,** Phenolic plant compounds functioning as reproduction inhibitors in *Microtus montanus, Science,* 195, 575, 1977.
19. **Jung, H. G. and Batzli, G. O.,** Nutritional ecology of microtine rodents: effects of plant extracts on the growth of arctic microtines, *J. Mammal.,* 62, 286, 1981.
20. **Jambernathan, R. and Mertz, E. T.,** Relationship between tannin levels, rat growth and distribution of proteins in sorghum, *J. Agric. Food. Chem.,* 21, 691, 1973.
21. **Labov, J. B.,** Phytoestrogens and mammalian reproduction, *Comp. Biochem. Physiol.,* 57A, 3, 1977.
22. **Sanders, E. H., Gardner, P. D., Berger, P. J. and Negus, N. C.,** 6-Methoxybenzoxazolinone: a plant derivative that stimulates reproduction in *Microtus montanus, Science,* 214, 67, 1981.
23. **Radwan, M. A.,** Differences between Douglas-fir genotypes in relation to browsing preference by black-tailed deer, *Can. J. For. Res.,* 2, 250, 1972.
24. **Dimock, E. J., Silen, R. E., and Allen, V. E.,** Genetic resistance in Douglas-fir to damage by snowshoe hare and black-tailed deer, *For. Sci.,* 22, 106, 1976.
25. **Myllymäki, A.,** Outbreaks and damage by field rodents and other harmful small mammals in Finland, *Ecol. Bull. Stockholm,* 19, 17, 1975.

26. **Christiansen, E.,** Field rodent damage in Norway, *Ecol. Bull. Stockholm,* 19, 37, 1975.
27. **Hansson, L.,** Natural repellence of plants towards small mammals, *Ecol. Bull. Stockholm,* 19, 213, 1975.
28. **Radwan, M. A., Crouch, G. L., Harrington, C. A., and Ellis, W. D.,** Terpenes of Ponderosa pine and feeding preferences by pocket gophers, *J. Chem. Ecol.,* 8, 241, 1982.
29. **Farentinos, R. C., Capretta, P. J., Kepner, R. E., and Littlefield, V. M.,** Selective herbivory in tassel-eared squirrels: role of monterpenes in Ponderosa pines chosen as feeding trees, *Science,* 213, 1273, 1981.
30. **Pehrson, Å,** Winter food consumption and digestibility in caged montain hares, in *World Lagomorph Conference,* Myers, K. and MacInnes, C. D., Eds., University of Guelph, Canada, 732, 1981.
31. **Palo, R. T., Pehrson, Å., and Knutsson, P.-G.,** Can birch phenolics be of importance in the defence against browsing vertebrates?, *Finn. Game Res.,* 41, 77, 1983.
32. **Rousi, A. and Häggman, J.,** Relationship between the total phenol content of Scots pines and browsing by the arctic hare, *Silvae Gen.,* 33, 95, 1984.
33. **Larsson, T.-B.,** Damage caused by small rodents in Sweden, *Ecol. Bull. Stockholm,* 19, 47, 1975.
34. **Rousi, M.,** Susceptibility of pine to mammalian herbivores in northern Finland, *Silva Fenn.,* 17, 301, 1983.
35. **Libby, W. J. and Hood, J. V.,** Juvenility in hedged Radiata pine, *Acta Hortic.,* 56, 91, 1976.
36. **Gustine, D. L., Shenk, J. S., Mayer, B. G., and Barnes, R. F.,** Isolation of β-nitropropionic acid from crownwetch, *Agron. J.,* 66, 636, 1974.
37. **Asay, K. H., Carlsson, I. T., and Wilsie, C. P.,** Genetic variability in forage yield crude protein percentage and palatability in reed grass, *Phalaris arundinacea* L., *Crop Sci.,* 8, 568, 1968.
38. **Kendall, W. A. and Sherwood, R. T.,** Palatability of leaves of tall fescue and reed canary grass and of some of their alkaloids to meadow voles, *Agron. J.,* 67, 667, 1975.
39. **Picman, J., Picman, A. K., and Towers, G. H. N.,** Effects of the sesquiterpene lactone, helenin, on feeding rates and survival of the tundra red-back vole *Clethrionomys rutilus, Biochem. Syst. Ecol.,* 10, 269, 1982.
40. **Miller, G. R.,** Evidence for selective feeding on fertilized plots by red grouse, hares and rabbits, *J. Wildl. Manage.,* 32, 849, 1968.
41. **Gessel, S. P. and Orians, G. H.,** Rodent damage to fertilized silver fir in western Washington, *Ecology,* 48, 649, 1967.
42. **Sullivan, T. P. and Sullivan, D. S.,** Influence of fertilization on feeding attacks to Lodgepole pine by snowshoe hares and red squirrels, *For. Chron.,* 263, 1982.
43. **Scholl, J. P., Kelsay, R. G., and Shafizadeh, F.,** Involvement of volatile compounds of *Artemisia* in browse preference by mule deer, *Biochem. Syst. Ecol.,* 5, 291, 1977.
44. **Schwartz, C. C., Nagy, J. G., and Regelin, W. L.,** Juniper oil yield, terpenoid concentration, and antimicrobial effects on deer, *J. Wildl. Manage.,* 44, 107, 1980.
45. **Schwartz, C. C., Regelin, W. L., and Nagy, J. G.,** Deer preference for juniper forage and volatile oil treated foods, *J. Wildl. Manage.,* 44, 114, 1980.
46. **Conolly, G. E., Ellison, B. O., Fleming, J. W., Geng, S., Kepner, R. E., Longhurst, W. M., Oh, J. H., and Russel, G. F.,** Deer browsing of Douglas fir trees in relation to volatile terpene composition and in vitro fermentability, *Forest Sci.,* 26, 179, 1980.
47. **Haukioja, E., Houpalahti, R., Kotiaho, J., and Nygrén, K.,** The kinds of pine preferred by moose, *Soumen Riista,* 30, 22, 1983.
48. **Oh, J. H., Sakai, T., Jones, M. B., and Longhurst, W. M.,** Effects of various essential oils isolated from Douglas-fir needles upon sheep and deer rumen microbial activity, *Appl. Microbiol.,* 15, 777, 1967.
49. **Radwan, M. A. and Crouch, G. L.,** Plant characteristics related to feeding preference by black-tailed deer, *J. Wildl. Manage.,* 38, 32, 1974.
50. **Lindgren, K.,** Provenances of *Pinus contorta* in northern Sweden, Ph. D. thesis, Agricultural University, Umeå, Sweden, 1983.
51. **Hansson, L.,** Damage by wildlife, especially small rodents, to North American *Pinus contorta* provenances introduced to Sweden, *Can. J. For. Res.,* 15, 1167, 1985.
52. **Hansson, L. and Lavsund, S.,** Damage by vertebrates to *Pinus contorta* plantations, *Sver. Skogsv. Förb. Tidskr.,* 80, 65, 1982.
53. **Rousi, M.,** Vole damage in tree species trials in northern Finland in the winter of 1981-82, *Folia For.,* 569, 1, 1983.
54. **Hansson, L., Gref, R., Lundgren, L., and Theander, O.,** Susceptibility to vole attacks due to bark phenols and terpenes in *Pinus contorta* provenances introduced into Sweden, *J. Chem. Ecol.,* 12, 1578, 1986.
55. **Rousi, M.,** personal communication, 1984.
56. **Shenk, J. S., Risius, M. L., and Barnes, R. F.,** Weanling meadow vole responses to crownvetch forage, *Agron. J.,* 66, 13, 1974.
57. **Markham, K. R.,** A chemotaxonomic approach to the selection of opposum resistant willows and poplars for use in soil conservation, *New Zealand J. Sci.,* 14, 179, 1971.

Chapter 25

CHEMICAL REPELLENTS

Adrian P. Meehan

TABLE OF CONTENTS

I. INTRODUCTION

There is little doubt that rodents are responsible for causing vast but unspecifiable damage to stored products (both in the "raw" and the packaged state), growing crops, and the fabric of buildings and structures. In addition they are known to be vectors in the spread of diseases to man, domestic stock, and wild animals. Most control methods involve the use of toxic compounds, but these have various attendant problems (for overview see Meehan[1]). A compound which is capable of repelling rodents from areas where they are not desired, or of protecting packaging and foodstuffs, but which is harmless and nonaversive to man would be of obvious interest.

In this account the term repellent is taken to mean a compound which can potentially repel by taste, although the primary effect in the author's opinion *should be* olfactory. If taste is the critical factor it may act too late to protect the structure or package as the rodent will need to gnaw, but not necessarily eat, before repellency occurs. Thus packaged goods may be destroyed, with consequent economic loss, before the animal is deterred. However, to the best of the author's knowledge there is no compound which repels by olfaction alone. Depending on the situation this definition may be changed so that gustation becomes the priority.[2]

Relatively little recent work has been carried out on repellents compared to other areas of rodent control. Most of the data presented here were collected in the 1950s and 1960s.

Earlier techniques for repelling rodents included the use of a deer skin on the floor or a circle of sawdust treated with creosote around the material to be protected.[36] Both of these were reputed to be effective! Even more unlikely was the use of hot-cross buns hung in barns to keep rats away.[4] However, many people, particularly in rural areas of the world, still believe in such folk-lore.

Early chemical repellents for rodents included sulphur, lime, lye, copperas, carbolic acid, kerosene, and oils of peppermint and wintergreen.[5] The active constituent of skunk odor, n-butyl mercaptan,[6] sodium fluosilicate,[7] and the bactericide, actidione[8,9] have also been suggested. For more details of the early work on rodent repellents the reader is referred to Welch.[10,11] Later work is reviewed by Drummond,[12] and Gutteridge[13] mentions many of the chemicals implicated as repellents.

II. COMPOUNDS TESTED AS REPELLENTS

Perhaps the most concerted effort to find a rodent repellent was carried out by the U.S. Army. Of more than 4600 compounds tested in 1 series of tests (involving addition to food), 123 were considered suitable for further testing.[14] In an attempt to determine structure/activity relationships it was found that acyclic and heterocyclic compounds containing tri- or pentavalent nitrogen groups were the most repellent. However there was no indication of repellency by olfaction.

Denver Wildlife Research Center staff have tested over 900 candidate compounds during the past 25 years and all the resulting data have been bought together in one publication.[15] More recently certain mercaptoketones have been evaluated as stress-inducing odorants in *Rattus norvegicus*.[16] It is claimed that the mercapto group is essential for an aversive effect and that certain of the compounds tested could play a role as area repellents. However, this last claim was not followed up. The subject of repellents can be conveniently divided into protection of (1) seeds and growing plants, (2) packaging, and (3) structure and fabric of buildings.

A. Protection of Seeds and Crops

Perhaps the major application of repellents applied to edible materials is the protection of seeds. To test possible repellents Schafer and Bowles[15] treated white wheat seeds (*Triticum aestivum*) and Douglas fir seeds (*Pseudotuga menziesii*) with 696 and 81 chemicals, respectively, at 2%. Known numbers of treated wheat seeds were made available to deer mice (*Peromyscus maniculatus*) and house mice (*Mus musculus*) in competition with unpoisoned laboratory diet for 3 days. A repellency index was established by counting numbers of treated grains eaten. Douglas fir seeds treated with the more promising chemicals were similarly evaluated. Other tests carried out involved longer feeding periods. Unfortunately the authors do not state which compounds, in their opinion, are worthy of further exploration although details of repellency are given for all the compounds tested.[15]

Two compounds which have received attention recently as seed protectants are methiocarb and thiram. The latter is commonly used as a fungicide and the former is an insecticide which is also used as a bird repellent. Corn (maize?) seed coated with thiram at 1.25% was found to repel the thirteen-lined ground squirrel (*Spermophilus tridecimlineatus*)[17] and apparently caused no phytotoxicity. Lower rates of application were ineffective. Methiocarb at 5.0 and 2.5% was repellent to the same species of squirrel but may have inhibited seed germination. Dry unplanted seeds were protected by 0.5% methiocarb but when planted little repellent/protective effect was seen. Preliminary laboratory trials also indicated that thiram at 0.31 and 1.25% and methiocarb at 0.031 and 0.5% were able to repel *Peromyscus maniculatus*.[18] Learned taste aversion apparently did not occur and untreated seeds continued to be taken.[19] Although both of these chemicals have potential as seed protectants their effect may be diminished by rain.[20]

Thiram has also been tested against *Rattus rattus mindanensis* and *R. argentiventer* in the Philippines. It was effective at reducing consumption of Benguet pine (*Pinus kesiya*) seeds when applied at concentrations down to 1.9%. Aldrin at 0.5% had a similar effect.[21] Thiram and ziram at 12% have also reduced damage to standing hard-woods in Canada by small populations of *Peromyscus pennsylvanicus* (meadow voles).[22] Crimidine and endrin have, however, failed to protect tree stems from voles when painted on.[23,24]

It seems that a common repellent for seeds is a mixture of "Arasan" (thiram) and endrin. Using this mixture as a standard, Weingartner and Cech[25] screened 27 nontoxic materials as rodent repellents on "field corn". Grafting wax, "Arasan" and red pepper (*Capsicum annum* var.) were all effective at repelling white footed mice (*Peromyscus sp.*), chipmunks (*Tamias striatus*), and grey squirrels (*Sciurus carolinensis*). Arasan-endrin and sodium selenate also performed well against the two former species (the latter was not tested). It was postulated that one or other of these five compounds might be effective at protecting northern red oak (*Quercus rubra*) acorns and wild black cherry (*Prunus serotina*) seeds.

Arasan-endrin mixtures are sometimes used in combination with a latex material but in this situation care has to be taken with the amount of latex. When applied to pine seeds greater amounts of latex allow more depredation by several rodent species and germination may also be decreased.[20]

Other compounds tested as seed repellents include R-55 (see later section), mestranol (a hormone), red squill (a rodenticide), and the rodenticide, alphanapthylthiourea (ANTU).[27] ANTU reduced seed consumption of bitterbrush (*Purskia tridentata*) by *Peromyscus maniculatus* significantly, but was not as effective as endrin.

Endrin has been mentioned several times in this section as being one of the better repellents when applied to seeds. However, it is extremely toxic to all wildlife and its use should be discouraged — it is in fact banned in many countries. Everett and Stevens[27] state that ANTU may be preferable, at least in Canada, but even this has its problems as it is a proven carcinogen. It has been banned from use as a rodenticide in some countries.

It is interesting, although perhaps not surprising, that most of the compounds tested and certainly most of the effective seed repellents are toxic in their own right. There is little question in the author's opinion that essentially nontoxic materials should be sought. If repellents are to be applied to building fabrics and packaging materials then this is a required property.

B. Protection of Packaging

A commonly used test to assess the effectiveness of repellents for use in packaging involves the placement of a treated barrier between the test animal and its food. Rodents will usually gnaw the treated barrier but will not necessarily taste the repellent because they possess a diastema through which nonedible materials can be ejected from the mouth without tasting. Thus if repellency is demonstrated it is likely to be by smell. It is the author's opinion that candidate materials for use on packaging can be better assessed by allowing rodents to attempt to penetrate the protected barrier. Nevertheless there is a claim that if a chemical repels rodents when given in food it will also be repellent when packaging is treated.[28] This has not always proven to be the case in the author's experience.[29]

The protection of packing material, whether it be paper, cardboard, or even wooden crates is of obvious concern to food and other producers. Once a commodity has been packaged it has a premium placed upon it and rodent damage at this stage becomes even more expensive. Thus it is not surprising that a number of studies to investigate repellents in this situation have been carried out — although most are not recent.[30-32] Attempts have also been made to incorporate repellents into the adhesives used in packing materials.[33]

Compounds tested include zinc dimethyl dithiocarbamate cyclohexylamine and tetraethyl thiamine sulphide.[31,33,34] Although these appear to be effective at keeping rats from packages there is some controversy over the best method of application. Certain trinitrobenzene compounds are claimed to be superior to the zinc dimethyl dithiocarbamate cyclohexamine complex when incorporated into packaging "film".[35] A trinitrobenzene aniline compound, which was apparently commercialized in the 1950s, is also claimed to be an effective rat repellent on asbestos and paper.[36]

Little research has been conducted on species other than *R. norvegicus* but asphalt emulsions and copper carbonate as well as certain wood preservatives containing metallic napthanates have been claimed to prevent squirrels gnawing on wood.[37]

DRC 5593, a compound similar to the irritant capsaicin, is effective in repelling *Mus musculus* when applied to burlap bags.[38] Malathion in a high viscosity oil can also deter *R. rattus* from treated burlap.[39] Khristov and Mateva[41] have described tests with "polymers" impregnated with many different potential repellents with varying results. It appears that the best results were obtained with polyvinylchloride and polyamide containing finely ground quartz glass. These materials showed at least a partial repellency to black rats, wood mice, and house mice.

Other compounds examined in the author's laboratory, and claimed to have significant repellency to rats, are Latakia tobacco and "Rodentifuge". Both failed to prevent rats feeding at treated hoppers as opposed to untreated ones.[29,40]

Perhaps the compound most thoroughly examined as a repellent for any application is Rotran R55. This compound, which to the best of the author's knowledge is no longer available, was manufactured by Philips Petroleum Company. Chemically it is tertiary butylsulfenyldimethyldithiocarbamate.

When wild brown rats and house mice are exposed to bags treated with 1% R55 and control bags treated with kerosene they tended to damage less of the R55 treated bags, but for some unknown reason the compound was less successful against albino and black rats.[42] In tests conducted in the author's laboratories, involving R55 emulsion or dust applied to food hoppers, the amount of take from the treated hoppers decreased somewhat in the first few days but increased with time. When incorporated into food strong repellency was shown, indicating that oral contact is necessary. In spite of this apparent repellency R55 cannot be used as a seed protectant as it significantly decreases germination, at least of Douglas fir and Western hemlock.[43]

C. Protection of Buildings

The results of unpublished field trials with R55 sprayed into rat burrows and harborage areas around grain silos showed that after treatment there was no significant level of infestation in a previously heavily populated area.[44] Similarly, R55 used in a food warehouse in the U.S. was claimed to have kept the area rodent free for at least 8 months after application to a ''Celotex'' wall.[45] Paul and Perti[46] also suggest that R55 may be a useful repellent when applied as a lacquer to PVC panelling. However details are lacking.

The fumigant, chloropicrin, has also been suggested as an area repellent.[47] It is no doubt effective against house mice but great care would be required in its use.

One of the most common questions asked the author — ''Is there a rat repellent I can put into my cables?''

There is a great need for an effective repellent for inclusion in cables, plastic pipes, and similar building materials. However, nothing suitable exists and Deighton[48] suggests that proofing and correct routing will be of more use.

III. CONCLUSIONS

This review is brief — more examples of candidate repellents exist, but very few have ever been commercialized or even taken past the first development stage. In fact no *effective* rodent repellent is currently available commercially, with the possible exception of some seed protectants, and these still suffer from many disadvantages. Not least, they tend to be toxic in their own right and perhaps are more likely to act as rodenticides than repellents. Similarly, repellents for the other uses mentioned are either too toxic (e.g., actidione, chloropicrin) or too aversive (e.g., R55) to humans and domestic animals to have any practical applications.

A glance at the references listed at the end of this section shows that much of the work outlined here is old. The author may have inadvertently missed some more recent relevant works (although a number of computer searches have been carried out), but there is no question that the amount of published work on rodent repellents has significantly decreased in recent years. None the less there is a definite need for repellents, particularly with the growing pressure against the use of toxic rodenticides and especially in the food industry. In Sweden some years ago it was stated that more attention was being paid to repellents than rodenticides.[49]

Should any reader wish to take up the challenge of developing rodent repellents then Salmon and Marsh[50] have produced a good review of some of the techniques available.

At a recent symposium an experienced rodent controller noted that '' . . . better chemical repellents are certainly needed''.[51] Perhaps the answer lies in repellent pheromones[52] or other ''natural'' chemicals; it is unlikely to lie in those compounds already evaluated.

REFERENCES

1. **Meehan, A. P.,** *Rats and Mice: Their Biology and Control,* Rentokil Ltd., East Grinstead, England, 1984.
2. **Rogers, J. G.,** Repellents to protect crops from vertebrate pests: some considerations for their use and development, in *Flavour Chemistry of Animal Foods,* ACS Symposium Series 67, Bullard, R. W. Ed., American Chemical Society, Washington, D.C., 1978, 150.
3. **Shorten, M.,** The reaction of the brown rat towards changes in its environment, in *Control of Rats and Mice,* Vol. 2, Clarendon Press, Oxford, 1954, 307.
4. **Baker, M.,** *Folklore and Customs of Rural England,* 4th ed., David & Charles, Newton Abbot, England, 1978, 54.
5. **Brians, F.,** Discussions on rodents — rat and mice control, *Exterm. Log,* 6(3), 15, 1938.

6. **Ford, L. A. and Clausen, D. F.,** Rodent repellent tests: n-butyl mercaptan against white and wild rats, *Ind. Eng. Chem. News Edn.* 19, 783, 1941.
7. Anonymous, Many new uses for sodium fluorosilicate, *Forecast,* June, 190, 1949.
8. **Traub, R., Dewitt, S., Welch, J., and Newman, D.,** Toxicity and repellency to rats of actidione, *J. Am. Pharm Assoc.,* 39(10), 552, 1950.
9. **Kuwabara, M., Shiraishi, A., and Tateda, H.,** New method for testing rat repellents, *Appl. Ent. Zool.,* 5(2), 84, 1970.
10. **Welch, J. F.,** A review of chemical repellents for rodents, *J. Agric. Food Chem.,* 2, 142, 1954.
11. **Nigam, B. S., Perti, S. L., and Agarwal, P. N.,** Susceptibility of packaging materials to insects and rodents. *Labdev J. Sci. Tech.,* 7-B(2), 89, 1969.
12. **Drummond, D. C.,** Repellents and attractants and their role in the control of rodents. Proc. Int. Symp. Bionomics and Control of Rodents, Kanpur, September 29 to Oct. 2, 1968, 101.
13. **Gutteridge, N. J. A.,** Chemicals in rodent control, *Chem. Soc. Rev.* 1(3), 381, 1972.
14. **Bowles, W. A., Dewitt, J. B., Adomaitis, V. A., and Pratt, J. J.,** Relationships between chemical structure and rat repellency. II. Compounds screened between 1950 and 1960. Report No. 75-11-FE1, U.S. Army Natick Laboratories, Mass. 1974.
15. **Schafer, E. W. and Bowles, W. A.,** Acute oral toxicity and repellency of 933 chemicals to house and deer mice, *Arch. Environ. Contam. Toxicol.,* 14, 111, 1985.
16. **Vernet-Maury, E., Polar, E. H., and Demael, A.,** Structure-activity relationships of stress-inducing odorants in the rat, *J. Chem. Ecol.* 10(7), 1007, 1984.
17. **Zurcher, N. J., Johnson, R. J., and Timm, R. M.,** Methiocarb and thiram as corn seed repellents for thirteen-lined ground squirrels. *Wildl. Soc. Bull.,* 11, 38, 1983.
18. **Johnson, R. J.,** Repellents for rodents in conservation-tillage agriculture, presented at Second Eastern Wildlife Damage Control Workshop, North Carolina State University, Raleigh, N.C., Sept. 22 to 85, 1985.
19. **Johnson, R. J. and Holm, K. E.,** Rodents in conservation-tillage fields: net loss or gain? in *Conservation Tillage for Row Crop Production,* Institute of Agriculture and Natural Resources, University of Nebraska, Lincoln, Tillage Conference Proceedings, 4, 161, 1985.
20. **Koehler, A. E. and Johnson, R. J.,** Rodent repellents for conservation agriculture, *Proc. Nebraska Academy of Sciences,* Nebraska Wesleyan University, Lincoln, Neb. 1983, 11.
21. **Dalmacio, M. V.,** Chemical protectants for Benguet pine seeds against rats, *Proc. Symp. Forest Pests and Diseases in Southeast Asia,* Bogor, Indonesia. April 20 to 23, 1976, BIOTROP Spec. Pub. No. 2.
22. **Martell, A. M.,** Protection of hardwood plantations from rodent damage with chemical repellents, *Acta Zool. Fenn.,* 173, 185, 1985.
23. **Myllymäki, A.,** Control of field rodents, in *Small Mammals: Their Productivity and Population Dynamics,* Gulley, F. B., Petrusewicz, K., and Ryszkowski, L., Cambridge University Press, London, 1975, chap. 14.
24. **Myllymäki, A.,** Outbreaks and damage by the field vole, *Microtus agrestis* (L.) since World War II in Europe, *EPPO Bull.,* 7, 177, 1977.
25. **Weingartner, D. H. and Cech, F. C.,** Screening materials for rodent repellency, *Proc. W. Va. Acad. Sci.,* 46(1), 27, 1974.
26. **Campbell, T. E.,** Increasing latex in pine seed coating impairs repellency and germination, *South J. Appl. For.,* 3(2), 61, 1979.
27. **Everett, R. L. and Stevens, R.,** Deer mouse *(Peromyscus maniculatus)* consumption of bitterbrush *(Purshia tridentata)* seed treated with 4 repellents, *J. Range Manage.,* 34(5), 393, 1981.
28. **Bellack, E. and Dewitt, J. B.,** Rodent repellent studies. III. Advanced studies in the evaluation of chemical repellents, *J. Am. Pharm. Assoc.,* 39, 197, 1950.
29. **Meehan, A. P. and Cole, S. G.,** Attempts to influence the feeding behaviour of the brown rat, *Int. Pest Cont.,* 16(1), 5, 1974.
30. **Tigner, J. R.,** Chemically treated multiwall tarps and bags tested for rat repellency, *J. Wildl. Mgmt.,* 30(1), 180, 1966.
31. **Welch, J. F., Dewitt, J. B., and Bellack, E.,** Rat deterrents for paper packages, *Soap Sanit. Chem.,* 26(4), 122, 1950.
32. **Welch, J. F., Dewitt, J. B., and Bellack, E.,** Rodent deterrents. Part II, *Soap. Sanit. Chem.,* 26(5), 147, 1950.
33. **Welch, J. F.,** Rat repellent findings, *Modern Packaging Manuf.,* 24, 138, 1951.
34. **Dewitt, J. B., Welch, J. F., and Bellack, E.,** Rodent repellency, *Modern Packaging* 23(9), 123, 1950.
35. **Welch, J. F. and Duggan, E. W.,** Rodent resistant vinyl films, *Modern Packaging,* 25(6), 130, 1952.
36. **Oderkirky, G. C.,** Do repellents stop animal damage, *Pest Cont.,* 23(12), 9, 1955.
37. **Jones, J. C.,** Squirrel problems and what to do about them, *Pest Cont.,* 29(8), 18, 1961.
38. **La Voie, K.,** Personal communications, 1978.

39. **Majumder, S. K., Krishnakumari, M. K., and Krishna, Rao, J. K.,** Malathion as a repellent for rats, *Curr. Sci.*, 33(7), 212, 1964.
40. **Bull, J. O.,** The influence of attractants and repellents on the feeding behaviour of *Rattus norvegicus,* Proc. 5th Vert. Pest. Conf., Fresno, Calif., March 7 to 9, 1972, 154.
41. **Khristov, L. and Mateva, M.,** Study on the repellent properties of polymers towards rodents, *Ekologiya (Sofia),* 2 (Abstr.), 61, 1976 (In Bulgarian).
42. **Khrishnakumari, M. K. and Jackson, W. B.,** Laboratory evaluation of Rotran (R55), an organic rodent repellent against some rodents, *Pest Control,* 41(12), 22, 1973.
43. **Edwards, D. G. W. and Olsen, P. E.,** R-55 Rodent repellent — Effect on germination in Douglas fir and western hemlock *Can. J. For. Res.*, 2, 256, 1972.
44. **Stansbury, R. E.,** Use of R-55 rodent repellent to move rodents from harbourage areas, Unpublished mimeo, Phillips Petroleum Company, Bartlesville, Ok., January 1969.
45. **Mace, E. P.,** Testing of R-55 against rats in food warehouse, Unpublished mimeo, Sur-Kil Chemical Co., Tulsa, Oklahoma, 1968.
46. **Paul, C. F. and Perti, S. L.,** Role of repellents in rodent control, Proc. *Int. Symp. Bionomics Control Rodents,* Kanpur, September 29 to October 2, 1968, 105.
47. **Tigner, J. R. and Bowles, W. A.,** Chloropicrin tested as an area repellent for house mice, *J. Wildl. Manage.*, 28(4), 748, 1964.
48. **Deighton, J. M.,** Damage to telephone cables and other transmission lines by pests, and means of preventing such damage, *Pest. Abstr. Sect. A,* 10, 445, 1964.
49. **Stenmark, A.,** Försök med avskräck ningsmedel mot sork, *Växtskyddsnotiser,* 34, 82, 1970.
50. **Salmon, T. P. and Marsh, R. E.,** Equipment and methods available for measuring the behavioural response of rodents to odors, in *Test Methods for Vertebrate Pest Control & Management Materials,* ASTM STP 625, Jackson, W. B. and Marsh, R. E., Eds., American Society for Testing & Materials, Philadelphia, Pa., 1977, 86.
51. **Howard, W. E.,** Research needs in rat control, Proc. 1st Symp. Recent Advances Rodent Control, Kuwait, February 8 to 11, 1982, 209.
52. **Christiansen, E. and Døving, K.,** Kommunikasjon has smagnagere viser nye veier til bekjempelse, *Tidsskr. Skogsbruk.*, 81, 293, 1973.

Chapter 26

ULTRASOUND DEVICES

M. Lund

TABLE OF CONTENTS

I. INTRODUCTION

For the last 20 years a method of rodent-proofing of premises has become increasingly common, especially in warehouses, food-processing industries, grain stores, etc. Instead of creating rodent-proof structures, electric devices emitting high-frequency sounds (i.e., frequencies ranging from 15 to 40 KHz) are installed in order to drive out already established rodents, or discourage others from invading the premises. A newer concept is that the devices would at least divert the rodents from their normal food supply to poison baits. The interest in such rodent control devices is elucidated by the fact that in U.S. alone estimated consumer expenditures on ultrasonics has increased from $75,000 in 1978 to $17 million in 1982.[6]

The method seems to have been developed on the basic observations that high-intensity, high-frequency sounds can cause harm to rodents, white mice being killed by just 1 min exposure to a sound field with a frequency of about 20 kHz and an acoustic level between 160 and 165 dB.[5] This is an extremely high intensity, however, also hazardous to man and other animals, and the devices marked for rodent control usually do not produce sounds with an intensity of more than 130 dB at a distance of 1 m from the loud speaker. It is a well-known fact that rodents can perceive sounds with frequences far above the hearing capacity of man, and even above that of a dog or a cat.[7] Rodents communicate to a considerable degree at frequences around 40 to 50 kHz, occasionally up to 100 kHz, and rats are known to produce ultrasounds as a method of echolocation.[8]

That high-frequency sounds at moderate intensities should in general provoke a flight response in rodents has apparently never been reported, specific distress calls may however elicit such a response in rats.[1]

From theoretical point of view the efficacy of such ultrasound devices can be doubted for several reasons:

1. Commensal rodents, especially rats, are highly adaptable and capable of habituating themselves to extremely different ecological situations, including noisy ones, such as mills and airports, or close to highways.
2. Rodents subjected to auditory stimuli at regular intervals are known to habituate quickly. They may show various initial responses but they all wane after 5 min to 2 days.[9]
3. High-frequency sounds are more directional than lower and audible frequencies, and they do not readily reflect around corners or solid obstacles. The higher the frequency the shorter the effective distance and the lower the probability that they can penetrate to rodent burrows and nests.[11]
4. If ultrasound devices are as harmful to rodents as claimed, they are likely to be so also to other mammals including man, pets, and domestic animals. If installed in a pigsty or stable they might reduce growth rate or milk production causing economic losses far beyond those inflicted by the mere presence of rodents.
5. Sound intensities above 110 to 120 dB are considered harmful, or at least unpleasant to man, if exposed for longer periods, and in numerous countries use of ultrasound devices is not permitted during working hours.

II. REPORTED EFFICACY OF ULTRASOUND DEVICES

Very few studies on the efficacy of ultrasound devices in rodent control are available from independent sources, and most are negative in their conclusions.

Based on laboratory tests Sprock et al.[11] state that continuous sounds, irrespective of frequency, are ineffective in deterring wild rats from entering and foraging in interconnected boxes; intermittent sounds, however, were found more effective. In-field tests ultrasonics did not repel rats or mice from any of the tested area.[11] Greaves et al.[2] conclude from laboratory tests that ultrasonic devices may prevent or reduce *invasion* of premises with few entry points.

In a large outdoor pen Meehan[5] tested four ultrasonic devices with largely negative results. Apart from an initial slight repellency of varying duration they failed to influence the behavior of the rats for any length of time. One device claimed to be effective over an area of 330 m^2 was not able to cover 1 m^2.[4]

Tests carried out at the Danish Pest Infestation Laboratory with 11 different devices, including such with varying frequencies and random intervals between signals, all indicate that apart from an initial repellency lasting from 30 min to 3 hr, no durable influence on rat behavior can be achieved. It is concluded that a maximum time of habituation of 3 hr in a situation where the rats had no particular motivation for venturing into the sound field, and where the sound waves could easily hit the target without being stopped by solid obstacles, was such a weak reaction that a practical effect in warehouses, in storage rooms, etc. seems to be out of question. It is further argued that the claim of some producers of ultrasound equipment that it should be used in combination with more conventional methods, i.e., traps and poison baits placed in the ''sound shadow areas'' is not valid. Rodents usually feed in these particular hidden areas where they feel safe, and the ultrasound devices in no way can change the situation.[3,4]

Recently the Denver Wildlife Research Center, has developed controlled laboratory and field protocols to assess the repellent efficacy of ultrasonic devices and in agreement with the U.S. Environmental Protection Agency (EPA) it is intended to evaluate the many different types of equipment marketed in the U.S.[10]

REFERENCES

1. **Frings, H. and Frings, M.,** *Animal Communication,* Blaisdell, New York, 1964, 204.
2. **Greaves, J. H. and Rowe, F. P.,** Responses of confined rodent populations to an ultrasound generator, *J. Wildl. Manage.,* 33(2), 409, 1969.
3. **Lund, M.,** Ultrasound disputed, *Pest Control,* 52(12), 16, 1984.
4. **Lund, M. and Lodal, J.,** Current work at the Danish Pest Infestation Laboratory, in *Proc. Conf. Organisation and Practice of Vertebrate Pest Control,* Dubock, A. C., Ed., ICI Plant Protection, Surrey, England, 1984, 253.
5. **Meehan, A. P.,** Attempts to influence the feeding behaviour of brown rats using ultrasonic noise generators, *Int. Pest Control,* 4, 12, 1976.
6. **Mix, J.,** Researchers dubunk controlling insects with ultrasonic devices, *Pest Control,* 52(2), 26, 1984.
7. **Prosser, C. L. and Brown, F. A.,** *Comparative Animal Physiology,* 2nd ed., W. B. Saunders Co., Philadelphia, Pa., 1961, 688.
8. **Riley, D. A. and Rosenzweig, M. R.,** Echolocation in rats, *J. Comp. Physiol. Psychol.,* 50(4), 323, 1957.
9. **Scourse, N. J. S. and Hinde, R. A.,** Habituation to auditory stimuli in mice, *Behaviour,* XLVII, 1, 1972.
10. **Shumake, S. A., LaVoie, G. K., and Crane, K.,** Efficacy test protocols for evaluation of ultrasonic rodent repellent devices, Proc. 11th Vertebrate Pest Conf., Sacramento, Calif., March 6 to 8, 1984, 85.
11. **Sprock, M., Howard, W. E., and Jacob, F. C.,** Sound as a deterrent to rats and mice, *J. Wildl. Manage.,* 31(4), 729, 1967.

Chapter 27

PREDATION

Sam Erlinge and Lennart Hansson

TABLE OF CONTENTS

I. INTRODUCTION

How much can predation help in rodent pest management? Are some of the problems of rodent pest populations consequences of a disturbed balance earlier achieved by predator-prey interactions? Do severe outbreaks occur in environments where predators are absent or greatly reduced? Are outbreaks absent in more natural, undisturbed environments with abundant predators? If so, what are the conditions under which predation can efficiently keep rodent populations under control? These are some of the questions encountered when discussing predation and rodent pest management.

II. PREDICTIONS FROM ECOLOGICAL THEORY

In theory, several attempts have been made to model predator-prey interactions in order to predict how predator and prey populations behave numerically.[42,43,67] Numerical changes over time were first modeled by means of differential equations by Lotka[36] and Volterra.[73] The growth rate of the prey population was considered to be determined by two factors; one includes the reproductive rate of each prey animal multiplied by the number of individuals in the population, causing an increase in numbers. The other factor represents losses of prey animals due to predation; the extent of predation was considered as a product of prey and predator densities and a factor that denotes the predators' efficiency at catching prey. Similarly, the growth of the predator population was presented by an equation including one factor increasing predator numbers; this factor contains the predation rate multiplied by a factor measuring the predator's efficiency at transforming prey animals into new predator individuals. The growth of the predator population was reduced by a given mortality rate in the predator population.

What predictions does this model yield? Graphic analysis suggests that predator and prey interactions result in an oscillating system; predator and prey populations continuously fluctuate out of phase with each other (Figure 1).

If a delay is built into this model, i.e., if the response of the predator population has a time lag, the system shows increased instability, the fluctuations increase in amplitude and one or both populations eventually become extinct.

III. EMPIRICAL EVIDENCE

Do predator-prey systems behave as predicted from the Lotka-Volterra equations? Tests with simple predator-prey systems in the laboratory have shown that predicted stable predator-prey oscillations occur only under certain conditions, i.e., when increased complexity of the system is built in, e.g., by periodic immigration of predators and prey[12] or by providing good hiding places for prey.[23] In very simple systems one of the species easily disappears. In nature, some rodent (Rodentia and Lagomorpha) populations perform regular oscillations (cycles) with 3 to 4 years between peak densities.[32,33] Also, predator numbers fluctuate synchronously with the rodents.[11,37,45,49,52] Is this cyclic pattern created by interactions between the predators and their prey? This was examined on a vole *(Microtus californicus)* population.[46-48] The extent of vole predation was measured during different phases of the vole population cycle. The highest predation rate (percentage of the vole population eaten by the predators) occurred during the decline and low phase of the cycle and the lowest predation rate during the increase phase. The conclusion was that predation had a decisive influence on vole numbers during the low phase but little influence during the increase, and that predation did not initiate the decline in the vole population. Some other studies on cyclic microtine populations have provided similar results.[10,40,41,51,56] Predation presumably increases the amplitude and influences the timing of a microtine cycle but predation does not

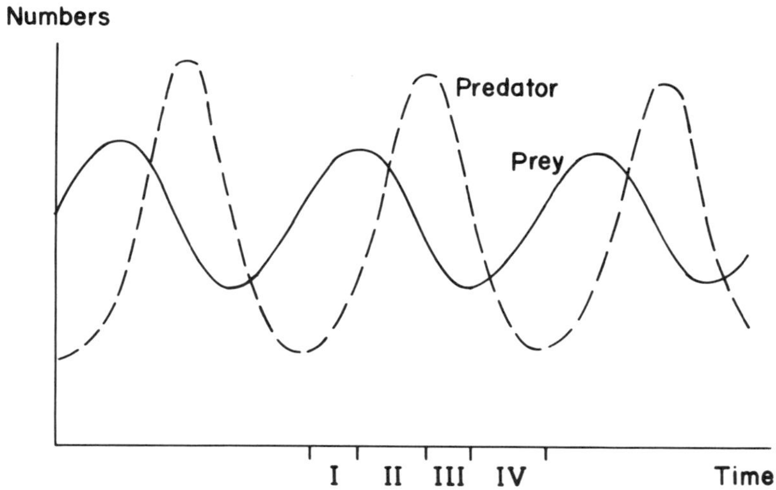

FIGURE 1. Changes in numbers with time of two interacting predator and prey species according to the Lotka-Volterra equations. (I) Both predator and prey increase in numbers; (II) prey decreases, predator goes on increasing; (III) both decrease; (IV) prey increases and predator decreases.

create the cycle. Other authors, however, suggest that the dynamics of microtine rodents and their predators are more closely interwoven throughout a cycle[68] and different kinds of predators could play the main role during different phases; avian predators truncating the top of peak numbers and mammalian predators being responsible for the great decline in numbers.[41]

Other rodent populations do not show any between-year cyclicity but fluctuate annually with peaks at the end of the breeding season.[14,31] Could this difference in rodent dynamics be explained by different influence of predation? Before reviewing studies designed to answer that question, we will return to some theoretical considerations.

IV. CONDITIONS PROMOTING STABILITY IN A PREDATOR-PREY SYSTEM

Graphical analysis illustrates that fluctuations in a predator-prey system according to the Lotka-Volterra equations can be reduced under certain conditions.[54] If predators have a reduced prey-capture efficiency, whereby prey are taken mainly when they are abundant or easily available, and if predators are able to switch to alternative food when prey becomes scarce or difficult to catch, the system will move towards stability. Also, if the number of predators is limited by some other factor than prey densities, e.g., by territorial behavior, this will have a similar effect.

Do any data support these theoretical predictions? In southern Sweden, the influence of predation on a field vole population dynamics was examined. This vole population showed few between-year variations in numbers. Instead, densities varied seasonally with peaks in autumn and lows in spring and early summer every year.[16] Thus, in an annual perspective this population showed a high degree of stability. Was this caused by predation? The annual vole consumption by avian and mammalian predators (eight species) was estimated from data on numbers of predators, their food requirements, and the proportion of voles in their diet. These estimates suggested that the annual vole consumption equaled the estimated annual production of voles in the area.[8] Also, the decline in vole numbers from autumn to spring (nonreproductive season) could be assigned to the predators' vole consumption during the same period. These and other data suggested that the noncyclic pattern in this population

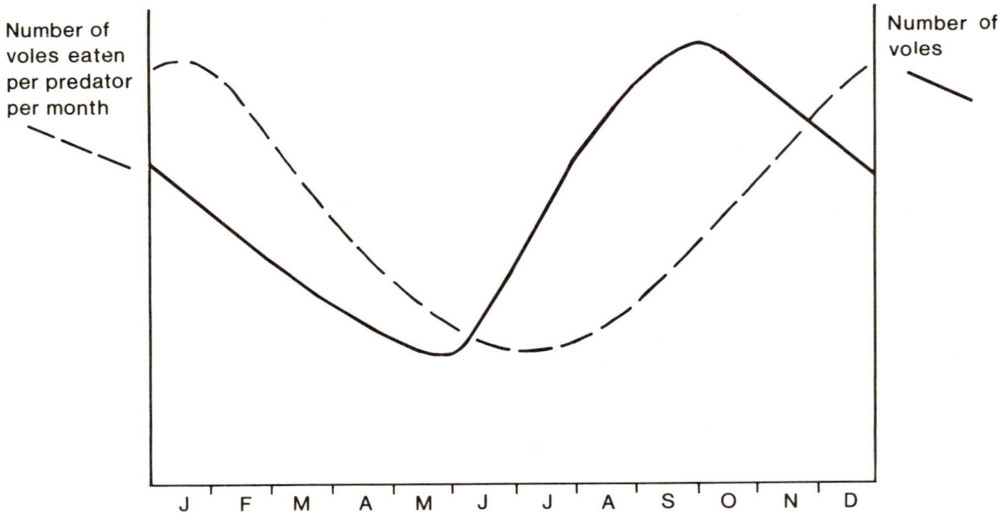

FIGURE 2. Functional response of the total assemblage of predators in the Revinge area (broken line) in relation to seasonal changes in vole numbers (continuous line). The functional response was measured as the mean number of field voles eaten by each predator during each month. (From Erlinge, S., Göransson, G., Hansson, L., Högstedt, G., Liberg, O., Nilsson, I. N., Nilsson, T., von Schantz, T., and Sylvén, M., *Oikos*, 40, 36, 1983. With permission.)

was caused by predation. Has this predator-prey system any characteristics which would promote stability? Firstly, have the predators in this system a relatively low prey-capture efficiency and do they catch the prey mainly when abundant or easily available? In the system examined, three predators were important, the common buzzard *(Buteo buteo)*, red fox *(Vulpes vulpes)*, and domestic cat *(Felis catus)*. They accounted for about 70% of the total annual vole consumption. These predators had a mixed diet; voles were important food during autumn, winter, and spring when voles were abundant or easily available due to reduced vegetation cover (Figure 2). When voles were scarce, these predators switched to other prey, especially young rabbits. Thus, the two prerequisites suggested for stability (low prey-capture efficiency and access to alternative food) were found in this system. Also, the common buzzard and red fox have a territorial organization at breeding time which probably limits population sizes.

High predation pressure has been observed in other studies on noncyclic rodents. In a Polish deciduous forest, the rodents declined markedly during the nonbreeding season every year. The impact by predators was high and estimated at a minimum of 70% of the losses.[55] Irregular fluctuations were observed in rodent populations (the bank vole, *Clethrionomys glareolus*, and wood mouse *Apodemus sylvaticus*), in a deciduous woodland in England.[58] A high predation rate was estimated for each of the two important predators, the tawny owl, *Strix aluco* (20 to 30% of the standing crop in any 2-month period)[58] and the weasel, *Mustela nivalis* (8 to 10% per month on each of the two rodents populations).[28] Both predators had small rodents as their favorite food but they changed to other prey, primarily earth worms (tawny owls), bird eggs, and nestlings (weasels) when small rodents became scarce. The highest predation rate probably occurred when rodents were scarce.[58]

Experimental studies where predation has been manipulated provide more conclusive evidence. In such studies the decline in spring in temperate areas which seems to be a characteristic feature in noncyclic rodent populations[57,64,66] is significantly lower in manipulated areas than on controls where rodents are exposed to predation.[64,65] Especially females have higher survival in predation-reduced areas, resulting in higher densities at the beginning of the breeding season.

V. GENERAL CONCLUSIONS REGARDING PREDATOR IMPACT

Predation is probably an important agent influencing dynamics of both cyclic and noncyclic rodent populations. The interactions, however, are performed differently in the two cases. In cyclic populations the reactions of predators to changing rodent numbers are primarily numerical and predator numbers are closely related to the densities of their prey. The numerical response generally has a considerable time delay and predation rate will be inversely density-dependent, i.e., high predation rate occurs at low prey densities and low rate during increasing and high densities. A pronounced functional response of predators (prey consumption rate by each individual predator as a reaction to varying prey densities) has sometimes been recorded on cyclic microtine populations.[56] This, however, has not changed the basic pattern described. Thus in nature, predation tends to increase the amplitude of the fluctuations by depressing rodent densities during the low phase. In such cases, predation has a destabilizing effect. This is, however, not to say that predators do not reduce rodents numbers at high densities. Estimates from studies on eruptive rodent population show that predators consume a substantial number of rodents at peak densities and this might reduce their damage.[56]

In predator-prey systems examined with noncyclic microtine populations, the response of the predators to varying prey densities is mainly functional, i.e., individual predators increase their vole consumption at high vole numbers (Figure 2). This response has a short time delay. On the other hand, the numerical response by the predators is weak and numbers of predators are little influenced by changeable vole numbers. This requires that alternative food is available to sustain predators during periods of rodent scarcity. If a rich supply of alternative prey maintains dense predator populations the effect of predation on a rodent population may be sufficiently substantial to prevent outbreaks and between-year cyclicity.[9] A characteristic feature of noncyclic microtine populations seems to be a marked spring decline causing low densities at the beginning of the breeding season each year. There is good evidence that this spring decline is — at least in some systems — caused by predation.

The answers to the questions posed at the beginning of this paper will be: Predators probably can contribute importantly in rodents pest management, their effect, however, will depend on several factors. Outbreaks in terms of peak numbers occurs in some natural ecosystems but not in others. Evidence suggests that long-term stability (noncyclic pattern) could be achieved by predation; the main prerequisite seems to be that predators should not be depending on rodents but sustained by alternative prey when rodents are scarce. By switching to small rodents when they increase in numbers, predators could keep rodents under control.

VI. POPULATION OUTBREAKS AND PREDATION

If predation is an important factor in limiting rodent numbers then outbreaks and damage levels of small rodents should often be associated with low numbers of predators. There is some evidence suggesting this.

Rodents introduced voluntarily or haphazardly on islands with impoverished faunas, with few or no predator species, often erupt.[7] Certain outbreaks of tropical small rodents have been related to temporarily relatively low predator numbers.[15,21,53] Late seasonal rains may produce more vegetation to protect the rodents especially from avian predators during the dry season. Predation has thus to be considered as one important variable in models predicting population development of noxious tropical or subtropical rodents.[53] Rabbits (*Oryctolagus cuniculus)* in their original mediterranean habitats in Spain are obviously regulated by predators while the deviating predator guild in the mediterranean habitats of Chile cannot prevent outbreaks of introduced rabbits.[24] Population increases in North American snowshoe hares (*Lepus americanus)* are related to a relaxation in predation pressure[26,27] and conditions seem similar in North Scandinavian small rodents.[18,20] Fox control in Denmark caused increases in lagomorph numbers[25] and a fox disease (*Sarcoptes* mange) in Sweden correlated with a marked increase in mountain hares (*Lepus timidus*).[19]

These observations demonstrate the potential, although not the practical, usefulness of affecting predator numbers or the predation process when faced with excessive rodent numbers and damage.

VII. INTRODUCTION OF NEW PREDATOR SPECIES

The most common, and probably most overlooked, example of predator introduction to control noxious rodents is the domestic cat. Still, there are very few estimates of the efficiency of cats in controlling mice or rats in human habitations. Elton[6] poisoned the rats on five farms in England and cats were then introduced on four of them. These four farms remained almost free from rats while the fifth farm suffered recurrent rat infestation. The author suggested that part of the food of the cats should consist of milk (as "alternative prey") and then the cats could keep the farm and its immediate surroundings rat-free. Feral cats make a great impact on field rodents.[35] and may, together with native predators, regulate such populations in certain areas.[8] Cats on islands have been shown to strongly affect introduced rabbit numbers[2] but have also been found to cause reduction or disappearance of native prey species.[74]

Monitor lizards (*Varanus indicus*) were introduced to some Pacific islands to control rats. However, their effect on one specified island (Ifoluk atoll) was lower than expected.[71] Instead, they caused declines in indigeneous crab species.

The Indian mongoose (*Herpestes auropunctatus*) has been introduced to several larger islands and initially it seemed to depress the numbers of *Rattus norvegicus* numbers, which may have caused increases in *Rattus rattus*.[50] An established Puerto Rican population had no longer any obvious effects on the rats, while the mongoose hunted poultry, causing declines in several reptilian species and transmitting rabies.[50]

Tame ferrets (*Putorius putorius*) are used for rabbit hunting in Europe and have been liberated on New Zealand to control rabbits.[74] The efficiency of this introduction is not well known but feral ferrets and cats together with food shortage halted a population expansion of European rabbits in New Zealand.[13]

Weasels have been employed as biological control agents especially by the Japanese. Several species, but especially *Mustela sibirica*, have been released on small islands around Japan and in the Pacific.[72] As an example 6843 weasels were stated to be released on 17 islands (up to 25 weasels per km^2) in 1967. At the same time rodenticides were often employed so it is difficult to evaluate the effect of the weasel predation alone. However, rat populations declined on some islands with introduced weasels but no rodenticides. Studies of feces showed preys from insects to birds with relatively few rodents being taken. Therefore, the efficiency of this weasel introduction has been considered doubtful.[74] Still, *Mustela erminea* is stated to have reduced the occurrence of both *R. norvegicus* and *R. rattus* on small New Zealand islands.[69] Stoats (*M. erminea*) were also recently introduced to a Danish island with outbreak numbers of water voles (*Arvicola terrestris*) and they were reported to halt the population development of the rodents.[38,39] As a precaution against future damage to, e.g., sea birds, only males were released in this case and no adverse effects were observed there on the resident population of ground-living birds. The effects on mice and vole populations of weasel introductions to a Canadian mainland clearcut area[63] are unclear but may have resulted in a prolonged low density phase in the vole population.

Summarizing, predator introductions have been applied mainly in human habitations and on islands. The effects vary widely and are often difficult to evaluate. On islands there is often a strong impact on the native wildlife[1] and this environmentally negative effect is often greater than that caused by introduction of competing species.[59]

VIII. PREDATION AND HABITAT ALTERATION

Habitat alteration is one of the best possibilities in integrated rodent pest management. Most alterations having a pronounced effect on rodent populations include the removal of cover,[15,17] which will not only affect the food supply but also, at least to the same extent, the shelter from predators. Less agile species such as voles and rabbits (generally folivores) avoid exposed areas where they may more easily be taken by predators. Thus, predation can be increased considerably by removal of weeds, bushes, or other shelter, either close to the vulnerable objects or in refuge areas for rodents making regular foraging sallies.[30,70]

The importance of keeping fruit tree and forest plantations free from excessive vegetation has been stressed many times. However, the effect varies geographically. It is clearly positive in temperate areas with little snow, such as Denmark[3] and Virginia,[4] but inefficient in more northernly areas such as in Finland[44] and north Sweden.[34] In the latter areas snow covers the ground for up to 7 months a year and permits voles, as well as other rodents, to invade fruit and forest seed orchards under a protective cover of snow in winter and cause considerable damage also in well-managed orchards. In tropical-subtropical areas the removal of sheltering vegetation is generally an efficient type of rodent management.[15]

There are thus obvious possibilities of exploiting the predation effects on the behavior of many rodent species by removing shelter in temperate to tropical regions. However, this is usually a very expensive method which might not only have to include the immediate crop areas but also large parts of the surrounding landscape. Foraging and dispersal behavior have to be considered and might render the total enterprise unsurmountable. Only local and fairly comprehensive cost-benefit analyses can decide whether or not it is economically profitable. Widespread habitat alterations will also profoundly affect the local fauna and flora.[22]

IX. POSSIBILITIES TO INCREASE PREDATION PRESSURE OR OTHER PREDATION EFFECTS

Nesting and denning sites are sometimes limiting for certain predator species; and raptors, for example, may be induced to breed in artificial brush nests and owls in nest boxes.[61] Many nomadic avian predators are not territorial[29] and respond numerically strongly to rodent numbers. Other more resident predators such as foxes and badgers are territorial and social behavior will limit their numbers irrespective of the number of breeding sites. Thus, use of artificial nests should be most profitable in heavily cultivated areas, in regions with strongly fluctuating rodent numbers, and for nonterritorial predator species. However, it is still not clear whether such measures would have any significant effect on rodent numbers.

Predators show various adaptations for hunting. Weasels, for example, might hunt in very dense vegetation where the rodents are usually unavailable to kestrels or buzzards. Formerly, many forest plantations in central Europe with dense vegetation were equipped with perch sites for raptors, but the use of such devices has disappeared during the last 20 to 30 years. Recent experiments in Norway on reforestation areas have, however, shown that perch sites are used by owls and in this way it is possible to direct predation to plantations vulnerable to vole damage.[60]

Odor from weasels has been found highly repellent to several rodent species.[62] Such factors were hoped to be exploited for deterring noxious rodents from vulnerable objects but results of practical tests did not come up to expectations.[38]

REFERENCES

1. **van Aarde, R. J.,** Population biology and the control of feral cats on Marion Island, *Acta Zool. Fenn.*, 172, 107, 1984.
2. **Apps, P.,** Cats on Dassen Island, *Acta Zool. Fenn.*, 172, 115, 1984.
3. **Bang, P.,** Damage by small mammals in Denmark, *Ecol. Bull. (Stockholm)*, 19, 13, 1975.
4. **Byers, R. E., Young, R. S., and Neely, R. D.,** Review of cultural and other control methods for reducing pine vole populations in apple orchards, *Proc. Vert. Pest. Conf.*, 7, 242, 1976.
5. **Davis, D. E.,** Advances in rodent control, *Z. Angew. Zool.*, 64, 193, 1977.
6. **Elton, C. S.,** The use of cats in farm rat control, *Br. J. Anim. Behav.*, 1, 151, 1953.
7. **Elton, C. S.,** *The Ecology of Invasions by Animals and Plants*, Methuen, London, 1958.
8. **Erlinge, S., Göransson, G., Hansson, L., Högstedt, G., Liberg, O., Nilsson, I. N., Nilsson, T., von Schantz, T., and Sylvén, M.,** Predation as a regulating factor on small rodent populations in southern Sweden, *Oikos*, 40, 36, 1983.
9. **Erlinge, S., Göransson, G., Högstedt, G., Jansson, G., Liberg, O., Loman, J., Nilsson, I. N., von Schantz, T., and Sylvén, M.,** Can vertebrate predators regulate their prey?, *Am. Nat.*, 123, 125, 1984.
10. **Fitzgerald, B. M.,** Weasel predation on a cyclic population of the montane vole *(Microtus montanus)* in California, *J. Anim. Ecol.*, 46, 367, 1977.
11. **Galushin, V. M.,** Synchronous fluctuations in populations of some raptors and their prey, *Ibis*, 116, 127, 1974.
12. **Gause, G. J.,** *The Struggle for Existence*, Williams and Wilkins, Baltimore, Md., 1934.
13. **Gibb, J. A.,** What determines the number of small herbivorous mammals, *New Zealand J. Ecol.*, 4, 73, 1981.
14. **Gliwicz, J.,** Island populations of rodents: their organization and functioning, *Biol. Rev.*, 55, 109, 1980.
15. **Green, M. G. and Taylor, K. D.,** Preliminary experiments in habitat alteration as a means of controlling field rodents in Kenya, *Ecol. Bull. (Stockholm)*, 19, 175, 1975.
16. **Hansson, L.,** Habitat, food and population dynamics of the field vole *Microtus agrestis* (L.) in south Sweden, *Viltrevy*, 8, 267, 1971.
17. **Hansson, L.,** Effects of habitat manipulation on small rodent populations, *Ecol. Bull. (Stockholm)*, 19, 163, 1975.
18. **Hansson, L.,** Predation as the factor causing extended low densities in microtine cycles, *Oikos*, 43, 255, 1984.
19. **Hansson, L.,** unpublished data, 1985.
20. **Henttonen, H.,** Predation causing extended low densities in microtine cycles: further evidence from shrew dynamics, *Oikos*, 45, 156, 1985.
21. **Hubert, B. and Adam, F.,** Outbreaks of *Mastomys erythroleucus* and *Taterillus gracilis* in the Sahelo-Sudanian zone in Senegal, *Acta Zool. Fenn.*, 173, 113, 1985.
22. **Howard, W. E.,** Biological control of vertebrate pests, *Proc. Vert. Pest Conf.*, 3, 137, 1967.
23. **Huffaker, C. B.,** Experimental studies on predation: dispersion factors and predator-prey oscillations, *Hilgardia*, 27, 343, 1958.
24. **Jaksic', F. M. and Soriguer, R. C.,** Predation upon the European rabbit *(Oryctolagus cuniculus)* in mediterranean habitats of Chile and Spain: a comparative analysis, *J. Anim. Ecol.*, 50, 269, 1981.
25. **Jensen, B.,** Effect of a fox control programme on the bag of some other game species, *Trans. Int. Congr. Game Biol.*, 9, 480, 1970.
26. **Keith, L. N., Cary, J. R., Rongstad, O. J., and Brittingham, M. C.,** Demography and ecology of a declining snowshoe hare population, *Wildl. Monogr.*, 90, 1, 1984.
27. **Keith, L. N. and Windberg, L. A.,** A demographic analysis of the snowshoe hare cycle, *Wildl. Monogr.*, 58, 1, 1978.
28. **King, C. M.,** The weasel *Mustela nivalis* and its prey in an English woodland, *J. Anim. Ecol.*, 49, 127, 1980.
29. **Korpimäki, E.,** Rapid tracking of microtine populations by their avian predators: possible evidence for stabilizing predation, *Oikos*, 45, 281, 1985.
30. **Kotler, B. P.,** Risk of predation and the structure of desert rodent communities, *Ecology*, 65, 689, 1984.
31. **Krebs, C. J.,** Dispersal, spacing behaviour and genetics in relation to population fluctuations in the vole *Microtus townsendii*, *Fortschr. Zool*, 25, 61, 1979.
32. **Krebs, C. J., Gaines, M. S., Keller, B. L., Myers, J. H., and Tamarin, R. H.,** Population cycles in small rodents, *Science*, 179, 35, 1973.
33. **Krebs, C. J. and Myers, J. H.,** Population cycles in small mammals, *Adv. Ecol. Res.*, 8, 267, 1974.
34. **Larsson, T.-B.,** Damage caused by small rodents in Sweden, *Ecol. Bull. (Stockholm)*, 19, 47, 1975.
35. **Liberg, O.,** Predation and social behaviour in a population of domestic cat: an evolutionary perspective, Thesis, Lund University, Lund, Sweden, 1981.

36. **Lotka, A. J.,** *Elements of Physical Biology,* Williams and Wilkins, Baltimore, Md., 1925.
37. **Luttich, S. N., Keith, L. B., and Stephenson, J. D.,** Population dynamics of the red-tailed hawk at Rochester, Alberta, *Auk,* 88, 75, 1971.
38. **Lund, M.,** Repellent effect of weasel odour on various wild rodents, *Danish Pest Inf. Lab. Annu. Rep. 1980,* 94, 1981.
39. **Lund, M.,** Biological control of water voles, *Danish Pest Inf. Lab. Annu. Rep. 1981,* 69, 1982.
40. **Maher, W. J.,** Predation by weasels on a winter population of lemmings, Banks Island, Northwest Territories, *Can. Field Nat.,* 81, 248, 1967.
41. **Maher, W. J.,** The pomarine jaeger as a brown lemming predator in northern Alaska, *Wilson Bull.,* 82, 130, 1970.
42. **May, R. M.,** Time-delay versus stability in population models with two and three trophic levels, *Ecology,* 54, 315, 1973.
43. **May, R. M.,** Theoretical ecology, principles and applications, Blackwell Scientific, Oxford, 1976.
44. **Myllymäki, A.,** Population ecology and its application to the control of the field vole, *Microtus agrestis* (L.)., *EPPO Publ. Ser.,* 58, 27, 1971.
45. **Parker, G. R.,** A population peak and crash of lemmings and snowy owls on Southampton Island, N. W. T., *Can. Field Nat.,* 88, 151, 1974.
46. **Pearson, O. P.,** Canivore-mouse predation: an example of its intensity and bioenergetics *J. Mammal.,* 45, 177, 1964.
47. **Pearson, O. P.,** The prey of carnivores during one cycle of mouse abundance, *J. Anim. Ecol.,* 35, 217, 1966.
48. **Pearson, O. P.,** Additional measurements of the impact of carnivores of California voles *(Microtus californicus).*, *J. Mammal.,* 52, 41, 1971.
49. **Phelan, F. J. S. and Robertson, R. J.,** Predatory responses of a raptor guild to changes in prey density, *Can. J. Zool.,* 56, 2565, 1978.
50. **Pimentel, D.,** Biology of the Indian mongoose in Puerto Rico, J., *Mammal.,* 36, 62, 1955.
51. **Pitelka, F. A.,** Cyclic pattern in lemming population near Barrow, Alaska, in *Alaskan Arctic Tundra,* Techn. Paper 25, Britton, M. E., Ed., Arctic Institute North America, 1973, 199.
52. **Pitelka, F. A., Tomich, P. Q., and Treichel, G. W.,** Ecological relations of jaegers and owls as lemming predators near Barrow, Alaska, *Ecol. Monogr.,* 25, 85, 1955.
53. **Poulet, A. R.,** The ecological basis of forecasting rodent outbreaks in a Sahelian agrosystem, *Acta Zool. Fenn.,* 173, 107, 1985.
54. **Ricklefs, R. E.,** *Ecology,* Chiron Press, Newton, Mass., 1973.
55. **Ryszkowski, L., Wagner, C. K., Goszczynski, J., and Truszkowski, J.,** Operation of predators in a forest and cultivated fields, *Ann. Zool. Fenn.,* 8, 160, 1971.
56. **Ryszkowski, L., Goszczynski, J., and Truszkowski, J.,** Trophic relationships of the common vole in cultivated fields, *Acta Theriol.,* 18, 125, 1973.
57. **Sadleir, R. M. F. S.,** The relationship between agonistic behaviour and population changes in deermouse, *Peromyscus maniculatus* (Wagner), *J. Anim. Ecol.,* 34, 331, 1965.
58. **Southern, H. N. and Lowe, V. P. N.,** Predation by tawny owls *(Strix aluco)* on bank voles *(Clethrionomys glareolus)* and wood mice *(Apodemus sylvaticus), J. Zool. (London),* 198, 83, 1982.
59. **Simberloff, D.,** Community effects of introduced species, in *Biotic Crises in Ecological and Evolutionary Time,* Nitecki, M. H., Ed., Academic Press, New York, 1981, 53.
60. **Sonerud, G.,** Jaktstrategier hos flyvende småpattedyrspredatorer, Thesis, Oslo University, Oslo, Sweden, 1980.
61. **Sonerud, G.,** Nest hole shift in Tengmalm's owl *Aegolius funereus* as defence against nest predation involving long term memory in the predator, *J. Anim. Ecol.,* 54, 179, 1985.
62. **Stoddart, D. M.,** Effect of the odour of weasels *(Mustela nivalis* L.) on trapped samples of their prey, *Oecologia,* 22, 439, 1976.
63. **Sullivan, Th.P. and Sullivan, D. S.,** The use of weasels for natural control of mouse and vole populations in a coastal coniferous forest, *Oecologia,* 47, 125, 1980.
64. **Taitt, M. J., Gipps, J. H. W., Krebs, C. J., and Dunjerski, Z.,** The effect of extra food and cover on declining populations of *Microtus townsendii, Can. J. Zool.,* 59, 1593, 1981.
65. **Taitt, M. J. and Krebs, C. J.,** Predation, cover, and food manipulations during a spring decline of *Microtus townsendii, J. Anim. Ecol.,* 52, 837, 1983.
66. **Tamarin, R. H.,** Dispersal in island and mainland voles, *Ecology,* 58, 1044, 1977.
67. **Tanner, J. T.,** The stability and the intrinsic growth rates of prey and predator populations, *Ecology,* 56, 855, 1975.
68. **Tapper, S.,** The effect of fluctuating vole numbers *(Microtus agrestis)* on a population of weasels *(Mustela nivalis)* on farmland, *J. Anim. Ecol.,* 48, 603, 1979.
69. **Taylor, R. H.,** Distribution and interactions of introduced rodents and carnivores in New Zealand, *Acta Zool. Fenn.,* 172, 103, 1984.

70. **Thompson, S. D.,** Structure and species composition of desert heteromyid rodent species assemblages: effects of a simple habitat manipulation, *Ecology,* 63, 1313, 1982.
71. **Uchida, T. A.,** Observations on the monitor lizard, *Varanus indicus* (Daudin), as a rat-control agent on Ifaluk, Western Caroline Islands, *Bull. W. H. O.,* 35, 976, 1966.
72. **Uchida, T. A.,** Observations on the efficiency of the Japanese weasel, *Mustele sibirica itatsi* Temminck & Schlegel, as a rat-control agent in the Ryukyus., *Bull. W. H. O.,* 39, 980, 1968.
73. **Volterra, V.,** *Variations and Fluctuations of the Number of Individuals of Animal Species Living Together,* McGraw-Hill, New York, 1926.
74. **Wodzicki, K.,** Prospects for biological control of rodent populations, *Bull. W. H. O.,* 48, 461, 1973.

Chapter 28

NONLETHAL METHODS IN RODENT CONTROL

William D. Fitzwater

TABLE OF CONTENTS

I. INTRODUCTION

It is difficult to justify to animal damage control specialists the use of ''nonlethal'' controls against such prolific, elusive, and aggressive competitors to human welfare as rodents. But besides a deep-rooted religious[7] distaste in some parts of the world for the destruction of lower vertebrate species there is increasing pressure from people, who should be intelligent enough to know better, who say we should ''live and let live''. This philosophy is espoused particularly by vegetarians and urbanites who have no realization of the need to save food and fiber crops from depredations by vertebrate animals.

While methods that kill outright like toxicants, fumigants, and deadfall traps can be clearly defined, most ''nonlethal'' methods will indirectly cause mortality by limiting survival success of a rodent population by such means as modification of the habitat. And as Walter E. Howard[9] has so aptly pointed out, ''Any habitat manipulation that effectively reduces the troublesome status of a population of vertebrates will most likely alter the entire ecosystem, i.e., effect the species composition and density of all other kinds of vertebrates and invertebrates living in that ecosystem, far more drastically than could result if the same degree of control was achieved by some relatively selective 'conventional' control method.''

Nonlethal approaches, semantically at least, should include antifertility chemicals and repellents which, by affecting one or more of the five senses, deter animals from occupying an area or eating a food material. However, chemosterilants and repellent chemicals/devices are discussed elsewhere in this volume.

II. MYTHS

An amazing amount of folklore has been generated around ''nonlethal'' methods. Religious incantations were thought to drive rats out of an area. In the 12th century the peninsula of Jumieges (France) was overrun with rats. The residents invoked the Spirit of St. Valentine who drove all the rats into the River Seine. This particular spot is pointed out to tourists as the ''Rat's Run''.[2] Along with incantations, the ancient Greeks believed that rodents could read. They even had a standard warning, ''I adjure ye, ye mice here present, that ye neither injure me nor suffer another mouse to do so. I give you yonder field (usually some neighbor) but if I ever catch you here again, by the Mother of God, I will rend you in seven pieces.'' This eviction notice was then stuck under a stone in the field, taking care that the written side was up.[22]

In more modern times a letter was found stuck in a crevice in an old basement in Sandwich, New Hampshire. The letter dated May 9, 1845 said, ''I have bourn with you till my patience is gone. I cannot find words bad enough to express how I feel . . . now spirits of the bottomless pit, depart from this place with all speed! Begone or you are ruined . . . We are preparing water (to) drown you; fire to roast you; cats to catch you; and clubs to maul you . . . This is for cellar rats. Please give notice to these in the chamber. There are many of us plotting against you . . . ''.[3]

Other common beliefs were to release ''different'' individuals into a rodent colony to chase out the others. Topsell[19] wrote, ''If mice are left in a pot without water they will kill one another. The winner if turned loose will hunt down and eat mice.'' ''In the early 1800s Norwegians singed the hair of a rat over a fire and let it loose. The smell was so offensive to his comrades that they quitted the house.''[2] And Lantz[13] mentions the recommendation of belling a rat and turning it loose to scare the others in the colony was totally ineffective.

Finally in this category we must include the Pied Piper of Hamelin who spirited away the rats of that community with his magic flute. Unfortunately, the belief in these mixed up memories of plague and the Children's Crusades still persists in wishful thinking that nonlethal methods could be a magic cure-all for our rodent problems.

III. HABITAT MODIFICATION (RURAL)

Habitat modification will be discussed as it applies to rural or agricultural situations as contrasted to urban environments or man-made structures where it is generally referred to as "sanitation and rodent proofing." In both instances it involves the removal of basic life needs — food, water, and shelter — from rodent environments making the animals move to more favorable habitats or be subjected to increased predation and other population depressants if they remain.

Clean cultivation of an area as quickly after the crop is harvested reduces this source of food supply on agricultural lands. The storage time of grain in field shocks should be kept to a minimum to prevent the buildup of rodent populations. Woodlots and brushy areas should be eliminated around orchards and cultivated crops. While hedgerows are recommended by the conservationists, clean farming to the field borders does reduce the incidence of vertebrate damage. An excellent example of effective habitat modification was the spraying of rangeland with 2,4-D (herbicide) for pocket gopher (*Thomomys talpoides)* control. This selectivity reduced forb density (favorite gopher food) resulting in 87 to 93% reduction in pocket gopher activity.[10,12]

Another method of reducing damage to field crops is the use of buffer crops or supplemental feeding. The writer is not aware of any studies on the effectiveness of buffer crops (a low choice food planted around a high choice crop that is to be protected) against rodents, but Lewis[14] found that a 5 m strip of rye reduced damage by jackrabbits (*Lepus sp.*) to a field of barley. Fitzwater[6] found that fall prunings piled around fruit trees would reduce meadow vole (*Microtus pennsylvanicus*) damage to the main stems except under severe weather conditions. Studies[18] showed that direct seeding a mixture of Douglas fir (*Pseudotsuga taxifolia*) and sunflower seeds with oats in late winter/early spring resulted in successful regeneration of cutover forest land as the seed-eating deer mice (*Peromyscus maniculatus*) were diverted by the more attractive seeds. Changes in agricultural practices have also been suggested. A Malaysian study[11] on rice rats (*Rattus argentiventer*) recommended that planting schedules be standardized over large areas and weeding efforts extended to cover not only rice paddies but adjacent ditches, fields, and waste lands as well. Elsewhere larger paddies/smaller bunds have been used to cut rodent numbers.[20] Sugihara et al.[17] found that herbicide treatments of crop borders reduced rat damage to sugarcane by 43% during the last half of the crop cycle.

Habitat manipulation by physical barriers to protect a structure from rodent invasion is known as "rodent proofing". While it is impractical to "proof" large acreages outdoors, individual plant protectors are available. Metal bands around palms and other trees have been used around the world to keep squirrels and rats from harvesting the crop. These bands should be placed high enough off the ground so rodents cannot jump over them, wide enough so they cannot reach over them, and isolated enough so they cannot jump to banded trees from nearby ones. Plastic netting developed to protect forest tree seedlings from deer and rabbit damage has shown some success in reducing pocket gopher (*Thomomys spp.*) damage to seedlings.[1] While fencing against rodents is impractical over large areas, some spectacular results have been achieved in the Phillipines around experimental rice plots[20] (these fences are electrified and probably should be considered lethal methods).

IV. SANITATION

Habitat manipulation is more practical and effective around man-made structures ("urban" environments). The processes here are generally known as "sanitation" and "rodent proofing". Sanitation involves cleaning up an area — the removal of materials such as garbage, piles of trash, improperly stacked supplies, free water sources, heavy vegetative ground cover, and excess poultry, livestock, and pet feed.

The above attractants for rodents can be reduced by storing garbage and other refuse left by careless humans in rodentproof containers until it can be removed to a proper disposal area. The storage containers should be of sufficient capacity to handle all the refuse generated between collection periods. Cleaning up an area should be taken literally to include the continuous sweeping up and proper disposing of loose grain and other food materials scattered around in the processing of foods. Waste materials should be disposed of by burning or burial or new rodent problems will be created. Piles of trash should also be collected at intervals and removed as in the case of garbage. Burnable trash may be collected at intervals and removed as in the case of garbage or on the spot. Accumulations of building materials, warehoused merchandise, or foodstuffs should be neatly stacked off the ground at a height of 45 cm so as to eliminate cover under which rodents might hide. These stacks should also be located away from walls and intersected frequently with passageways. The distance from walls should be 45 cm and the distance between alleyways in the stacked materials should be 1.25 m. While these practices, in themselves, are not deterrents to rodents they give easier access for rodent control personnel to use lethal measures more effectively in the area.

Reducing the sources of water available to rodents in outside environments is usually a practical impossibility, but some sources of standing water can be heavily polluted with petroleum products. With the better control possible indoors, leaky faucets and other plumbing repairs can be made, drains fitted with heavy perforated covers through which rodents cannot pass, free-standing water drained, or access blocked to some water sources by mesh wire. The removal of heavy vegetative cover from around buildings is an excellent deterrent to rodent invasion. The role played by dense vegetative cover in encouraging rodent invasions has been documented by Brooks[4] in his studies of roof rat movements into densely landscaped housing developments in California. And of course in semi-urban situations like food processing plants, the extensive use of paving around the plant discourages rodent invasion.

Careless exposure of feed for poultry, livestock, and even pets can be a source of sustenance for a rodent colony. Exposure should be confined and timed so as to be inaccessible during periods of high rodent numbers.

V. RODENT PROOFING

The second basic approach to limiting rodent access to food/water/shelter rich areas is known as "rodent proofing". In planning rodent proofing one must realize the physical abilities of the species of rodents concerned. Take the cosmopolitan Norway rat (*Rattus norvegicus*) as an example. The majority of burrows made by this rodent are only 30 cm below ground level.[16] However, as under some circumstances, they will go deeper, it is advisable to protect a building foundation down to 90 cm below ground level. Norway rats have been known to jump 76 cm directly up in the air and can go higher if able to scramble up a rough surface.[8] Thus 90 cm is considered a reasonably safe distance that should be protected above ground level. The equally ubiquitious roof rat (*Rattus rattus*) while not a good burrower is considerably more agile and thus protection above ground level may necessarily be more extensive where this species is a problem.

Rodents have very flexible bodies and generally can enter openings only large enough to admit their skulls. Thus small house mice (*Mus musculus*) can squeeze through openings larger than 6 mm, while young rats are capable of getting through openings over 12 mm. These then are the minimum diameters of openings that can be permitted in rodent-proof structures. While roof rats are the most agile, all of the rodents are able to climb vertically up rough surfaces and walk along wires and ropes. Rats can reach about 33 cm (safety factor should be 45 cm), scramble up inside pipes with 4 to 10 cm inside diameters, and climb pipes up to 7.5 cm in outside diameter. They can jump about 1.22 m horizontally.[15]

In designing rodent-proof structures remember rodent teeth are as hard as the steel blade in a pocket knife. Thus the materials used must be rodent resistant. This includes such building materials as brick, cement block, concrete (at least 75 mm thick), galvanized sheet iron (at least 26 ga), metal mesh (openings no larger than 6 mm), glass (unbroken, at least 6 mm thick), and similar materials. Wood will keep them out too providing there are no "gnawing edges" available. Gnawing edges are places where small holes, open joints, or other irregularities in the wood surface enable the animals to get a grip on the wood and gradually splinter it away to enlarge the opening to a size they can enter.

Finally, there must be a program of inspection and continual maintenance of these devices. Human carelessness in propping doors open, banging into mesh covered windows, or new construction around existing structures may permit reinvasions to occur.

VI. LIVE TRAPPING

Live trapping could be considered "nonlethal" in comparison to deadfall trapping in that the animals are taken unharmed. Designs for live traps include pitfalls, drop doors, and false floors. The pitfalls are very primitive types of traps consisting of holes dug in the ground along paths used by rodents. Animals fall into them or are attracted by bait in the bottom. The steep vertical sides prevent them from scrambling back out. Plastic or metal containers are used to line these pits to make it even more difficult for the rodents to get out.

"Drop door" traps are those that drop down and close off the entryway when the animal upsets the trigger mechanism. Generally only one animal will be caught at a time. "False floor" traps generally give way to the weight of the animal dropping it into a holding cage. The floor is counterbalanced so that when the animal is dropped down, it moves back into position acting as a multicatch trap. A modification of this is a spring-wound compartmented trap known as the "KETCH-ALL", widely used in the U.S. When a mouse enters the trap and upsets the trigger mechanism, he is thrust into a holding cage as one compartment revolves and a new one takes its place.

The effectiveness of all of the above types of traps is dependent upon proper location. As rodents are creatures of habit and travel along regular travel routes, traps should be set in these heavily-used pathways. If possible, funnel the direction of the rodent's movements so they converge on the trap entrance by using wings. In buildings this can be accomplished by setting the traps tight against walls across recognizable travel routes.

While the use of "nonlethal" measures seems a futile gesture in reducing rodent numbers and damage to most animal damage control specialists, there is no denying the practicality of it when applied to a man-made environment. As David E. Davis[5] so graphically illustrated in his classic study of rat populations in Baltimore, Maryland, education of the public (with a little push from public health regulations enforcement) to use good sanitation and rodentproofing practices provided longer and more effective control of rat populations than the use of toxicants.

REFERENCES

1. **Anthony, R. M., Barnes, V. G., Jr., and Evans, J.,** "Vexar" plastic netting to reduce pocket gopher depredation of conifer seedlings, in *Proc. 8th Vert. Pest Contr. Conf.,* Howard, W. E. and Marsh, R. E., Eds., University of California, Davis, Calif., 1978, 138.
2. **Boelter, W. R.,** *The Rat Problem,* John Bale, Ed., Danielsson, London, 1909.
3. **Botkin, B. A.,** *A Treasury of New England Folklore,* Crown Publ., New York, 1947.
4. **Brooks, J. E.,** Roof rats in residential areas — The ecology of invasion, *Calif. Vector News,* 13, 69, 1966.
5. **Davis, D. E.,** Rodent control strategy, in *Pest Control Strategies for the Future,* National Academy of Sciences, Washington, D.C., 1972, 157.
6. **Fitzwater, W. D.,** Meeting the meadow mouse menace, in *Proc. Vert. Pest Contr. Conf.,* National Pest Controllers Association, Elizabeth, N. J., 1962, 67.
7. **Fitzwater, W. D. and Prakash, I.,** *Handbook of Vertebrate Pest Control,* Indian Council of Agricultural Research, New Delhi, 1973, 96.
8. **Hansgen, K. H.,** Distances jumped by Norway rats during competition, *Calif. Vector News,* 19, 67, 1972.
9. **Howard, W. E.,** Biological control of vertebrate pests, in *Proc. 3rd Vert. Pest. Conf.,* Cummings, M. W., Ed., University of California Davis, Calif., 1967, 137.
10. **Hull, A. C., Jr.,** Effect of spraying with 2,4-D upon abundance of pocket gophers in Franklin Basin, Idaho, *J. Range Manage.,* 24, 230, 1971.
11. **Majid, T. B. and Chye, O. S.,** The organisation and practice of rice field rat control in Malaysia, in *Proc. Organisation and Practice of Vert. Pest Contr.,* Dubock, A. C., Ed., ICI Plant Protect. Div., Surrey, England, 1984, 465.
12. **Keith, J. O., Hansen, R. M., and Ward, A. L.,** Effect of 2,4-D on abundance and foods of pocket gophers, *J. Wildl. Manage.,* 23, 137, 1959.
13. **Lantz, D. E.,** How to destroy rats, *U.S. Dept. Agric. Farmer's Bull.,* 369, 1909.
14. **Lewis, J. H.,** Planting practice to reduce crop damage by jackrabbits, *J. Wildl. Manage.,* 10, 277, 1946.
15. **Peterson, G. N.,** *Rodent Proof Design/Construction,* Wavary Press, Kent, 1978.
16. **Pisano, R. G. and Storer, T. I.,** Burrows and feeding of the Norway rat, *J. Mammal.,* 29, 374, 1948.
17. **Sugihara, R. T., Pank, L. F., Fellows, D. P., Hirata, D. N., Stott, R. S., Hilton, H. W., and Kaya, H.,** Noncrop habitat manipulation as a means of controlled rats and reducing damage to sugarcane, *Hawaii. Sugar Techol. 1977 Rep.,* 1978, 84.
18. **Sullivan, T. P.,** Biological control of conifer seed damage by the deer mouse *(Peromyscus maniculatus),* in *Proc. 8th Vert. Pest Conf.,* Howard, W. E. and Marsh, R. E., Eds., University of California, Davis, Calif., 1978, 237.
19. **Topsell, E.,** *The Historie of Foure-Footed Beastes,* W. Iaggard, London, 1607.
20. **Uhler, L. D.,** Seasonal abundance of *(Rattus rattus mindanensis)* at the International Rice Research Institute College, Laguna, Phillipines, *The Phillip. Agric.,* 51, 581, 1967.
21. **Wood, B. J.,** Rat pests of tropical crops — A review of practical aspects, in *Proc. The Organisation and Practice of Vert. Pest Contr.,* Dubock, A. C., Ed., ICI Plant Protect. Div., Surrey, England, 1984, 265.
22. **Zagoren, R.,** Post a letter to the mice, *Our Dumb Anim.,* 83, 12, 1950.

Chapter 29

INTEGRATED PEST MANAGEMENT, PRINCIPLES IN RODENT CONTROL

Heimo Posamentier

TABLE OF CONTENTS

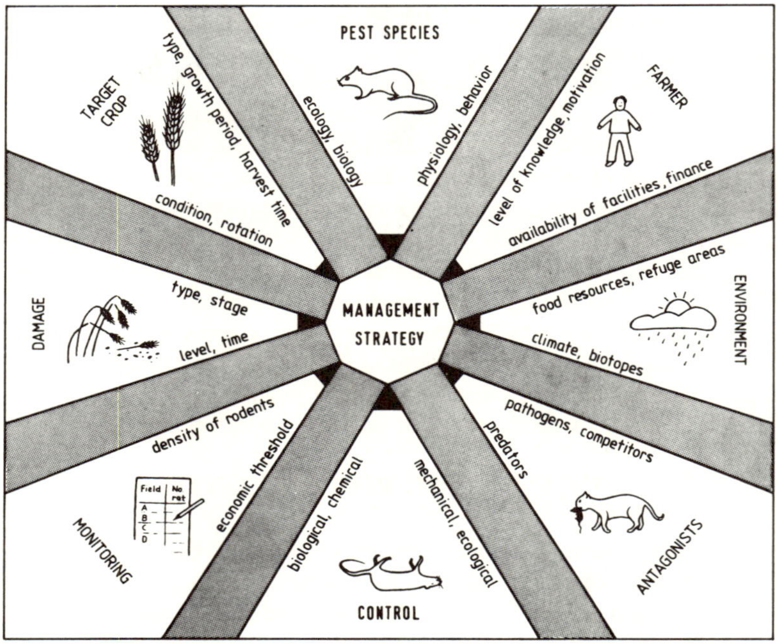

FIGURE 1. Parameters that need to be integrated into management strategies.

I. INTRODUCTION

Integrated pest management is a strategy blending various control techniques with the characteristics of the crop, its management and pest organisms, environmental factors, and in some cases the scope and means of those applying the measures (Figure 1). The ultimate aim of such a strategy is to keep the damages through pest organisms below economically relevant levels. At the same time the cost to the users, for example, farmers, and the detrimental effects on the environment should be kept to a minimum.

To obtain this goal the following steps should be taken.[12]

1. Information has to be obtained regarding the respective ecosystem and its regulatory mechanisms.
2. Criteria must be set for determining the need for plant protection measures.
3. Methods for corrective measures must be selected.

Initially these concepts were applied to insect pests and diseases. It recently became obvious that they are also relevant in rodent pest management, although some obvious characteristics distinguishing rodent from insect pests should be given before discussing the factors that need to be considered.

1. In most cases rodent pests can be potentially responsible for damage independent of any season.
2. Many species of rodents cause damage to different types of standing crops and all types of stored produce and structures.
3. Most rodent pests are commensal and can live in very different habitat types, produced through human activities.
4. Most rodent pests can survive under different climatic conditions.
5. Most rodent control measures are indirect (i.e. baiting, hygiene).

The criteria for determining the need for plant protection measures refer to threshhold levels of infestation or damage at which control becomes economic or necessary. This concept is less significant where rodents are concerned, because the mode of damage is different to that of insects and diseases. Rodents in most cases feed directly on the crop and a single animal can cause considerable economic damage. Therefore steps for control need to be initiated at the first sign of infestation, and certain preventive actions even earlier.

On the other hand, it is very important to have information on the respective ecosystem and its regulatory mechanisms. With specific regard to rodents it refers to preventive measures, migration patterns, time of expected infestation or damage incidence, climatic factors, as well as the biology, and ecology of the pest species involved.

Since rodents and even individual species, many being commensal, can survive in totally different areas ranging from sewerage systems to field crops, for example *Rattus norvegicus*, the type of area infested must be considered in any management system as well. Finally, replacement of one species with another can be an additional complicating factor.

II. PARAMETERS INFLUENCING RODENT CONTROL TECHNIQUES

Preventive measures, such as habitat management, are discussed in a different chapter. Points of recolonization, often serving as refuge areas, such as permanent swamps, creek sides, or weedy fields and dikes[21-23,42,56] should be reduced or eliminated, and can be included under habitat management as well. However parallel action has to be taken if the points of recolonization are houses, particularly in tropical countries with relatively open structures. This parallel action may involve simple hygienic practices or certain control activities. In addition some species have regular migration patterns. Tast[54] found that the migration of *Microtus oeconomus* from winter and summer habitats is related to seasonal floods caused by the melting of ice and snow. The initiation of migration is probably related to the length of day. If habitat changes seasonally, animals can and do adapt by migrating to more suitable areas nearby.[7,15,20] Migration is also caused by drastic alterations of habitat, whether harvesting agricultural fields or natural influences such as wildfires,[22,44] floods,[33,35,45,48,55] or landslips.[11] Finally, in some specific cases predators may play an important role in keeping rodent pest populations at acceptable levels. For example, trapping results in Bangladesh have shown that households which keep cats had a much lower incidence of rodents than others.

Regular migrational patterns and microclimatic conditions are to some extent predictable and can therefore be incorporated into rodent control operations. This is very helpful in countries with less efficient communication systems, if certain requirements, such as the distribution of information and facilities, can be organized ahead of the expected infestation periods.

The choice of methods has to be considered. These again depend to a large extent on the factors mentioned above, and in addition on the applicators of control measures. In some countries applicators are specifically trained in pest management and can be expected to use relatively complex management techniques. Furthermore, the total amount of finance involved is not a limiting factor if the overall operation is economic. In other countries, like Bangladesh, the techniques that can be utilized are more restricted by:

1. Low level of technical knowledge and general education
2. Absolute cost involved
3. Lower level of motivation
4. Inadequate facilities (i.e., rodenticides)
5. Inaccessibility of facilities.

III. THE EXAMPLE OF BANGLADESH

Using Bangladesh as an example, management techniques specifically adapted to local conditions, and the information required to develop these, are described below.

In the case of Bangladesh a considerable amount of information has been obtained since 1978 on the distribution of rodent damage to agricultural crops,[3,4,6,13,14,16-18,26,30-38] and on the biology and ecology of the major pest species, *Bandicota bengalensis*.[2,19,27-41,43,46,47,52,53] This boost of information came mainly from two projects. One, concerned primarily with applied research, was financed by the U.S. Agency for International Development (U.S. AID) while the second was integrated into the local plant protection service and financed by the Federal Republic of Germany through its Agency, the Gesellschaft für Technische Zusammenarbeit (GTZ) GmbH and administrated by the Bangladesh-German Plant Protection Program. Further information has been taken from the literature referring to neighboring countries.[5,8-10,24,25,42,50,51] In addition, the situation and problems of the people applying the measures, in this case farmers, were ascertained by qualitative observations and opinion surveys[1,37] along with availability of facilities.

With this information in hand some general and specific recommendations were drawn up. Subsequently, technical and extension staff were trained on rodents, their biology and control, and with the help of these services as well as national information and motivation campaigns, farmers were made aware of the appropriate techniques of control and facilities available. These activities resulted in a considerable increase in the number of farmers carrying out rodent control by themselves.

From the works quoted above, the following important facts were considered for the recommendations on rodent management.

1. The grain crops mainly damaged in Bangladesh are wheat and deep water rice; the transplanted rice varieties are hardly attacked.
2. Serious damage to grain crops starts around booting stage.
3. The only species of real economic significance in the fields of Bangladesh is *Bandicota bengalensis*.
4. In most cases each burrow of *Bandicota bengalensis* is inhabited by a single adult.
5. The activity radius of *Bandicota bengalensis* is small and restricted to the immediate vicinity of its burrow system.
6. The distribution of *Bandicota bengalensis* within a given area is very patchy.
7. The baiting periods in wheat and deep water rice need only be short.
8. Active burrows of *Bandicota bengalensis* are obvious and cannot be missed in agricultural fields.
9. *Bandicota bengalensis* readily accepts most baits and removes these from burrow openings shortly after application if placement is in the late afternoon.
10. Shyness to acute rodenticide baits is not a serious problem in Bangledesh fields.
11. At present most farmers consider $0.15 per packet of rodenticide bait acceptable. One packet should be sufficient to control rodents in about 2 acres for one season.
12. At present almost 70% of farmers involved in rodent control use rodenticides and 74% of all farmers consider the use of rodenticides the best way to control rats.
13. Only 11% favored mechanical methods other than trapping (14%).
14. Farmers practice direct baiting of burrow systems.
15. Traps are fairly expensive and could be used only in houses, due to pilferage.
16. Mechanical control on dikes, the major harborage for *Bandicota bengalensis* is impractical and in some cases impossible.
17. Good rodenticide or bait was not accessible according to the opinion of farmers and a market survey.

This general and specific information led to the following considerations:

1. To make farmers accept any recommendations on control, measures must incorporate the use of rodenticides. Mechanical methods are not favored, mainly because they are labor intensive, expensive, and often not very effective.
2. In most cases acute rodenticide bait (i.e., zinc phosphide, 2 to 3% a.i.) of proper formulation would achieve the desired results. Its advantages are low cost, farmers have used it in the past, and it is easy and simple to apply. With respect to the behavior of *Bandicota bengalensis* it can be applied in a way that is safe to humans and nontarget animals, and presents neglible bait shyness problems.
3. Applying bait directly to burrows has several advantages. Most farmers use this method. Active burrow systems are easily identified and located and as a rule only one adult inhabits one system dictating the exact quantity of bait required. Furthermore this mode of application is relatively safe with respect to nontarget organisms, which is particularly important when using acute rodenticides. Due to the patchy distribution of *Bandicota bengalensis,* area baiting would be wasteful and therefore not acceptable to the local farmers.

Additional general suggestions and information to the above points were incorporated into the extension advices. These included:

1. That major or economical damage starts around booting stage.
2. That dense stands of grain crops are more susceptible to damage by rodents than poor stands.
3. That acute rodenticides should only be used for 2 to 3 days every 4 to 6 weeks.
4. That hygienic practices in the house and field are a major factor in keeping infestation incidence at a low level and making control activities far more efficient.

Finally the importance of monitoring, interpreted in Bangladesh as "vigilance", was stressed.

As mentioned earlier, strategies for rodent control may differ between crops or places, and more specific management sequences have been drawn up. Three examples, incorporating the above criteria and further information characteristic to the respective crop, are presented below.

A. Rodent Control in Wheat

Among grain crops, wheat is heavily attacked by rats in Bangladesh. Nationwide surveys conducted between 1980 and 1984 indicate that the annual loss to wheat stands is between 1.8 and 2.5% of tillers cut, which can be equated to about 20,000 tons before harvest.[41]

The distribution of damage between areas and even fields can vary considerably. While about 50% showed no infestation in 1980 and 1981, the percentage of plots with more than 2% damage was 31 and 24%, respectively.[34,38] As a result of a survey in 4 out of 21 districts, overall damage of 12.1% in mature wheat fields of the 1979 season was found.[30] According to various animal surveys it could be established that the species almost exclusively responsible for this damage is *Bandicota bengalensis*.[30] Consequently various studies in Bangladesh on ecology and behavior concentrated on this species. On the basis of these results the following strategy was suggested (Figure 2).

1. Inspect fields regularly, about once a week, particularly close to booting stage.
2. Keep fields and surrounding areas clean of weeds and redundant vegetation.
3. At first infestation use mechanical methods to control rats. If not possible, apply rodenticide bait directly to the burrow systems late in the afternoon, and inform other

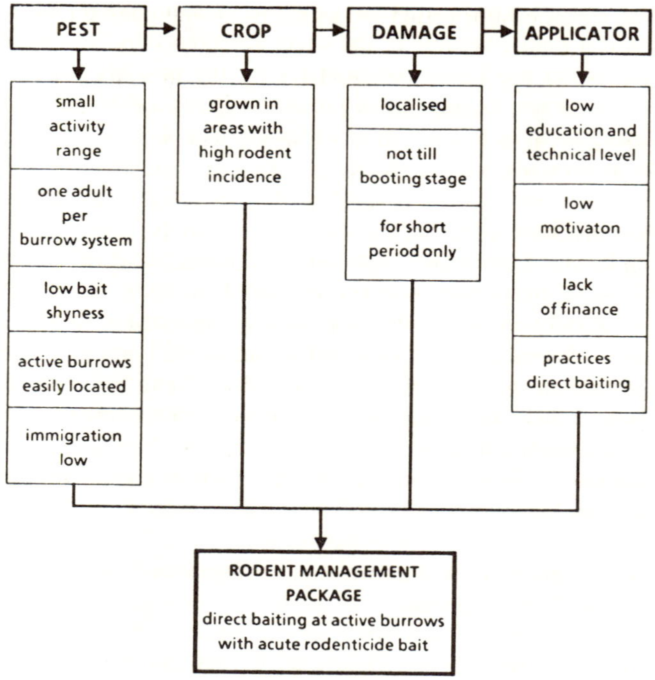

FIGURE 2. Major factors considered in recommending a rodent management package in wheat.

farmers if surrounding fields are also infested. In the case of acute rodenticide bait (i.e., zinc phosphide 2 to 3% a.i.) use about 5 g of bait per burrow system, and collect bait not taken on the following morning; apply for 2 to 3 days every 4 to 6 weeks only. If an anticoagulant bait is used apply about 35 g (coumatetralyl). In most fields situations 1 to 2 days of application is sufficient; repeat as required.

4. Animals not taking acute poison bait should be controlled by nonchemical methods or anticoagulant rodenticide bait.
5. Continue checking fields for immigration.

The advantage of placing bait directly into burrow systems has been discussed already. In most cases baits are immediately taken by animals, and stored underground if not consumed, thereby reducing the chance of poisoning nontarget animals. For simple and easy application the use of rodenticide bait in block form (2 × 2 × 0.5 cm) is recommended, formulated as suggested by Smythe and Khan[49] and Posamentier and van Elsen.[41]

Trials using zinc phosphide baits of 2 to 3% concentration in Bangladesh have shown to be effective against *Bandicota bengalensis* and efficiencies of 90% could be obtained if applied by technicians. A reduction of some 40 to 60% in damage in wheat fields was recorded where this bait was applied by farmers on a wide scale.[1,37]

The major problem with acute rodenticides is the formulation, particularly if mixed by farmers themselves in these countries. Therefore, local manufactures were found and motivated to produce and distribute ready made zinc phosphide bait blocks as mentioned above. The costs could be kept low enough, at about $0.20 per ha, and farmers readily accepted and used this formulation.[1,37]

The traditional draw-back of acute rodenticides, i.e., bait shyness, was not critical in Bangladesh. *Bandicota bengalensis* does not live in groups and the population cannot build up again before the wheat harvest. At present the use of acute rodenticides is favored over anticoagulants in wheat fields, because of the wider acceptance by farmers, even among the poorer section. The wider use easily compensates for the possibly lower effectiveness of acute rodenticides. As farmers become more familiar with the use of anticoagulants, perhaps through initially using them in houses, their wider use would certainly be advisable and in most cases still be economical.

B. Rodent Control in Pineapple

Damaged and undamaged pineapples were surveyed when harvested and it was found that between 8 and 9% of all fruits were attacked by rats.[34] Trapping results revealed that two species, *Rattus rattus* and *Bandicota bengalensis*, were responsible for this damage. *Rattus rattus* probably moved between nearby huts and gardens, and immigration into gardens had to be expected from all sides. In spite of low population density the damages, and hence financial losses, were relatively high. Since pineapples ripen continuously the harvesting period is spread over about 3 months.

Mechanical methods could not be applied because burrows were difficult to locate and plants may be damaged. Since no reliable rodenticide baits were available at the time, rodent control was hardly practiced by garden owners despite heavy losses.

Keeping in mind the problem of continuous immigration and the long harvesting period it becomes obvious that rodent control had to be carried out over an extended period. Furthermore, since not all the animals reside in the garden, immigration can be from any side and burrows are not easily located in pineapple gardens, so baiting points have to be laid out in a regular pattern. Control should start just before the first fruit begin to ripen (Figure 3).

Field tests have shown that about 12 bait stations should be set per acre, initially with about 50 g of anticoagulant bait each. Refilling and relocating of stations should be adapted to the bait take by rodents to save costs. This is relatively important, since most stations are not visited by rats, due to the relatively low population density and the bait quickly deteriorates since ripening time is during the monsoon period. In addition, it would be advantageous to conduct control in nearby houses as well. The concentration of baiting points makes overall costs fairly high; however, pineapple is an expensive crop and 2 to 3 damaged fruits compensate for a single bait station including contents. During the harvesting period, because of theft, each garden has a guard, and in practice bait stations could be checked each day. This means that labor costs for rodent control can be disregarded, and it would be possible to reduce the amount of bait still further, depending on the diligence of the respective guard.

Relatively good results have also been obtained with zinc phosphide bait. Bait (5 to 7 g) should be laid at each of 30 points per acre, evenly spaced. It should be laid and checked for 2 or 3 days every 4 weeks starting just prior to the ripening of the fruit.

Finally, it was stressed that the vegetation between rows should be kept as low as possible. In most cases gardens are not cared for outside the harvesting period and are overgrown with weeds and unncessary vegetation.

Garden owners normally have a higher than average level of knowledge, and are capable of learning the more complex technique described. Moreover, since this group resided within a relatively small area, it was simple to directly train most owners in a short period of time (Figure 3).

C. Rodent Control in Deep Water Rice

Deep water rice is grown in the low lying areas of Bangladesh. The flood waters start rising towards the end of March, and by June most fields are flooded (Figure 4A). Maximum water level is normally reached by mid-August when the level drops again and most fields are dry by November.

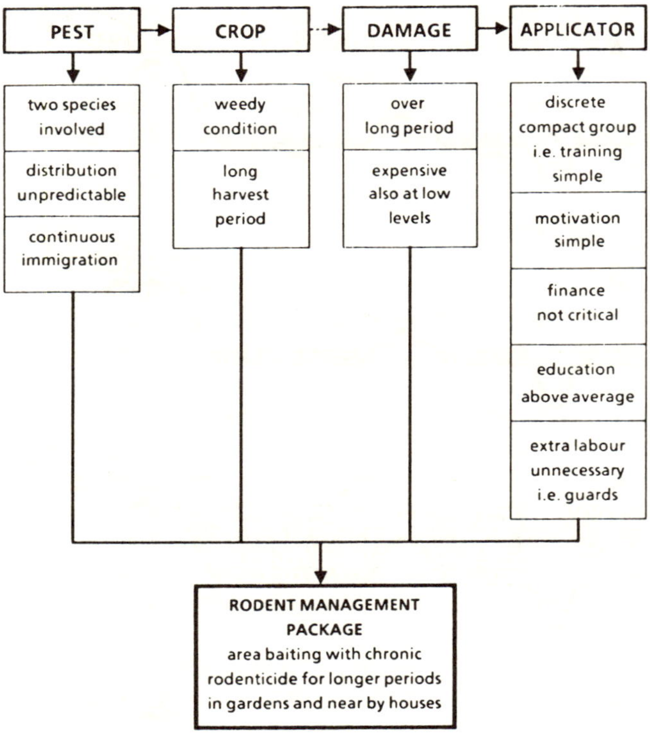

FIGURE 3. Major factors considered in recommending a rodent management package in pineapple.

Deep water rice is sown on to dry fields in April with the first rains. It grows and tillers (Figure 4B) but proper growth does not start until the fields are flooded. Tillers start elongating with the rising water level. Around maximum flooding, tillers, which reach lengths of up to 6 m, have collapsed and grown in different directions with part of the tiller floating on the surface (Figure 4C). These criss-crossing tillers form a matting at the surface with sufficient bouyancy to support a rat. At about the same time, the booting stage has been reached. The rice continues to develop as the waters recede and is harvested between October and November when most fields are dry again (Figure 4E).

The rats, *Bandicota bengalensis* and less frequently *Bandicota indica*, have adapted to this sequence. As fields are initially flooded, the rats leave the fields, which at that time do not provide a suitable habitat in terms of food and shelter, and colonize higher ground (Figure 4B). At this time the population is concentrated in small areas above the water level, mainly communication dikes and village islands. This is a very suitable time to initiate control actions. Bait acceptance is fairly high, because of a relative shortage of food.

At about the time of maximum water level the flooded fields provide suitable habitat in terms of shelter and food (i.e., the rice itself, invertebrates, and weeds). At this point part of the population returns to the flooded fields to take up residence (Figure 4C). This could be established by noting damage, floating rat nests and trapped animals in excess of 400 m from the next dry land.[35]

As the water level drops, reappearing land is immediately settled and these rats start breeding (Figure 4D). This would be the second control period, but since part of the rat population remains in the flooded fields this control action cannot be very effective. Finally water leaves the fields, the rice ripens, rats start building burrows, and breeding activity reaches its peak.

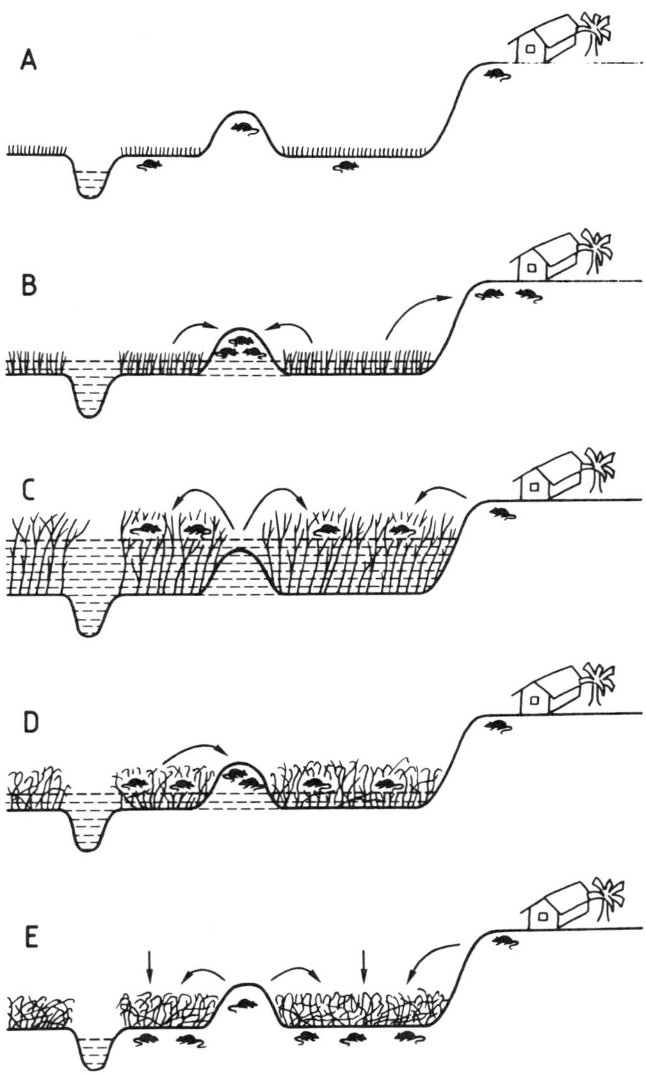

FIGURE 4. Migration pattern of *Bandicota* in deep water rice. Arrows show direction of movement. (A) Seedling to tillering stage, (B) Vegetative stage, fields just covered by floods, rats are forced on to higher ground, concentrated on a small area. Best time for control action. (C) Booting stage, near maximum flooding, economic damage starts, control only on very high ground possible. (D) Near ripening stage, water level drops, land reappears, and is settled by some rats, breeding starts. Good time for second control. (E) Harvest stage, rats migrate into fields, heavy breeding.

Rats and therefore damage is distributed throughout the large expanses, possibly related to the position of higher grounds and the rate of flooding. As access into and through these expanses during the flood period is extremely difficult, it is practically impossible to carry out rodent control in flooded fields.

Rat damage in deep water rice has been estimated at 3 to 4% of tillers cut according to national surveys.[4,35] However, another study[57] indicates that this estimate may be four times too low, possibly because of the difficulty in locating cut tillers under flood conditions. This level of damage certainly warrants control measures, even if difficult. To date farmers have not been involved in rodent control in deep water rice fields.

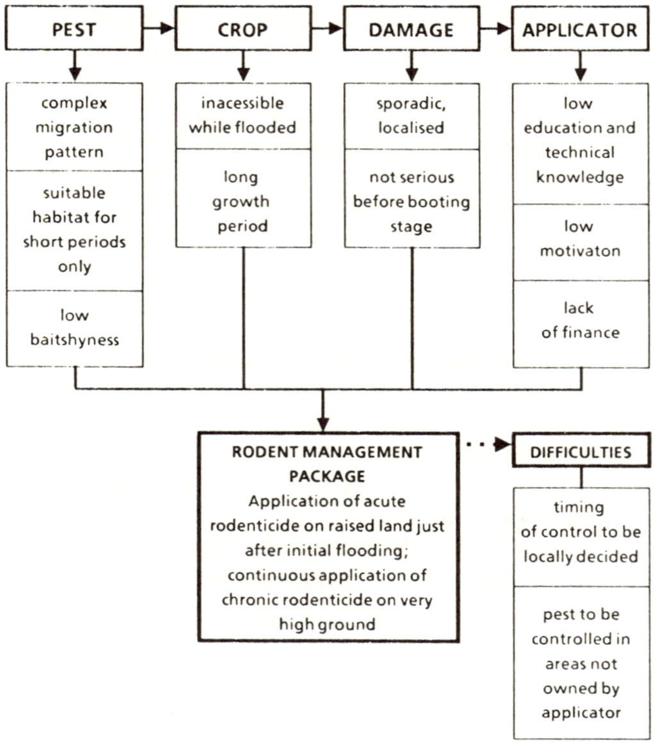

FIGURE 5. Major factors considered in recommending a rodent management package in deep water rice fields.

The effective time for control has already been mentioned, and to reduce the chance of immigration and build-up of a population on high grounds it is advisable to conduct control in these areas, mainly island villages, throughout the flood period.

The first and best period of control is very short, may be 3 to 5 weeks at most, and therefore the use of acute rodenticides would produce acceptable results, further improved by habitat management of these restricted areas. On ground never flooded, where control should be carried out over an extended period, the use of anticoagulants would be advisable.

The above suggestions can be very effective (Figure 5). One difficulty is due to the fact that in most cases the highland is not owned by any specific person and therefore nobody feels responsible for carrying out control in these areas, even if fields are owned nearby. This process to inform farmers about the advantages and the need to take action in this case is difficult and will take time.

IV. CONCLUDING REMARKS

The three examples, though primarily dependent on rodenticides, differed in the form and timing, the period and place of bait application, and the type of rodenticide used. In each case the methodology took into account the growth habit of the crop, the time and mode of damage, and restrictions imposed by the crop situation, and the absolute and relative costs involved. It also took advantage of the ecology of the species involved. In simple situations, i.e., wheat fields, the method is straightforward and quickly accepted by farmers, while in more complex situations, i.e., deep water rice fields, the method is complex and acceptance will take time.

These strategies could only be arrived at following the availability of a considerable amount of information on the respective situations. In each one discrete techniques and activities, sometimes concurrent in different localities, were integrated into a unified package appropriate for the specific situation. In addition, these packages or management procedures attempted to optimize the control effort and effect rather than maximizing it. It should be noted that biological systems are dynamic and these management techniques should not be viewed as final, but as basic ideas, which will have to be modified as some factors, including the educational level of the person applying the methods, will change with time.

ACKNOWLEDGMENTS

In this chapter I have drawn heavily on the information made available by Mr. Richard Poché and Joe Brooks from Denver Wildlife Research Center and their colleagues at the Bangladesh Agricultural Research Center, Vertebrate Pest Section, for which I am grateful. I would also like to thank the staff of the Vertebrate Pest Control Section, Plant Protection, Bangladesh for their devoted cooperation, and Professor W. Baeumler and Mr. W. Flache for preparing Figures 1 and 4, respectively.

The work was undertaken within the framework of the Bangladesh-German Plant Protection Program, a bilateral project of the People's Republic of Bangladesh and the Federal Republic of Germany, administered through the Gesellschaft für Technische Zusammenarbeit (GTZ) GmbH., Eschborn, Germany.

REFERENCES

1. **Adhikarya, R. and Posamentier, H.,** Motivating farmers for action. How strategic multi-media campaigns can help, Deutsche Gesellschaft fuer Technische Zusammenarbeit (GTZ) GmbH, Eschborn, Federal Republic of Germany, Schriftenreihe No. 185, 1987, 209.
2. **Ahmed, A., Mannan, M. A., and Mian, A.,** Evaluation of anticoagulant rodenticides for the control of house rat, *Bangladesh J. Agric.*, 5, 24, 1979.
3. **Alam, S. and Posamentier, H.,** National Rodent Damage Assessment in Wheat '81, Technical Rep., Bangladesh/German Plant Protection Programme, Khamar Bari, Dhaka, Bangladesh (Mimeo), 5, 1981.
4. **Alam, S., Huq, M., and Posamentier, H.,** National Rodent Damage Assessment in Deep Water Rice — 1980 Season, *Technical Rep.*, Bangladesh/German Plant Protection Programme, Khamar Bari, Dhaka, Bangladesh (Mimeo), 2, 1981.
5. **Brooks, J. E., Htun, P. T., Naing, H., and Walton, D. W.,** Food requirements of the lesser bandicoot rat, *Acta Theriol.*, 26, 373, 1981.
6. **Brooks, J. E. and Siddique, M. A. B.,** Rat Damage to the 1982 Wheat Crop in Bangladesh, Technical Rep. No. 16, Bangladesh Agricultural Research Institute, Joydebpur, Bangladesh (Mimeo), 1, 1982.
7. **Curry-Lindahl, K.,** Notes on the Ecology and periodicity of some rodents and shrews in Sweden, *Mammalia*, 23, 389, 1959.
8. **Fulk, G. W.,** Food hoarding of *Bandicota bengalensis* in a rice field, *Mammalia*, 41, 539, 1977.
9. **Fulk, G. W., Akhtar, M. T., Alam, S. A., and Smiet, A. C.,** Rodent damage to the 1978 and 1979 wheat crops in Pakistan, *Pak. J. Agric. Res.*, 1, 19, 1980.
10. **Fulk, G. W., Lathiya, S. B., and Khokar, A. R.,** Rice-field rats of lower Sind: abundance, reproduction and diet, *J. Zool. London*, 192, 371, 1981.
11. **Grodzinski, W.,** The succession of small mammal communities on an overgrown clearing and landslip mountain in the Beskid Sredni (Western Carpathians), *Ekol. Pol.*, 7, 83, 1959.
12. **Hoffmann, G. M. F., Nienhaus, F., Schonbeck, F., Weltzien, H. C., and Wilbert, H.,** *Lehrbuch der Phytomedizin*, Paul Parey, Berlin, 1976.
13. **Huq, S. M.,** National rodent damage assessment in wheat 1982 harvest, *Plant Prot. Newslett.*, Bangladesh, 4, 15, 1982.
14. **Huq, S. M., Alam, S., and Posamentier, H.,** National Rodent Damage Assessment in transplanted Rice Crops (T-Aman '80; Boro '81; Aus '81), Technical Rep., Bangladesh/German Plant Protection Programme, Khamar Bari, Dhaka, Bangladesh (Mimeo), 8, 1981.

15. **Kalela, O.,** Seasonal change of habitat in the Norwegian Lemming, *Lemmus lemmus* (L.), *Ann. Acad. Sci. Fenn.*, 34, 1, 1961.
16. **Khan, M. A. R.,** On the distribution of mammalian fauna of Bangladesh, Forestry Conf., Bangladesh, 1982, 13.
17. **Kock, D. and Posamentier, H.,** *Cannomys badius* (Hodgson, 1842) in Bangladesh (Rodentia: Rhizomyidae), *Z. Saugetierkd.*, 48, 314, 1983.
18. **Kock, D. and Posamentier, H.,** *Vulpes bengalensis* (Shaw, 1800) in Bangladesh (Carnivora: Canidae), *Z. Saugetierkd.*, 49, 374, 1984.
19. **Mian, M. D. Y., Ahmed, M. D. S., and Brooks, J. E.,** Rodent control on island villages in deepwater rice in Bangladesh, Technical Rep. No. 19, Bangladesh Agricultural Research Institute, Joydebpur, Bangladesh, *(Mimeo)*, 10, 1983.
20. **Myllymaki, A.,** Bedeutung and Ursachen der Mausefrasschaden in Finnland, *Publ. Finn. State Agric. Res. Board*, 178, 75, 1959.
21. **Newsome, A. E.,** A Population Study of House-mice permanently inhabitating a Reed-bed in South Australia, *J. Anim. Ecol.*, 38, 361, 1969.
22. **Newsome, A. E., Mcilroy, J., and Catling, P.,** The effects of an extensive wildfire on populations of twenty ground vertebrates in south-east Australia, *Proc. Ecol. Soc. Austr.*, 9, 107, 1975.
23. **Otto, R. A.,** Control of rice rats in west Malaysia, *Mardi (Crop Prot. Res. Bull.*, Kuala Lumpur), 1(4), 1, 1971.
24. **Parrack, D. W.,** A note on the loss of food to the lesser Bandicoot rat, *Bandicota bengalensis* (Gray and Hardwicke), *Curr. Sci.*, 38, 93, 1969.
25. **Parrack, D. W. and Thomas, J.,** The behavior of the lesser Bandicoot rat, *Bandicota bengalensis* (Gray and Hardwicke), *J. Bombay Nat. Hist. Soc.*, 67, 67, 1970.
26. **Poche, R. M.,** Range extension of *Rattus exulans* in South Asia, *Mammalia*, 44, 272, 1980.
27. **Poche, R. M., Haque, MD. E., Mian MD. Y., Sultana, P., and Karim, MD. A.,** Rice yield reduction by simulated Rat damage in Bangladesh, Technical Rep. No. 2, Bangladesh Agricultural Research Institute, Vertebrate Pest Division, Joydebpur, Bangladesh (Mimeo), 14, 1980.
28. **Poche, R. M., Haque, MD. E., Siddique, MD. A., Sultana, P., Mian, MD. Y., Degrazio, J., and Thompson, D.,** Rodent control in wheat fields, Technical Rep. No. 10, Bangladesh Agricultural Research Institute, Joydebpur, Bangladesh (Mimeo), 17, 1980.
29. **Poche, R. M. and Mian, MD. Y.,** Comparison of two methods for monitoring population indices of small mammals in Bangladesh deep water rice, Technical Rep. No. 6, Bangladesh Agricultural Research Institute, Joydebpur, Bangladesh, (Mimeo), 8, 1980.
30. **Poche, R. M., Mian, MD. Y., Haque, MD. E., and Sultana, P.,** Rodent damage and burrowing characteristics in Bangladesh wheat fields, *J. Wildl. Manage.*, 46, 139, 1982.
31. **Poche, R. M., Mian, MD. Y., Sterner, R., Haque, MD. E., and Sultana, P.,** Rodent movements in wheat fields, Technical Rep. No. 8, Bangladesh Agricultural Research Institute, Joydebpur, Bangladesh (Mimeo), 13, 1980.
32. **Poche, R. M., Sultana, P., Mian, MD. Y., and Haque, M.,** Studies with zinc phosphide on *Bandicota bengalensis* (Gray) and *Rattus rattus* (Linnaeus) in Bangladesh, *Bangladesh J. Zool.*, 7, 117, 1979.
33. **Posamentier, H.,** Habitat requirements of small Mammals in coastal health lands of New South Wales, M.Sc. thesis, Sydney University, Australia, 1975.
34. **Posamentier, H.,** Rodent damage and evaluation of rodent control in pineapple, Bangladesh, *Bangladesh J. Zool.*, 9, 41, 1981.
35. **Posamentier, H.,** Observations on three species of rodents in deep water rice in Bangladesh, *Z. Angew. Zool.*, 68, 155, 1981.
36. **Posamentier, H.,** Rodent damage assessment method in sown grain crops, *Bangladesh J. Zool.*, 9, 87, 1981.
37. **Posamentier, H.,** *Evaluation Report of the National Rat Control Campaign in Wheat 1982/83*, Bangladesh/German Plant Prot. Programme, Khamar Bari, Dhaka, Bangladesh, 1983, 127.
38. **Posamentier, H. and Alam, S.,** Rodent damage and control in wheat in Bangladesh during the 1980 season, *Bangladesh J. Zool.*, 8, 96, 1980.
39. **Posamentier, H. and Alam, S.,** Food choice with preconditioning between wheat and rice of *Bandicota bengalensis* and *Rattus rattus*, *Bangladesh J. Zool.*, 9, 87, 1981.
40. **Posamentier, H., Alam, S., and Haq, M.,** Interaction of stem density and growth stage on rodent damage distribution in wheat in Bangladesh, *J. Agric.*, in press.
41. **Posamentier, H. and Van Elsen, A.,** *Rodent Pests, their Biology and Control in Bangladesh*, Bangladesh/German Plant Protection Programme, Khamar Bari, Farmgate, Dhaka, Bangladesh, 1984.
42. **Prakash, I.,** *Rodent Pest Management, Principles and Practices*, Central Arid Zone Res. Inst., Jodhpur, India, Monograph 4, 28, 1976.

43. **Rahman, S. A. M. M. and Brooks, J.,** Comparative field trials of two rodenticides in small farm housing clusters, Technical Rep. No. 14, Bangladesh Agricultural Research Institute, Joydebpur, Bangladesh, (Mimeo), 10, 1982.
44. **Recher, H. F., Lunney, D., and Posamentier, H.,** A grand natural experiment: the Nadgee Wildfire, *Austr. Nat. Hist.,* 18, 153, 1975.
45. **Redhead, T. D.,** On the demography of *Rattus sordidus colletti* in monsoonal Australia, *Austr. J. Ecol.,* 4, 115, 1979.
46. **Sarker, S. K.,** Efficacy trials in fields of Bangladesh with the acute Rodenticide Zinc Phosphide, *Plant Prot. Newslett. Bangladesh,* 4, 11, 1982.
47. **Sarker, S. K., Posamentier, H., and Alam, S.,** Preference tests between different rice textures for wild *Bandicota bengalensis* and *Rattus rattus, Bangladesh J. Agric.,* 9, 4, 1986.
48. **Sheppe, W. and Osborne, T.,** Patterns of use of a flood plain by Zambian mammals, *Ecol. Monogr.,* 1, 179, 1971.
49. **Smythe, W. R. and Khan, A. A.,** Instruction for the preparation and use of zinc phosphide rat cakes for the protection of agricultural crops in Pakistan, Vert. Pest Centre Rep., Karachi, Pakistan, 1978, 4.
50. **Spillet, J. J.,** *The Ecology of the Lesser Bandicoot Rat in Calcutta,* Bombay Natural History Society and Johns Hopkins University Center for Medical Research and Training, Calcutta, 1968, 223.
51. **Sridhara, S. and Krishnamoorthy, R. V.,** Cereal preferences and intake of four species of field rodents, *J. Bombay Nat. Hist. Soc.,* 75, 129, 1978.
52. **Sultana, P.,** A comparison of three methods to determine food preferences in the lesser Bandicoot rat *(Bandicota bengalensis* GRAY), Technical Rep. No. 7, Bangladesh Agricultural Research Institute, Vertebrate Pest Division, Joydebpur, Bangladesh (Mimeo), 1980, 14.
53. **Sultana, P., Brooks, J. E., Siddique, MD. A., and Rahman, MD. M.,** Comparative Toxicities of several anticoagulant rodenticides to *Bandicota bengalensis,* Technical Rep. No. 13, Bangladesh Agricultural Research Institute, Joydebpur, Bangladesh (Mimeo), 5, 1981.
54. **Tast, J.,** the root vole, *Microtus oeconomus* (Pallas), as an inhabitant of seasonally flooded land, *Ann. Zool. Fenn.,* 3, 127, 1966.
55. **Temme, M.,** Beitrag zum Verhalten von Ratten *(Rattus rattus mindanensis* und *Rattus exulans)* wahrend und nach einer Uberschwemmung in Zentral Luzon, Philippinen, *Z. Angew. Zool.,* 60, 385, 1973.
56. **Tietjem, H. P., Halvorson, C. H., Hegdal, P. L., and Johnson, A. M.,** Gopher relationships in Black Mesa, Colorado, *Ecology,* 48, 635, 1967.
57. **Brooks, J.,** personal communication.

Chapter 30

TRANSFER OF RODENT PEST CONTROL TECHNOLOGY

M. M. Hoque and R. C. Saxena

TABLE OF CONTENTS

I. INTRODUCTION

Rodent damage to agricultural crops, stored products, and health is widely recognized as economically important. Awareness of the problem has stimulated research activity on these topics: the innate capacity of the rodents to reproduce, their ability to cope with stress induced by internal and external factors, their movement, birth rate, and survival of individuals in the population. Quantification of rodent damage, invention of new rodenticides, and development of new control techniques are necessary to establish a sound and efficient rodent control. A variety of methods for controlling rats has been used. They either protect crops or exclude the pest from human habitation and stored products by rodent proofing. The methods are broadly categorized as physical, biological, and chemical. Physical methods, such as traps when used intensively, can abruptly reduce the rat populations. Compared with chemical control, biological control of rodents has received little attention. The introduction of predators, parasites, and disease agents against rodents would have serious practical and theoretical difficulties.

The advent of anticoagulant rodenticides (e.g., warfarin, coumachlor, coumatetralyl, and indandiones) and the development of a delivery system for them have provided better techniques for their use in croplands and for urban rat control. But the development of anticoagulant-resistant populations has prompted a search for alternative compounds. Rodenticides that have been tested as alternatives range from the acute to the chronic type; some can be used as both.

Monitoring programs that periodically sample and test rat populations to detect resistance is another feature of the management of rodent pests in developed countries. A monitoring program provides the feedback necessary to determine when alternative control methods should be used.

In spite of the availability of rodent pest control technology backed by years of scientific investigations, technology users are still very few. The two major constraints are the absence of rodent control programs adequately supported by the government and a backlog in the transfer of rodent control technology. This paper highlights the developments in rodent control technology and constraints in its transfer. A case is presented on how a rodent control program can be instrumental in the transfer of rodent control technology and suggestions are given on how future control methods might be developed for greater use.

II. DEVELOPMENT IN RODENT CONTROL TECHNOLOGY

In the last two or three decades the use of rodenticides has become effective and economically feasible. Success is attributed to understanding of the pest's ecology, population dynamics, and behavior. Integration of control strategies such as habitat manipulation and use of rodenticides proved effective in urban rodent control. Similar principles were used in reducing crop depredations by the pest. Anticoagulant rodenticides, discovered in the early 1950s have greatly revolutionized chemical rodent control. These materials are blood anticoagulants which require repeated ingestion before a lethal dose builds up. Warfarin and other compounds such as racumin, diphacinone, and chlorophacinone were among the first-generation anticoagulants. Their use exploits the most important biological attribute of anticoagulants: symptoms of poisoning develop slowly so that bait shyness is avoided. By the time rodents stop feeding, they have already ingested a lethal dose.

Anticoagulant compounds allow continuous baiting and provide 100% control of rodent infestation. They practically eliminated many perennial rodent problems in temperate countries and made possible the attainment of rat-free towns and cities in Europe and the U.S.[2,9,24] The rat-free town approach was later modified for use in tropical croplands.

In 1974, the sustained baiting method, a modified form of continuous baiting, was developed to protect growing rice in the Philippines. This method requires a definite but relatively short period of activity on the part of the farmer, usually about 10 weeks during the crop season when damage would be greatest.[30] Its acceptance in combination with other methods has significantly reduced damage to crops where rodents were a problem.

The drawback of anticoagulant rodenticides is the development of resistant populations after a decade or so of continuous use of the anticoagulants.[1,17,18] Experience in Europe and the U.S. prompted researchers to develop ways of prolonging the use of these effective rodenticides. Methods of screening resistant rats and programs to prevent their spread were developed.[3,8] In the U.K. and the U.S., suspending the use of anticoagulants until the proportion of resistant rats became negligible suppressed the build-up of pockets of resistant rats.[3,13] Acute poisons and mechanical and physical methods were used for such cases. Researchers in other countries developed their own monitoring schemes.[15]

Another development in the area of anticoagulant resistance was the renewed interest in the search for new rodenticides,[14] that are still anticoagulant but potent against resistant rats. Some such rodenticides were difenacoum, bromadiolone, and broadifacoum.[14,23,28] However, cross-resistance for the first two compounds was later reported. Of the three, brodifacoum is reportedly the most potent. A delivery system for this rodenticide has been developed for use in controlling rodents in urban areas and croplands. The technique, known as pulsed baiting, takes advantage of the delayed action and acute potency of the compound.[10] Rodents ingest a lethal dose of brodifacoum bait at a single feed, whereas multiple-feed anticoagulants usually require 5 to 10 days of feeding.

Quantification of losses to rodent depredation is not yet well developed; however, various techniques for assessing damage done to the pest are available. The techniques aid in monitoring infestation and in making decisions for appropriate action. In the Philippines, for example, rat damage to rice is regularly assessed shortly before harvest.[31] Survey data for several years have indicated that damage at this period reflected the level of damage that will occur in the next season. When rat-damaged tillers exceed 3%, massive rodent campaign is conducted, followed by chronic poisoning during the crop period. Results of damage assessment studies in Malaysia indicate that level of damaged tillers may be used as a basis for treatment during the crop period.[3]

Damage survey data needed for decision-making cannot be obtained without sufficient manpower support. This points to the importance of rodent control organization. Most countries have their own rodent control programs.[16,20,22,26] The degree of success varies, however, and greatly depends on available financial and manpower resources as well as on government policies.

The number of international organizations funding rodent research indicate increasing international awareness of the significance of rodent pests. The World Health Organization's (WHO) major research area is control of rodent vectors of disease-causing microorganisms. WHO also responds to government requests by providing assistance of highly experienced, short-term consultants for rodent vector control and by holding training courses for personnel in developing countries.[11] WHO has surveillance units all over the world for rodent-borne diseases. The information is disseminated and becomes a basis for appropriate action. The Food and Agriculture Organization of the United Nations (FAO UN) provides grants to developing countries and technical assistance in developing local expertise on rodent research and control. The U.S. Agency for International Development (U.S. AID) has been assisting several U.S. Fish and Wildlife Service projects on control of vertebrate pests in Africa, Asia, and the Pacific. Several European organizations also help in solving rodent problems in both their respective countries and in others.[25] These are the European Plant Protection Organization, the Overseas Development Administration of U.K., the German Agency for Technical Cooperation, and several private chemical industries. In addition, international conferences and symposia held regularly allow scientists from all over the world to discuss their interest in rodent management.

III. CONSTRAINTS IN THE TRANSFER OF CONTROL TECHNOLOGY

Sociocultural characteristics, e.g., belief systems, customs, traditions, and values of a society or people can be problems in the effective management of rodent pests.

A common belief in Southeast Asia is that, like domestic animals, rats deserve decent treatment. Offering unpoisoned bait instead of rodenticide bait around crop fields is an old practice in some area of the Philippines. The belief stems from the practitioner's notion that befriending the pest would keep it from damaging his crop.

Religious beliefs about rats can be another problem. A typical example is in India where devotees of the Hindu goddess Bhagwati advocate tolerance for rats; killing the rodent is taboo.[5] Such belief systems are problems that confront a rat control advocate, hindering his operation.

Customs and traditions often dictate people's response to the rodent crisis. People adhere to their traditional practices of rodent control, unwilling to take the risks of using a new one. This is specially so when the expected result — seeing dead rats — is not immediate. For instance, people are still more interested in using acute poison that affects only about 80% of the rodent population and teaches surviving animals to associate their illness with the bait material, instead of using slow-acting rodenticides that affect almost 100% of the population and do not induce bait shyness.

The value system of people can also be a hindrance to introducing rat control innovations. A farmer may know that protecting a crop from rodent attack is one way of increasing yield without increasing area; however, he may aim for another goal — such as buying something for the family to enjoy now, not later — instead of using his resources to buy a rodenticide.[21] Traditional practices are not likely to be abandoned unless it can be demonstrated that new rodent control technology will be effective and will not result in intolerable social pressure or change in tradition.

Economic status can influence people's outlook on rodent pests. In an affluent society, for example, what is considered a disaster may not be so among the poor. For example, the presence of a single rodent hair or dropping in a bag of flour may drive a housewife in the U.S. to file a complaint against a supermarket for selling contaminated food. In the U.S., 5 million pounds of grain are seized or condemned yearly because of rodent contamination.[19] Such action is unthinkable in nations where food is scarce.

The characteristics of agricultural enterprises and people engaged in the trade also influence the adoption of new production technology. Farmers in developed countries adopt a new technology faster than their counterpart in the developing countries. The former are big landholders, with capital and modern equipment at their disposal, whereas the latter are small landholders with low income, are not profit-oriented and treat their farm as a single source of family livelihood.

Two types of farmers — the modern and the traditional — also use available resources and production technology differently. Government programs to induce the traditional farmer to get modernized are plenty, yet few farmers do so and many lag behind. One typical example of such a program is the Philippine food production program initiated in 1973. Farmer participants were provided with the package of production technology including rodent control and credits and every extension effort was made to promote the program. Nevertheless, many farmers later dropped out. Many were low-income, small landholders who were not sold to the idea of obtaining credit for capital. Such experience in the developing countries is less likely in the developed countries. This shows that not only economic status but also the attitude of people toward obtaining credit for capital creates constraints in transferring a technology such as that for rodent control.

The absence of rodent control programs adequately supported by the government greatly hampers the rate at which rodent control technology is disseminated. This situation is more acute in the developing nations where limited financial support causes a shortage of extension workers. Where farmers are poor and have minimum education, the extension worker is greatly needed.[6] He is the principal conduit for agricultural technology; however, the high farmer to technician ratio in less developed countries generally results in a backlog in the transfer of rodent control technology. Without rodent control programs therefore, available rodent control technology could hardly get to the user except for the contributions of the private chemical suppliers or equipment manufacturers. Moreover, with inadequate support of the program, communication technology could not be best utilized. These factors mentioned above would consequently affect communication systems, resulting in slow transfer of rodent control technology.

The inherent characteristic of a rodent control innovation may itself be a constraint. While scientists base innovations on important rodent control principles, what is often lacking is a consideration of the social and cultural implications of the technology. Researchers develop the technology as they perceive the problem but rarely, if ever, from the clientele's perspective. This is one major factor why some rat control innovations either are not consistently used or are not at all used. Communication experts say that what is accepted depends upon what is communicated and acceptance depends upon the attributes of innovations as perceived by the receiver.[29] In other words, like beauty, innovations exist only in the eyes of the beholder, and it is the beholder's perceptions which influence his behavior and decision to adopt or not to adopt a certain new approach. With this communication principle let us analyze the characteristics of a rat control strategy, whether it would possess the necessary attributes (i.e., relative advantage, compatibility, simplicity, trialability, and observability) as perceived by our clientele.[29]

The following example illustrates the perceived attributes of a rat control innovation aimed at protection growing crops. The rat control strategy is continuous baiting with anticoagulants. Some farmers could not perceive that the method is better than what they have been using earlier, such as single baiting with acute rodenticides. In this instance the farmer perceived more work activity than the advantage he would get by adopting the method. The same farmer may also perceive the innovation as incompatible with his existing farming practices, such as the control of insect pests. He may not understand why there is a need to start baiting even when damage is not yet apparent, whereas for insects he applies the control only when a certain level of damage or insect population occurs. The farmer may also find the innovation too complex, and as such, difficult to use.

One important rodent control principle that the farmer should accept is that it is cheaper to prevent an agricultural rodent problem than to stop losses once rodents have started to do serious damage to the crop. This means that the farmer should start rodent control activity as early as planting. Such an activity may conflict with his other activities of higher priority, or may not appreciate its use if there is no current evidence of rodent damage. To effectively protect his growing crop from rodent depredation, continuous baiting should be carried out and multiple dose anticoagulant rodenticides that require longer baiting should be used. Problems arise here because the farmer may think that continuous baiting means that the rats are not dying, specifically since he does not see dead rats around.

We emphasize the advantage of executing a rodent control measure by groups in a wide area than by single individuals. Because of the individualistic behavior of some people, they may try the technology by themselves. This often results in failure of the trial and eventually the farmer's fear of trying again. Because in this instance the control method failed, the farmer may associate risks with adoption of innovation.

The fate of a technology greatly depends upon the designers themselves. Often researchers tend to develop technologies purely by discipline, expecting that the clientele would have something to use to solve their problem. Rodent pest specialist are no exceptions. We believe that our responsibility is to develop rodent control technology, evaluate it, and hand it over to the extension worker. Having done that, we sit back hoping that it would be used. Sooner or later we receive comments or complaints that our technology does not work. We put the blame on the extension workers or even on the stupidity of the user. However, the problem might just have been our failure to follow up and monitor the performance of our method as used by our clientele. We then face the problem of developing another method which we think will perform better instead of probing the possible causes of failure of the technology.

IV. RODENT CONTROL PROGRAM IN TECHNOLOGY TRANSFER

Most rodent control technologies awaiting transfer are indigenously developed. Some could be transferred between neighboring countries, especially those with the same problem species and habitat. Transfer of pest control technology in the developed countries, like the U.S., is through government agencies, e.g., university extension personnel, regulatory personnel, and researchers. The private agencies involved in the transfer are professional pest control operators, chemical suppliers, and equipment manufacturers.[27] Most developing countries have their technologies transferred through ministries of the government with minor contributions from chemical and equipment suppliers and private pest control operators. In spite of extension efforts by those involved in the transfer, much remains to be done for rodent control technology to be used continuously. Deliberate transfer efforts are being done through organized rodent control programs. Many countries use this approach as seen in Taiwan's field and commensal rodent control program initiated in 1979. Taiwan has a three-level task force organized to handle all activities related to the program, with definite functions at each level (Figure 1).[20] A strong coordination between the agencies involved is evident as shown by the amount of time spent in carrying out the program. Full mobilization of farmers and city dwellers under the different agencies takes 6 months of preparation and 2 months post control evaluation. Preparation includes meetings between member agencies, publicity campaigns, and survey of areas to be baited; it culminates with the distribution of prepared baits to city dwellers and farmers. Island-wide baiting was done 1 to 2 weeks before the designated annual rat extermination week. Representatives from different agencies conduct a follow-up rat population density survey. Then, the village people are again convened and informed of the results of the campaign. Awards are given to the counties and townships that have done the best job. Results of the 1981 campaign reduced rodent populations all over the country. However, the population built up again a year later. This points to the need for a sustained rodent control program, not just a periodic one or an activity that is instituted only when population reaches devastating levels.

The success of the Taiwan program was primarily due to the cooperation of all sectors involved. Success did not come overnight, but evolved through painstaking efforts invested in organizing the program which started as a 6-year field rodent control project in 1970. A strong government support, like free baits for the public and incentives to those who have done their best, contributed much to the success. It demonstrates what an organized rodent control program could do in reducing pest problems and strongly supports the argument of rodent specialists that establishment of rodent control programs would serve as vehicles for technology transfer. Many countries, especially in the tropics, have instituted control programs within the last two decades. Some have included regular rat damage surveillance. One limitation of their programs, however, was the shortage of manpower to work with farmers as consultants with respect to production technology, including rodent control. The Philippines' food production program in the 1970s is another program that was used as a vehicle in transferring a rodent control technology. The innovation was incorporated into the production program, a package of rice production technologies that included loans to farmer participants. Such strategy, i.e., use of authority or compulsion for the use of the rat control innovation, was not readily accepted by farmers. The result of a sample survey

CENTRAL TASK FORCE

Member Agencies	Funtions
Council for Agricultural Planning and Development Ministry of Interior National Health Administration Provincial Department of Agriculture and Forestry Taipei Municipal Government Kaohsiung Municipal Government Provincial Health Department Provincial Environmental Sanitation Institute Provincial Food Bureau Provincial Forestry Bureau Taiwan Sugar Corporation	Plan the entire campaign, budget, and fund for country government; decide the acreage for baiting and the coverage of both rural and urban households; supply and distribute rodenticides, brown rice, and other bait materials; impart training; compile and print educational materials; publicize; supervise evaluate effectiveness of campaign

COUNTY CITY TASK FORCE

Member Agencies	Functions
County/City Governments County City Farmers' associations Sugar Mills District Agricultural Improvement Stations Provincial Food Bureau Branch Offices	Plan county/city campaigns; provide financial support, provincial subsidies, and help townships to raise funds; check acreage of baiting and coverage of farms in respective townships; education, supervison, and publicity; survey rodent population densities before and after control; oversee bait preparation for townships; compile data and report

TOWNSHIP TASK FORCE

Member	Functions
Township Offices Townships Farmers' Association and Farming Section Chiefs Village Office Chiefs and Neighbourhood Leaders Farmers and City Dwellers	Plan township campaigns; survey baited acreage, farms and city dwellings; mix bait materials (loose grain) and distribute bait; education and publicity; field baiting; dispose off dead rats; compile data and report

FIGURE 1. Organization of Field Rodent Control in Taiwan. (After Ku, T. Y., Proc. Conf. on the Organ. and Prac. of Vert. Pest Cont., August 30 to September 3, 1982, Elvetham Hall, Hampshire, England, 1984.)

revealed that 1 year after its introduction in one of the program areas, adoption of the rat control innovations was very low (12.5%).[7] It seems that programs are not always sure guarantees that the transfer of rat control innovation will be faster. Exceptions are probably in countries with better strategies of enforcing the adoption of an innovation. This may also depend on how extensive the innovation is promoted through communication.

V. RECOMMENDATION

A common problem in having rodent control programs is sustaining them. This is most often felt when the program is instituted by reason of politics. It is therefore highly desirable that one or two representatives from the technology designers become members of the policy-making body which would formulate guidelines in program implementation. The designers'

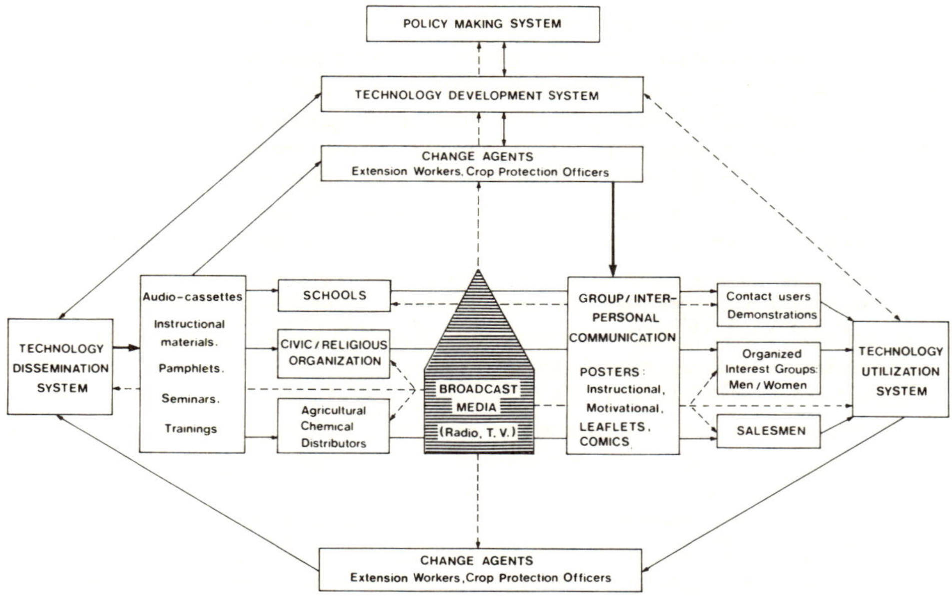

FIGURE 2. A paradigm of the technology transfer process in rodent control.

inputs would be valuable as advice on what control strategies could be integrated for specific situations. The technologists could also help sustain the interest of top officials in the program. Officials should be made to realize that a continuing effort to keep rodent population low must be the long-range objective of the program.

Often the most difficult problem is getting the people to carry out a sustained control program. When people are no longer bothered by rodents, they lose interest and abandon preventive measures that they should follow. This problem may be lessened through improved communication technology. The design of the communication strategy should be directed to the major decision makers, for they are the ones who should be convinced that the rat control innovation is worthy of adoption. The goal of dissemination efforts should be not just adoption but *continuing* adoption of the rodent control innovation.

Ideally, technology development should be a dynamic process with the involvement of the technology utilization system, technology dissemination system, and change agents (Figure 2). The technology development system must have a large input with respect to the content and clarity of the message as being translated into appropriate extension materials for diffusion. Such materials could be in the form of audio-cassettes, instructional materials, and pamphlets that are used for seminars and trainings. These could be used by change agents, either officially designated ones, as represented by government extension workers, crop protection officers, or indigenous community workers, and civic or religious organizations. The use of broadcast media is effective in creating awareness among target groups and all sectors of the society. The rat problem message and available control techniques, or where the technology could be obtained, may be aired over the radio in the form of spots and plugs. The technology dissemination system has the responsibility of translating technical materials into forms that are easily understood by the change agents and the technology users.

A rural sociologist says that what is accepted depends on what is communicated, and who brings the message in turn determines the effectiveness of the technology transfer.[6] This statement points to two things: the technology itself and those who do the extension work. Many studies have shown the critical role that change agents play in the transfer and adoption of a technology. The change agents can determine whether the whole technology can be transferred without some of its components getting lost in the process. This depends to some extent on the simplicity or complexity of the innovation as perceived by the change agent and clientele. A seemingly simple procedure from the designer's perspective may be perceived as too complex by the receiver. One example is a delivery system for anticoagulant rodenticides called sustained baiting, often perceived as difficult by rice growers in the Philippines. The procedure has three key features, namely: (1) continuous baiting from at most 2 weeks after transplanting to maturity, (2) increase of the number of bait holders and baits as rat pressure mounts, and (3) baiting points located at each corner and center of a hectare of rice field inside paddies, not atop dikes. All these key features are essential if one is to succeed. Many farmers who have used sustained baiting report failure because they failed to follow the correct procedures.

The preceding example points to an important consideration when designing a rodent control technology. It should be made even simpler than what is currently practiced. Ideally, the procedures should be self-explanatory so that even without the aid of extension workers, the user could easily follow the instructions. We technologists should closely observe the practices of the prospective technology users before we even attempt to change them. As much as possible, the design should not deviate so far from present practice; otherwise, its adoption would meet problems. What good would a highly efficient technology be if it were not used, compared with a less efficient one which has a wider adoption. Let us take the rice growers' practices in anticoagulant baiting in rice fields in the Philippines. Although the recommendation calls for baiting only at 5 baiting points per hectare, the farmers would place bait holders all over the rice fields. What they see is the advantage of more bait holders to intercept the rats. They could not see the logic of training the animals to feed at the baiting points, so that only the five baiting places need to be visited when checking baits. If the technologist had not revised the bait placement practice of the farmers, perhaps the sustained baiting technique might have been adopted.

In technology development, it is important that farmers/clientele participate in designing rodent control technology to ensure its future adoption. Furthermore, designers should be involved in the evaluation of the technology as used by the clientele. Even if new rodent control techniques could be handed over to efficient monitoring and extension services for implementation, the exchange between the clientele and the scientists would still be essential for the process of technology development itself.[11] The discipline in bringing technology fully to the level at which farmers can use it should be central to all research in technology development.

REFERENCES

1. **Boyle, C. M.,** Case of apparent resistance of *Rattus norvegicus* Berkenhout to anticoagulant poisons, *Nature,* 188, 517, 1960.
2. **Brooks, J. E.,** A review of rodent control programs in New York State, Proc. 6th Vert. Pest Conf., March 5 to 7, 1974, Anaheim, Calif., 1974, 132.
3. **Brooks, J. E., Bowerman, A. M., Jackson, W. B., and Kaukeinen, D.,** in *Anticoagulant Resistance Studies,* N. Y. State Department of Health and Ohio Environmental Studies Center, BGSU, 1974, 3.
4. **Buckle, A. P., Huson, L. W., and Rowe, F. P.,** The frequency distribution of rat damage in rice fields and its implication for the cost-benefit of control programmes, in *Proc. Conf. on the Organ. and Pract. of Pest Cont.,* August 30 to September 3, 1982, Elvetham Hall, Hampshire, England, 1984, 289.
5. **Canby, T. Y. and Stanfield, J. L.,** The rat: lapdog of the devil, *Natl. Geo.,* 152(1), 60, 1977.
6. **Castillo, G. T., Ed.,** Technology transfer in agriculture and family planning, in *Beyond Manila: Philippine Rural Problems in Perspective,* International Development Research Center, Ottawa, Canada, 1980, 420.

7. **Dizon, R. C.,** Duffusion of a rat control innovation among *Masagana* 99 farmers in Laguna, Philippines, M.S. Thesis, University of the Philippines at Los Banos, College, Laguna, Philippines, 1978, 69.
8. **Drummond, D. C. and Bentley, E. W.,** Laboratory investigation of resistance to warfarin in England and Wales, Rep. Int. Conf. Rodent and Rodenticides (Dec. 1965, Paris), *EPPO Publ. Ser. A,* 41, 57, 1967.
9. **Drummond, D. C.,** An outline of the current status of large scale urban rat control in England and Wales, *Int. Symp. Fragen Rattenvertilgung Budapest/Ungarn,* Dunker and Humbolt, Berlin, 1973.
10. **Dubock, A. C.,** Pulse baiting — a new technique for high potency, slow acting rodenticides, in *Proc. Conf. on the Organ. and Pract. of Bert. Pest Control,* August 30 to September 3, 1982, Elvetham Hall, Hampshire, England, 105, 1974.
11. **Goodell, G. E., Kenmore, P. E., Litsinger, J. A., Bandong, J. P., dela Cruz, C. G., and Lumaban, M. D.,** Rice insect pest management and its transfer to small-scale farmers in the Philippines, in *Report of anthropologists and other social scientists in interdisciplinary team developing improved food production technology,* International Rice Research Institute, Los Banos, Philippines, 1982, 25.
12. **Gratz, N. G.,** The role of W.H.O. in the study and control of rodent-borne diseases, in *Proc. 6th Vert. Pest Conf.,* March 5 to 7, 1974. Anaheim, Calif., 1974.
13. **Greaves, J. H.,** personal communication, 1984.
14. **Hadler, M. R., Redfern, R., and Rowe, F. P.,** Laboratory evaluation of difenacoum as a rodenticide, *J. Hyg. Cambridge,* 74, 641, 1975.
15. **Hoque, M. M.,** Development of anticoagulant resistance monitoring program for Philippine rats, *Phillip. Agric.,* 66, (1), 36, 1983.
16. **Indrarto, N.,** Five year rat control programme in Indonesia, in Proc. Conf. on the Organ. and Prac. of Vert. Pest Conf. August 30 to September 3, 1982, Elvetham Hall, Hampshire, England, 1984, 487.
17. **Jackson, W. B., Spear, P. J., and Wright, C. G.,** Resistance of Norway rats to anticoagulant rodenticides confirmed in the United States, in *Report from Pest Control,* The Harvest Publ. Co., 1971.
18. **Jackson, W. B. and Kaukeinen, D. E.,** Resistance of wild Norway rats in North Carolina to warfarin rodenticides, *Science,* 176, 1343, 1972.
19. **Jackson, W. B.,** Evaluation of rodent depredations to crops and stored products, *EPPO Bull.,* 7(2), 439, 1977.
20. **Ku, T. Y.,** Distribution and population fluctuations of field rodents and their control in Taiwan, in *Proc. Conf. on the Organ. and Prac. of Vert. Pest Cont.,* August 30 to September 3, 1982, Elvetham Hall, Hampshire, England, 1984, 487.
21. **Legans, J. P. and Loomis, C. P., Eds.,** *Behavioral Change in Agriculture,* Cornell University Press, Ithaca, 506 p., 1971, 506.
22. **Majid, T. B. and Chye, O. S.,** The organization and practice of rice field rat control in Malaysia, in *Proc. Conf. on the Organ. and Prac. of Vert. Pest Cont.,* August 30 to September 3, 1982, Elvetham Hall, Hampshire, England, 1984, 465.
23. **Marsh, R. E.,** Bromadiolone, a new anticoagulant rodenticide, *EPPO Bull.,* 7(2), 495, 1977.
24. **Myllymaki, A. A.,** Programme for the extermination of rats at minimal cost in Tampere Finland, in *Int. Symp. Fragen der Rattenvertilgung in Budapest/Ungarn,* Dunker and Humbolt, Berlin, 1973.
25. **Myllymaki, A.,** Activities of the rodent branch of the European and Mediterranean Plant Protection Organization (EPPO), undated report.
26. **Omar, M. T.,** The National rat control campaign in Egypt, in *Proc. Conf. on the Organ. and Prac. of Vert. Pest Cont.,* August 30 to September 3, 1982, Elvetham Hall, Hampshire, England, 1984, 443.
27. **Rambo, G. W.,** Professional pest control in the U.S. in *Proc. Conf. on the Organ. and Prac. of Vert. Pest Cont.,* August 30 to September 3, 1982, Elvetham Hall, Hampshire, England, 1984, 459.
28. **Redfern, R., Gill, J. E., and Hadler, M. R.,** Laboratory evaluation of WBA 8119 as a rodenticide for use against warfarin-resistant and non-resistant rats and mice, *J. Hyg. Cambridge,* 77, 419, 1976.
29. **Rogers, E. M. and Shoemaker, F. M.,** *Communication of Innovations: Across-Cultural Approach,* The Free Press of Glencoe, New York, 1972, 476.
30. **Sanchez, F. F. et al.,** Rodent Research Center 1975 Annual Progress Report, University of the Philippines at Los Baños, 1975.
31. **Sanchez, F. F. et al.,** Rodent Research Center 1974 Annual Progress Report, College, Laguna, Philippines, 1974.

Chapter 31

AREAS OF FURTHER RESEARCH

Walter E. Howard

TABLE OF CONTENTS

I. INTRODUCTION

There have not been many attempts to identify the areas of rodent control in need of further research.[14,18] Because of space limitation no attempt has been made to review the literature, instead only a few examples are provided of the research needs.

II. AREAS OF FURTHER RESEARCH

A. Evaluation and Choice of Rodenticides

Research needs concerning the efficaciousness of rodenticides are many.[4] The literature is full of improper laboratory testing procedures, especially when comparisons are made between acute and first and second generation anticoagulants. Most anticoagulants are very toxic in pure form. They are more effective if consumed by rodents in small amounts over several days, with the sum of the multiple chronic dose being lower than that required in a single acute lethal dose. This becomes less applicable the more potent the anticoagulant, e.g., with the second generation anticoagulants brodifacoum, bromadiolone, and difenacoum. There is insufficient data concerning what the most efficacious concentrations of rodenticides are to use for the many different pest rodent species, and concerning proper comparisons between different rodenticides.[27]

To determine average lethal dose (ALD), we need to known LD_{95} or LD_{99}, not LD_{50}, a term useful only for comparing a rodenticide's general toxicity with other compounds. LD_{50} does not tell you the concentration to be used. The dose linear regression lines are not straight lines. Toxicity, as such, only gives some indication of how much a compound will need to be diluted in a bait formulation. What control people are really concerned with is not how toxic the pure compound is (most anticoagulants are more toxic to man than 1080), but what are the hazards associated with how the material is going to be used.

More research is needed to determine what the minimum effective concentration of various rodenticides is to use with different rodent species, since the concentration often depends on the method of application and application rate.[30] When baiting in the field, success depends not only upon the type of bait formulation and its degree of acceptance by the target animal, but on the time of day when applied, baiting techniques practiced, the weather, and other conditions of the habitat.

Little is known about synergistic effects of a simultaneous application of baits of similar or different food bases that have been treated with different rodenticides where bait shyness may develop with one of the baits. We need to develop ways of fooling those few rodents that too often survive conventional rodent baiting.

Research on resistance has been investigated.[25] If permanent preventive baiting with anticoagulants is being practiced, what if some animals do not die for a month or so? The point is, if there are known survivors, don't go to the expense of testing for resistance, just use another rodenticide until all the rodents are eliminated. It doesn't matter whether it is genetic resistance, bait, or toxicant aversion or acceptability, just run a few simple field tests in bait boxes on acceptance of bait material, of rodenticide-treated bait, and of survival. This will give prompt results and not be costly like the time-consuming laboratory evaluations. Let major government rodent control laboratories do the expensive resistance and LD_{50} tests, if they so desire. Except for anticoagulants, few tests have been made to measure different degrees of resistance to acute rodenticides.[22]

Since pharmaceutical and chemical companies will not permit most of the potential chemosterilants to be used in rodent control, and in some countries they may also be difficult to register, we do not place antifertility studies in high priority, although there are numerous instances where such rodenticides would be helpful.[8,34] There is still a greater role for tracking powders.[33]

B. Bait Formulation

Much research is needed to find ways of improving bait acceptance by the target rodents. To know your rodenticide well and the intricacies of bait formulation is as important to rodent control as the knowledge of rodent biology.[30,44] Surely there are many yet untested innovative techniques and procedures for safeguarding nontarget species other than the questionable values of ''tamper-proof'' bait boxes.

Questions needing better answers for the great variety of rodent control problems include durable[48] and degradable baits, slow vs. fast kill toxic baits, size, texture, type, and kinds of bait, role of additives,[29] to improve and simplify laboratory and field testing techniques, test dilution baiting incorporating single lethal baits, determine the quantity of bait needed depending upon weather, season, and time of day of baiting, frequency of baiting, prebaiting, broadcast vs. spot baiting, use of bait boxes (stations), aerial baiting, and ground machine baiting.[17]

C. Baiting

With regard to baiting, research is needed concerning relative effectiveness of baiting depending upon weather, season, time of day, type and amount of food available to the rodents, what they are currently feeding upon, amount of bait needed depending on above factors, and frequency the control operations are needed, if not done on a permanent basis. Many questions stll exist concerning the cost effectiveness of prebaiting where acute rodenticides are to be used. When should bait be placed in bait boxes (stations) and when applied by hand, with hand-carried spreading equipment, or by airplane? What is the cost effectiveness of using a helicopter instead of a fixed wing when no airport is close?

D. Integrated Pest Management

More analyses of the integrated pest management (IPM) approach to rodent control, especially agriculture[10] and forestry rodents, are needed. As Marsh[31] points out, this is because methods used to control animal damage have historically been quite varied because no single approach was universally effective or acceptable. Because of plague, one study was conducted on integrated management of rodents in campgrounds.[37]

E. Aversion and Bait Shyness

More research is needed to distinguish between bait shyness and toxicant shyness with the different rodenticides.[39,40] How long will the various developed aversions to baits and/or toxicants persist?[20] How do bait shyness and neophobia relate to different rodent species?

F. Chemical Repellents

Much laboratory research has been undertaken to find chemical repellents[26] that can be used to prevent rodents from gnawing on packages, wires, and cables, but little evaluation has been made of duration of repellency, i.e., frequency of habitation, use of natural plant compounds such as tanins, and use of predator odors.

G. Frightening Devices

Devices which create an electromagnetic field to control rodents have been proven ineffective.[19] Since rodents are first exposed to the ultrasounds from places of security, it usually takes the rodents only a few hours or days for them to lose fear of such devices. We have not found examples where they can equal the proper use of rodenticides. Low frequency sounds possibly need more researching. Electric shock is too dangerous and costly. Soil vibrations have not been effective so far. Flickering lights and air hoses with numerous small openings need further investigation.

H. Electrocution

For a long time various rodent control electrocuting devices have periodically appeared on the market. We have not seen any that could be considered cost effective. Electric barriers around rice research plots have proven effective, but they are costly with high maintenance requirements.[41]

I. Physical Barriers

There has always been room for more information on how to physically rodent-proof structures and how to establish rodent-proof barriers in the field.

J. Predation

It has been shown that even though household predators, such as cats and dogs, cannot eliminate rodents, they can help prevent commensal rodent populations from exceeding the carrying capacity of the rodent harborage that is available. More research is needed on how to encourage natural predators in agricultural areas as part of integrated pest control programs. For example, it has been shown that raptors will promptly use temporary perches established in crops, although to date no significant rodent control benefit has been detected.[11] Any natural predation that will delay rodents from becoming established in a new crop would be beneficial.

K. Epizootics

Even though the introduced myxoma virus (myxomatosis) has been very dramatic in reducing European rabbits *(Oryctologus cuniculus)*, the intentional introduction of specific contagious or infectious diseases to control rodents does not appear very promising. The use of serious diseases that affect rodents provides too much hazard to man and other nontarget species, and conventional rodent control methods seem to be more efficacious.

L. Environmental Manipulation

With commensal rodents, general sanitation is helpful in holding down the density of pest rodents, but this method by itself usually cannot eliminate pest rodents. Habitat manipulation, though usually not feasible, can help keep pest rodents out of agricultural crops. It must always be remembered, however, that whenever the habitat is modified to control a field rodent pest, it will also be deleterious to many other kinds of vertebrates that suffer when the habitat is modified.

M. Trapping

It is amazing that so little research has been done on how to best use break-back kill traps, live traps,[28] and neck snares to control the various species of pest rodents.

N. Pheromones, Odor, and Taste Attractants

If we can make it easier for rodents to find a bait, increase specificity of attractants to target species, or encourage rodents to more quickly sample an unfamiliar bait, less rodenticide would be required and the effectiveness of baiting improved. Attractants would also be useful on traps and in bait boxes. However, we cannot expect rodents to respond reflexively to pheromones as do some insects, as rodents tend to evaluate the whole scene before responding.

A body capacitor-lickometer proved useful for testing responses to odors.[23] Research indicated that a small forest rodent (*Peromyscus* sp.) easily located conifer seeds by the volatiles the seeds produced.[42] Takada et al.[47] showed that acceptability of sodium warfarin above threshold for discrimination was improved by the additions of 8% sucrose. Other examples of odor and taste tests include olfaction.[1,3,21,35,36]

O. Permanent Preventive Control

Not enough research has been done on how to locally eliminate rodent pests on a permanent basis.[16,24] From conservation and safety points of view, permanent control of rodent pests is desirable because then so little rodenticide will be needed and nontoxic bait can be used to monitor the appearance of invading rodents.

P. Monitoring with Nontoxic Bait

Only by careful monitoring the effectiveness of a control program can you detect needed subtle changes in the operation that may greatly improve the program.[5,6] Even though agricultural rodent control usually consists of periodic control operations when deemed necessary, year-around monitoring research can make the timing of operations more effective and often at considerable savings. One technique needing more testing in the monitoring of rodent reinvasion is by using nontoxic wax blocks or other kinds of clean baits, usually in bait boxes.

Q. Damage Assessment and Economics

There is a great need for more sophisticated data concerning the economic damage caused by rodents and the cost-benefit values of various control programs.[45] Preventive control may prove more economically sound than damage assessments designed to determine some threshold when control should be undertaken since it is better to prevent the problem rather than try to reduce the degree of further loss to the crop.

R. Animal Movements

We need to understand better how the social status of rodent populations influence the pattern of movements of rodents.[2] More information is needed concerning distances which various species of pest rodents are likely to disperse due to population pressure and when programmed genetically to make extensive innate dispersal movements when they first mature sexually.[12]

S. Population Dynamics of Pest Rodents and Forecasting

This article does not have the space to discuss the great amount of information still needed concerning population dynamics of the many species of pest rodents and how to better forecast irruptions among cyclic species of pest rodents.[38,43,46,49,50]

T. Species Interactions

We have relatively little information on how control measures selective for one species may indirectly affect the movements and density of other species. Surprisingly, there is little information on the direct interaction of different species, such as *Rattus norvegicus* and *R. rattus*.

U. Imprinting of Young

Howard and Marsh[51] have demonstrated how adult rats made bait-shy can imprint this aversive behavior on their young if they are exposed to the bait continuously prior to weaning.

V. Social Attitudes and Humaneness

There is need to appraise the changing social attitudes which affect rodent control, both beneficially and detrimentally, and to analyze what actions can be taken to constructively modify some of the negative attitudes. A realistic balance of nature is no longer understood by most people.[15]

It is a law of nature that large numbers of all animals born must die prematurely, and they don't die nicely. Consequently, from an ethical and humane point of view man has an obligation in environments he has modified of both managing and controlling various rodent populations. More attention is needed on what the most humane yet effective ways are of controlling pest rodents.

W. Safety and Protection of Nontarget Species

It is paramount that more analyses be made on the safety and protection of nontarget species.[9,30,33]

X. Secondary Hazards

The biological half-life of rodenticides in bodies of rodents needs further researching. More research should be done with "pulse-baiting",[7] when using second generation anticoagulants, to reduce secondary hazards that can result from an overkill of the target rodents if they continue to eat the toxic bait until they die.

Y. Organization and Planning

A comparison of the effectiveness of different types of organizations and planning schemes for rodent control operations would be helpful since an effective organization structure is essential for successful rodent control.[13]

REFERENCES

1. **Barnett, S. A., Cowan, P. E., Radford, G. G., and Prakash, I.,** Peripheral ansomia and the discrimination of poisoned food by *Rattus rattus* L. *Behav. Ecol.*, 13, 183, 1975.
2. **Barnett, S. A., Dickson, R. G., and Warth, K. G.,** Social status, activity and preputial glands of wild and domestic house mice, *Zool. J. Linnean Soc.*, 70, 421, 1980.
3. **Drummond, D. C.,** Repellents and attractants and their role in the control of rodents, in *Proc. of Intern. Symp. on Bionomics and Control of Rodents,* September 29 to October 2, 1968, Kanpur, India, 1968, 101.
4. **Drummond, D. C.,** The role of biological efficacy evaluation and methodology in registration-rodenticides, Plant Protection Service, Food and Agricultural Organization, AGP:PRR/77/BP.7, 1977.
5. **Drummond, D. C.,** Rodent control programmes and the need for monitoring, in *Proc. of Symp. on Plant Protection,* Gesellschaft für Technische Zusammenarbeit, Münster, 1978.
6. **Drummond, D. C.,** Developing and monitoring urban rodent control programmes, *Acta Zool. Fenn.*, 173, 145, 1985.
7. **Dubock, A. D.,** Pulsed-baiting — a new technique for high potency, slow-acting rodenticides, in *Proc. 10th Vertebrate Pest Conf.,* University of California at Davis, Davis, Calif., 1982, 123.
8. **Ericsson, R. J., Downing, H. E., Marsh, R. E., and Howard, W. E.,** Bait acceptance by rats of microencapsulated male sterilant alphachlorohydrin, *J. Wildl. Manage.*, 35, 573, 1971.
9. **Godfrey, M. E. R.,** Non-target and secondary poisoning hazards of "second generation" anticoagulants, *Acta Zool. Fenn.*, 173, 209, 1985.
10. **Greaves, J. H.,** Rodent control in agriculture, Plant Production and Protection Paper 40, Food and Agriculture Organization, 1982.
11. **Hall, T. R., Howard, W. E., and Marsh, R. E.,** Raptor use of artificial perches, *Wildl. Soc. Bull.*, 9, 1981, 296.
12. **Howard, W. E.,** Innate and environmental dispersal of individual vertebrates, Reprinted in *Dispersal and Migration,* W. Z. Lidicker, Jr., and R. L. Caldwell, Eds., Benchmark Papers in Ecol., II, Hutchison Ross, 1982, 33.
13. **Howard, W. E.,** An effective organization is essential for successful rodent control, in *Proc. Conf. on Organization and Practice of Vertebrate Pest Control,* A. C. Dubock, Ed., ICI Plant Protection Division, Surrey, England, 1984, 437.
14. **Howard, W. E.,** Research needs in rat control, in *Proc. 1st Symp. on Recent Advances in Rodent Control,* A. H. H. Mohammad, T. M. Zaghloul, and M. Zakaria, Eds., February 8 to 11, 1982, Kuwait Ministry of Public Health, Kuwait, 1984, 209.
15. **Howard, W. E.,** Balance of nature: fiction or reality, in *Trans. 49th N.A. Wildl. and Natural Resources Conf.,* Wildlife Management Institute, 1984, 469.

16. **Howard, W. E.,** Rodent-free using permanent bait stations, in *Control of Mammal Pests,* Richards, C. G. J. and Ku, T. Y., Eds., Fourth International Theriological Congress, Edmonton, Canada, 1987, 147.
17. **Howard, W. E. and Marsh, R. E.,** Method of selecting preferred baits and rodenticides, *EPPO Bull.,* 7, 459, 1977.
18. **Howard, W. E. and Marsh, R. E.,** Improvements needed in rodent-control research, in *Proc. All India Rodent Seminar,* Ahmedabad, India, September 23 to 26, 1975, Rodent Control Project, 1977, 285.
19. **Howard, W. E. and Marsh, R. E.,** Ultrasonics and electromagnetic control of rodents, *Acta Zool. Fenn.,* 173, 1985, 187.
20. **Howard, W. E., Marsh, R. E., and Cole, R. E.,** Duration of associative memory to toxic bait in deer mice, *J. Wildl. Manage.,* 41, 1977, 484.
21. **Howard, W. E., Marsh, R. E., and Palmateer, S. D.,** Rat acceptance of different sugar concentrations in water baits, *Intern. Pest Control,* 14(6), 17, 1972.
22. **Howard, W. E., Marsh, R. E., and Palmateer, S. D.,** Selective breeding of rats for resistance to sodium monofluoroacetate, *J. Appl. Ecol.,* 10, 731, 1973.
23. **Howard, W. E., Palmateer, S. D., and Marsh, R. E.,** A body capacitor-olfactometer for squirrels and rats, *J. Mammal.,* 50, 771, 1969.
24. **Howard, W. E., Park, J. S., Cho, W. S., and Kim, S. I.,** A safe and inexpensive way of making South Korean villages rodent-free, in *Vertebrate Pest Control and Management Materials,* Spec. Tech. Publ. 680, Beck, J. R., Ed., American Society for Testing and Materials, Philadelphia, Pa., 1979, 50.
25. **Jackson, W. B., Ashton, A. D., Frantz, S., and Padula, C.,** Present status of rodent resistance to warfarin in the United States, *Acta Zool. Fenn.,* 173, 163, 1985.
26. **Johnson, R. J., Koehler, A. E., and Burnside, O. C.,** Rodent repellents for planted corn, in *Proc. 10th Vertebrate Pest Conf.,* Marsh, R. E., Ed., University of California at Davis, Davis, Calif., 1982, 205.
27. **Lorgue, G., Rampaud, M., Keck, C., and Jaussaud, P.,** Toxicological data on wild and domestic animals poisoned with rodenticides, *Acta Zool. Fenn.,* 173, 213, 1985.
28. **Lund, M.,** Rat traps, in *Danish Pest Infestation Lab. Ann. Rep.,* Lyngby, Denmark, 1981, 64.
29. **Mainardi, M., Pasquali, A., and Pasquale, E.,** Food choice in *Mus musculus:* effects of experience on aversion to an unpleasant flavour, *Biol. Behav.,* 10, 67, 1985.
30. **Marsh, R. E.,** Rodenticide selection and bait composition to minimize potential primary hazard to nontarget species when baiting field rodents, in *Proc. 1st East. Wildl. Damage Control Conf.,* Decker, D. J., Ed., Cornell University, Ithaca, New York, 1983, 155.
31. **Marsh, R. E.,** IPM concepts in vertebrate pest control relative to forestry, in *Proc. Integrated Forest Pest Manage. Symp.,* Branham, S. J. and Hertel, G. O., Eds., University of Georgia, Athens, Georgia, 1984, 25.
32. **Marsh, R. E.,** Tracking powders as a rodent control technique, *Acta Zool. Fenn.,* 173, 1985, 181.
33. **Marsh, R. E.,** Methods of safeguarding nontarget wildlife when controlling rodents in *Proc. Seminar on the Control of Squirrel Damage to Forest Trees,* Kuo, P. C., Ed., COA, Forestry Service No. 2, Taiwan, Republic of China, 1985, 169.
34. **Marsh, R. E. and Howard, W. E.,** Testing rodent chemosterilants, *EPPO Bull.* 7, 485, 1977.
35. **Marsh, R. E. and Howard, W. E.,** Pheromones (odors) for rodent control, *Pest Control Tech.,* 7(6), 22, 1979.
36. **Marsh, R. E., Howard, W. E., and Palmateer, S. D.,** Effects of odors of rodenticides and adherents on attractiveness of oats to ground squirrels, *J. Wildl. Manage.,* 34, 821, 1970.
37. **Marsh, R. E., Salmon, T. P., and Howard, W. E.,** Integrated management of rodents and other wildlife in campgrounds, Pacific Southwest Region, U.S. Forest Service, Rep. No. 81-39 2140 IPM, 1982, 126.
38. **Myllymäki, A., Hanson, L., and Christiansen, E.,** Models for forecasting population trends in two species of microtine rodent, *Microtus agrestis and Clethrionomys glareolus, Acta Zool. Fenn.,* 173, 93, 1985.
39. **Prakash, I. and Ojha, P.,** The phenomenon of poison-bait aversion and its attenuation in the Indian gerbil, *Tatera indica indica, Z. Angew. Zool.,* 65, 385, 1978.
40. **Prakash, I., Rana, B. D., and Jain, A. P.,** Bait shyness in three species of *Rattus, Z. Angew. Zool.,* 62, 89, 1975.
41. **Ramos, F. V.,** Construction and operation of the electric rat fence, Tech. Paper 3, International Rice Research Institutes, Los Banos, Philippines, 1967.
42. **Record, C. R., Marsh, R. E., Howard, W. E., and Stern, D. J.,** Olfactory responses of deer mice to Douglas-fir seed volatiles, in *Proc. 8th Vertebrate Pest Conf.,* Siebe, C. C., Ed., University of California at Davis, Davis, Calif., 1976, 116.
43. **Redhead, T. D., Enright, N., and Newsome, A. E.,** Causes and predictions of outbreaks of *Mus musculus* in irrigated and non-irrigated cereal farms, *Acta Zool. Fenn.,* 173, 123, 1985.
44. **Richards, C. G. J. and Huson, L. W.,** Towards the optimal use of anticoagulant rodenticides, *Acta Zool. Fenn.,* 173, 159, 1985.

45. **Saunders, G. R. and Robards, G. E.,** Economic considerations of mouse-plague control in irrigated sunflower crops, *Crop Prot.*, 2, 153, 1983.
46. **Spitz, F.,** Further development of the forecasting model for *Microtus orvalis, Acta Zool. Fenn.*, 173, 89, 1985.
47. **Takada, Y., Kusano, T., and Sugawara, R.,** Behavioral and physiological approach to the evaluation of rodenticidal effectiveness of sodium warfarin, Rep. Spec. Res. Project Trop. Agric. Resources, Japan, 2, 1984, 91.
48. **Tunberg, A. D., Howard, W. E., and Marsh, R. E.,** A new concept on pocket gopher control, in *Proc. 11th Vertebrate Pest Conf.*, Clark, D. O., Ed., University of California, Davis, Calif., 1984, 7.
49. **Wood, B. J. and Lian, S. S.,** A long-term study of *Rattus tiomanicus* populations in Johore, Malaysia. II. Recovery from control and economic aspects, *J. Appl. Ecol.*, 21, 465, 1984.
50. **Xia, W., Wei, S., Zhou, L., and Yang, K.,** Simulation and forecasting of oscillations in numbers of *Clethrionomys rufocanus, Acta Zool. Fenn.*, 173, 103, 1985.
51. **Howard, W. E. and Marsh, R. E.,** personal communication.

Chapter 32

ECOLOGY AND ECONOMICS OF RODENT PEST MANAGEMENT: THE NEED FOR RESEARCH

S. A. Barnett

TABLE OF CONTENTS

I. INTRODUCTION

Research on the management of rodent (and other) pests is of two main kinds: (1) technological, and (2) ecological, in its widest sense.

The leading technological aims are the discovery and development of new poisons and other devices. The anticoagulants illustrate both the value and the limitations of this approach: on the one hand, they provide a relatively safe and often effective alternative to acute poisons; on the other, resistant populations are likely to arise wherever they are used on a large scale. Technological research is certain to continue. A quick and well defined outcome is always possible, and it can be profitable: a firm that markets a new, successful product benefits financially.

A comprehensive ecological program should seek reliable statements on the economics of control. "Control" here means all measures designed to prevent loss of food intended for consumption by people or domestic animals. Such programs are expensive, yet do not yield immediate or direct profits. Hence the need for them requires special emphasis; for in the long run applying their findings can lead to vast reduction in losses due to rodent pests. Indeed, major advances in the war against the rodents of fields and forests may depend on ecological studies.

Funding must therefore come principally from central or local government, or from farmers' cooperatives. But industry may contribute, especially if the proposed research entails using or testing a potentially profitable product.

The criterion of economic success depends on the type of agriculture. (1) At one extreme is the "agrobusiness" that produces exclusively for the market. For that, the ideal is to calculate the cost of different measures of control, and to set them against the market value of the food saved. Such figures would allow a decision on the kind and extent of control operations. (2) At the other extreme is a purely subsistence economy. For that one needs to ask whether the labor expended on saving food could have been more usefully employed in producing it. Such questions are, of course, very difficult to answer; but *they should be asked,* however painful it is to do so.

The proposals that follow concern pests of agriculture. They are to some extent relevant to the management of rodents in cities; but the control of urban rodents is, to a much larger extent than in fields and villages, a matter of hygiene, proofing buildings, and similar measures. More important, rodents in cities may be less significant as food pests than as disease vectors. Where there is serious rodent-borne disease, economic criteria do not apply.

II. THE COMPONENTS OF A PROGRAM

A. The Human Population

In planning large-scale field research, a first step is to ensure the help of farmers and others. Some members of the research team should belong to the local community. It is not appropriate for a university graduate, unacquainted with the people, to descend on a rural community and propose research on the ecology, bionomics, and economic implications of the population dynamics of the sympatric species of the local Rodentia.

Each human group presents its own problems. Scepticism and disinclination to change traditional ways may be widespread, but can be overcome by matter-of-fact statements on the benefits of protecting food from pests. Some communities, however, have specific taboos concerning "vermin" such as rodents: for Hindus they are unclean, and should therefore be dealt with only by "tribals" outside the Hindu system. In such a case, familiarity with local requirements is essential.

A perhaps more general hazard is suspicion. Strangers who wish to record crop yields may be supposed to be agents of tax collectors. (See also Barnett and Prakash, p. 67[3].)

Scientific and technical workers need to establish a *rapport* not only with farmers but also with the office workers who lay down policy and provide funds. A good acquaintance with the regulations that restrict administrative action (and the anxieties that beset administrators) is helpful. Similarly, the managers themselves should be willing to understand scientific attitudes. Here, as an illustration, is a perhaps over-simple parable. An administrator (A) asks a biologist (B) what is the best design for a mouse trap. B replies that, given time and resources, she can tell him; but why does he want to know? A replies, perhaps irritably, that the idea is to kill mice. B retorts that there are other ways of killing mice. A concedes that he should have asked for the best way of killing mice, and no doubt expects B to demand substantial funds for research on this question. But B again asks why A wants to kill mice; and to A's no doubt still more exasperated explanation that he wants to preserve food B points out that food might be better protected by preventing mice from breeding or . . .

A scientific worker is likely to concentrate on acquiring knowledge and asking critical questions, while a manager is preoccupied with what is customary, with keeping to a budget and with adhering to regulations. Neither has a monopoly of the right or the good. Both need to recognize the need for *reliable, quantitative information on the effects of control measures.*

B. The Rodent Population

Before experiments begin, the species to be controlled must be identified. Rodent systematics have been well studied, but some species, for instance of the genus *Mus*, may present problems.

Most major agricultural regions support several sympatric species. Unfortunately, much of the early rigorous research was in northern Europe or North America, where there are only three species; often, only one, *Rattus norvegicus*, had to be dealt with. Methods and findings from such simple ecosystems are not always applicable to the vast areas of the tropics and subtropics with more complex ecosystems. Here are some examples.

Happold[9] has described the rodents of Nigerian farmlands. The most prominent is the so-called multimammate mouse, *Mastomys (Rattus) natalensis*; other numerous species are the rusty-bellied rat, *Lophuromys sikapusi*, the brush-furred mouse, *Uranomys ruddi*, and the pygmy mouse, *Mus minutoides*. In the villages, *M. natalensis* is again common, often in company with the roof rat, *Rattus rattus*. (There is an unexplained absence of house mice, *Mus musculus*, in most of Africa.)

For East Africa we have an account by Taylor[11] of a devastating outbreak of rats in Kenya and Tanzania. *Mastomys natalensis* was again conspicuous, but was accompanied by seven other species regarded as pests. Of these, the Nile rat, *Arvicanthis niloticus*, (which is also common in West Africa) and the zebra mouse, *Rhabdomys pumilio*, were especially important. A pygmy mouse, *Mus triton*, represented the smallest category of rodents, but its prevalence was not established. Such unobtrusive species are widespread, but often little studied.

A corresponding group in rice-growing regions of India is led by two very widespread and numerous species, the Indian gerbil, *Tatera indica*, and the mole rat, *Bandicota bengalensis*. Also present in the fields may be the soft-furred field rat, *Rattus meltada*, the common Indian field mouse, *Mus booduga*, and other species of *Mus*. In the villages are *R. rattus*, *M. musculus* and the very large bandicoot rat, *B. indica*.[3]

In the sugar plantations of Hawaii, the most important rodents are the introduced species, *Rattus rattus* and *R. norvegicus*. But *R. hawaiiensis* is also a significant pest.[6]

Ideally, a control program should cover all the species in such an association. Hence their distributions should be known. Most are burrowers, but just where they burrow varies with the species (and, sometimes, with the season). Some important species, notably *R. rattus* and the Polynesian rat, *R. exulans*, rarely burrow. The former nests in trees, such as coconut palms, or in roofs; the Polynesian rat is small enough to nest in bamboo furniture. Corresponding differences in feeding habits, and the effects of poisons, when known, must influence the control methods to be used.

Important but rarely asked questions concern the interactions of species. As an example, in the natural pastures of Rajasthan, India, the abundant species are the gerbils, *Meriones hurrianae* and *Tatera indica*. When most of these have been removed by poisoning, *Gerbillus gleadowi* and *Rattus meltada* are still present and can become more numerous. In each region there is therefore a general question: if the most prominent species are removed, will they be quickly replaced by others?

C. Assessing the Effects of Control

The outcome of a control program can, in principle, be assessed (1) by estimating the change in the numbers of rodents, and (2) by recording the effect on food production.

An accurate estimate (say, within 10%) of the absolute numbers of a large rodent population is rarely practicable, but a sound knowledge of the natural history of a species usually makes possible an indirect census (for a summary and references see also Barnett and Prakash, p. 136[3]). Traces that can be used include a count of the number of "active" burrows and sample counts of fecal pellets. Two much used, and sometimes valuable, techniques require trapping: in one, rodents are caught, marked, released, and caught again, to provide a "Lincoln index"; in the other, a trap line provides a trap index. Different species respond to traps in different ways. Hence if trapping is to be used on a little studied species (of which there are many) preliminary research may be needed to determine the exact procedure.

For cost-effective control it is necessary to apply knowledge of population dynamics. There is a useful general principle: if a population, at or near the maximum allowed by the environment, is reduced by, say, 50 or 60%, the fertlity of the survivors rises, and numbers are quickly restored; but if only 5% remain, recovery is disproportionately slower (see also Barnett and Prakash, pp. 26, 60[3]). The numbers of some species vary with the season, and reach a maximum soon after the harvest. The kill from an operation at that time may be visually impressive, but most of the animals would have died soon in any case. The time to act is when the population is already low, and can be reduced to a level from which it cannot quickly recover. An apparently expensive campaign that reduces rodent numbers to a minimum allows a long interval before the next operation is needed; it is then, in the long run, cheaper than a perfunctory skimming of the surface of the population.[5]

The difficulties of estimating rodent populations suggest that control operations should be assessed by some measure of food saved; but estimates of crop yields may be equally difficult to make or, if made, to interpret. It is desirable to have figures for several years before experiments on control begin. Yields vary from year to year with the weather. The conditions in which the crops are grown, or even the crops themselves, may be changed without regard for the needs of a properly controlled experiment. There are other sources of uncontrolled variation. Losses due to rodents occur both in the fields and after harvest; hence estimates should, if possible, cover the amounts harvested and also the amounts ultimately sold or consumed. Very thorough planning in advance may be needed to ensure that information of this kind will be available.

It is hardly surprising that most of the published statements on rodent numbers, on losses due to rodents, and on the effects of control are little more than inventions. A few research programs have, however, shown, if only imperfectly, how the results of control can be related to changes in rodent populations or in crop yields. Wood[12] has compared alternative control methods for *Rattus* of three species in Malaysian rice fields. The preferred method is said to give a three-fold increase in yield. He has also described a series of studies in oil palm estates, where the most prominent rodent is *R. tiomanicus*.[13] Here too control was evidently economically effective. Wood's method included the use of an anticoagulant in an unusual base, and so illustrates the need for innovations to meet special needs. His papers may be studied for their critical account both of achievements and of unsolved problems.

Saunders and Robards[10] have made an economic assessment of aerial poisoning during a plague of house mice, *Mus musculus*, in Australia. The crop was sunflower, but their conclusions are held to apply also to cereal crops. The most successful treatment is estimated to have killed 90% of the mice, and to have increased yield by 20%.

The need for careful scrutiny of accepted routine is shown in an investigation of two English villages heavily infested with *Rattus norvegicus*.[2] Partial clearances of the rats were followed within 6 months by a restoration of rat numbers to their previous level. In contrast, more thorough clearances, which required at least two successive poisonings, had a much more lasting effect which was still evident after a year. At the time, it was usual for rat treatments on English farms to be carried out at intervals of 3 months; but a more economical procedure would have been a really efficient operation once a year.

D. The Control Operation

To render valid conclusions on the economic effects of control, experiments should compare areas in which rodents are killed or otherwise treated with others left untreated. A research program has therefore been designed, in which 9 areas would be studied over a period of at least 3 years.[3] In three, the rodent populations and crop yields would be recorded, but the rodent populations would be left untreated. In another three, as many rodents as possible would be killed, but nothing further would be done against them: but the return of the rodents, by breeding or reinvasion, would be observed. In the last three, a similar initial strike would be followed by attempts, by all practicable means, to keep the rodents to a minimum. Where such a program is too much for the resources available, it can of course be reduced. Even experiments of the third type by themselves can yield useful findings.

The methods used for the initial strike must depend on the species present and on the availability (including the price) of poisons or other devices. Often, it would require the use of both an acute poison, such as zinc phosphide, Zn_3P_2, and an anticoagulant; but other substances may be appropriate.

Whatever the scale of the experiments, regular monitoring of the rodent population is essential. Sooner or later numbers will rise again; and since spontaneous generation (often assumed to occur by the untutored) can be excluded, breeding and immigration must be recorded. Hence it is desirable to record pregnancy rates and to map changes at the periphery of each experimental area. The importance of reinvasion has been illustrated in experimental treatments of sewers heavily infested with *Rattus norvegicus*.[1,4,8] Sewer rats, left to themselves, are a constant source of infestation of the surface; but if they are killed while those outside remain, their numbers are quickly restored from above.

III. FUNDAMENTAL RESEARCH

The proposals above, for large-scale operational research in the field, may seem absurdly grandiose, even utopian. While this chapter was being written, funds for the arts of peace were steadily diminishing throughout the world. Human resources are being directed, on a vast scale, toward killing, but not toward killing pests. Yet research on the scale proposed is far from excessive in relation to the size of the problem.

The proposals are, indeed, incomplete. Their emphasis on economically significant projects should not disguise the need for more fundamental inquiry. The biology of many of the important rodent pests, in much of the world, is little known. At the same time, there is a dearth of zoologists with relevant experience. Both deficiencies can be partly remedied by combining operational research with general investigations supervized from an agricultural university or similar institution. Any rodent population, but especially one that is changing during attempts at control, presents plenty of ecological, ethological, and physiological problems. Much of the research should be done by graduate students working for a higher degree (Ph.D. or M.Sc. or the equivalents). More knowledge of pest species would then be gathered. At the same time, young zoologists would be prepared for independent work in the management of mammalian pests: some would take up operational research, others would train pest control workers and advise farmers and warehouse-keepers. (For a brief account of such an organization, see Elton.[7])

The management of rodent pests is not a temporary problem. No panacea is to be expected. New hazards will always arise. Human societies and their agricultural practices change, and the environments of associated rodents change with them. The rodents themselves vary also, often unpredictably, but — as a rule — adaptively. For the foreseeable future our food supplies and, often, our health, will be threatened by these omnipresent trespassers. Only unremitting, large-scale research at all levels will enable us to counter this threat.

REFERENCES

1. **Barnett, S. A. and Bathard, A. H.,** Population dynamics of sewer rats, *J. Hyg., Cambridge,* 51, 483, 1953.
2. **Barnett, S. A., Bathard, A. H., and Spencer, M. M.,** Rat populations and control in two English villages, *Ann. Appl. Biol.,* 38, 444, 1951.
3. **Barnett, S. A. and Prakash, I.,** *Rodents of Economic Importance in India,* Heinemann-Educational, London, 1975.
4. **Bentley, E. W., Bathard, A. H., and Riley, J. D.,** The rates of recovery of sewer rats after poisoning, *J. Hyg., Cambridge,* 57, 291, 1959.
5. **Davis, D. E.,** The characteristics of global rat populations, *Am. J. Public Health,* 41, 158, 1951.
6. **Doty, R. E.,** Rat control on a Hawaiian sugar cane plantation, *Hawaii. Planters' Rec.,* 49, 71, 1945.
7. **Elton, C. S.,** Research on rodent control by the Bureau of Animal Population September 1939 to July 1947, in *Control of Rats and Mice,* Chitty, D., Ed., Clarendon Press, Oxford, 1954, 1.
8. **Greaves, J. H., Hammond, L. E., and Bathard, A. H.,** The control of re-invasion by rats of part of a sewer network, *Ann. Appl. Biol.,* 62, 341, 1968.
9. **Happold, D. C. D.,** *The Mammals of Nigeria,* Oxford University Press, Oxford, 1986.
10. **Saunders, G. R. and Robards, G. E.,** Economic considerations of mouse-plague control in irrigated sunflower crops, *Crop Prot.,* 2, 153, 1983.
11. **Taylor, K. D.,** An outbreak of rats in agricultural areas of Kenya in 1962, *E. Afr. Agric. J.,* 34, 66, 1968.
12. **Wood, B. J.,** Investigations of rats in ricefields demonstrating an effective control method giving substantial yield increase, *PANS,* 17, 180, 1971.
13. **Wood, B. J. and Nicol, I.,** Rat control on oil palm estates, in *Advances in Oil Palm Cultivation,* Wastie, R. L. and Earp, D. A., Ed., Incorporated Society of Planters, Kuala Lumpur, 1972, 380.

INDEX

A

B

C

N

O

P

Q

R

S

T

U

V

W